T0301016

Derived Langlands

Monomial Resolutions of Admissible Representations

Series on Number Theory and Its Applications Vol. 15

Derived Langlands

Monomial Resolutions of Admissible Representations

Victor Snaith
University of Sheffield, UK

 **World Scientific**

NEW JERSEY · LONDON · SINGAPORE · BEIJING · SHANGHAI · HONG KONG · TAIPEI · CHENNAI · TOKYO

Published by

World Scientific Publishing Co. Pte. Ltd.
5 Toh Tuck Link, Singapore 596224
USA office: 27 Warren Street, Suite 401-402, Hackensack, NJ 07601
UK office: 57 Shelton Street, Covent Garden, London WC2H 9HE

Library of Congress Cataloging-in-Publication Data
Names: Snaith, Victor P. (Victor Percy), 1944–   author.
Title: Derived Langlands : monomial resolutions of admissible representations /
    by Victor Snaith (University of Sheffield, UK).
Description: New Jersey : World Scientific, 2019. | Series: Series on number theory and
    its applications ; volume 15 | Includes bibliographical references and index.
Identifiers: LCCN 2018048216 | ISBN 9789813275744 (hardcover : alk. paper)
Subjects: LCSH: Algebraic number theory. | Representations of groups. | Galois modules (Algebra)
Classification: LCC QA247 .S59725 2019 | DDC 512.7/4--dc23
LC record available at https://lccn.loc.gov/2018048216

British Library Cataloguing-in-Publication Data
A catalogue record for this book is available from the British Library.

For any available supplementary material, please visit
https://www.worldscientific.com/worldscibooks/10.1142/11142#t=suppl

Printed in Singapore

Contents

Preface

The purpose of this monograph is to describe a functorial embedding of the category of admissible $k[G]$-representations of a locally profinite topological group G into the derived category of the additive category of the admissible $k[G]$-monomial module category[1], based on the family of compact open modulo the centre subgroups.

By virtue of the Langlands Programme (see, for example, [1] and [2]) the representation theory of locally profinite topological groups is related in a very important manner to modern number theory and arithmetic-algebraic geometry ([3], [52]). There are many facets to this relationship (for example, there are more than 40 sources in the biography of this monograph which deal with some feature of this relationship).

I shall concentrate, for simplicity, on the locally profinite groups associated with GL_n. However in Appendix IV I briefly indicate how the main construction generalises to an admissible representation of an arbitrary locally profinite group.

After the local field case, motivated by the Langlands Programme, also one is interested in monomial resolutions of a number of other settings:

(i) The admissible representations of the semi-direct product of GL_n with a Galois group occur in the phenomenon of Galois base change (aka Galois descent; [7] and [91]).

(ii) The restricted tensor product of admissible representations in the local field case occur in the construction of adèlic automorphic representations and their connection with modular forms and Hecke operators ([67] and [51]).

(iii) The local Langlands correspondence involves Deligne representations of the local Weil group [40].

---

[1]Experts in the Langlands Programme may be interested to learn, before going any further, that when G is a locally p-adic Lie group the monomial category is closely related to the category of topological modules over a sort of enlarged Hecke algebra with generators $((K, \psi), g, (H, \phi))$ with H, K compact open modulo the centre and $(K, \psi) \leq (gHg^{-1}, (g^{-1})^*(\phi))$ subject to the multiplication and relations which are described (for G finite) in Chapter One, §1.

(iv) The local correspondence is characterised in terms of invariants such as ϵ-factors and L-functions which participate in the local functional equation as developed in Tate's thesis ([**86**] and [**142**]).

In each of (i)–(iv) I have attempted to give at least an example of how monomial resolutions are constructed and fit in to the overall picture. Sometimes these examples are just given for GL_2 — perhaps out of lack of time and, more often, an indication of the threadbare state of my expertise.

This monograph contains a partial fulfilment of a mathematical ambition which I have harboured since 1986, which is my cue for a brief scrap of long-forgotten history! In the 1940's Richard Brauer proved his famous Induction Theorem for representations in characteristic zero of a finite group. In particular, explicit induction theorems (e.g. Artin's Induction Theorem [**126**]) are important in the derivation of Brauer relations between class numbers and orders of units [**31**]. Around 1946 Brauer posed the problem of deriving an explicit formula for his induction theorem analogous to that of Artin's induction theorem (see the footnote [**113**] p. 71). In a series of results Dwork, Langlands and Deligne derived results which essentially solve Brauer's problem for solvable groups ([**57**], [**48**] and [**89**]).

Using a topological construction which originated in terms of formulae in the stable homotopy category, I gave the first explicit formula to solve Brauer's problem ([**121**] and [**122**]). Monomial resolutions for finite-dimensional complex representations of compact Lie groups were implicit in my original formula because it was the Euler characteristic of a topologically constructed chain complex of monomial modules (i.e. sums of modules induced from lines). This point of view was particularly stressed in [**121**] and [**124**]. At that time my ambition was to construct explicit monomial resolutions for (a) finite-dimensional Galois representations and (b) for admissible representations of $GL_n K$ when K was a local field and thence to attempt to go back and forth (as then predicted by the Langlands correspondence, since proved by Mike Harris and Richard Taylor) capitalising on the fact that monomial resolutions are built from one-dimensional representations to which the local class field theory correspondence applies.

The two main obstacles to this ambition were (a) insufficient expertise concerning admissible representations of locally profinite groups and (b) complete ignorance of the correct categorical setting. The crucial advance made by Robert Boltje in [**19**] was to overcome obstacle (b) by describing the (additive) category $_{k[G]}$**mon** in whose derived category monomial resolutions naturally live and to develop all the techniques for working there when G is finite.

It is straightforward to extend $_{k[G]}$**mon** to the case where G is a locally profinite group which is compact open modulo the centre such as $K^* \cdot GL_n \mathcal{O}_K$ in $GL_n K$ when K is a local field. By good fortune Ian Leary had, around 2001, pointed out to me the properties of the Baum-Connes

space $\underline{E}(G,\mathcal{C})$ (see Appendix IV) and that the Bruhat-Tits building for
GL_nK almost equals $\underline{E}(GL_nK,\mathcal{C})$ when K is a local field. This fact implies, if $\mathcal{C}$ is the family of compact open modulo the centre subgroups,
that given a sheaf of functorial monomial resolutions for groups in $\mathcal{C}$ on
$\underline{E}(GL_nK,\mathcal{C})$ one may construct a double complex to give a $_{k[GL_nK]}$mon-
monomial resolution of any admissible k-representation V of GL_nK having
a fixed central character.

The construction of monomial resolutions for arbitrary admissible k-
representations V of arbitrary locally profinite groups follows once one has
the sheaf. This is given by the functorial bar-monomial resolution for the
restrictions of V to subgroups of G which lie in $\mathcal{C}$, the family of compact
open modulo the centre subgroups. The difficulty of verifying that the bar-
monomial resolution is indeed a monomial resolution is overcome by using
extensions of the recognition criteria of [19].

Without going into technical details, a monomial resolution $M_* \longrightarrow V$
gives rise to an exact "resolution" sequence of k-vector spaces of the form

$$\cdots \longrightarrow M_i^{((H,\phi))} \longrightarrow M_{i-1}^{((H,\phi))} \longrightarrow \cdots \longrightarrow M_0^{((H,\phi))} \longrightarrow V^{(H,\phi)} \longrightarrow 0$$

for each continuous character $\phi : H \longrightarrow k^*$ where $H \in \mathcal{C}$ and k is an
algebraically closed field. When G is, for example, adèlic GL_2 and V is an
automorphic representation then $V^{(H,\phi)}$'s (the subspace of V where H acts
via ϕ) include the classical spaces of modular forms. Hence the interest in
setting (ii) mentioned above. Hecke operators

$$[JgH] : V^{(H,\phi)} \longrightarrow V^{(H',\phi')}$$

famously operate on spaces of modular forms. Hence the question arises
whether $[JgH]$ may be extended to the above "resolution" for each (H,ϕ).

One of my favourite mathematical discoveries is the Shintani correspondence of [117] which is a bijection between Galois invariant complex
irreducibles of $GL_n\mathbb{F}_q$ and irreducibles of the Galois-fixed subgroup $GL_n\mathbb{F}_{q'}$.
For classical algebraic groups this is a consequence of Lang's Theorem, as
explained in [53] (see also [54] and Chapter Eight, §3). In the case of
local fields the analogue is Galois base change for admissible irreducibles
of GL_nK as mentioned in setting (i) above. By virtue of a theorem of
Tate, a Galois invariant admissible irreducible representation V is the same
as one which extends to an admissible irreducible representation of the
semi-direct product of the Galois group with GL_nK. This extension is
unique up to twists by one-dimensional characters. Hence the question
arises of constructing monomial resolutions of admissible representations of
such semi-direct products.

For finite-dimensional representations of a finite group G monomial resolutions were defined and shown to exist, unique in the derived category of
$_{k[G]}$mon in [19]. However, even in this case the functorial bar-monomial

resolution of Chapter One was previously unknown. It immediately extends the monomial resolutions to finite dimensional representations of absolute Galois groups and Weil groups. In the Langlands correspondence, as mentioned in setting (iii) above, Deligne representations of Weil groups are important in order to complete the correspondence on the Galois side. A Deligne representation is a finite dimensional representation of the Weil group together with a nilpotent endomorphism. Hence the question arises of constructing monomial resolutions of Deligne representations.

The local functional equation, as described in [86] and [142], is important in the characterisation of the Langlands correspondence, as mentioned in setting (iv) above. It uses the Fourier transform on vector spaces of eigendistributions. Hence the question arises of deriving functional equations in spaces of eigendistributions for each term in a monomial resolution.

This monograph is organised in the following manner. Details of the contents of each chapter are given in the chapter's introduction.

In Chapter One we shall recapitulate the theory of the category of $k[G]$-monomial modules and monomial resolutions of finitely generated $k[G]$-modules. When G is a finite group this material is due to Robert Boltje [19]. We shall be concerned (with a view eventually to treating the case of G a locally p-adic Lie group in later chapters) with the extension to the case where G is finite modulo the centre.

In Chapter Two we shall consider, in the local field case, the existence and structure of the monomial resolution of an admissible k-representation V of GL_2K with central character $\underline{\phi}$. The monomial resolution constructed in this case is unique in the derived category of $_{k[GL_2K],\underline{\phi}}\mathbf{mon}$.

There are two (possibly important) incongruous sections which I have included in Chapter Two. These are §7 and §8 concerning a "descent construction" which is a quotient monomial complex that one may construct from a monomial resolution. Appendix I was written several years before the majority of this monograph and contains a tediously lengthy, explicit analysis of the example of Shintani descent afforded by the Galois group $\mathrm{Gal}(\mathbb{F}_4/\mathbb{F}_2)$ acting on $GL_2\mathbb{F}_4$. These calculations allow one to calculate (in §6 and §7 of Appendix I) the Euler characteristic of the monomial resolutions of an extension of the Galois invariant irreducibles to the semi-direct product of the Galois group with $PGL_2\mathbb{F}_4$. In §8 of Appendix I the Euler characteristic of the descent construction is calculated when mapped to the representation ring and its data compared with the Shintani descent formula for this example. Even though the relevance and utility of the descent construction is highly speculative, I thought I should attempt to point out what it might be good for in the context of settings (i)–(iv). Accordingly Chapter Two §7 describes the descent construction in general and §8 gives the $(-)^{((J,\lambda))}$-data of the descent construction monomial complex in an example of an involutory outer automorphism of the dihedral group of order eight and in the Shintani descent example of Appendix I.

In Chapter Three one encounters the profound relation between automorphic representations and modular forms in [[51], [62], [67], [80]], for example. The topic is a breathtaking mathematical story of local-global flavour which has proved so important in number theory and arithmetic-algebraic geometry. Having already introduced monomial resolutions in the admissible local case, in this chapter I shall give a brief sketch of their introduction for global automorphic representations via the Tensor Product Theorem.

In Chapter Four I shall verify Conjecture 3.3 for $GL_n K$ for all $n \geq 2$ where K is a p-adic local field. For $GL_2 K$ this was accomplished (in Chapter Two, Theorem 4.9 and Corollary 4.10) by means of explicit formulae, in order to introduce the ideas of the general proof gradually. In this chapter I shall adopt a similar gradual approach, going into considerable detail in the $GL_3 K$ case before giving the general case.

For $GL_2 K$ the proof of Chapter Two Conjecture 3.3 was accomplished by constructing a double complex in $_{k[GL_2 K], \underline{\phi}}\mathbf{mon}$ using several bar-monomial resolutions together with a simplicial action on the tree for $GL_2 K$. For $GL_2 K$, by some low-dimensional good fortune, the construction of the differential in the double complex was made particularly easy (see the introduction to Chapter Two). For $GL_n K$ with $n \geq 3$ we have to use in a crucial way the naturality of the bar-monomial resolutions in order to apply the construction of the monomial complex given in Chapter Two §3. This requires a simplicial action on a space Y which, for $GL_n K$ with $n \geq 2$, we take to be the Bruhat-Tits building. Such buildings are constructed from BN-pairs.

In Chapter Five we recall the definition and properties of Deligne representations of the Weil group. In Conjecture 2.4 we describe the bar-monomial resolution resolution for a finite-dimensional Deligne representation $(\rho, V, \mathbf{n})$. The verification of Conjecture 2.4 should be straightforward but for the time being, out of laziness, I have left it unproved.

In [85] a Gauss sum is attached to each finite-dimensional complex irreducible representation V of $GL_n \mathbb{F}_q$. The Kondo-Gauss sum is a scalar $d \times d$-matrix where $d = \dim_{\mathbb{C}}(V)$. In Appendix III, §3 I recapitulate the construction of [85] but using the formulae in terms of character values, which simultaneously removes the irreducibility condition and reveals the functorial properties (e.g. invariance under induction; see Appendix III, Theorem 3.2).

In Chapter Six the theme is the association of ϵ-factors, L-functions and Kondo-style invariants to the terms in a monomial resolution of an admissible representation V of $GL_n K$ when K is a p-adic local field. The examples here suggest that eventually one may be able to construct the ϵ-factors and

L-functions of [66] by merely applying variations of my constructions to the monomial modules which occur in the monomial resolution of V and taking the Euler characteristic.

Chapter Seven recalls how Hecke operators

$$[JgH] : V^{(H,\phi)} \longrightarrow V^{(H',\phi')}$$

are defined in terms of the Double Coset Formula and explains how they fit in with the exact sequences

$$M_*^{((H,\phi))} \longrightarrow V^{(H,\phi)} \longrightarrow 0$$

which originate from a monomial resolution of the representation V. The chapter explains the conditions under which $[JgH]$ extends to the entire chain complex and gives a solitary illustrative example. In particular the latter may apply to the adèlic case of an automorphic representation. Then, if J, H are the usual congruence subgroups $\Gamma_0(N), \Gamma_1(N)$, the $[JgH]$'s are the classical Hecke operators and the $V^{(H,\phi)}$'s are spaces of modular forms ([51] §11.2).

Throughout this monograph Galois base change keeps being mentioned (particularly in Appendix I). Analogues of base change for representations are known in the context of modular forms ([55], [56], [81], [107], [91]; see ([39] pp. 84–88 and pp. 90–103)) and are predicted in the global Langlands Programme. As explained in Chapter Eight §2, functoriality of automorphic Galois base change would lead quickly to base change for modular forms. Therefore, Chapter Eight §1 is concerned with examples to illustrate the possibility of functoriality of Galois base change in the simpler context of [117]. To establish functoriality of base change in the case of Shintani descent one would need a different approach to the main result of [117]. With this in mind, Chapter Eight §3 sketches the original proof and then establishes the equivalence with a family of integrality conditions. Chapter Eight §4 introduces some curious polynomials with integer coefficients which were suggested by the discussion of §3. Chapter Eight §5 is a reminder for homotopy theorists and stable homotopy theorists of the functorial topological constructions which should exist as a consequence of (and as evidence for) functoriality of Shintani base change. Chapter Eight §6 examines an example of the inverse Shintani correspondence, where one starts with an irreducible for $GL_s\mathbb{F}_q$ and receives one for $GL_s\mathbb{F}_{q^n}$. The section finishes by posing a question, related to base change functoriality, about constructing a resolution of the target irreducible from the monomial resolution of the input irreducible.

Appendix I contains more explicit detail than any reader might conceivably want concerning Shintani descent from Galois invariant complex irreducible representations of $GL_2\mathbb{F}_4$ to $GL_2\mathbb{F}_2$. On the other hand, inter alia, it introduces a "descent algorithm" which may be of importance in connection with Galois descent. The "descent algorithm" is a procedure to construct an approximation to a monomial complex for $G \times H^G$ from a monomial resolution for an admissible representation of $G \propto H$, the semi-direct product of G acting on H.

As I have mentioned earlier, Appendix I was written several years before most of this monograph. This is not strictly true in relation to Appendix I, §11 which derives tables of $(-)^{((J,\lambda))}$-data for use in the descent construction examples of Chapter Two §7 and §8.

Appendix II consists of a version of the calculation, by Deligne and Henniart, of wildly ramified local root numbers modulo p-primary roots of unity where p is the residue characteristic. In this monograph it is relevant to the setting (iv) above. My hope is that a similar argument will allow one to construct ϵ-factors for admissible complex representations of $GL_n K$ (at least modulo p-primary roots of unity when K is local) term by term in the monomial resolution. In the case of complex admissible representations (as proved in Chapter One, §6) the monomial resolution has a "finite type" $\mathcal{P}_K$-adic filtration whose associated graded Euler characteristics are therefore finite. Included here because of its relevance, Appendix II has been gathering dust on my home page for several years, which accounts for its abstract!

Appendix III recalls in §1 the characterisation of irreducible complex representations of the symmetric groups and finite general linear groups, together with the construction of their zeta functions. Also the formulae for and the functorial properties of Kondo-Gauss sums [85] are explained.

Appendix IV assures the reader, without going into a single detail, that replacing $GL_n K$ and its Bruhat-Tits building by any locally p-adic Lie group and its Tammo tom Dieck space (aka its Baum-Connes space) $\underline{E}(G, \mathcal{C})$, where $\mathcal{C}$ is the family of compact modulo the centre subgroups $H \subseteq G$, results in a construction of functorial monomial resolutions for any admissible representation V of G with a fixed central character ϕ. The construction is accomplished by a direct imitation of that of Chapter Four.

As far as I know the embedding of the category of admissible representations into the monomial derived category is new, even in the case of finite groups. Apologies for my ignorance (characteristic of out of touch retirees, particularly in the UK) if it is not.

The research in this monograph was partially supported by a Leverhulme Emeritus Professorial Fellowship. I am very grateful to the Leverhulme Foundation without whose contribution I would have had no chance of travel to seek the advice of experts of this sort of representation theory.

I am especially very grateful to Jim Arthur, Paul Baum, Tobias Berger, Ken Brown, Paul Buckingham, Gerald Cliff, Ivan Fesenko, Guy Henniart, Florian Herzig, Steve Kudla, Ian Leary, Rob Kurinczuk, Jayanta Manoharmayum, Tom Oliver, Roger Plymen, Peter Schneider, Alexander Stasinski, Al Weiss for their interest and for their suggestions. As mentioned earlier, this monograph is intended to introduce the embedding of representations of locally p-adic Lie groups into the derived category of $_{k[G]}\mathbf{mon}$ and to give some nascent examples. In a further article I hope to explain the "Bernstein centre" [**14**] through the eyes of $_{k[G]}\mathbf{mon}$. The reader might also find this project interesting — it may not work out. (I have not had the opportunity to consult a haruspex![2])

Victor Snaith, FRSC, FFI, University of Sheffield,
September 2018

---

[2]An augury who foretells with the aid of the sacrifice of medium-sized farm animals.

CHAPTER 1

Finite modulo the centre groups

In this chapter we shall recapitulate the theory of the category of $k[G]$-monomial modules and monomial resolutions of finitely generated $k[G]$-modules. When G is a finite group this material is due to Robert Boltje [19]. Boltje's paper was the culmination of a series of articles concerning explicit (or canonical) versions of Brauer's induction theorem for finite groups[1], details of which are to be found in the series ([17], [18], [20], [102], [121], [122], [126]).

In this chapter we shall be concerned (with a view eventually to treating the case of G a locally p-adic Lie group in later chapters) with the extension to the case where G is finite modulo the centre.

Monomial resolutions for finite-dimensional complex representations of compact Lie groups were implicit in the original, topological construction of an explicit (or canonical) Brauer induction formula in [122]. This is because the formula was the Euler characteristic of a topologically constructed chain complex (this point of view was particularly stressed in [121] and [124]).

The crucial advance made by Robert Boltje in [19] was to describe the (additive) category $_{k[G]}\mathbf{mon}$ in whose derived category monomial resolutions naturally live and to develop all the techniques for working there when G is finite.

This chapter is arranged in the following manner. §1 sets up the notation and §2 defines a monomial resolution of a $k[G]$-module, which is a chain complex in the category $_{k[G]}\mathbf{mon}$. The category $_{k[G]}\mathbf{mon}$ is additive but not abelian so §3 introduces some functor categories which are used in §4 (in the style of what I imagine must have been the proof of the classical Freyd-Mitchell Theorem) to obtain a full embedding of the monomial category into a module category. This enables one to recognise a monomial resolution by mapping it to the module category where it becomes a projective resolution. In §5 I introduce a new canonical and functorial monomial resolution called the bar-monomial resolution. It is recognisable as a monomial resolution because it becomes the familiar bar resolution in the module category. The functoriality of the bar-monomial resolution is

---

[1]This was a classical problem (see the footnote [113] p. 71) of Richard Brauer from the 1940's. The first solution, using homotopy theory, appears in [121], [122].

essential because it permits an extension to the case where G is a locally p-adic Lie group which is compact, open modulo the centre. Thereafter the functoriality allows one to construct a sheaf of monomial resolutions on the Baum-Connes space $\underline{E}G$ when G is a locally p-adic Lie group. From this in Chapters Two and Four we obtain $_{k[G]}\mathbf{mon}$-monomial resolutions for admissible $k[G]$-representations in general.

1. Notation

1.1. The first difference between this chapter and [**19**] is that we shall assume that G is a finitely generated group with centre $Z(G)$ and finite quotient group $G/Z(G)$.

Fix a commutative Noetherian ring k and write $\hat{G}$ for the group of character homomorphisms $\mathrm{Hom}(G, k^*)$ from G to k^*, the multiplicative group of units of k. In addition we shall fix a central character $\underline{\phi} \in \mathrm{Hom}(Z(G), k^*)$. Let H be a subgroup of G which contains $Z(G)$ and denote by $\hat{H}_{\underline{\phi}}$ the finite subset of $\hat{H}$ consisting of characters which are equal to $\underline{\phi}$ when restricted to $Z(G)$.

For an arbitrary k-algebra A we write $_A\mathbf{mod}$ (resp. $\mathbf{mod}_A$) for the category of left (resp. right) A-modules. We denote by $_A\mathbf{lat}$ (resp. $\mathbf{lat}_A$) the category of left (resp. right) A-lattices i.e. the subcategory of $_A\mathbf{mod}$ consisting of those A-modules which are finitely generated and A-projective. The rank of a free k-module M will be denoted by $\mathrm{rk}_k(M)$. When $A = k[G]$ we have subcategories $_{k[G],\underline{\phi}}\mathbf{mod} \subset {}_{k[G]}\mathbf{mod}$ and $_{k[G],\underline{\phi}}\mathbf{lat} \subset {}_{k[G]}\mathbf{lat}$ whose objects are those on which $Z(G)$ acts via $\underline{\phi}$.

Let $\mathcal{M}_{\underline{\phi}}(G)$ denote the finite poset consisting of all pairs $(H, \phi) \in \hat{H}_{\underline{\phi}}$ where $(K, \psi) \leq (H, \phi)$ in the partial ordering if and only if $K \leq H$ and the restriction of ϕ to K is equal to ψ. Moreover $\mathcal{M}_{\underline{\phi}}(G)$ admits a left G-action by conjugation. That is, for $g \in G$, $g(H, \phi) = (gHg^{-1}, (g^{-1})^*(\phi))$ where $(g^{-1})^*(\phi)(ghg^{-1}) = \phi(h)$ for all $h \in H$. We shall write $N_G(H, \phi)$ for the G-stabiliser of (H, ϕ)

$$N_G(H, \phi) = \{g \in G \mid g(H, \phi) = (H, \phi)\}.$$

The G-orbit of (H, ϕ) will be denoted by $(H, \phi)^G$.

For $(H, \phi) \in \mathcal{M}_{\underline{\phi}}(G)$ we denote by k_ϕ the $k[H]$-module given by $h \cdot v = \phi(h)v$ for all $v \in k, h \in H$.

DEFINITION 1.2.[2]

A finite $(G, \underline{\phi})$-Line Bundle[3] over k is a left $k[G]$-module M together with a fixed finite direct sum decomposition

$$M = M_1 \oplus \cdots \oplus M_m$$

where each of the M_i is a free k-module of rank one on which $Z(G)$ acts via ϕ and the G-action permutes the M_i. The M_i's are called the Lines of M. For $1 \leq i \leq m$ let H_i denote the subgroup of G with stabilises the Line M_i. Then there exists a unique $\phi_i \in \hat{H}_{i\phi}$ such that $h \cdot v = \phi_i(h)v$ for all $v \in M_i, h \in H_i$. The pair $(H_i, \phi_i) \in \mathcal{M}_{\underline{\phi}}(G)$ is called the stabilising pair of M_i.

The k-submodule of M given by

$$M^{((H,\phi))} = \oplus_{1 \leq i \leq m, \ (H,\phi) \leq (H_i, \phi_i)} \ M_i$$

is called the (H, ϕ)-fixed points of M.

A morphism from M to the finite $(G, \underline{\phi})$-Line Bundle $N = N_1 \oplus \cdots \oplus N_n$ is defined to be a $k[G]$-module homomorphism $f : M \longrightarrow N$ such that

$$f(M^{((H,\phi))}) \subseteq N^{((H,\phi))}$$

for all $(H, \phi) \in \mathcal{M}_{\underline{\phi}}(G)$. The (left) finite $(G, \underline{\phi})$-Line Bundles and their morphisms define an additive category denoted by $_{k[G],\underline{\phi}}\mathbf{mon}$.

By definition each $(G, \underline{\phi})$-Line Bundle is a k-free $k[G]$-module so there is a forgetful functor

$$\mathcal{V} : \ _{k[G],\underline{\phi}}\mathbf{mon} \longrightarrow \ _{k[G],\underline{\phi}}\mathbf{mod}.$$

1.3. *Some natural operations*

There are several operations which are obvious lifts to $_{k[G],\underline{\phi}}\mathbf{mon}$ of well-known operations in $_{k[G],\underline{\phi}}\mathbf{mod}$. This means that the resulting functors commute with the forgetful functor $\mathcal{V} : \ _{k[G],\underline{\phi}}\mathbf{mon} \longrightarrow \ _{k[G],\underline{\phi}}\mathbf{mod}$.

(i) Direct sum: If $M = M_1 \oplus \cdots \oplus M_m$ and $N = N_1 \oplus \cdots \oplus N_n$ are objects in $_{k[G],\underline{\phi}}\mathbf{mon}$ then so is

$$M \oplus N = M_1 \oplus \cdots \oplus M_m \oplus N_1 \oplus \cdots \oplus N_n.$$

---

[2]When we come to the bar-monomial resolution in Chapter One §5 and its subsequent passage from finite groups to locally compact modulo the centre subgroups of locally p-adic Lie groups in Chapter Two and later it becomes clear that I should have set up the monomial category by stipulating that we are given the $M^{((H,\phi))}$'s with the property that they are "Lineable" rather than that we are given the Lines. This would require a straightforward but extensive revision which, at my age and with my resources, is unlikely to happen! Apologies!

[3]The capital letters are chosen there to distinguish the Line Bundle from the familiar vector bundle terminology.

(ii) Tensor product: If M belongs to $_{k[G],\underline{\phi_M}}\mathbf{mon}$ and N belongs to $_{k[G],\underline{\phi_N}}\mathbf{mon}$ then the tensor product

$$M \otimes_k N = \oplus_{1 \leq i \leq m, 1 \leq j \leq n} M_i \otimes_k N_j$$

belongs to $_{k[G],\underline{\phi_M \cdot \phi_N}}\mathbf{mon}$.

(iii) Homomorphisms: The $k[G]$-module $\mathrm{Hom}_k(M,N)$ with the decomposition into Lines of the form $\oplus_{1 \leq i \leq m, 1 \leq j \leq n} \mathrm{Hom}_k(M_i,N_j)$ is an object of $_{k[G],\underline{\phi_M \cdot \phi_N}^{-1}}\mathbf{mon}$ if G acts via the usual formula $(g \cdot f)(m) = g \cdot (f(g^{-1} \cdot m))$ for $g \in G, m \in M$. As a special case we have the dual of M given by $\mathrm{Hom}_k(M,k)$.

(iv) Restriction: Let $f : G' \longrightarrow G$ be a homomorphism of finitely generated groups which are both finite modulo the centre and such that $f(Z(G')) \subseteq Z(G)$. Then we have a restriction map Res_f from $_{k[G],\underline{\phi}}\mathbf{mod}$ to $_{k[G'],\underline{\phi} \cdot f}\mathbf{mod}$ and similarly Res_f from $_{k[G],\underline{\phi}}\mathbf{mon}$ to $_{k[G'],\underline{\phi} \cdot f}\mathbf{mon}$.

(v) Induction: Suppose that $H \subseteq G$ are finitely generated groups which are both finite modulo the centre with $Z(G) \subseteq H$. Then the index of H in G is finite and the usual induced $k[G]$-module $\mathrm{Ind}_H^G(P)$ for $P \in {}_{k[H],\underline{\phi}}\mathbf{mod}$ is the object of $_{k[G],\underline{\phi}}\mathbf{mod}$ given by $k[G] \otimes_{k[H]} P$. If $P = P_1 \oplus \cdots \oplus P_s$ lies in $_{k[H],\underline{\phi}}\mathbf{mon}$ then $\mathrm{Ind}_H^G(P)$ with the Line-decomposition given by $\oplus_{g,1 \leq i \leq s} g \otimes_{k[H]} P_i$, as g runs through a set of coset representatives for G/H, is the object of $_{k[G],\underline{\phi}}\mathbf{mon}$ denoted by $\underline{\mathrm{Ind}}_H^G(P)$.

Note that, if the stabilising pair for P_i is (H_i, ϕ_i) then the stabilising pair of the Line $g \otimes_{k[H]} P_i$ is $g(H_i, \phi_i)$. Also for $(H, \phi) \in \mathcal{M}_{\underline{\phi}}(G)$ then we have

$$\oplus_{(J,\psi) \in (H,\phi)^G} L_{(J,\psi)} \cong \underline{\mathrm{Ind}}_{N_G(H,\phi)}^G(k_\phi)$$

where (J, ψ) runs through the G-conjugates of (H, ϕ) and $L_{(J,\psi)}$ is the J-module k_ψ.

(vi) Canonical isomorphisms: Analogues of the usual distributivity isomorphism of direct sums over tensor products, the Frobenius reciprocity isomorphism and the Mackey decomposition isomorphism all hold in the Line Bundle context (see [19] §1.5(f)–(h)).

PROPOSITION 1.4. ([19] §1.6 and §1.7)
If $M = M_1 \oplus \cdots \oplus M_m$ and $N = N_1 \oplus \cdots \oplus N_n$ are objects in $_{k[G],\underline{\phi}}\mathbf{mon}$ the following statements are equivalent:

(i) M and N are isomorphic in $_{k[G],\underline{\phi}}\mathbf{mon}$.

(ii) For all $(H, \phi) \in \mathcal{M}_{\underline{\phi}}(G)$ the $(N_G(H, \phi), \underline{\phi})$-Line Bundles over k

$$M(H, \phi) = \oplus_{\text{stabilising pair of } M_i \text{ equals } (H,\phi)} M_i$$

and

$$N(H, \phi) = \oplus_{\text{stabilising pair of } N_j \text{ equals } (H,\phi)} N_j$$

are isomorphic in $_{k[N_G(H,\phi)],\underline{\phi}}\mathbf{mod}$.

(iii) For all $(H, \phi) \in \mathcal{M}_{\underline{\phi}}(G)$ the $(N_G(H, \phi), \underline{\phi})$-Line Bundles over k

$$M(H, \phi) = \oplus_{\text{stabilising pair of } M_i \text{ equals } (H, \phi)} M_i$$

and

$$N(H, \phi) = \oplus_{\text{stabilising pair of } N_j \text{ equals } (H, \phi)} N_j$$

are isomorphic in $_{k[N_G(H,\phi)],\underline{\phi}}\mathbf{mon}$.

(iv) For all $(H, \phi) \in \mathcal{M}_{\underline{\phi}}(G)$ the $(G, \underline{\phi})$-Line Bundles over k

$$M((H, \phi)^G) = \oplus_{\text{stabilising pair of } M_i \in (H,\phi)^G} M_i$$

and

$$N((H, \phi)^G) = \oplus_{\text{stabilising pair of } N_j \in (H,\phi)^G} N_j$$

are isomorphic in $_{k[G],\underline{\phi}}\mathbf{mon}$.

In fact M is isomorphic in $_{k[G],\underline{\phi}}\mathbf{mon}$ to the direct sum over the distinct G-orbits on $\mathcal{M}_{\underline{\phi}}(G)$ of the $M((H, \phi)^G)$'s.

1.5. We call a finite $(G, \underline{\phi})$-Line Bundle $M = M_1 \oplus \cdots \oplus M_m$ over k indecomposable if it is not isomorphic to a non-trival direct sum $N \oplus P$ in $_{k[G],\underline{\phi}}\mathbf{mon}$. If we form the direct sum of the lines of a single G-orbit then we obtain a finite $(G, \underline{\phi})$-Line Bundle over k and every M may be written as the direct sum of these. Therefore, if M is indecomposable, then G acts transitively on the Lines of M. In this case $M \cong \underline{\mathrm{Ind}}_{H_i}^G(k_{\phi_i})$ for any $1 \leq i \leq m$ where (H_i, ϕ_i) is the stabilising pair of the Line M_i. Explicitly, the isomorphism is given by sending $g \otimes_{H_i} v \in \underline{\mathrm{Ind}}_{H_i}^G(k_{\phi_i})$ to $g \cdot v \in M$.

Therefore for each object M in $_{k[G],\underline{\phi}}\mathbf{mon}$ we have a sum over the G-orbits of $\mathcal{M}_{\underline{\phi}}(G)$ and uniquely determined integers $r_{(h,\phi)}(M) \geq 0$ such that

$$M \cong \oplus_{G \backslash \mathcal{M}_{\underline{\phi}}(G)} r_{(h,\phi)}(M) \cdot \underline{\mathrm{Ind}}_H^G(k_\phi)$$

where $r_{(h,\phi)}(M) \cdot P$ denotes the $r_{(h,\phi)}(M)$-fold direct sum of copies of P.

Note that, in the notation of Proposition 1.4(iv),

$$\mathrm{rk}_k(M((H, \phi)^G)) = [G : H] \cdot r_{(h,\phi)}(M).$$

PROPOSITION 1.6. ([19] §1.9)
The set of finite $(G, \underline{\phi})$-Line Bundles over k given by

$$\{\underline{\mathrm{Ind}}_H^G(k_\phi) \mid (H, \phi) \in G \backslash \mathcal{M}_{\underline{\phi}}(G)\}$$

is a full set of pairwise non-isomorphic representatives for the isomorphism classes of indecomposable objects in $_{k[G],\underline{\phi}}\mathbf{mon}$. Moreover each finite $(G, \underline{\phi})$-Line Bundle M over k is isomorphic to the direct sum of objects $\underline{\mathrm{Ind}}_H^G(k_\phi)$ with $(H, \phi) \in G \backslash \mathcal{M}_{\underline{\phi}}(G)$ and uniquely determined multiplicity $r_{(h,\phi)}(M) = \mathrm{rk}_k(M((H, \phi)^G))/[G : H]$.

COROLLARY 1.7. *([19] §1.10)*

Let M, N be objects of $_{k[G],\phi}\mathbf{mon}$. In the notation of Proposition 1.4 the following are equivalent:

(i) M and N are isomorphic in $_{k[G],\phi}\mathbf{mon}$.

(ii) For all $(H, \phi) \in \mathcal{M}_\phi(G)$, $rk_k(M((H,\phi)^G)) = rk_k(N((H,\phi)^G))$.

(iii) For all $(H, \phi) \in \mathcal{M}_{\underline{\phi}}(G)$, $rk_k(M(H,\phi)) = rk_k(N(H,\phi))$.

(iv) For all $(H, \phi) \in \mathcal{M}_\phi(G)$, $rk_k(M^{((H,\phi))}) = rk_k(N^{((H,\phi))})$.

1.8. Let $(K, \psi), (H, \phi) \in \mathcal{M}_\phi(G)$ and let $g \in G$. Define a morphism

$$f_g \in \mathrm{Hom}_{k[G],\underline{\phi}\mathbf{mon}}(\mathrm{Ind}_K^G(k_\psi), \mathrm{Ind}_H^G(k_\phi))$$

by the formula

$$f_g(g' \otimes_K v) = \begin{cases} g'g \otimes_H v & \text{if } (K, \psi) \le (gHg^{-1}, (g^{-1})^*(\phi)) \\ 0 & \text{otherwise.} \end{cases}$$

This is well-defined because, for $k' \in K$,

$$f_g(g'k' \otimes_K \psi(k')^{-1}v) = g'k'g \otimes_H \psi(k')^{-1}v$$

$$= g'gg^{-1}k'g \otimes_H \psi(k')^{-1}v$$

$$= g'g \otimes_H \phi(g^{-1}k'g)\psi(k')^{-1}v$$

$$= g'g \otimes_H \psi(k')\psi(k')^{-1}v$$

$$= f_g(g' \otimes_K v).$$

The composition of morphisms

$$\mathrm{Hom}_{k[G],\underline{\phi}\mathbf{mon}}(\mathrm{Ind}_K^G(k_\psi), \mathrm{Ind}_H^G(k_\phi)) \times \mathrm{Hom}_{k[G],\underline{\phi}\mathbf{mon}}(\mathrm{Ind}_H^G(k_\phi), \mathrm{Ind}_U^G(k_\mu))$$

$$\downarrow$$

$$\mathrm{Hom}_{k[G],\phi\mathbf{mon}}(\mathrm{Ind}_K^G(k_\psi), \mathrm{Ind}_U^G(k_\mu))$$

is given by $(f_g, f_{g_1}) \mapsto f_{gg_1} = f_{g_1} \cdot f_g$.

If $h \in H$ and $f_g \in \mathrm{Hom}_{k[G],\underline{\phi}\mathbf{mon}}(\mathrm{Ind}_K^G(k_\psi), \mathrm{Ind}_H^G(k_\phi))$ then so does f_{gh} and $f_{gh} = \phi(h)f_g$ since

$$f_{gh}(g' \otimes_K v) = g'gh \otimes_H v = \phi(h)g'g \otimes_H v = \phi(h)f_g(g' \otimes_K v).$$

Similarly if $k \in K$ then $f_{kg} = \psi(k)f_g$. In particular f_{kg}, f_{gh} and f_g generate the same line

$$\langle f_{kg} \rangle = \langle f_{gh} \rangle = \langle f_g \rangle \subset \mathrm{Hom}_{k[G],\underline{\phi}\mathbf{mon}}(\mathrm{Ind}_K^G(k_\psi), \mathrm{Ind}_H^G(k_\phi)).$$

LEMMA 1.9. ([19] §1.11)

(i) Let $(K, \psi) \in \mathcal{M}_{\underline{\phi}}(G)$ and let N be an object of $_{k[G],\underline{\phi}}\mathbf{mon}$. Then there is a k-linear isomorphism

$$\mathrm{Hom}_{k[G],\underline{\phi}\mathbf{mon}}(\underline{\mathrm{Ind}}_K^G(k_\psi), N) \xrightarrow{\cong} N^{((K,\psi))}$$

given by $f \mapsto f(1 \otimes_K 1)$. The inverse isomorphism is given by

$$n \mapsto ((g \otimes_K v \mapsto vg \cdot n)).$$

(ii) Let $(K, \psi), (H, \phi) \in \mathcal{M}_{\underline{\phi}}(G)$. In the notation of §1.8 there is a k-linear isomorphism

$$\mathrm{Hom}_{k[G],\phi\mathbf{mon}}(\underline{\mathrm{Ind}}_K^G(k_\psi), \underline{\mathrm{Ind}}_H^G(k_\phi))$$

$$\cong \frac{k \langle f_g \mid (K,\psi) \leq (gHg^{-1}, (g^{-1})^*(\phi)) \rangle}{\langle f_{gh} - \phi(h)f_g, \langle f_{kg} - \psi(k)f_g \mid h \in H, k \in K \rangle}.$$

(iii) $\mathrm{Hom}_{k[G],\underline{\phi}\mathbf{mon}}(\underline{\mathrm{Ind}}_K^G(k_\psi), \underline{\mathrm{Ind}}_H^G(k_\phi))$ is a free k-module of finite rank with basis given by $f_{g_1}, \ldots, f_{g_t}$ where g_i runs through the subset of double coset representatives of $K \backslash G / H$ such that

$$(K, \psi) \leq (g_i H g_i^{-1}, (g_i^{-1})^*(\phi)).$$

LEMMA 1.10. ([19] §1.12)

Consider the diagram

$$M \xrightarrow{h} N \xleftarrow{f} P$$

in which $M, P \in_{k[G],\underline{\phi}} \mathbf{mon}$ and $N \in_{k[G],\underline{\phi}} \mathbf{mod}$ with h, f being morphisms in $_{k[G],\underline{\phi}}\mathbf{mod}$. In particular we include the situation where $N' \in_{k[G],\underline{\phi}} \mathbf{mon}$ with h, f being morphisms to N' in $_{k[G],\underline{\phi}}\mathbf{mon}$ and the diagram above being the result of applying the forgetful functor $\mathcal{V}$ with $N = \mathcal{V}(N')$. Assume, for all $(H, \phi) \in \mathcal{M}_{\underline{\phi}}(G)$, that

$$f(P^{((H,\phi))}) \subseteq h(M^{((H,\phi))}).$$

Then there exists $j \in \mathrm{Hom}_{k[G],\underline{\phi}\mathbf{mon}}(P, M)$ such that $h \cdot j = f$.

REMARK 1.11. *Partial central characters* $\underline{\phi}'$

There is an obvious analogous version of this section with partial central characters $\underline{\phi}'$. That is, one fixes a central subgroup $H' \subseteq Z(G)$ such that G/H' is finite and fixes $\underline{\phi}' \in \hat{H}'$. Then one repeats the section with $(Z(G), \underline{\phi})$ replaced by $(H', \underline{\phi}')$.

The case of finite G, which is treated in [19] is the case in which $H' = \{1\}$.

2. Monomial resolutions

2.1. Let $G, k, \underline{\phi}$ be as in §1.1. Let V be a finitely generated $k[G]$-module on which $Z(G)$ acts via the central character $\underline{\phi}$. That is, V is an object of $_{k[G],\underline{\phi}}\mathbf{mod}$. In this section we shall shall define the notion of a $_{k[G],\underline{\phi}}\mathbf{mon}$-resolution of V. This is a chain complex of morphisms in $_{k[G],\underline{\phi}}\mathbf{mon}$ with certain properties which will ensure that it is exists and is unique up to chain homotopy in $_{k[G],\underline{\phi}}\mathbf{mon}$.

2.2. For $V \in {}_{k[G],\underline{\phi}} \mathbf{mod}$ and $(H, \phi) \in \mathcal{M}_{\underline{\phi}}(G)$ define the (H, ϕ)-fixed points of V by

$$V^{(H,\phi)} = \{v \in V \mid h \cdot v = \phi(h)v \text{ for all } h \in H\}.$$

Clearly $g(V^{(H,\phi)}) = V^{g(H,\phi)}$, $V^{(Z(G),\underline{\phi})} = V$ and $(K, \psi) \leq (H, \phi)$ implies that $V^{(H,\phi)} \subseteq V^{(K,\psi)}$. Note that $f \in \mathrm{Hom}_{k[G],\underline{\phi}\mathbf{mod}}(V, W)$ satisfies $f(V^{(H,\phi)}) \subseteq W^{(H,\phi)}$ for all $(H, \phi) \in \mathcal{M}_{\underline{\phi}}(G)$. In addition, if $M \in {}_{k[G],\underline{\phi}} \mathbf{mon}$ then $M^{((H,\phi))} \subseteq M^{(H,\phi)}$ so that $f \in \mathrm{Hom}_{k[G],\underline{\phi}\mathbf{mod}}(\mathcal{V}(M), V)$ satisfies $f(M^{((H,\phi))}) \subseteq V^{(H,\phi)}$ for all $(H, \phi) \in \mathcal{M}_{\underline{\phi}}(G)$.

DEFINITION 2.3. ([19] §2.2)
Let $V \in {}_{k[G],\underline{\phi}} \mathbf{mod}$. A $_{k[G],\underline{\phi}}\mathbf{mon}$-resolution of V is a chain complex

$$M_* : \quad \cdots \xrightarrow{\partial_{i+1}} M_{i+1} \xrightarrow{\partial_i} M_i \xrightarrow{\partial_{i-1}} \cdots \xrightarrow{\partial_1} M_1 \xrightarrow{\partial_0} M_0$$

with $M_i \in {}_{k[G],\underline{\phi}} \mathbf{mon}$ and $\partial_i \in \mathrm{Hom}_{k[G],\underline{\phi}\mathbf{mon}}(M_{i+1}, M_i)$ for all $i \geq 0$ together with $\epsilon \in \mathrm{Hom}_{k[G],\underline{\phi}\mathbf{mod}}(\mathcal{V}(M_0), V)$ such that

$$\cdots \xrightarrow{\partial_i} M_i^{((H,\phi))} \xrightarrow{\partial_{i-1}} \cdots \xrightarrow{\partial_1} M_1^{((H,\phi))} \xrightarrow{\partial_0} M_0^{((H,\phi))} \xrightarrow{\epsilon} V^{(H,\phi)} \longrightarrow 0$$

is an exact sequence of k-modules for each $(H, \phi) \in \mathcal{M}_{\underline{\phi}}(G)$. In particular, when $(H, \phi) = (Z(G), \underline{\phi})$ we see that

$$\cdots \xrightarrow{\partial_i} M_i \xrightarrow{\partial_{i-1}} \cdots \xrightarrow{\partial_1} M_1 \xrightarrow{\partial_0} M_0 \xrightarrow{\epsilon} V \longrightarrow 0$$

is an exact sequence in $_{k[G],\underline{\phi}}\mathbf{mod}$.

PROPOSITION 2.4.
Let $V \in {}_{k[G],\underline{\phi}} \mathbf{mod}$ and let

$$\cdots \longrightarrow M_n \xrightarrow{\partial_{n-1}} M_{n-1} \xrightarrow{\partial_{n-2}} \cdots \xrightarrow{\partial_0} M_0 \xrightarrow{\epsilon} V \longrightarrow 0$$

be a $_{k[G],\underline{\phi}}\mathbf{mon}$-resolution of V. Suppose that

$$\cdots \longrightarrow C_n \xrightarrow{\partial'_{n-1}} C_{n-1} \xrightarrow{\partial'_{n-2}} \cdots \xrightarrow{\partial'_0} C_0 \xrightarrow{\epsilon'} V \longrightarrow 0$$

a chain complex where each ∂_i' and C_i belong to $_{k[G],\underline{\phi}}\mathbf{mon}$ and ϵ' is a $_{k[G],\underline{\phi}}\mathbf{mod}$ homomorphism such that $\epsilon'(C_0^{((H,\phi))}) \subseteq V^{(H,\phi)}$ for each $(H,\phi) \in \mathcal{M}_\phi(G)$.

Then there exists a chain map of $_{k[G],\underline{\phi}}\mathbf{mon}$-morphisms $\{f_i : C_i \longrightarrow M_i, i \geq 0\}$ such that

$$\epsilon \cdot f_0 = \epsilon', \ f_{i-1} \cdot \partial_i' = \partial_i \cdot f_i \text{ for all } i \geq 1.$$

In addition, if $\{f_i' : C_i \longrightarrow M_i, i \geq 0\}$ is another chain map of $_{k[G],\underline{\phi}}\mathbf{mon}$-morphisms such that $\epsilon \cdot f_0 = \epsilon \cdot f_0'$ then there exists a $_{k[G],\underline{\phi}}\mathbf{mon}$-chain homotopy $\{s_i : C_i \longrightarrow M_{i+1}, \text{ for all } i \geq 0\}$ such that $\partial_i \cdot s_i + s_{i-1} \cdot \partial_i' = f_i - f_i'$ for all $i \geq 1$ and $f_0 - f_0' = \partial_0 \cdot s_0$.

Proof

This is the usual homological algebra argument using Lemma 1.10. □

REMARK 2.5. Needless to say, Proposition 2.4 has an analogue to the effect that every $_{k[G],\underline{\phi}}\mathbf{mod}$-homomorphism $V \longrightarrow V'$ extends to a $_{k[G],\underline{\phi}}\mathbf{mon}$-morphism between the monomial resolutions of V and V', if they exist, and the extension is unique up to $_{k[G],\underline{\phi}}\mathbf{mon}$-chain homotopy.

3. Some functor categories

3.1. The category $_{k[G],\underline{\phi}}\mathbf{mon}$ is additive but not abelian. Homological algebra (e.g. a projective resolution) is more conveniently accomplished in an abelian category. To overcome this difficulty we shall embed $_{k[G],\underline{\phi}}\mathbf{mon}$ into more convenient abelian categories. This is reminiscent of the Freyd-Mitchell Theorem which embeds every abelian category into a category of modules.

3.2. *The functor category* $funct_k^o(_{k[G],\underline{\phi}}\mathbf{mon},_k \mathbf{mod})$

Let $funct_k^o(_{k[G],\underline{\phi}}\mathbf{mon},_k \mathbf{mod})$ denote the category of contravariant functors, $\mathcal{F}, \mathcal{G}$ etc., from $_{k[G],\underline{\phi}}\mathbf{mon}$ to the category of finitely generated k-modules whose morphisms are k-linear natural transformations $\alpha : \mathcal{F} \longrightarrow \mathcal{G}$ etc.

Let $_{k[G],\underline{\phi}}\mathbf{mod}$ denote the category of finite rank $k[G]$-modules with central character $\underline{\phi}$ (see §1.1). Consider the functor

$$\mathcal{I} :_{k[G],\underline{\phi}} \mathbf{mod} \longrightarrow funct_k^o(_{k[G],\underline{\phi}}\mathbf{mon},_k \mathbf{mod})$$

given on objects by

$$\mathcal{I}(V) = \mathrm{Hom}_{k[G],\underline{\phi}\mathbf{mod}}(\mathcal{V}(-),V)$$

with $\mathcal{I}(\beta : V \longrightarrow W) = (f \mapsto \beta \cdot f)$.

Now we shall consider morphisms. In order to keep track not only of $g \in G$ but also of K and H we shall write f_g of §1.8 as a triple

$$((K, \psi), g, (H, \phi)) : \operatorname{Ind}_K^G(k_\psi) \longrightarrow \operatorname{Ind}_H^G(k_\phi).$$

Then the composition

$$\operatorname{Hom}_{k[G], \underline{\phi}\mathbf{mon}}(\operatorname{Ind}_K^G(k_\psi), \operatorname{Ind}_H^G(k_\phi)) \times \operatorname{Hom}_{k[G], \underline{\phi}\mathbf{mon}}(\operatorname{Ind}_H^G(k_\phi), \operatorname{Ind}_U^G(k_\mu))$$

$$\downarrow$$

$$\operatorname{Hom}_{k[G], \underline{\phi}\mathbf{mon}}(\operatorname{Ind}_K^G(k_\psi), \operatorname{Ind}_U^G(k_\mu))$$

is given by $(((K, \psi), g, (H, \phi)), ((H, \phi), g_1, (U, \mu))) \mapsto (((K, \psi), gg_1, (U, \mu)))$.
The tautological equality $(gHg^{-1}, (g^{-1})^*(\phi)) = (gHg^{-1}, (g^{-1})^*(\phi))$ yields an isomorphism

$$((gHg^{-1}, (g^{-1})^*(\phi)), g, (H, \phi)) : \operatorname{Ind}_{gHg^{-1}}^G(k_{(g^{-1})^*(\phi)}) \overset{\cong}{\longrightarrow} \operatorname{Ind}_H^G(k_\phi)$$

given by $g' \otimes_{gHg^{-1}} v \mapsto g'g \otimes_H v$, which is an automorphism if and only if $g \in N_G(H, \phi)$, the normaliser of (H, ϕ).

The morphism $((K, \psi), g, (H, \phi))$ induces

$$((K, \psi), g, (H, \phi))^* : \mathcal{I}(V)(\operatorname{Ind}_H^G(k_\phi)) \longrightarrow \mathcal{I}(V)(\operatorname{Ind}_K^G(k_\psi))$$

given by the pre-composition $((K, \psi), g, (H, \phi))^*(f) = f \cdot ((K, \psi), g, (H, \phi))$.
There is an isomorphism of k-modules (analogous to §1.9(i))

$$\mathcal{I}(V)(\operatorname{Ind}_H^G(k_\phi)) \cong V^{(H, \phi)}$$

given by sending $v \in V^{(H, \phi)}$ to the $k[G, \underline{\phi}]$-mod morphism

$$\operatorname{Ind}_H^G(k_\phi)) \longrightarrow V \qquad g_1 \otimes_H \alpha \mapsto \alpha g_1 v$$

for $\alpha \in k$. Therefore $((K, \psi), g, (H, \phi))^*$ corresponds to a k-linear map

$$V^{(H, \phi)} \longrightarrow V^{(K, \psi)}$$

given by $v' \mapsto gv'$ since

$$g_1 \otimes_K \alpha \longrightarrow g_1 g \otimes_H \alpha \longrightarrow \alpha g_1 g v'$$

is the map which corresponds to $gv' \in V^{(K, \psi)}$. This makes sense because, if $z \in K$, then $zgv' = gg^{-1}zgv' = \phi(g^{-1}zg)gv' = \psi(z)gv'$.

3.3. The functor $\mathcal{I}$ on morphisms

Let $V, W \in_{k[G], \underline{\phi}} \mathbf{mod}$ and set $\mathcal{F}(-) = \mathcal{I}(V)(-)$, $\mathcal{G}(-) = \mathcal{I}(W)(-)$.
Given a natural transformation $\alpha : \mathcal{F} \longrightarrow \mathcal{G}$ for each $M \in_{k[G], \underline{\phi}} \mathbf{mon}$ we have

$$\alpha(M) : \mathcal{F}(M) \longrightarrow \mathcal{G}(M)$$

such that if $\beta : M \longrightarrow M'$ is a morphism in $_{k[G], \underline{\phi}}\mathbf{mon}$ we have a commutative diagram

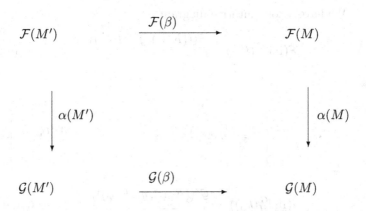

If we have a homomorphism $\gamma : V \longrightarrow W$ in $_{k[G],\underline{\phi}}\mathbf{mod}$ we obtain a natural transformation

$$\gamma_* : \mathcal{F} \longrightarrow \mathcal{G}$$

given by $\gamma_*(M)(f) = \gamma \cdot f \in \mathcal{G}(M)$ for all $f : M \longrightarrow V$ in $\mathcal{F}(M)$. However, given any natural transformation α there is a unique homomorphism γ in $_{k[G],\underline{\phi}}\mathbf{mod}$ such that $\alpha = \gamma_*$. This is seen by the following discussion. We have a morphism in $_{k[G],\underline{\phi}}\mathbf{mon}$

$$((K,\psi),g,(H,\phi)) : \underline{\mathrm{Ind}}_K^G(k_\psi) \longrightarrow \underline{\mathrm{Ind}}_H^G(k_\phi).$$

For example, when $(K,\psi) = (Z(G),\underline{\phi}) = (H,\phi)$ we have

$$((Z(G),\underline{\phi}),g,(Z(G),\underline{\phi})).$$

We have a commutative diagram

$$\mathcal{F}(\underline{\mathrm{Ind}}_H^G(k_\phi)) \xrightarrow{\ \mathcal{F}(((K,\psi),g,(H,\phi)))\ } \mathcal{F}(\underline{\mathrm{Ind}}_K^G(k_\psi))$$

$$\downarrow \alpha(\underline{\mathrm{Ind}}_H^G(k_\phi)) \qquad\qquad \alpha(\underline{\mathrm{Ind}}_K^G(k_\psi)))\downarrow$$

$$\mathcal{G}(\underline{\mathrm{Ind}}_H^G(k_\phi)) \xrightarrow{\ \mathcal{G}(((K,\psi),g,(H,\phi)))\ } \mathcal{G}(\underline{\mathrm{Ind}}_K^G(k_\psi))$$

Taking $(K,\psi) = (Z(G),\underline{\phi}) = (H,\phi)$ the commutative square may be identified with

$$V \xrightarrow{\ (g\cdot -)\ } V$$

$$\downarrow \alpha(\underline{\mathrm{Ind}}_{Z(G)}^G(k_{\underline{\phi}})) \qquad\qquad \alpha(\underline{\mathrm{Ind}}_{Z(G)}^G(k_{\underline{\phi}}))\downarrow$$

$$W \xrightarrow{\ (g\cdot -)\ } W$$

which shows that $\alpha(\underline{\mathrm{Ind}}_{Z(G)}^G(k_{\underline{\phi}}))$ is a homomorphism in $_{k[G],\underline{\phi}}\mathbf{mod}$. Now setting $(K,\psi) = (Z(\overline{G}),\underline{\phi})$ we have a morphism

$$((Z(G),\underline{\phi}),g,(H,\phi)): \underline{\mathrm{Ind}}_{Z(G)}^G(k_{\underline{\phi}}) \longrightarrow \underline{\mathrm{Ind}}_H^G(k_\phi).$$

This yields the commutative diagram

$$
\begin{array}{ccc}
\mathcal{F}(\underline{\mathrm{Ind}}_H^G(k_\phi)) & \xrightarrow{\ \mathcal{F}(((Z(G),\underline{\phi}),g,(H,\phi)))\ } & \mathcal{F}(\underline{\mathrm{Ind}}_{Z(G)}^G(k_{\underline{\phi}})) \\[2em]
\Big\downarrow{\scriptstyle\alpha(\underline{\mathrm{Ind}}_H^G(k_\phi))} & & \Big\downarrow{\scriptstyle\alpha(\underline{\mathrm{Ind}}_{Z(G)}^G(k_{\underline{\phi}}))} \\[2em]
\mathcal{G}(\underline{\mathrm{Ind}}_H^G(k_\phi)) & \xrightarrow{\ \mathcal{G}(((Z(G),\underline{\phi}),g,(H,\phi)))\ } & \mathcal{G}(\underline{\mathrm{Ind}}_{Z(G)}^G(k_{\underline{\phi}}))
\end{array}
$$

which in turn may be identified with

$$
\begin{array}{ccc}
V^{(H,\phi)} & \xrightarrow{\ \mathcal{F}(((Z(G),\underline{\phi}),g,(H,\phi)))\ } & V \\[2em]
\Big\downarrow{\scriptstyle\alpha(\underline{\mathrm{Ind}}_H^G(k_\phi))} & & \Big\downarrow{\scriptstyle\alpha(\underline{\mathrm{Ind}}_{Z(G)}^G(k_{\underline{\phi}}))} \\[2em]
W^{(H,\phi)} & \xrightarrow{\ \mathcal{G}(((Z(G),\underline{\phi}),g,(H,\phi)))\ } & W
\end{array}
$$

The horizontal maps are injective. In fact, for example, the upper horizontal map sends $v \in V^{(H,\phi)}$ to $g \cdot v \in V$. For $v \in V^{(H,\phi)}$ corresponds to the homomorphism $g_1 \otimes_H \alpha \mapsto \alpha g_1 \cdot v$. Since $\mathcal{F}(((Z(G),\underline{\phi}),g,(H,\phi)))$ corresponds to pre-composition with f_g we see that the image of v corresponds to the homomorphism $g_1 \otimes_{Z(G)} \alpha \mapsto g_1 g \otimes_H \alpha \mapsto \alpha g_1 g \cdot v$ which is identified with $g \cdot v \in V$.

Hence $\alpha(\underline{\mathrm{Ind}}_H^G(k_\phi))$ is uniquely determined by $\alpha(\underline{\mathrm{Ind}}_{Z(G)}^G(k_{\underline{\phi}}))$ and so we have established the following proposition, which follows from the previous discussion together with additivity of the functors and natural transformations.

PROPOSITION 3.4.

In the notation of §3.3 given any natural transformation

$$\alpha : \mathcal{F} \longrightarrow \mathcal{G}$$

there is a unique homomorphism $\gamma : V \longrightarrow W$ in $_{k[G],\underline{\phi}}\mathbf{mod}$ such that $\alpha = \gamma_*$, which completely determines α.

PROPOSITION 3.5. ([19] §3.2)

Let $\mathcal{I}$ denote the functor of §3.3 and define a functor

$$\mathcal{J} :_{k[G],\underline{\phi}} \mathbf{mon} \longrightarrow funct_k^o(_{k[G],\underline{\phi}}\mathbf{mon},_k \mathbf{mod})$$

by $\mathcal{J}(M) = \mathrm{Hom}_{k[G],\underline{\phi}\mathbf{mon}}(-, M)$.

Then the category $funct_k^o(_{k[G],\underline{\phi}}\mathbf{mon},_k \mathbf{mod})$ is abelian. Furthermore both $\mathcal{I}$ and $\mathcal{J}$ are full embeddings (i.e. bijective on morphisms and hence injective on isomorphism classes of objects).

Proof

By Yoneda's Lemma $\mathcal{J}$ is a full embedding. The result for $\mathcal{I}$ follows from Proposition 3.4. □

PROPOSITION 3.6. ([19] §3.3)

For $M \in_{k[G],\underline{\phi}} \mathbf{mon}$ the functor $\mathcal{J}(M)$ in $funct_k^o(_{k[G],\underline{\phi}}\mathbf{mon},_k \mathbf{mod})$ is projective.

Proof

Suppose that we have a diagram in $funct_k^o(_{k[G],\underline{\phi}}\mathbf{mon},_k \mathbf{mod})$ of the form

$$\mathcal{G} \xrightarrow{\alpha} \mathcal{H} \xleftarrow{\beta} \mathcal{J}(M)$$

in which α is surjective. Therefore for every $N \in_{k[G],\underline{\phi}} \mathbf{mon}$ the homomorphism $\alpha(N) : \mathcal{G}(N) \longrightarrow \mathcal{H}(N)$ is surjective. By Yoneda's Lemma natural transformations from $\mathcal{J}(M)$ to $\mathcal{G}$ correspond bijectively to the elements of $\mathcal{G}(M)$. Similarly the natural transformations from $\mathcal{J}(M)$ to $\mathcal{H}$ correspond bijectively to the elements of $\mathcal{H}(M)$. The fact that $\alpha(M)$ is surjective is therefore equivalent to the fact that there exists a natural transformation γ from $\mathcal{J}(M)$ to $\mathcal{G}$ such that $\alpha \cdot \gamma = \beta$. □

REMARK 3.7. In general $funct_k^o(_{k[G],\underline{\phi}}\mathbf{mon},_k \mathbf{mod})$ has more projectives than just the $\mathcal{J}(M)$'s. A complete analysis of all the projectives in $funct_k^o(_{k[G],\underline{\phi}}\mathbf{mon},_k \mathbf{mod})$ may be given along the lines of the finite group case, which is given in ([19] §3.4 and §3.8).

DEFINITION 3.8.

Let $M \in_{k[G],\phi} \mathbf{mon}, V \in_{k[G],\phi} \mathbf{mod}$. Define a k-linear isomorphism $\mathcal{K}_{M,V}$ of the form

$$\mathrm{Hom}_{k[G],\phi\mathbf{mod}}(\mathcal{V}(M),V) \xrightarrow{\mathcal{K}_{M,V}} \mathrm{Hom}_{funct_k^o(k[G],\phi\mathbf{mon},k\mathbf{mod})}(\mathcal{J}(M),\mathcal{I}(V))$$

by sending $f : \mathcal{V}(M) \longrightarrow V$ to the natural transformation

$$\mathcal{K}_{M,V}(N) : \mathcal{J}(M)(N) \longrightarrow \mathcal{I}(V)(N)$$

given by $h \mapsto f \cdot \mathcal{V}(h)$ for all $N \in_{k[G],\phi} \mathbf{mon}$

$$\mathrm{Hom}_{k[G],\phi\mathbf{mon}}(N,M) \longrightarrow \mathrm{Hom}_{k[G],\phi\mathbf{mod}}(\mathcal{V}(N),V).$$

The inverse isomorphism is given by $\mathcal{K}_{M,V}^{-1}(\phi) = \phi(M)(1_M)$ where 1_M denotes the identity morphism on M.

In fact $\mathcal{K}$ is a functorial equivalence of the form

$$\mathcal{K} : \mathrm{Hom}_{k[G],\phi\mathbf{mod}}(\mathcal{V}(-),-) \xrightarrow{\cong} \mathrm{Hom}_{funct_k^o(k[G],\phi\mathbf{mon},k\mathbf{mod})}(\mathcal{J}(-),\mathcal{I}(-)).$$

THEOREM 3.9. ([19] §3.6)

Let

$$\ldots \xrightarrow{\partial_i} M_i \xrightarrow{\partial_{i-1}} \ldots \xrightarrow{\partial_1} M_1 \xrightarrow{\partial_0} M_0 \xrightarrow{\epsilon} V \longrightarrow 0$$

be a chain complex with $M_i \in_{k[G],\phi} \mathbf{mon}$ for $i \geq 0$, $V \in_{k[G],\phi} \mathbf{mod}$, $\partial_i \in \mathrm{Hom}_{k[G],\phi\mathbf{mon}}(M_{i+1},M_i)$ and $\epsilon \in \mathrm{Hom}_{k[G],\phi\mathbf{mod}}(\mathcal{V}(M_0),V)$. Then the following are equivalent:

(i) $M_* \longrightarrow V$ is a $k[G],\phi\mathbf{mon}$-resolution of V.

(ii) The sequence

$$\ldots \xrightarrow{\mathcal{J}(\partial_i)} \mathcal{J}(M_i) \xrightarrow{\mathcal{J}(\partial_{i-1})} \ldots \xrightarrow{\mathcal{J}(\partial_1)} \mathcal{J}(M_1) \xrightarrow{\mathcal{J}(\partial_0)} \mathcal{J}(M_0) \xrightarrow{\mathcal{K}_{M_0,V}(\epsilon)} \mathcal{I}(V) \longrightarrow 0$$

is exact in $funct_k^o(k[G],\phi\mathbf{mon},k\ \mathbf{mod})$.

REMARK 3.10. Theorem 3.9 together with Proposition 3.5 and Proposition 3.6 imply that the map

$$(M_* \xrightarrow{\epsilon} V) \mapsto (\mathcal{J}(M_*) \xrightarrow{\mathcal{K}_{M_0,V}(\epsilon)} \mathcal{I}(V))$$

is a bijection between the $k[G],\phi\mathbf{mon}$-resolutions of V and the projective resolutions of $\mathcal{I}(V)$ consisting of objects from the subcategory $\mathcal{J}(k[G],\phi\mathbf{mon})$.

4. From functors to modules

4.1. The functor Φ_M

Let $M \in_{k[G],\phi} \mathbf{mon}$ and let $\mathcal{A}_M = \mathrm{Hom}_{k[G],\phi\mathbf{mon}}(M,M)$, the ring of endomorphisms on M under composition. By Lemma 1.9 $\mathcal{A}_M$ is a finitely generated k-algebra.

In this section we shall show that there is an equivalence of categories between $funct_k^o({}_{k[G],\underline{\phi}}\mathbf{mon},{}_k\mathbf{mod})$ and the category of right modules $\mathbf{mod}_{\mathcal{A}_M}$ for a suitable choice of M.

We have a functor

$$\Phi_M : funct_k^o({}_{k[G],\underline{\phi}}\mathbf{mon},{}_k\mathbf{mod}) \longrightarrow \mathbf{mod}_{\mathcal{A}_M}$$

given by $\Phi(\mathcal{F}) = \mathcal{F}(M)$. Right multiplication by $z \in \mathcal{A}_M$ on $v \in \mathcal{F}(M)$ is given by

$$v\#z = \mathcal{F}(z)(v)$$

where $\mathcal{F}(z) : \mathcal{F}(M) \longrightarrow \mathcal{F}(M)$ is the left k-module morphism obtained by applying $\mathcal{F}$ to the endomorphism z. This is a right-$\mathcal{A}_M$ action since

$$v\#(zz_1) = \mathcal{F}(zz_1)(v) = (\mathcal{F}(z_1) \cdot \mathcal{F}(z))(v) = \mathcal{F}(z_1)(\mathcal{F}(z)(v)) = (v\#z)\#z_1.$$

In the other direction define a functor

$$\Psi_M : \mathbf{mod}_{\mathcal{A}_M} \longrightarrow funct_k^o({}_{k[G],\underline{\phi}}\mathbf{mon},{}_k\mathbf{mod}),$$

for $P \in \mathbf{mod}_{\mathcal{A}_M}$, by

$$\Psi_M(P) = \mathrm{Hom}_{\mathcal{A}_M}(\mathrm{Hom}_{{}_{k[G],\underline{\phi}}\mathbf{mon}}(M,-),P).$$

Here, for $N \in {}_{k[G],\underline{\phi}}\mathbf{mon}$, $\mathrm{Hom}_{{}_{k[G],\underline{\phi}}\mathbf{mon}}(M,-)$ is a right $\mathcal{A}_M$-module via pre-composition by endomorphisms of M. For a homomorphism of $\mathcal{A}_M$-modules $f : P \longrightarrow Q$ the map $\Psi_M(f)$ is given by composition with f.

Next we consider the composite functor

$$\Phi_M \cdot \Psi_M : \mathbf{mod}_{\mathcal{A}_M} \longrightarrow \mathbf{mod}_{\mathcal{A}_M}.$$

This is given by $P \mapsto \mathrm{Hom}_{\mathcal{A}_M}(\mathrm{Hom}_{{}_{k[G],\underline{\phi}}\mathbf{mon}}(M,M),P) = \mathrm{Hom}_{\mathcal{A}_M}(\mathcal{A}_M,P)$ so that there is an obvious natural transformation $\eta : 1 \xrightarrow{\cong} \Phi_M \cdot \Psi_M$ such that $\eta(P)$ is an isomorphism for each module P.

Now consider the composite functor

$$\Psi_M \cdot \Phi_M : funct_k^o({}_{k[G],\underline{\phi}}\mathbf{mon},{}_k\mathbf{mod}) \longrightarrow funct_k^o({}_{k[G],\underline{\phi}}\mathbf{mon},{}_k\mathbf{mod}).$$

For a functor $\mathcal{F}$ we shall define a natural transformation

$$\epsilon_{\mathcal{F}} : \mathcal{F} \longrightarrow \mathrm{Hom}_{\mathcal{A}_M}(\mathrm{Hom}_{{}_{k[G],\underline{\phi}}\mathbf{mon}}(M,-),\mathcal{F}(M)) = \Psi_M \cdot \Phi_M(\mathcal{F}).$$

For $N \in {}_{k[G],\underline{\phi}}\mathbf{mon}$ we define

$$\epsilon_{\mathcal{F}}(N) : \mathcal{F}(N) \longrightarrow \mathrm{Hom}_{\mathcal{A}_M}(\mathrm{Hom}_{{}_{k[G],\underline{\phi}}\mathbf{mon}}(M,N),\mathcal{F}(M))$$

by the formula $v \mapsto (f \mapsto \mathcal{F}(f)(v))$.

THEOREM 4.2. ([19] §3.8)

Let $S \in_{k[G], \underline{\phi}}$ **mon** be the finite $(G, \underline{\phi})$-Line Bundle over k given by

$$S = \oplus_{(H, \phi) \in \mathcal{M}_{\phi}(G)} \underline{\mathrm{Ind}}_H^G(k_\phi).$$

Then, in the notation of §4.1,

$$\Phi_S : funct_k^o({}_{k[G], \underline{\phi}} \textbf{mon}, _k \textbf{mod}) \longrightarrow \textbf{mod}_{A_S}$$

and

$$\Psi_S : \textbf{mod}_{A_S} \longrightarrow funct_k^o({}_{k[G], \underline{\phi}} \textbf{mon}, _k \textbf{mod})$$

are inverse equivalences of categories. In fact, the natural transformations η and ϵ are isomorphisms of functors when $M = S$.

REMARK 4.3. Theorem 4.2 is true when S is replaced by any M which is the direct sum of $\underline{\mathrm{Ind}}_H^G(k_\phi)$'s containing at least one pair (H, ϕ) from each G-orbit of $\mathcal{M}_{\phi}(G)$. That is, for any $(G, \underline{\phi})$-Line Bundle containing

$$\oplus_{(H, \phi) \in G \backslash \mathcal{M}_{\phi}(G)} \underline{\mathrm{Ind}}_H^G(k_\phi)$$

as a summand. This remark is established by Morita theory ([88] p. 636).

5. The bar-monomial resolution

5.1. *The bar resolution*

We begin this section by recalling the two-sided bar-resolution for A-modules. Let k be a commutative Noetherian ring and let A be a (not necessarily commutative) k-algebra. For each integer $p \geq 0$ set

$$B_p(M, A, N) = M \otimes_k A \otimes_k A \otimes_k \ldots \otimes_k A \otimes_k N$$

in which there are p copies of A and $M \in \textbf{mod}_A, N \in_A \textbf{mod}$. Define

$$d : B_1(M, A, N) \longrightarrow B_0(M, A, N)$$

by $d(m \otimes a \otimes n) = m \cdot a \otimes n - m \otimes a \cdot n$. For $p \geq 2$ define

$$d : B_p(M, A, N) \longrightarrow B_{p-1}(M, A, N)$$

by

$$d(m \otimes a_1 \otimes \ldots \otimes a_p \otimes n)$$

$$= m \cdot a_1 \otimes \ldots \otimes a_p \otimes n$$

$$+ \sum_{i=1}^{p-1} (-1)^i m \otimes a_1 \otimes \ldots \otimes a_i \cdot a_{i+1} \otimes \ldots \otimes a_p \otimes n$$

$$+ (-1)^p m \otimes a_1 \otimes \ldots \otimes a_p \cdot n.$$

Setting $N = A$ we define

$$\epsilon : B_0(M, A, A) = M \otimes_k A \longrightarrow M$$

by $\epsilon(m \otimes a) = m \cdot a$ and

$$\eta : M \longrightarrow B_0(M, A, A) = M \otimes_k A$$

by $\eta(m) = m \otimes 1$. Finally define, for $p \geq 0$,

$$s : B_p(M, A, A) \longrightarrow B_{p+1}(M, A, A)$$

by $s(m \otimes a_1 \otimes \ldots \otimes a_p \otimes a) = (-1)^{p+1} m \otimes a_1 \otimes \ldots \otimes a_p \otimes a \otimes 1$.

With these definitions, if $p \geq 2$, we have, for $i \geq 0$,

$$dd = 0 : B_{i+2}(M, A, A) \longrightarrow B_i(M, A, A) \text{ and } \epsilon.d = 0 : B_1(M, A, A) \longrightarrow M.$$

Also we have

$$1 = \epsilon \cdot \eta : M \longrightarrow B_0(M, A, A) \longrightarrow M$$

and for $p \geq 0$

$$ds + sd = 1 : B_p(M, A, A) \longrightarrow B_p(M, A, A).$$

Finally we have

$$ds + \eta\epsilon = 1 : B_0(M, A, A) \longrightarrow B_0(M, A, A).$$

All the d's and ϵ are right A-module maps if the A-multiplication is given by multiplication on the right-hand factor only.

Therefore we have established the following well-known result concerning the bar resolution for a right A-module.

PROPOSITION 5.2.

In the situation of §5.1 the chain complex

$$\ldots \longrightarrow B_p(M, A, A) \xrightarrow{d} B_{p-1}(M, A, A) \xrightarrow{d} \ldots$$

$$\ldots \longrightarrow B_1(M, A, A) \xrightarrow{d} B_0(M, A, A) \xrightarrow{\epsilon} M \longrightarrow 0$$

is a free right-A-module resolution of M.

5.3. As in §3.8 and §4.1, let $M \in_{k[G],\phi}$ mon, $V \in_{k[G],\phi}$ mod and let $\mathcal{A}_M = \text{Hom}_{k[G],\underline{\phi}\text{mon}}(M, M)$, the ring of endomorphisms on M under composition. For $i \geq 0$ define $\tilde{M}_{M,i} \in {}_k\text{mod}$ by (i copies of $\mathcal{A}_M$)

$$\tilde{M}_{M,i} = \text{Hom}_{k[G],\underline{\phi}\text{mod}}(\mathcal{V}(M), V) \otimes_k \mathcal{A}_M \otimes_k \ldots \otimes_k \mathcal{A}_M$$

and set

$$\underline{M}_{M,i} = \tilde{M}_{M,i} \otimes_k \text{Hom}_{k[G],\underline{\phi}\text{mon}}(-, M).$$

Hence $\underline{M}_{M,i} \in funct_k^o({}_{k[G],\underline{\phi}}\text{mon}, {}_k\text{mod})$ and in fact the values of this functor are not merely objects in ${}_k\text{mod}$ because they have a natural right $\mathcal{A}_M$-module structure, defined as in §4.1.

If $i \geq 1$ we defined natural transformations $d_{M,0}, d_{M,1}, \ldots, d_{M,i}$ in the following way. Define

$$d_{M,0} : \underline{M}_{M,i} \longrightarrow \underline{M}_{M,i-1}$$

by
$$d_{M,0}(f \otimes \alpha_1 \otimes \ldots \otimes \alpha_i \otimes u) = f(- \cdot \alpha_1) \otimes \alpha_2 \ldots \otimes \alpha_i \otimes u.$$
The map $f(- \cdot \alpha_1) : \mathcal{V}(M) \longrightarrow V$ is a $_{k[G],\underline{\phi}}\mathbf{mod}$- homomorphism since α_i acts on the right of M.

For $1 \leq j \leq i - 1$ we define
$$d_{M,j} : \underline{M}_{M,i} \longrightarrow \underline{M}_{M,i-1}$$
by
$$d_{M,j}(f \otimes \alpha_1 \otimes \ldots \otimes \alpha_i \otimes u) = f \otimes \alpha_1 \ldots \otimes \alpha_j\alpha_{j+1} \otimes \ldots \otimes \alpha_i \otimes u.$$
Finally
$$d_{M,i} : \underline{M}_{M,i} \longrightarrow \underline{M}_{M,i-1}$$
is given by
$$d_i(M)(f \otimes \alpha_1 \otimes \ldots \otimes \alpha_i \otimes u) = f \otimes \alpha_1 \otimes \ldots \otimes \alpha_{i-1} \otimes \alpha_i \cdot u.$$
Since u is a $_{k[G],\underline{\phi}}\mathbf{mon}$-morphism so is $\alpha_i \cdot u$ because
$$(\alpha_i \cdot u)(\alpha m) = \alpha_i(u(\alpha m)) = \alpha_i(\alpha u(m)) = \alpha\alpha_i(u(m)) = \alpha(\alpha_i \cdot u)(m)$$
since α_i is a $_{k[G],\underline{\phi}}\mathbf{mon}$ endomorphism of M.

Next we define a natural transformation
$$\epsilon_M : \underline{M}_{M,0} \longrightarrow \mathcal{I}(V) = \mathrm{Hom}_{k[G],\underline{\phi}\mathbf{mod}}(\mathcal{V}(-), V)$$
by sending $f \otimes u \in \underline{M}_{M,0}$ to $f \cdot \mathcal{V}(u) \in \mathcal{I}(V)$.

Finally we define
$$d_M = \sum_{j=0}^{i} (-1)^j d_{M,j} : \underline{M}_{M,i} \longrightarrow \underline{M}_{M,i-1}.$$

THEOREM 5.4.

The sequence
$$\ldots \xrightarrow{d_M} \underline{M}_{M,i}(M) \xrightarrow{d_M} \underline{M}_{M,i-1}(M) \ldots \xrightarrow{d_M} \underline{M}_{M,0}(M) \xrightarrow{\epsilon_M} \mathcal{I}(V)(M) \longrightarrow 0$$
is the right $\mathcal{A}_M$-module bar resolution of $\mathcal{I}(V)(M)$.

5.5. *The functorial monomial resolution of V*

Let V be a finite rank k-lattice with a left G-action. Let $M \in {}_{k[G],\underline{\phi}}\mathbf{mon}$ and $W \in {}_k\mathbf{lat}$. Define another object $W \otimes_k M \in {}_{k[G],\underline{\phi}}\mathbf{mon}$ by letting G act only on the M-factor, $g(w \otimes m) = w \otimes gm$, and defining the Lines of $W \otimes_k M$ to consist of the one-dimensional subspaces $\langle w \otimes L \rangle$ where $w \in W$, runs through a k-basis of W, and L is a Line of M. Therefore, if $M' \in {}_{k[G],\underline{\phi}}\mathbf{mon}$ we have an isomorphism
$$\mathrm{Hom}_{k[G],\underline{\phi}\mathbf{mon}}(W \otimes_k M, M') \xrightarrow{\cong} W \otimes_k \mathrm{Hom}_{k[G],\underline{\phi}\mathbf{mon}}(M, M')$$

providing that W is finite-dimensional. Similarly we have an isomorphism

$$\mathrm{Hom}_{k[G],\underline{\phi}\mathbf{mon}}(M, W \otimes_k M') \xrightarrow{\cong} W \otimes_k \mathrm{Hom}_{k[G],\underline{\phi}\mathbf{mon}}(M, M')$$

when W is finite dimensional.

As in Theorem 4.2, let $S \in_{k[G],\underline{\phi}} \mathbf{mon}$ be the finite $(G, \underline{\phi})$-Line Bundle over k given by

$$S = \oplus_{(H,\phi) \in \mathcal{M}_{\underline{\phi}}(G)} \underline{\mathrm{Ind}}_H^G(k_\phi).$$

As in §5.3, for $i \geq 0$ we have $\tilde{M}_{S,i} \in {}_k\mathbf{mod}$ by (i copies of $\mathcal{A}_S$)

$$\tilde{M}_{S,i} = \mathrm{Hom}_{k[G],\underline{\phi}\mathbf{mod}}(\mathcal{V}(S), V) \otimes_k \mathcal{A}_S \otimes_k \ldots \otimes_k \mathcal{A}_S,$$

which is a finite dimensional k-lattice. As a k-basis for $\tilde{M}_{S,i}$ we take the tensor product of the direct sum of bases for each $V^{(H,\phi)}$ and a basis for each $\mathcal{A}_S$-factor given by the f_g's of Lemma 1.9. Note that, conveniently, the product of two f_g's is either zero or an f_g. Therefore we may form $\tilde{M}_{S,i} \otimes_k S \in {}_{k[G],\underline{\phi}}\mathbf{mon}$.

Recall from §3.2 that

$$\mathrm{Hom}_{k[G],\underline{\phi}\mathbf{mod}}(\mathcal{V}(S), V) = \mathcal{I}(V)(S) \cong \oplus_{(H,\phi) \in \mathcal{M}_{\underline{\phi}}(G)} V^{(H,\phi)},$$

which we shall assume is a finite dimensional k-lattice. In our principal application where k is a field this will be fulfilled automatically.

We have morphisms in ${}_{k[G],\underline{\phi}}\mathbf{mon}$ for $i \geq 1$

$$d_0, d_1, \ldots, d_i : \tilde{M}_{S,i} \otimes_k S \longrightarrow \tilde{M}_{S,i-1} \otimes_k S$$

defined on

$$f \otimes \alpha_1 \otimes \ldots \otimes \alpha_i \otimes s \in \mathrm{Hom}_{k[G,\underline{\phi}]-mod}(\mathcal{V}(S), V) \otimes_k \mathcal{A}_S^{\otimes^i} \otimes_k S$$

by

$$d_0(f \otimes \alpha_1 \otimes \ldots \otimes \alpha_i \otimes s) = f \cdot \mathcal{V}(\alpha_1) \otimes \alpha_2 \otimes \ldots \otimes \alpha_i \otimes s,$$

and for $1 \leq j \leq i - 1$

$$d_j(f \otimes \alpha_1 \otimes \ldots \otimes \alpha_i \otimes s) = f \otimes \alpha_1 \otimes \ldots \otimes \alpha_j\alpha_{j+1} \ldots \otimes \alpha_i \otimes s$$

$$d_i(f \otimes \alpha_1 \otimes \ldots \otimes \alpha_i \otimes s) = f \otimes \alpha_1 \otimes \ldots \otimes \alpha_{i-1} \otimes \alpha_i(s).$$

Setting $d = \sum_{j=0}^i (-1)^j d_j$ gives a morphism in ${}_{k[G],\underline{\phi}}\mathbf{mon}$

$$d : \tilde{M}_{S,i} \otimes_k S \longrightarrow \tilde{M}_{S,i-1} \otimes_k S$$

for all $i \geq 1$. In addition we define a homomorphism in ${}_{k[G],\underline{\phi}}\mathbf{mod}$

$$\epsilon : \tilde{M}_{S,0} \otimes_k S = \mathrm{Hom}_{k[G],\underline{\phi}\mathbf{mod}}(\mathcal{V}(S), V) \otimes_k S \longrightarrow V$$

by $\epsilon(f \otimes s) = f(s)$.

The chain complex

$$\cdots \xrightarrow{d} \tilde{M}_{S,i} \otimes_k S \xrightarrow{d} \cdots \xrightarrow{d} \tilde{M}_{S,1} \otimes_k S \xrightarrow{d} \tilde{M}_{S,0} \otimes_k S \xrightarrow{\epsilon} V \longrightarrow 0$$

satisfies the conditions of Theorem 3.9. Therefore, abbreviating $\tilde{M}_{S,i} \otimes_k S$ to M_i pro tem, this chain complex is a $_{k[G],\underline{\phi}}$**mon**-resolution of V if and only if the sequence

$$\cdots \xrightarrow{\mathcal{J}(d)} \mathcal{J}(M_i) \xrightarrow{\mathcal{J}(d)} \cdots \xrightarrow{\mathcal{J}(d)} \mathcal{J}(M_1) \xrightarrow{\mathcal{J}(d)} \mathcal{J}(M_0) \xrightarrow{\mathcal{K}_{M_0,V}(\epsilon)} \mathcal{I}(V) \longrightarrow 0$$

is exact in $funct^o_k(_{k[G],\underline{\phi}}\mathbf{mon},_k \mathbf{mod})$. By Theorem 4.2 this chain complex of functors is exact if and only if the result of applying Φ_S to it is exact in the category $\mathbf{mod}_{A_S}$. However, by Theorem 5.4 with $M = S$ the resulting chain complex in $\mathbf{mod}_{A_S}$ is the bar resolution, which is exact.

Therefore, taking k to be a field in order to ensure the lattice conditions, we have proved the following result.

THEOREM 5.6. *Existence of the bar-monomial resolution*
Let k be a field. Then, in the notation of §5.5, The chain complex

$$\cdots \xrightarrow{d} \tilde{M}_{S,i} \otimes_k S \xrightarrow{d} \cdots \xrightarrow{d} \tilde{M}_{S,1} \otimes_k S \xrightarrow{d} \tilde{M}_{S,0} \otimes_k S \xrightarrow{\epsilon} V \longrightarrow 0$$

is a $_{k[G],\underline{\phi}}$**mon**-resolution of V.

REMARK 5.7. Using Theorem 3.9 and Theorem 4.2 to prove Theorem 5.6 had the advantage that it guided us directly from the classical bar resolution for rings and modules to a description of the bar-monomial resolution. However, now that we have its description, it is presumably straightforward to construct the correct type of "contracting homotopy" which would immediately show that the complex of Theorem 5.6 is a monomial resolution.

5.8. *Naturality - inclusions of subgroups*
Suppose we have an inclusion homomorphism $i : G \subseteq J$ of finite modulo the centre groups with $i(Z(G)) \subseteq Z(J)$. Suppose that G and J have central characters $\underline{\phi}_G$ and $\underline{\phi}_J$, respectively, which satisfy $\underline{\phi}_G = \underline{\phi}_J \cdot i$.
As in Theorem 4.2 define $S_G \in_{k[G],\underline{\phi}} \mathbf{mon}$ and $S_J \in_{k[J],\underline{\phi}} \mathbf{mon}$ to be

$$S_G = \oplus_{(H,\phi)\in\mathcal{M}_{\underline{\phi}}(G)} \underline{\mathrm{Ind}}_H^G(k_\phi)$$

and

$$S_J = \oplus_{(H',\phi')\in\mathcal{M}_{\underline{\phi}}(J)} \underline{\mathrm{Ind}}_{H'}^J(k_{\phi'}).$$

We have a $_{k[G],\underline{\phi}}$**mon**-morphism, as in §3.2,

$$((K,\psi),g,(H,\phi)) : S_G \longrightarrow S_G$$

associated to each triple $((K,\psi),g,(H,\phi))$ which satisfies the condition that $(K,\psi) \leq (gHg^{-1},(g^{-1})^*(\phi))$. It maps $\underline{\mathrm{Ind}}_K^G(k_\psi) \longrightarrow \underline{\mathrm{Ind}}_H^G(k_\phi)$ via

$g' \otimes_K v \mapsto g'g \otimes_H v$ and is zero on the other summands. Varying over the set of such triples, we have

$$\mathcal{A}_{S_G} = \mathrm{Hom}_{k[G],\underline{\phi}\mathbf{mon}}(S_G, S_G) \cong k\langle((K,\psi), g, (H,\phi))\rangle / \simeq$$

where $\simeq$ is the subspace generated by

$$((K,\psi), gh, (H,\phi)) - \phi(h)((K,\psi), g, (H,\phi)) \text{ for } h \in H.$$

Then $\mathcal{A}_{S_G}$ is a ring under composition which is given in terms of generators by

$$((K,\psi), g, (H,\phi)) \cdot ((H,\phi), u, (U,\mu)) = ((K,\psi), gu, (U,\mu)).$$

The inclusion homomorphism $G \subseteq J$ means that we have a map on triples which sends $((K,\psi), g, (H,\phi))$, a triple for G, to $((K,\psi), g, (H,\phi))$ considered as a triple for J. This preserves composition and the relation $\simeq$ so we have a ring hommorphism

$$i_{G,J} : \mathcal{A}_{S_G} \longrightarrow \mathcal{A}_{S_J}.$$

Next, if V is a representation of J then Frobenius reciprocity gives an isomorphism, also $i_{G,J}$,

$$\mathrm{Hom}_{k[G],\underline{\phi}\mathbf{mod}}(\mathcal{V}(\underline{\mathrm{Ind}}_H^G(k_\phi)), \mathrm{Res}_G^J(V)) \xrightarrow{\cong} \mathrm{Hom}_{k[J],\underline{\phi}\mathbf{mod}}(\mathcal{V}(\underline{\mathrm{Ind}}_H^J(k_\phi)), V)$$

which sends f to $i_{G,J}(f) : j \otimes_H v \mapsto jf(1 \otimes_H v)$.

We have two routes from $\mathrm{Hom}_{k[G],\underline{\phi}\mathbf{mod}}(\mathcal{V}(\underline{\mathrm{Ind}}_H^G(k_\phi)), \mathrm{Res}_G^J(V)) \otimes \mathcal{A}_{S_G}$ to $\mathrm{Hom}_{k[J],\underline{\phi}\mathbf{mod}}(\mathcal{V}(\underline{\mathrm{Ind}}_H^J(k_\phi)), V)$. Starting with $f : \underline{\mathrm{Ind}}_H^G(\phi)) \longrightarrow \mathrm{Res}_G^J(V)$ and $((K,\psi), g, (H,\phi))$ we may form $i_{G,J}(f \cdot \mathcal{V}(((K,\psi), g, (H,\phi))))$ given by

$$g_1 \otimes_K v \mapsto g_1 g \otimes_H v \mapsto g_1 g(f(1 \otimes_H v) \in V$$

or we can form $i_{G,J}(f) \cdot i_{G,J}(\mathcal{V}((K,\psi), g, (H,\phi)))$ which is given by

$$j \otimes_K v \mapsto jg \otimes_H v \mapsto jgf(1 \otimes_H v).$$

Therefore the two routes agree.

Also we have two routes $\mathcal{A}_G \otimes S_G \longrightarrow S_J$ given by evaluation followed by $i_{G,J}$ or $i_{G,J} \otimes i_{G,J}$ followed by evaluation. Both are given by

$$((K,\psi), g, (H,\phi)) \otimes (g_1 \otimes_K v) \mapsto g_1 g \otimes_H v.$$

We define a G-map $i_{G,J} : S_G \longrightarrow S_J$ by sending $g \otimes_H v \in \underline{\mathrm{Ind}}_H^G(k_\phi)$ to $g \otimes_H v \in \underline{\mathrm{Ind}}_H^J(k_\phi)$.

Therefore it is easy to see that $i_{G,J}$ induces a canonical homomorphism of bar-monomial resolutions of Theorem 5.6

$$i_{G,J} : (\tilde{M}_{S_G,*} \otimes_k S_G \xrightarrow{\epsilon} \mathrm{Res}_G^J(V)) \longrightarrow (\tilde{M}_{S_J,*} \otimes_k S_J \xrightarrow{\epsilon} V)$$

which commutes with the augmentations to V and with the left-action by the subgroup G.

In addition, if $G \subseteq J \subseteq H$ is a chain of groups the canonical homomorphisms are transitive in the sense that $i_{J,H} \cdot i_{G,J} = i_{G,H}$.

5.9. Naturality - surjections onto quotient groups

Let $N \lhd G$ be a normal subgroup which acts trivially on V. In this case the central character $\underline{\phi}$ factorises through G/N and we shall also denote by $\underline{\phi}$ the resulting central character on G/N. If $\phi \in \hat{H}$ is trivial on $H \bigcap N$ then ϕ induces a unique character $\tilde{\phi}$ on HN/N. Let $\pi : G \longrightarrow G/N$ denote the quotient map. Define

$$\pi_{G,G/N} : S_G \longrightarrow S_{G/N}$$

to be zero on summands $\underline{\mathrm{Ind}}_H^G(\phi)$ unless $N \subseteq H$ and $\mathrm{Res}_N^H(\phi)$ is trivial. If $N \subseteq H$ and $\mathrm{Res}_N^H(\phi)$ is trivial then there is an $_{k[G],\underline{\phi}}$mon-isomorphism

$$\pi_{G,G/N} : \underline{\mathrm{Ind}}_H^G(k_\phi) \xrightarrow{\cong} \underline{\mathrm{Ind}}_{H/N}^{G/N}(k_{\tilde{\phi}})$$

given by $g \otimes_H v \mapsto \pi(g) \otimes_{H/N} v$.

There is a ring homomorphism $\pi_{G,G/N} : \mathcal{A}_G \longrightarrow \mathcal{A}_{G/N}$ given by sending $((K,\psi), g, (H,\phi))$ to zero unless N is a subgroup of $H \bigcap K$, which is equivalent to $N \subseteq K$, and $\mathrm{Res}_N^H(\phi)$ (hence also $\mathrm{Res}_N^H(\psi)$) is trivial. Otherwise

$$\pi_{G,G/N}((K,\psi), g, (H,\phi)) = ((K/N, \tilde{\psi}), \pi(g), (H/N, \tilde{\phi})).$$

There is an isomorphism

$$\mathrm{Hom}_{k[G],\underline{\phi}\mathbf{mod}}(\mathcal{V}(\underline{\mathrm{Ind}}_H^G(k_\phi)), V) \xrightarrow{\pi_{G,G/N}} \mathrm{Hom}_{k[G/N],\underline{\phi}\mathbf{mod}}(\mathcal{V}(\underline{\mathrm{Ind}}_{H/N}^{G/N}(k_{\tilde{\phi}})), V)$$

if $N \subseteq H$ and $\mathrm{Res}_N^H(\phi)$ is trivial and we define $\pi_{G,G/N}$ to be zero on $\mathrm{Hom}_{k[G],\underline{\phi}\mathbf{mod}}(\mathcal{V}(\underline{\mathrm{Ind}}_H^G(k_\phi)), V)$ otherwise.

The maps $\pi_{G,G/N}$ induce a chain map of bar-monomial resolutions

$$\pi_{G,G/N} : (\tilde{M}_{S_G,*} \otimes_k S_G \xrightarrow{\epsilon} V) \longrightarrow (\tilde{M}_{S_{G/N},*} \otimes_k S_{G/N} \xrightarrow{\epsilon} V)$$

which commutes with the augmentations to V and with the left-action by G.

In addition, if $N \subseteq M$ is an inclusion of normal subgroups of G the canonical homomorphisms are transitive in the sense that $\pi_{G/N,G/M} \cdot \pi_{G,G/N} = \pi_{G,G/M}$.

The chain map $i_{G,J}$ of §5.8 together with $\pi_{G,G/N}$ define a canonical chain map of bar-monomial resolutions associated to any homomorphism $\lambda : G \longrightarrow G_1$

$$\lambda_* : (\tilde{M}_{S_G,*} \otimes_k S_G \xrightarrow{\epsilon} V) \longrightarrow (\tilde{M}_{S_{G_1},*} \otimes_k S_{G_1} \xrightarrow{\epsilon} V)$$

which commutes with the augmentations to V and with the left-action by the subgroup G. In addition $\lambda_* \cdot \tau_* = (\lambda \cdot \tau)_*$.

To see this one verifies the following property. Suppose that we have an inclusion $G \subseteq J$ and a surjection $J \longrightarrow J/N$ as in §5.8 and §5.9. The composition of these two homomorphisms is equal to the surjection $G \longrightarrow GN/N$ followed by the inclusion $GN/N \subseteq J/N$. However

$$i_{GN/N,J/N} \cdot \pi_{G,GN/N} = \pi_{J,J/N} \cdot i_{G,J}.$$

5.10. *Naturality - inclusions of sub-representations*

Let $j : V \subseteq V_1$ be the inclusion of a sub-representation of G such that each representation has the same central character ϕ. Then post-composition with j induces

$$\mathrm{Hom}_{k[G],\underline{\phi}\mathbf{mod}}(\mathcal{V}(S_G), V) \xrightarrow{j_*} \mathrm{Hom}_{k[G],\underline{\phi}\mathbf{mod}}(\mathcal{V}(S_G), V_1)$$

and a canonical chain map of bar-monomial resolutions

$$j_* : (\tilde{M}_{V,S_G,*} \otimes_k S_G \xrightarrow{\epsilon} V) \longrightarrow (\tilde{M}_{V_1,S_G,*} \otimes_k S_G \xrightarrow{\epsilon} V_1)$$

which commutes with the augmentations and with the left-action by G. Here the suffices V and V_1, which will usually be suppressed, have been included to stress which representation is being resolved.

6. Finiteness of monomial resolutions in characteristic zero

6.1. Suppose that G is a locally p-adic group which is finite modulo the centre and suppose that V is a finite dimensional, irreducible complex representation of G with central character ϕ on $Z(G)$. We shall construct a finite length, finite type monomial resolution of V by modifying the proof for finite groups which is given in ([**19**] §6). In this section, temporarily, we shall suppress the mention of $\underline{\phi}$ and merely write $\mathcal{M}_G$ for $\mathcal{M}_\phi(G)$.

This is an involved induction of the "homological algebra" type which produces a monomial resolution unique up to chain homotopy equivalence.

Eventually I hope to be able to construct a proof which proceeds directly by modification of the bar-monomial resolution, for example by constructing a contracting homotopy which respects "depth", the filtration on which the following proof is based.

6.2. We say that $(H, \phi) \in \mathcal{M}_G$ is V-admissible if $V^{(H,\phi)} \neq 0$ and $(Z(G), \underline{\phi}) \leq (H, \phi)$. Let $\mathcal{S}(V)$ denote the set of non-zero subspaces of V and let $A(V) \subseteq \mathcal{M}_G$ dente the set of V-admissible pairs. Define maps

$$F_V : A(V) \longrightarrow \mathcal{S}(V) \text{ and } P_V : \mathcal{S}(V) \longrightarrow A(V)$$

by the formulae

$$F_V(H, \phi) = V^{(H,\phi)} \text{ and } P_V(W) = \sup\{(H, \phi) \mid W \subseteq V^{(H,\phi)}\}.$$

Usually suprema do not exist in $\mathcal{M}_G$ but $P_V(W)$ exists in this context. Firstly $W \subseteq V^{(Z(G),\underline{\phi})}$. On the other hand, if $W \subseteq V^{(H,\phi)}$ and $W \subseteq V^{(H',\phi')}$ then $W \subseteq V^{(H'',\phi'')}$ where H'' is the subgroup generated by H

and H' and ϕ'' is a character which extends both ϕ' and ϕ. This extension exists since $H/Z(G)$ and $H'/Z(G)$ are both finite and $\mathbb{C}^*$ is an injective abelian group.

Both $S(V)$ and $A(V)$ are posets with G-action with $W \subseteq F_V(P_V(W))$ and $(H, \phi) \leq P_V(F_V(H, \phi))$. Define the V-closure $\mathrm{cl}_V(H, \phi)$ by

$$\mathrm{cl}_V(H, \phi) = P_V(F_V(H, \phi))$$

and say that (H, ϕ) is V-closed if $(H, \phi) = \mathrm{cl}_V(H, \phi)$. Hence $\mathrm{cl}_V(H, \phi)$ is the largest pair (H', ϕ') such that $V^{(H,\phi)} = V^{(H',\phi')}$. Closure commutes with the G-action, is idempotent and order-increasing. Let $\mathrm{Cl}(V) \subseteq A(V)$ denote the subset of closed pairs.

For $(H, \phi) \in \mathrm{Cl}(V)$ define the V-depth $d_V(H, \phi)$ to be the largest integer $n \geq 0$ such that there exists a strictly increasing chain in $\mathrm{Cl}(V)$ of length n of the form

$$(H, \phi) = (H_0, \phi_0) < (H_1, \phi_1) < \cdots < (H_{n-1}, \phi_{n-1}) < (H_n, \phi_n).$$

Therefore $d_V(H, \phi) = 0$ if and only if (H, ϕ) is maximal in $\mathrm{Cl}(V)$.

THEOREM 6.3.

In the situation and notation of §6.1 and §6.2 there exists a finite type $\mathbb{C}[G]$-monomial resolution

$$M_* \xrightarrow{\epsilon} V \longrightarrow 0$$

such that:

(i) For $i \geq 0$, M_i has no Line with stabiliser pair $(H, \phi) \notin \mathrm{Cl}(V)$.

(ii) For $i \geq 0$, M_i has no Line with stabiliser pair $(H, \phi) \in \mathrm{Cl}(V)$ and $d_V(H, \phi) < i$.

In particular $M_i = 0$ for all $i > \max\{d_V(H, \phi) \mid (H, \phi) \in \mathrm{Cl}(V)\}$.

Proof

By induction on n we shall show that there exists a chain complex

$$M_n \xrightarrow{\partial_{n-1}} M_{n-1} \xrightarrow{\partial_{n-2}} \cdots \xrightarrow{\partial_0} M_0 \xrightarrow{\epsilon} V \longrightarrow 0$$

in which each M_i is a $\mathbb{C}[G]$-Line Bundle, each ∂_i is a morphism and ϵ is a homomorphism of $\mathbb{C}[G]$-modules such that the following conditions are satisfied:

(A_n) For $0 \leq i \leq n$, M_i has no Line with stabiliser pair $(H, \phi) \notin \mathrm{Cl}(V)$.

(B_n) For $0 \leq i \leq n$, M_i has no Line with stabiliser pair $(H, \phi) \in \mathrm{Cl}(V)$ and $d_V(H, \phi) < i$.

(C_n) The sequence of vector spaces

$$M_n^{((H,\phi))} \xrightarrow{\partial_{n-1}} M_{n-1}^{((H,\phi))} \xrightarrow{\partial_{n-2}} \cdots \xrightarrow{\partial_0} M_0^{((H,\phi))} \xrightarrow{\epsilon} V^{(H,\phi)} \longrightarrow 0$$

is exact for all $(H, \phi) \in \mathcal{M}_G$.

(D_n) The sequence of vector spaces

$$0 \longrightarrow M_n^{((H,\phi))} \xrightarrow{\partial_{n-1}} M_{n-1}^{((H,\phi))} \xrightarrow{\partial_{n-2}} \ldots \xrightarrow{\partial_0} M_0^{((H,\phi))} \xrightarrow{\epsilon} V^{(H,\phi)} \longrightarrow 0$$

is exact for all $(H, \phi) \in \mathrm{Cl}(V)$ with $d_V(H, \phi) \leq n$.

Note that for $n > \max(d_V(H, \phi) \mid (H, \phi) \in \mathrm{Cl}(V))$ the properties (A_n) and (B_n) imply that $M_n = 0$ and property (C_n) implies that

$$M_* \xrightarrow{\epsilon} V \longrightarrow 0$$

is a $\mathbb{C}[G]$-monomial resolution satisfying conditions (i) and (ii) of Theorem 6.3. By $\mathbb{C}[G]$-equivariance it suffices to prove (A_n)-(D_n) for one pair (H, ϕ) in each G-orbit.

Step (a): We show that if (A_n) , (B_n) and (D_n) hold then it suffices to prove (C_n) only for $(H, \phi) \in \mathrm{Cl}(V)$. Let $(H, \phi) \in \mathcal{M}_G$ such that $(Z(G), \underline{\phi}) \leq (H, \phi)$. If $(H, \phi) \notin A(V)$ then no larger (H', ϕ') is V-admissible and since $Cl(V) \subseteq A(V)$ we have $M_i^{((H,\phi))} = 0$ for $0 \leq i \leq n$ by (A_n). Also $V^{(H,\phi)} = 0$. Now suppose that $(H, \phi) \in A(V)$ then we must prove that the sequence in (C_n) is exact, assuming that it is exact for all $(H, \phi) \in \mathrm{Cl}(V)$. We have $V^{(H,\phi)} = V^{\mathrm{cl}_V(H,\phi)}$ by the definition of closure. In addition, $M_i^{((H,\phi))} = M_i^{\mathrm{cl}_V(H,\phi)}$ for all $0 \leq i \leq n$. This is seen as follows: $(H, \phi) \leq \mathrm{cl}_V(H, \phi)$ implies that $M_i^{\mathrm{cl}_V(H,\phi)} \subseteq M_i^{((H,\phi))}$ and for all $(H', \phi') \geq (H, \phi)$ we have no Lines in M_i with stabiliser pair (H', ϕ') unless $(H', \phi') \in \mathrm{Cl}(V)$, by (A_n), but in this case we have $(H', \phi') = \mathrm{cl}_V(H', \phi') \geq \mathrm{cl}_V(H, \phi) = (H, \phi)$ so that $M_i^{((H',\phi'))} \subseteq M_i^{\mathrm{cl}_V(H,\phi)}$. Therefore the sequences in (C_n) for (H, ϕ) and for its closure coincide but the latter is exact by assumption.

Step (b): We start the induction on n by defining $\epsilon : M_0 \longrightarrow V$. For each (H, ϕ) we need to define the set of Lines in M_0 whose stabiliser pair is G-conjugate to (H, ϕ). If $(H, \phi) \notin \mathrm{Cl}(V)$ we shall define the set of such Lines to be empty. If $(H, \phi) \in \mathrm{Cl}(V)$ define this set of lines to be given by the Line Bundle $\underline{\mathrm{Ind}}_H^G(V^{(H,\phi)})$ where $V^{(H,\phi)}$ is viewed as a $\mathbb{C}[H]$-Line Bundle with any choice of decomposition into Lines with the H-action $hv = \phi(h) \cdot v$, which was given on $V^{(H,\phi)}$ already. Define ϵ on this sub-Line Bundle of M_0

$$\underline{\mathrm{Ind}}_H^G(V^{(H,\phi)}) \longrightarrow V$$

by $\epsilon(g \otimes_{\mathbb{C}[H]} v) = g \cdot v$. This satisfies both (A_0) and (B_0).

Next we show that (D_0) holds. Suppose that (H, ϕ) is a maximal element in $\mathrm{Cl}(V)$, which implies that (H, ϕ) is maximal in $A(V)$ because if there were a larger pair in $A(V)$ its closure would be in $\mathrm{Cl}(V)$ and larger than (H, ϕ). Therefore the normaliser of (H, ϕ) must equal H. For the normaliser is $\mathrm{stab}_G(H, \phi)$ and so if it is greater than H the character ϕ may be extended ϕ' on to $H' > H$ with an abelian quotient group H'/H

and $V^{(H,\phi)} \neq 0$. By Clifford theory $V^{(H',\phi')} \neq 0$ and therefore $(H',\phi') >$ (H,ϕ) in $A(V)$ — a contradiction. Maximality in $Cl(V)$, (A_0) and the condition $H = stab_G(H,\phi)$ imply that $M_0^{((H,\phi))} = 1 \otimes_{\mathbb{C}[H]} V^{(H,\phi)}$ so that $\epsilon : M^{((H,\phi))} \longrightarrow V^{(H,\phi)}$ is an isomorphism. Finally, by step (a), we must show that $\epsilon : M^{((H,\phi))} \longrightarrow V^{(H,\phi)}$ is surjective for any $(H,\phi) \in Cl(V)$, which is clear by construction.

Step (c): Next we assume that we have already constructed a chain complex

$$M_n \xrightarrow{\partial_{n-1}} M_{n-1} \xrightarrow{\partial_{n-2}} \ldots \xrightarrow{\partial_0} M_0 \xrightarrow{\epsilon} V \longrightarrow 0$$

such that (A_n), (B_n), (C_n) and (D_n) hold. When $n = 0$ we interpret (M_n, ∂_{n-1}) as (V, ϵ). We shall define $\partial_n : M_{n+1} \longrightarrow M_n$ by defining the Lines of M_{n+1} whose stabiliser pair is G-conjugate to (H,ϕ) and specifying ∂_n on the direct sum of those Lines. If $(H,\phi) \notin Cl(V)$ we set the sum of these Lines to be zero, which assures that (A_{n+1}) holds. If $(H,\phi) \in Cl(V)$ and $d_V(H,\phi) \leq n$ we also set the sum of these Lines to be zero, so that (B_{n+1}) follows from (B_n). Moreover (D_n) and (A_{n+1}) implies (C_{n+1}) for all (H,ϕ) with $d_V(H,\phi) \leq n$. Also (D_n) implies (D_{n+1}) for all $(H,\phi) \in Cl(V)$ with $d_V(H,\phi) \leq n$.

If $(H,\phi) \in Cl(V)$ with $d_V(H,\phi) > n+1$ we define the direct sum of Lines in M_{n+1} with stabiliser pair conjugate to (H,ϕ) to be $\underline{Ind}_H^G(\Omega_n^{(H,\phi)})$ with $\Omega_n^{(H,\phi)} = Ker(\partial_{n-1} : M_n^{((H,\phi))} \longrightarrow M_{n-1}^{((H,\phi))})$ considered as an H-Line Bundle with any chosen decomposition and we set $\partial_n(g \otimes_{\mathbb{C}[H]} v) = gv$. Clearly $\partial_{n-1} \cdot \partial_n = 0$ as defined so far and ∂_n is a morphism by Frobenius reciprocity. Also

$$Ker(\partial_{n-1} : M_n^{((H,\phi))} \longrightarrow M_{n-1}^{((H,\phi))}) = \Omega_n^{(H,\phi)} = \partial_n(1 \otimes_{\mathbb{C}[H]} \Omega_n^{(H,\phi)}) \subseteq Im(\partial_n)$$

shows that (C_{n+1}) holds for this pair (H,ϕ) (and its G-conjugates) and (D_{n+1}) is vacuously true for it.

It remains to deal with the case when $d_V(H,\phi) = n+1$ and only (D_{n+1}) requires to be proved since (C_{n+1}) is vacuously satisfied in this case. We shall show in Lemma 6.4 that

$$\Omega_n^{(H,\phi)} = Ker(\partial_{n-1} : M_n^{((H,\phi))} \longrightarrow M_{n-1}^{((H,\phi))}) \cong Ind_H^{stab_G(H,\phi)}(L_{(H,\phi)})$$

as $\mathbb{C}[stab_G(H,\phi)]$-modules for some $\mathbb{C}[H]$-submodule $L_{(H,\phi)} \subseteq \Omega_n^{(H,\phi)}$. Since H acts on $L_{(H,\phi)}$ via multiplication by ϕ we may choose a decomposition for $L_{(H,\phi)}$ as a direct sum of (H,ϕ)-Lines. Then we define the direct sum of Lines in M_{n+1} whose stabilisers are G-conjugate to (H,ϕ) to be given by $\underline{Ind}_H^G(L_{(H,\phi)})$ and define ∂_n on $\underline{Ind}_H^G(L_{(H,\phi)})$ by $\partial_n(g \otimes_{\mathbb{C}[H]} v) = gv$. Then ∂_n is a morphism, by Frobenius reciprocity, and (D_{n+1}) holds because, by (A_{n+1}) and (B_{n+1}),

$$M_{n+1}^{((H,\phi))} = (\underline{Ind}_H^G(L_{(H,\phi)}))^{((H,\phi))} = \oplus_{g \in stab_G(H,\phi)/H} \, s \otimes_{\mathbb{C}[H]} L_{(H,\phi)},$$

which shows that ∂_n induces an isomorphism $M_{n+1}^{((H,\phi))} \longrightarrow \Omega_n^{(H,\phi)}$.
The proof will be completed by Lemma 6.4. $\square$

LEMMA 6.4.

Suppose we have a chain complex as at the start of the proof of Theorem 6.3

$$M_n \xrightarrow{\partial_{n-1}} M_{n-1} \xrightarrow{\partial_{n-2}} \dots \xrightarrow{\partial_0} M_0 \xrightarrow{\epsilon} V \longrightarrow 0$$

such that conditions (A_n), (B_n), (C_n) and (D_n) hold. Let $(H, \phi) \in Cl(V)$ with $d_V(H, \phi) = n+1$. Then the class $\theta \in K_0(\mathbb{C}[N])$ of the $\mathbb{C}[\mathrm{stab}_G(H, \phi)]$-module

$$\mathrm{Ker}(\partial_{n-1} : M_n^{((H,\phi))} \longrightarrow M_{n-1}^{((H,\phi))}) = \Omega_n^{(H,\phi)}$$

is a (possibly zero) multiple of the character of $\mathrm{Ind}_H^{\mathrm{stab}_G(H,\phi)}(\phi)$.

Proof

Set $N = \mathrm{stab}_G(H, \phi)$. Then, by (C_n) for (H, ϕ) we have an exact sequence of $\mathbb{C}[N]$-modules

$$0 \longrightarrow \Omega_n^{(H,\phi)} \longrightarrow M_n^{((H,\phi))} \xrightarrow{\partial_{n-1}} M_{n-1}^{((H,\phi))} \xrightarrow{\partial_{n-2}}$$

$$\dots \xrightarrow{\partial_0} M_0^{((H,\phi))} \xrightarrow{\epsilon} V^{(H,\phi)} \longrightarrow 0.$$

Denote by $\chi_i \in K_0(\mathbb{C}[N])$ the character of $M_i^{((H,\phi))}$ and ν the class of $V^{(H,\phi)}$. Therefore we obtain

$$\nu = (-1)^{n+1}\theta + \sum_{i=0}^{n} (-1)^i \chi_i \in K_0(\mathbb{C}[N]).$$

At this point the proof of ([19] §6) uses the existence of the explicit Brauer induction maps a_N and b_N (in the notation of [126]) which could be established for finite modulo the centre groups (with a fixed, possibly non-finite central character) by the algebraic argument of [17] which is reproduced in [126]. However, it is easier to use the topological construction of these maps by the method of Peter Symonds construction [134], which uses the action of G on the projective space of V and works more general for finite modulo the centre groups.

Now consider the class of the $\mathbb{C}[N]$-Line Bundle $M_i^{((H,\phi))}$ which satisfies $\chi = b_N(M_i^{((H,\phi))})$, by definition. We also have $b_N(a_N(\nu)) = \nu$ so that we obtain

$$(-1)^{n+1}\theta = b_N(a_N(\nu) - \sum_{i=0}^{n} (-1)^i \chi_i \in K_0(\mathbb{C}[N]).$$

Observe next that all the stabiliser pairs of the Lines of $M_i^{((H,\phi))}$ for $0 \le i \le n$ have the form $(H' \bigcap N, \mathrm{Res}_{H' \cap N}^{H'}(\phi'))$ for some $(H', \phi') \in \mathcal{M}_G$ such that $(H, \phi) \le (H', \phi')$. Hence in the free abelian group $R_+(N)$ on the N-conjugacy classes of $\mathcal{M}_N$ the class of $M_i^{((H,\phi))} \in R_+(N)$ may have non-zero coefficients only at basis elements $(K, \psi) \in \mathcal{M}_N/N$ with $(H, \phi) \le (K, \psi)$. By a basic property of a_N, the same is true of $a_N(\nu)$ since ν restricts to a multiple of ϕ on H. Therefore we may write

$$a_N(\nu) - \sum_{i=0}^{n} (-1)^i M_i^{((H,\phi))} = \sum_{\substack{(K,\psi)_N \in \mathcal{M}_N/N \\ (H,\phi) \le (K,\psi)}} \alpha_{(K,\psi)_N} \cdot (K,\psi)_N \in R_+(N)$$

where each $\alpha_{(K,\psi)_N}$ is an integer.

We shall show that $\alpha_{(K,\psi)_N} = 0$ for all $(H, \phi) < (K, \psi)$, which concludes the proof.

Assume that $\alpha_{(K_0,\psi_0)_N} \ne 0$ for some $(H, \phi) < (K_0, \psi_0)$ and assume also that (K_0, ψ_0) is maximal amongst pairs satisfying this condition.

Recall that there is a (non-symmetric) bilinear form on $R_+(N)$ and maximality of (K_0, ψ_0) yields

$$\left((K_0, \psi_0)_N, \sum_{\substack{(K,\psi)_N \in \mathcal{M}_N/N \\ (H,\phi) \le (K,\psi)}} \alpha_{(K,\psi)_N} \cdot (K,\psi)_N \right)_N$$

$$= \sum_{\substack{(K,\psi)_N \in \mathcal{M}_N/N \\ (H,\phi) \le (K,\psi)}} \alpha_{(K,\psi)_N} ((K_0, \psi_0)_N, (K, \psi)_N)_N$$

$$= \alpha_{(K_0,\psi_0)_N} ((K_0, \psi_0)_N, (K_0, \psi_0)_N)_N$$

$$= \alpha_{(K_0,\psi_0)_N} [\mathrm{stab}_N(K_0, \psi_0) : K_0] \ne 0.$$

On the other hand, adjointness properties of a_G and the bilinear form yield

$$\left((K_0, \psi_0)_N, a_N(\nu) - \sum_{i=0}^{n} (-1)^i M_i^{((H,\phi))} \right)_N$$

$$= ((K_0, \psi_0)_N, a_N(\nu))_N - \sum_{i=0}^{n} (-1)^i ((K_0, \psi_0)_N, M_i^{((H,\phi))})_N$$

$$= (\mathrm{Ind}_{K_0}^{N}(\psi_0), \nu)_N - \sum_{i=0}^{n} (-1)^i \dim_{\mathbb{C}}(\mathrm{Hom}_{\mathbb{C}[N]-mon}(\underline{\mathrm{Ind}}_{K_0}^{N}(\psi_0), M_i^{((H,\phi))}))$$

$$= (\psi_0, \mathrm{Res}_{K_0}^{N}(\nu))_{K_0} - \sum_{i=0}^{n} (-1)^i \dim_{\mathbb{C}}(M_i^{((K_0,\psi_0))})$$

$$= \dim_{\mathbb{C}}(V^{(K_0,\psi_0)}) - \sum_{i=0}^{n} (-1)^i \dim_{\mathbb{C}}(M_i^{((K_0,\psi_0))}).$$

The lemma will be proved by showing that this last expression is zero. If (K_0, ψ_0) is not $V^{(H,\phi)}$-admissible then (A_n) implies that every term in this sum vanishes. If (K_0, ψ_0) is $V^{(H,\phi)}$-admissible then we have $cl_V(H, \phi) = (H, \phi) < (K_0, \psi_0) \le cl_V(K_0, \psi_0)$. This implies that

$$d_V(cl_V(K_0, \psi_0)) < d_V(H, \phi) = n + 1.$$

However (D_n) implies that the chain complex

$$0 \longrightarrow M_n^{(cl_V(K_0,\psi_0))} \xrightarrow{\partial_{n-1}} M_{n-1}^{(cl_V(K_0,\psi_0))} \xrightarrow{\partial_{n-2}} \cdots$$

$$\xrightarrow{\partial_0} M_0^{(cl_V(K_0,\psi_0))} \xrightarrow{\epsilon} V^{cl_V(K_0,\psi_0)} = V^{(K_0,\psi_0)} \longrightarrow 0$$

is exact. In addition, by an argument used in the proof of Theorem 6.3(a) we have $M_i^{(cl_V(K_0,\psi_0))} = M_i^{((K_0,\psi_0))}$ for $0 \leq i \leq n$, which completes the proof. $\square$

COROLLARY 6.5.

When k is an algebraically closed field of characteristic zero in Theorem 5.6 the bar-monomial resolution of V is chain homotopy equivalent in $_{k[G],\underline{\phi}}\mathbf{mon}$ to a finite length, finitely generated $_{k[G],\underline{\phi}}\mathbf{mon}$-resolution of V.

CHAPTER 2

GL_2 of a local field

In this chapter we shall consider, in the local field case, the existence and structure of the monomial resolution of an admissible k-representation V of $GL_2 K$ with central character $\underline{\phi}$. The monomial resolution constructed in this case is unique in the derived category of $_{k[GL_2 K],\underline{\phi}}\mathbf{mon}$.

In §1 we recall the definition and properties of compactly supported (modulo the centre) induction of an admissible representation (and the $_{k[G]}\mathbf{mon}$-analogue) of a locally profinite Lie group such as $GL_n K$. In §2 the finite modulo the centre monomial resolutions of Chapter One are extended, using the functoriality of the bar-monomial resolution, to the case of compact open modulo the centre groups. In §3 we construct a $_{k[GL_n K],\underline{\phi}}\mathbf{mon}$-monomial double complex made from a compact open modulo the centre bar-monomial resolution for each such orbit stabiliser of a $GL_n K$-simplicial complex Y and the natural monomial morphisms between them. In §4 with $n = 2$ we take Y to be a simplicial subdivision of the Bruhat-Tits building (the classical tree of [115]) of $GL_2 K$. Since the tree is one-dimensional one can make the construction without recourse to the naturality properties — homological algebra with chain complexes will suffice. For $GL_2 K$ we adopt this simplification in order better to illustrate the basic construction without the extra technicalities. In §5 I describe what becomes of a monomial resolution of an admissible representation of $GL_2 K$ when one takes the part "fixed by level-n units". Not surprisingly the result is a finite modulo the centre monomial resolution. §5 concludes with some remarks about the relevance of this to local ϵ-factors and L-functions. In §6 for $GL_2 K$ (although the results hold for arbitrary $GL_n K$) I describe the monomial resolution of an admissible representation of the semi-direct product of a Galois group with $GL_2 K$ which extends a given Galois invariant $GL_2 K$ admissible representation. This construction should be related to Galois base change (or Galois descent) of admissible irreducibles (see [91] and [7]). With this sort of application in mind I have given in Appendix I, an extended discussion of a particular Shintani descent example. Shintani descent [117] is the finite group (possibly motivating) analogue of base change.

1. Induction

In this section we are going to study admissible representations of GL_2K and its subgroups, where K is a p-adic local field. These representations will be given by left-actions of the groups on vector spaces over k, which is an algebraically closed field of arbitrary characteristic. Let us begin by recalling induced and compactly induced smooth representations.

DEFINITION 1.1. ([40] p. 17)
Let G be a locally profinite group and $H \subseteq G$ a closed subgroup. Thus H is also locally profinite. Let

$$\sigma : H \longrightarrow \mathrm{Aut}_k(W)$$

be a smooth representation of H. Set X equal to the space of functions $f : G \longrightarrow W$ such that
   (i) $f(hg) = hf(g)$ for all $h \in H, g \in G$,
   (ii) there is a compact open subgroup $K_f \subseteq G$ such that $f(gk) = f(g)$ for all $g \in G, k \in K_f$.
   The (left) action of G on X is given by $(g \cdot f)(x) = f(xg^{-1})$ and

$$\Sigma : G \longrightarrow \mathrm{Aut}_k(X)$$

gives a smooth representation of G.
   The representation Σ is called the representation of G smoothly induced from σ and is usually denoted by $\Sigma = \mathrm{Ind}_H^G(\sigma)$.

   1.2. Definition 1.1 does make sense since, if $g \in G, h \in H$ and $f \in X$, then

$$(g \cdot f)(hg_1) = f(hg_1g^{-1}) = hf(g_1g^{-1}) = h(g \cdot f)(g_1)$$

so that $(g \cdot f)$ satisfies condition (i) of Definition 1.1.
   Also

$$(gg_1 \cdot f)(x) = f(x(gg_1)^{-1}) = f(xg_1^{-1}g^{-1}) = g \cdot (x \mapsto f(xg_1^{-1})) = (g \cdot (g_1 \cdot f))(x)$$

so Σ is a left representation, providing that $g \cdot f \in X$ when $f \in X$. However, condition (ii) asserts that there exists a compact open subgroup K_f such that $k \cdot f = f$ for all $k \in K_f$. The subgroup gK_fg^{-1} is also a compact open subgroup and, if $k \in K_f$, we have

$$(gkg^{-1}) \cdot (g \cdot f) = (gkg^{-1}g) \cdot f = (gk) \cdot f = (g \cdot (k \cdot f)) = (g \cdot f)$$

so that $g \cdot f \in X$, as required.
   The smooth representations of G form an abelian category Rep(G).

PROPOSITION 1.3. ([40] p. 18)
   The functor

$$\mathrm{Ind}_H^G : \mathrm{Rep}(H) \longrightarrow \mathrm{Rep}(G)$$

is additive and exact.

PROPOSITION 1.4. *(Frobenius Reciprocity; ([40] p. 18))*
There is an isomorphism

$$\operatorname{Hom}_G(\pi, \operatorname{Ind}_H^G(\sigma)) \overset{\cong}{\longrightarrow} \operatorname{Hom}_H(\pi, \sigma)$$

given by $\phi \mapsto \alpha \cdot \phi$ where α is the H-map

$$\operatorname{Ind}_H^G(\sigma) \longrightarrow \sigma$$

given by $\alpha(f) = f(1)$.

1.5. In general, if $H \subseteq Q$ are two closed subgroups there is a Q-map

$$\operatorname{Ind}_H^G(\sigma) \longrightarrow \operatorname{Ind}_H^Q(\sigma)$$

given by restriction of functions. Note that α in Proposition 1.4 is the special case where $H = Q$.

1.6. *The c-Ind variation ([40] p. 19)*
Inside X let X_c denote the set of functions which are compactly supported modulo H. This means that the image of the support

$$\operatorname{supp}(f) = \{g \in G \mid f(g) \neq 0\}$$

has compact image in $H \backslash G$. Alternatively there is a compact subset $C \subseteq G$ such that $\operatorname{supp}(f) \subseteq H \cdot C$.

The Σ-action on X preserves X_c, since $\operatorname{supp}(g \cdot f) = \operatorname{supp}(f)g \subseteq HCg$, and we obtain $X_c = c - \operatorname{Ind}_H^G(W)$, the compact induction of W from H to G.

This construction is of particular interest when H is open. Then there is a canonical H-map

$$\alpha^c : W \longrightarrow c - \operatorname{Ind}_H^G(W)$$

given by $w \mapsto f_w$ where f_w is supported in H and $f_w(h) = h \cdot w$ (so $f_w(g) = 0$ if $g \notin H$).
For $g \in G$ we have

$$(g \cdot f_w)(x) = f_w(xg^{-1}) = \begin{cases} 0 & \text{if } xg^{-1} \notin H, \\ (xg^{-1}) \cdot w & \text{if } xg^{-1} \in H, \end{cases}$$

$$= \begin{cases} 0 & \text{if } x \notin Hg, \\ (xg^{-1}) \cdot w & \text{if } x \in Hg. \end{cases}$$

LEMMA 1.7. *([40] p. 19)*
Let H be an open subgroup of G. Then
(i) $\alpha^c : w \mapsto f_w$ is an H-isomorphism onto the space of functions $f \in c - \operatorname{Ind}_H^G(W)$ such that $\operatorname{supp}(f) \subseteq H$.
(ii) If $w \in W$ and $h \in H$ then $h \cdot f_w = f_{h^{-1}w}$.

(iii) If $\mathcal{W}$ is a k-basis of W and $\mathcal{G}$ is a set of coset representatives for $H\backslash G$ then
$$\{g \cdot f_w \mid w \in \mathcal{W}, \ g \in \mathcal{G}\}$$
is a k-basis of $c - \operatorname{Ind}_H^G(W)$.

Proof

If $\operatorname{supp}(f)$ is compact modulo H there exists a compact subset C such that
$$\operatorname{supp}(f) \subseteq HC = \bigcup_{c \in C} Hc.$$
Each Hc is open so the open covering of C by the Hc's refines to a finite covering and so
$$C = Hc_1 \bigcup \ldots \bigcup Hc_n$$
and so
$$\operatorname{supp}(f) \subseteq HC = Hc_1 \bigcup \ldots \bigcup Hc_n.$$

For part (i), the map α^c is an H-homomorphism to the space of functions supported in H with inverse map $f \mapsto f(1)$.

For part (ii), from §1.6 we have
$$(h \cdot f_w)(x) = f_w(xh^{-1}) = \begin{cases} 0 & \text{if } x \notin H, \\ xh^{-1}w & \text{if } x \in H. \end{cases}$$
so that, for all $x \in G$, $(h \cdot f_w)(x) = f_{h^{-1}w}(x)$, as required.

For part (iii), the support of any $f \in c - \operatorname{Ind}_H^G(W)$ is a finite union of cosets Hg where the g's are chosen from the set of coset representatives $\mathcal{G}$ of $H\backslash G$. The restriction of f to any one of these Hg's also lies in $c - \operatorname{Ind}_H^G(W)$. If $\operatorname{supp}(f) \subseteq Hg$ then $(g^{-1} \cdot f)(z) \neq 0$ implies that $zg \in Hg$ so that $g^{-1} \cdot f$ has support contained in H. Hence $g^{-1} \cdot f$ on H is a finite linear combination of the functions f_w with $w \in \mathcal{W}$. Therefore f is a finite linear combination of $g \cdot f_w$'s where $w \in \mathcal{W}, g \in \mathcal{G}$. Clearly the set of functions $g \cdot f_w$ with $g \in \mathcal{G}$ and $w \in \mathcal{W}$ is linearly independent. $\square$

EXAMPLE 1.8. Let K be a p-adic local field with valuation ring $\mathcal{O}_K$ and π_K a generator of the maximal ideal of $\mathcal{O}_K$. In GL_nK if H is compact, open modulo K^* then there is a subgroup H' of finite index in H such that $H' = K^*H_1$ with H_1 compact, open in SL_nK. This can be established by studying the simplicial action of GL_nK on a suitable barycentric subdivision of the Bruhat-Tits building of SL_nK (see §4.12 and Chapter Four §1).

To show that H is both open and closed it suffices to verify this for H'. Firstly H' is open, since it is $H' = \bigcup_{z \in K^*} zH_1 = \bigcup_{s \in \mathbb{Z}} \pi_K^s H_1$.

Also $H' = K^*H_1$ is closed. Suppose that $X' \notin K^*H_1$. K^*H_1 is closed under mutiplication by the multiplicative group generated by π_K so that $\pi_K^m X' \notin K^* H_1$ for all m. By conjugation we may assume that H_1 is a subgroup of $SL_n \mathcal{O}_K$, which is the maximal compact open subgroup of $SL_n K$, unique up to conjugacy. Choose the smallest non-negative integer m such that every entry of $X = \pi_K^m X'$ lies in $\mathcal{O}_K$. Therefore we may write $0 \neq \det(X) = \pi_K^s u$ where $u \in \mathcal{O}_K^*$ and $1 \leq s$. Now suppose that V is an $n \times n$ matrix with entries in $\mathcal{O}_K$ such that $X + \pi_K^t V \in K^* H_1$. Then

$$\det(X + \pi_K^t V) \equiv \pi_K^s u \ (\text{modulo } \pi_K^t).$$

So that if $t > s$ then s must have the form $s = nw$ for some integer w and $\pi_K^{-w}(X + \pi_K^t V) \in GL_n \mathcal{O}_K \bigcap K^* H_1 = H_1$. Therefore all the entries in $\pi_K^{-w} X$ lie in $\mathcal{O}_K$ and $\pi_K^{-w} X \in GL_n \mathcal{O}_K$. Enlarging t, if necessary, we can ensure that $\pi_K^{-w} X \in H_1$, since H_1 is closed (being compact), and therefore $X' \in K^* H_1$, which is a contradiction.

Since H is both closed and open in $GL_n K$ we may form the admissible representation $c - \mathrm{Ind}_H^{GL_n K}(k_\phi)$ for any continuous character $\phi : H \longrightarrow k^*$ and apply Lemma 1.7.

If $g \in GL_n K, h \in H$ then $(g \cdot f_1)(x) = \phi(xg^{-1})$ if $xg^{-1} \in H$ and zero otherwise. On the other hand, $(hg \cdot f_1)(x) = \phi(xg^{-1}h^{-1}) = \phi(h^{-1})\phi(xg^{-1})$ if $xg^{-1} \in H$ and zero otherwise. Therefore as a left $GL_n K$-representation $c - \mathrm{Ind}_H^{GL_n K}(k_\phi)$ is isomorphic to

$$k[GL_n K]/(g - \phi(h)hg \mid g \in GL_n K, \ h \in H)$$

with left action induced by $g_1 \cdot g = gg_1^{-1}$.

This vector space is isomorphic to the k-vector space whose basis is given by k-bilinear tensors over H of the form $g \otimes_H 1$ as in the case of finite groups. The basis vector $g \cdot f_1$ corresponds to $g^{-1} \otimes_H 1$ and $GL_n K$ acts on the tensors by left multiplication, as usual. This is well-defined because $\phi(h)(hg \cdot f_1)$ corresponds to $g^{-1}h^{-1} \otimes_H \phi(h) = g^{-1} \otimes_H 1$.

PROPOSITION 1.9. ([40] p. 19)
The functor

$$c - \mathrm{Ind}_H^G : \mathrm{Rep}(H) \longrightarrow \mathrm{Rep}(G)$$

is additive and exact.

PROPOSITION 1.10. ([40] p. 20)
Let $H \subseteq G$ be an open subgroup and (σ, W) smooth. Then there is a bi-functorial isomorphism

$$\mathrm{Hom}_G(c - \mathrm{Ind}_H^G(\sigma), \pi) \xrightarrow{\cong} \mathrm{Hom}_H(\sigma, \pi)$$

given by $f \mapsto f \cdot \alpha^c$.

EXAMPLE 1.11. *The Line Bundle* $c - \underline{\mathrm{Ind}}_H^{GL_n K}(M)$
In the situation of Example 1.8, suppose that M is a Line Bundle in $k[K^* H_1], \phi \mathbf{mon}$. Then $c - \mathrm{Ind}_H^{GL_n K}(M)$ can be given the structure of a Line

Bundle in $_{k[GL_nK],\phi}\mathbf{mon}$ denoted by $c - \mathrm{Ind}_H^{GL_nK}(M)$. Choose for $\mathcal{W}$, a basis for M, a set consisting of one non-zero vector from each Line. The Lines of $c - \mathrm{Ind}_H^{GL_nK}(M)$ are then given by

$$\{\langle g \cdot f_w\rangle \mid w \in \mathcal{W}, g \in \mathcal{G}\}$$

in the notation of Lemma 1.7.

This vector space is isomorphic to the k-vector space whose basis is given by k-bilinear tensors over H of the form $g \otimes_H w$ as in the case of finite groups. The basis vector $g \cdot f_w$ corresponds to $g^{-1} \otimes_H w$ and GL_nK acts on the tensors by left multiplication, as usual. This is well-defined because $\phi(h)(hg \cdot f_w) = \phi(h)g \cdot f_{h^{-1}w}$ corresponds to $g^{-1}h^{-1} \otimes_H \phi(h)w = g^{-1} \otimes_H w$.

2. From finite to compact open

2.1. Let K be a p-adic local field with valuation ring $\mathcal{O}_K$ and π_K a generator of the maximal ideal of $\mathcal{O}_K$. In GL_nK let H be compact, open modulo K^* of the form $H = K^*H_1$ with H_1 compact, open, as in Example 1.8.

For $1 \leq m$ let U_K^m denote the subgroup of GL_nK given by

$$U_K^m = \{X \in GL_n\mathcal{O}_K \mid X \equiv I \ (\text{modulo } \pi_K^m)\}.$$

Let V be a (left) admissible k-representation of GL_nK with central character ϕ. Since every vector of V lies in the fixed subspace $V^J = V^{(J,1)}$ for some open subgroup J, we have

$$V = \bigcup_{m \geq 1} V^{(U_K^m,1)} = \bigcup_{m \geq m_0} V^{(K^*U_K^m,\phi)}$$

where m_0 is the least integer such that the central character is trivial on $U_K^{m_0}$. Also each $V^{(K^*U_K^m,1)}$ is a finite-dimensional representation of H via a left action which factorises through the quotient $H/H \cap U_K^m = K^* \cdot H_1/H_1 \cap U_K^m$, which is finite modulo the centre providing that m is large enough. This fact is established by observing that every maximal compact open subgroup of GL_nK is conjugate to $GL_n\mathcal{O}_K$. For example, when $n = 2$ and $\alpha, \beta, \gamma.\delta \in \mathcal{O}_K$, the conjugate

$$\begin{pmatrix} \alpha & \beta \\ \gamma & \delta \end{pmatrix} GL_2\mathcal{O}_K \begin{pmatrix} \alpha & \beta \\ \gamma & \delta \end{pmatrix}^{-1}$$

$$= \begin{pmatrix} \alpha & \beta \\ \gamma & \delta \end{pmatrix} GL_2\mathcal{O}_K \begin{pmatrix} \frac{\delta}{\alpha\delta-\gamma\beta} & \frac{-\beta}{\alpha\delta-\gamma\beta} \\ \frac{-\gamma}{\alpha\delta-\gamma\beta} & \frac{\alpha}{\alpha\delta-\gamma\beta} \end{pmatrix}$$

contains

$$\begin{pmatrix} \alpha & \beta \\ \gamma & \delta \end{pmatrix} U_K^m \begin{pmatrix} \frac{\delta}{\alpha\delta-\gamma\beta} & \frac{-\beta}{\alpha\delta-\gamma\beta} \\ \frac{-\gamma}{\alpha\delta-\gamma\beta} & \frac{\alpha}{\alpha\delta-\gamma\beta} \end{pmatrix}$$

which contains U_K^{m-r} for some integer depending on the K-adic valuation of $\alpha\delta - \gamma\beta$.

Write $V(m)$ for the k-vector space $V^{(K^*U_K^m, 1)}$ which we consider as a k-representation, via inflation, of each of the finite modulo the centre quotients $G(m+r) = K^* \cdot H_1/H_1 \cap U_K^{m+r}$, for r large.

For each sufficiently large integer r we have a bar-monomial resolution of $V(m)$ in $_{k[G(m+r)], \underline{\phi}}\mathbf{mon}$ denoted by

$$M_{V(m), *, G(m+r)} \xrightarrow{\epsilon} V(m) \longrightarrow 0.$$

By inflation we shall construct a monomial resolution of $V(m)$ in $_{k[K^* \cdot H_1], \underline{\phi}}\mathbf{mon}$ denoted by

$$\mathrm{Inf}_{G(m+r)}^{K^* \cdot H_1}(M_{V(m), *, G(m+r)}) \xrightarrow{\epsilon} V(m) \longrightarrow 0.$$

Let $\pi : K^* \cdot H_1 \longrightarrow G(m+r)$ denote the canonical projection homomorphism. Since $H_1/H_1 \cap U_K^{m+r}$ is finite, there is an isomorphism of $_{k[K^* \cdot H_1], \underline{\phi}}\mathbf{mon}$ Line Bundles of the form

$$c - \mathrm{Ind}_{\pi^{-1}(H)}^{K^* \cdot H_1}(\pi^*(\phi)) \cong \mathrm{Inf}_{G(m+r)}^{K^* \cdot H_1}(\underline{\mathrm{Ind}}_H^G(k_\phi))$$

for $(H, \phi) \in \mathcal{M}_{\underline{\phi}(G(m+r))}$.

Define a $_{k[K^* \cdot H_1], \underline{\phi}}\mathbf{mon}$ Line Bundle $S_{K^* \cdot H_1, m+r}$ by

$$S_{K^* \cdot H_1, m+r} = \oplus_{(H, \phi) \in \mathcal{M}_{\underline{\phi}}(G(m+r))} c - \mathrm{Ind}_{\pi^{-1}(H)}^{K^* \cdot H_1}(\pi^*(\phi)).$$

For any triple $((K, \psi), g, (H, \phi)) \in \mathcal{A}_{G(m+r)}$ we may construct its inflation

$$((\pi^{-1}(K), \pi^*(\psi)), gH_1 \bigcap U_K^{m+r}, (\pi^{-1}(H), \pi^*(\phi)))$$

which represents a well-defined $_{k[K^* \cdot H_1], \underline{\phi}}\mathbf{mon}$-endomorphism of $S_{K^* \cdot H_1, m+r}$ given by the same formulae as for the finite modulo the centre case in Lemma 1.9. It only depends on the $H_1 \bigcap U_K^{m+r}$-coset of g because, as in Lemma 1.9, for $h \in H_1 \bigcap U_K^{m+r} \subseteq \pi^{-1}(H)$, we have

$$((\pi^{-1}(K), \pi^*(\psi)), gh, (\pi^{-1}(H), \pi^*(\phi)))$$

$$= \phi(h)((\pi^{-1}(K), \pi^*(\psi)), g, (\pi^{-1}(H), \pi^*(\phi)))$$

$$= ((\pi^{-1}(K), \pi^*(\psi)), g, (\pi^{-1}(H), \pi^*(\phi)))$$

because $\phi(h) = 1$.

The formulae of Lemma 1.9 give a ring structure on the k-vector space of spanned by the inflated triples

$$((\pi^{-1}(K), \pi^*(\psi)), gH_1 \bigcap U_K^{m+r}, (\pi^{-1}(H), \pi^*(\phi))).$$

We denote this ring of $_{k[K^* \cdot H_1], \underline{\phi}}$**mon**-endomorphism of $S_{K^* \cdot H_1, m+r}$ by $\mathcal{A}_{K^* \cdot H_1, m+r}$.

As in §3.2 we have isomorphisms

$$\mathrm{Hom}_{k[K^* \cdot H_1], \underline{\phi}\mathbf{monmod}}(\mathcal{V}(S_{K^* \cdot H_1, m+r}), V(m))$$

$$\cong \mathrm{Hom}_{k[G(m+r)], \underline{\phi}\mathbf{mod}}(\mathcal{V}(S_{G(m+r)}), V(m))$$

$$\cong \oplus_{(H,\phi) \in \mathcal{M}_{\underline{\phi}}(G(m+r))} V(m)^{(H,\phi)}.$$

THEOREM 2.2.

The chain complex

$$\mathrm{Inf}_{G(m+r)}^{K^* \cdot H_1}(M_{V(m),*,G(m+r)}) \xrightarrow{\epsilon} V(m) \longrightarrow 0$$

is defined by replacing $\mathrm{Hom}_{k[G], \underline{\phi}\mathbf{mod}}(\mathcal{V}(S), V)$, $\mathcal{A}_S$ and S, in the construction of §5.5, respectively by $\mathrm{Hom}_{k[K^* \cdot H_1], \underline{\phi}\mathbf{mon}\mathbf{mod}}(\mathcal{V}(S_{K^* \cdot H_1, m+r}), V(m))$, $\mathcal{A}_{K^* \cdot H_1, m+r}$ and $S_{K^* \cdot H_1, m+r}$.

It is a monomial resolution of $V(m)$ in $_{k[K^* \cdot H_1], \underline{\phi}}$**mon**.

Proof

As a chain complex the inflated complex and the bar-monomial resolution of $V(m)$ in $_{k[G(m+r)], \underline{\phi}}$**mon** are isomorphic. Therefore we have only to verify monomial exactness. Suppose that $(J, \lambda) \in \mathcal{M}_{\underline{\phi}}(K^* \cdot H_1)$. In order for $V(m)^{(J,\lambda)}$ to be non-zero we must have $\mathrm{Res}_{H_1 \cap U_K^{m+r} \cap J}^J(\lambda) = 1$. In this case we may extend λ trivially on $H_1 \cap U_K^{m+r}$ to give $\tilde{\lambda}$ on

$$\langle J, H_1 \cap U_K^{m+r} \rangle = \tilde{J} = \pi^{-1}(\pi(J)).$$

Also, by construction, $V(m)^{(J,\lambda)} = V(m)^{(\tilde{J} \ \tilde{\lambda})} = V(m)^{(\pi(J),\lambda')}$, where $\lambda' \cdot \pi = \lambda$. Also the $_{k[K^* \cdot H_1], \underline{\phi}}$**mon**-Lines of the inflated resolution whose stabiliser pair is greater than or equal to (J, λ) are the same as the $_{k[G(m+r)], \underline{\phi}}$**mon**-Lines whose stabiliser pair is greater than or equal to $(\pi(J), \lambda')$. On the other hand, if $V(m)^{(J,\lambda)} = 0$ there are no $_{k[K^* \cdot H_1], \underline{\phi}}$**mon**-Lines of the inflated resolution whose stabiliser pair is greater than or equal to (J, λ), which completes the verification of monomial exactness. $\square$

REMARK 2.3. The discussion of Chapter One §5.9 shows that if we fix m we may form the direct limit as r varies. In addition the discussion of Chapter One §5.10 shows that we may form the direct limit over m also. The net result is the following.

THEOREM 2.4.

The chain complex

$$\varinjlim_{m} \varinjlim_{r} \mathrm{Inf}_{G(m+r)}^{K^* \cdot H_1} (M_{V(m),*,G(m+r)}) \xrightarrow{\epsilon} \varinjlim_{m} V(m) = V \longrightarrow 0$$

is a monomial resolution of V in $_{k[K^* \cdot H_1], \phi}\mathbf{mon}$. This will be called the $_{k[K^* \cdot H_1], \phi}\mathbf{mon}$-bar monomial resolution of V and denoted by $W_{V,*,K^* \cdot H_1}$.

REMARK 2.5. The $_{k[K^* \cdot H_1], \phi}\mathbf{mon}$-bar monomial resolution of V of Theorem 2.4 inherits from Chapter One §5.8–§5.10 naturality properties analogous to those which hold in the finite module the centre case.

3. The admissible monomial double-complex

3.1. *Monomial complexes for $GL_n K$*

In this section we are going to study (left) admissible k-representations of $GL_n K$ with central character ϕ. Here K continues to be a p-adic local field. As usual k is an algebraically closed field of arbitrary characteristic. If V is such an admissible k-representation we shall begin by applying Theorem 2.4 to the restrictions of V to compact open modulo the centre subgroups.

Let Y be a simplicial complex upon which $GL_n K$ acts simplicially and in which the stabiliser $H_\sigma = \mathrm{stab}_G(\sigma)$ is compact, open modulo the centre, K^*, of $GL_n K$.

An example of such a Y is given by $GL_n K$ acting on (a suitable subdivision of) its Bruhat-Tits [35] building.

For each simplex σ of Y, by Theorem 2.4, we have a $_{k[H_\sigma], \phi}\mathbf{mon}$-bar monomial resolution of V

$$W_{V,*,H_\sigma} \longrightarrow V \longrightarrow 0.$$

Form the graded k-vector space which in degree m is equal to

$$\underline{M}_m = \oplus_{\alpha+n=m} W_{V,\alpha,H_{\sigma^n}}.$$

If σ^{n-1} is a face of σ^n there is an inclusion $H_{\sigma^n} \subseteq H_{\sigma^{n-1}}$. Therefore there is a canonical monomial chain map

$$i_{H_{\sigma^n}, H_{\sigma^{n-1}}} : W_{V,*,H_{\sigma^n}} \longrightarrow W_{V,*,H_{\sigma^{n-1}}}$$

such that

$$i_{H_{\sigma^{n-1}}, H_{\sigma^{n-2}}} i_{H_{\sigma^n}, H_{\sigma^{n-1}}} = i_{H_{\sigma^n}, H_{\sigma^{n-2}}}.$$

If σ^{n-1} is a face of σ^n let $d(\sigma^{n-1}, \sigma^n)$ denote the incidence degree of σ^{n-1} in σ^n; this is ± 1. In the simplicial chain complex of Y

$$d(\sigma^n) = \sum_{\sigma^{n-1} \text{ face of } \sigma^n} d(\sigma^{n-1}, \sigma^n)\sigma^{n-1}.$$

For $x \in W_{V,\alpha,H_{\sigma^n}}$ write

$$d_Y(x) = \sum_{\sigma^{n-1} \text{ face of } \sigma^n} d(\sigma^{n-1},\sigma^n)\, i_{H_{\sigma^n},H_{\sigma^{n-1}}}(x).$$

Let $d_{\sigma^n} : W_{V,\alpha,H_{\sigma^n}} \longrightarrow W_{V,\alpha-1,H_{\sigma^n}}$ denote the differential in the $k[H_\sigma],_\phi$**mon**-bar monomial resolution of V.

Define $\underline{d} : \underline{M}_m \longrightarrow \underline{M}_{m-1}$ when $m = \alpha + n$ by

$$\underline{d}(x) = d_Y(x) + (-1)^n d_{\sigma^n}(x).$$

Therefore we have

$$\underline{d}(\underline{d}(x))$$

$$= \underline{d}(\sum_{\sigma^{n-1} \text{ face of } \sigma^n} d(\sigma^{n-1},\sigma^n)\, i_{H_{\sigma^n},H_{\sigma^{n-1}}}(x)) + \underline{d}((-1)^n d_{\sigma^n}(x))$$

$$= \sum_{\substack{\sigma^{n-2} \text{ face of } \sigma^{n-1} \\ \sigma^{n-1} \text{ face of } \sigma^n}} d(\sigma^{n-2},\sigma^{n-1})\, i_{H_{\sigma^{n-1}},H_{\sigma^{n-2}}}$$

$$(d(\sigma^{n-1},\sigma^n)\, i_{H_{\sigma^n},H_{\sigma^{n-1}}}(x))$$

$$+ \sum_{\sigma^{n-1} \text{ face of } \sigma^n} d(\sigma^{n-1},\sigma^n)\, i_{H_{\sigma^n},H_{\sigma^{n-1}}}((-1)^n d_{\sigma^n}(x))$$

$$+ (-1)^{n-1} \sum_{\sigma^{n-1} \text{ face of } \sigma^n} d(\sigma^{n-1},\sigma^n)\, d_{\sigma^{n-1}}(i_{H_{\sigma^n},H_{\sigma^{n-1}}}(x))$$

$$+ (-1)^n d_{\sigma^n}((-1)^n d_{\sigma^n}(x))$$

$$= \sum_{\substack{\sigma^{n-2} \text{ face of } \sigma^{n-1} \\ \sigma^{n-1} \text{ face of } \sigma^n}} d(\sigma^{n-2},\sigma^{n-1})\, d(\sigma^{n-1},\sigma^n)\, i_{H_{\sigma^n},H_{\sigma^{n-2}}}(x)$$

$$+ (-1)^n \sum_{\sigma^{n-1} \text{ face of } \sigma^n} d(\sigma^{n-1},\sigma^n)\, i_{H_{\sigma^n},H_{\sigma^{n-1}}}(d_{\sigma^n}(x))$$

$$+ (-1)^{n-1} \sum_{\sigma^{n-1} \text{ face of } \sigma^n} d(\sigma^{n-1},\sigma^n)\, i_{H_{\sigma^n},H_{\sigma^{n-1}}}(d_{\sigma^n}(x))$$

$$+ d_{\sigma^n}(d_{\sigma^n}(x))$$

$$= \sum_{\substack{\sigma^{n-2} \text{ face of } \sigma^{n-1} \\ \sigma^{n-1} \text{ face of } \sigma^n}} d(\sigma^{n-2},\sigma^{n-1})\, d(\sigma^{n-1},\sigma^n)\, i_{H_{\sigma^n},H_{\sigma^{n-2}}}(x)$$

$$= 0$$

because, as is well-known, for each pair (σ^n,σ^{n-2}) the sum

$$\sum_{\substack{\sigma^{n-2} \text{ face of } \sigma^{n-1} \\ \sigma^{n-1} \text{ face of } \sigma^n}} d(\sigma^{n-2},\sigma^{n-1})\, d(\sigma^{n-1},\sigma^n) = 0.$$

Note that $\underline{M}_*$ has an obvious structure of a $_{k[GL_nK],\underline{\phi}}\mathbf{mon}$-Line Bundle since the GL_nK-action permutes the summands $W_{V,*,H_{\sigma^n}}$, each of which is a $_{k[H_{\sigma^n}],\underline{\phi}}\mathbf{mon}$-Line Bundle.

THEOREM 3.2.

If Y is the Bruhat-Tits building for GL_nK, suitably subdivided to make the GL_nK-action simplicial, then $(\underline{M}_*,\underline{d})$ is a chain complex in $_{k[GL_nK],\underline{\phi}}\mathbf{mon}$.

In addition, this complex has a canonical augmentation homomorphism in $_{k[GL_nK],\underline{\phi}}\mathbf{mod}$ of the form $\underline{M}_0 \xrightarrow{\epsilon} V$.

CONJECTURE 3.3. For $n \geq 2$, K local and $G = GL_nK$

$$\cdots \longrightarrow \underline{M}_i \xrightarrow{d} \underline{M}_{i-1} \xrightarrow{d} \cdots \xrightarrow{d} \underline{M}_0 \xrightarrow{\epsilon} V \longrightarrow 0$$

is a monomial resolution in $_{k[GL_nK],\underline{\phi}}\mathbf{mon}$. That is, for each $(H,\phi) \in \mathcal{M}_{G,\underline{\phi}}$

$$\cdots \longrightarrow \underline{M}_i^{((H,\phi))} \xrightarrow{d} \underline{M}_{i-1}^{((H,\phi))} \xrightarrow{d} \cdots \xrightarrow{d} \underline{M}_0^{((H,\phi))} \xrightarrow{\epsilon} V^{(H,\phi)} \longrightarrow 0$$

is an exact sequence of k-vector spaces.

REMARK 3.4. When we come to the proof of Conjecture 3.3 for GL_2K in §4.12 it will become clear that a "suitable" simplical action on the Bruhat-Tits building of GL_nK must have the property that every compact open modulo the centre subgroup of GL_nK must be contained in some simplex-stabiliser. In general this property is a consequence of the Bruhat-Tits fixed point theorem for group actions on CAT(0) spaces (see [4]). For GL_2K we prove it, for the specific simplicial structure used of §4.12, in Proposition 4.8.

4. Monomial resolutions for GL_2K

Let K be a p-adic local field. In this section I shall use the well-known action of GL_2K on its tree ([115] p. 69) to verify Conjecture 3.3 for GL_2K. The resulting monomial resolution is unique up to chain homotopy in $_{k[GL_nK],\underline{\phi}}\mathbf{mon}$.

I shall begin with a detailed recapitulation of the tree (also known as the Bruhat-Tits building for GL_2K [35] pp. 130–131). The notation is that of §2.1.

4.1. The GL_2K-action on its tree

A lattice in $K \oplus K$ is any finitely generated $\mathcal{O}_K$-submodule which generates $K \oplus K$ as a K-vector space. If $x \in K^*$ and L is a lattice then so also is xL. The homothety class of L is the orbit of L in the set of lattices under this K^*-action. The set of classes of lattices gives rise to a tree ([115] Chapter II) with a right GL_2K-action.

Let $H_1 = GL_2 \mathcal{O}_K$ and

$$H_2 = \begin{pmatrix} \pi_K & 0 \\ 0 & 1 \end{pmatrix} H_1 \begin{pmatrix} \pi_K & 0 \\ 0 & 1 \end{pmatrix}^{-1}$$

which are two of the maximal compact subgroups. All other maximal compact subgroups of $GL_2 K$ are conjugate to H_1. Explicitly we have

$$H_2 = \{ \begin{pmatrix} a & b\pi_K \\ c\pi_K^{-1} & d \end{pmatrix} \mid a, b, c, d \in \mathcal{O}_K, \ ad - bc \in \mathcal{O}_K^* \}.$$

By ([115] p. 69 et seq) if $L = \mathcal{O}_K \oplus \mathcal{O}_K$ and $L' = \mathcal{O}_K \oplus \mathcal{O}_K \pi_K$ then $\mathrm{Stab}_{GL_2 K}(L) = H_1 \cdot K^*$ and $\mathrm{Stab}_{GL_2 K}(L') = H_2 \cdot K^*$ where $GL_2 K$ acts by right multiplication on the vector space $V = K \oplus K$. This fact will enable us to calculate some normalisers.

If $XH_1 X^{-1} = H_1$ then $((L)X)H_1 = (L)H_1 X = (L)X$ but from the tree structure each homothety class of a lattice is stabilised by a different maximal compact subgroup so that $H_1 \cdot K^*$ stabilises L and $(L)X$ and so $(L)X = L$ and $X \in H_1 \cdot K^*$. This shows that $N_{GL_2 K} H_1 = H_1 \cdot K^*$. Similarly for H_2.

If $YH_1 \bigcap H_2 Y^{-1} = H_1 \bigcap H_2$ then $(L)Y = (L)YH_1 \bigcap H_2$ and $(L')Y = (L')YH_1 \bigcap H_2$. Also $(H_1 \bigcap H_2) \cdot K^* \subseteq \mathrm{Stab}_{GL_2 K}(L) \bigcap \mathrm{Stab}_{GL_2 K}(L')$. Now the distance from L to L' is one [115], so they are adjacent in the graph, and the (pointwise) stabiliser of the edge they define is precisely $(H_1 \bigcap H_2) \cdot K^*$. Furthermore this is the only edge that this subgroup stabilises. But $(L)Y$ and $(L')Y$ are also adjacent and $H_1 \bigcap H_2 \cdot K^*$ also stabilises this edge so the edges coincide. This coincidence can happen in two ways. If the ordered pair $((L)Y, (L')Y)$ is equal to (L, L') then $Y \in H_1 \bigcap H_2 \cdot K^*$. On the other hand it is possible that $((L)Y, (L')Y)$ is equal to (L', L). In fact, the matrix calculation

$$\begin{pmatrix} 0 & 1 \\ \pi_K^{-1} & 0 \end{pmatrix} \begin{pmatrix} \alpha & \pi_K \beta \\ \gamma & \delta \end{pmatrix} \begin{pmatrix} 0 & \pi_K \\ 1 & 0 \end{pmatrix} = \begin{pmatrix} \delta & \pi_K \gamma \\ \beta & \alpha \end{pmatrix}$$

shows that $H_1 \bigcap H_2$ is normalised by the matrix

$$\begin{pmatrix} 0 & 1 \\ \pi_K^{-1} & 0 \end{pmatrix},$$

which does not belong to $H_1 \bigcap H_2$. Therefore

$$N_{GL_2 K}(H_1 \bigcap H_2) = \langle (H_1 \bigcap H_2) \cdot K^*, \begin{pmatrix} 0 & 1 \\ \pi_K^{-1} & 0 \end{pmatrix} \rangle.$$

This group stabilises the edge $\{L, L'\}$ but only the subgroup of index two $(H_1 \cap H_2) \cdot K^*$ maps the ordered pair (L, L') to itself by the identity, other matrices interchange the order.

Therefore the Weyl groups of H_1, H_2 given by $N_{GL_2K}H_i/H_i$ are both isomorphic to $K^*/\mathcal{O}_K^* \cong \mathbb{Z}$ generated by the scalar matrix π_K. The Weyl group $N_{GL_2K}(H_1 \cap H_2)/H_1 \cap H_2$ is isomorphic to the infinite cyclic group generated by

$$u = \begin{pmatrix} 0 & 1 \\ \pi_K^{-1} & 0 \end{pmatrix}$$

which contains a subgroup of index two given by $\langle u^2 \rangle = K^*/\mathcal{O}_K^* \cong \mathbb{Z}$.

Now suppose that $Z(H_1 \cap H_2)Z^{-1} \subset H_1$ then $(L)Z = (L)ZH_1$ and the preceding argument shows that $Z \in H_1 \cdot K^*$. Hence the coset space

$$\frac{H_1 \cdot K^*}{H_1 \cap H_2} \cong \frac{H_1}{H_1 \cap H_2} \times \mathbb{Z}$$

and $\frac{H_1}{H_1 \cap H_2}$ is in one-one correspondence with $\mathbb{P}^1(\mathcal{O}_K/(\pi_K))$ because this coset is isomorphic to the orbit of the edge LL' under the action of $\mathrm{Stab}_{GL_2K}(L)$ ([**115**] p. 72).

The correspondence between $\mathbb{P}^1(\mathcal{O}_K/(\pi_K))$ and the set of lattices of distance one from L is described as follows in ([**115**] p. 72). Let $L'' \subset L$ be such that $L/L'' \cong \mathcal{O}_K/(\pi_K) \cong k$. Then we have a short exact sequence

$$0 \longrightarrow k \cong L''/\pi_K L \longrightarrow L/\pi_K L \cong k \oplus k \longrightarrow k \longrightarrow 0$$

which associates to L'' a linear subspace in $k \oplus k$ and hence a point in $\mathbb{P}^1(k)$.

Also, via the left action on lattices, since $H_1 \cap H_2$ stabilises the edge through L and L' we get a bijection

$$H_1 \cap H_2 \backslash GL_2\mathcal{O}_K \leftrightarrow \mathbb{P}^1(k).$$

The transpose of this bijection is given explicitly as follows. Suppose, for $a, b, c, d, \alpha, \beta, \gamma, \delta \in \mathcal{O}_K$, that

$$X = \begin{pmatrix} a & b \\ c & d \end{pmatrix}, Y = \begin{pmatrix} \alpha & \beta\pi_K \\ \gamma & \delta \end{pmatrix}$$

with $X \in GL_2\mathcal{O}_K$, $Y \in H_1 \cap H_2$. Then

$$XY = \begin{pmatrix} a & b \\ c & d \end{pmatrix}\begin{pmatrix} \alpha & \beta\pi_K \\ \gamma & \delta \end{pmatrix} = \begin{pmatrix} a\alpha + b\gamma & a\beta\pi_K + b\delta \\ c\alpha + d\gamma & c\beta\pi_K + d\delta \end{pmatrix}$$

and because $\alpha, \delta \in \mathcal{O}_K^*$ we have a well-defined element

$$\left(\frac{\overline{b}}{\overline{d}} \right) = \left(\frac{\overline{b\delta}}{\overline{d\delta}} \right) \in \mathbb{P}^1(k)$$

depending only on the coset $XH_1 \bigcap H_2$. Hence representatives of this coset are given by

$$\left\{ \begin{pmatrix} 1 & b \\ 0 & 1 \end{pmatrix} \right\} \quad \text{and} \quad \begin{pmatrix} 0 & 1 \\ -1 & 0 \end{pmatrix},$$

where $b \in \mathcal{O}_K$ runs through a set of representatives for k.

4.2. The simplicial complex of the tree

Now we are ready to calculate the simplicial chain complex of the tree together with its GL_2K-action. I am going to transpose to a left action on the tree by GL_2K.

The cellular 1-chain group of the tree, with coefficients in k, is the k-vector space whose basis consists of the 1-cells of the tree. This is clearly given by the induced representation

$$C_1 = c - \text{Ind}_{N_{GL_2K}(H_1 \cap H_2)}^{GL_2K}(k_\tau) = k[GL_2K] \otimes_{k[N_{GL_2K}(H_1 \cap H_2)]} k_\tau$$

where k_τ is a copy of k on which $(H_1 \bigcap H_2) \cdot K^*$ acts trivially and

$$\begin{pmatrix} 0 & 1 \\ \pi_K^{-1} & 0 \end{pmatrix}$$

acts like -1. Here, as described in Example 1.8, we are depicting the c-induction as the "crude" algebraic induction in terms of tensor product over group rings. This is algebraically more convenient and it will emphasise that individual simplices are far apart, as they are with respect to the distance function on lattice classes in ([**115**] pp. 69–70).

The 0-cells are given by the induced representation

$$C_0 = c - \text{Ind}_{N_{GL_2K}H_1}^{GL_2K}(k)$$

where k has the trivial action. The simplicial differential

$$d : C_1 \longrightarrow C_0$$

is a GL_2K-map and so, by Proposition 1.10, is determined by a $N_{GL_2K}(H_1 \cap H_2)$-map from k_τ to C_0. This map is easily seen to be given by

$$d(1) = 1 \otimes_{N_{GL_2K}H_1} 1 - \begin{pmatrix} 0 & 1 \\ \pi_K^{-1} & 0 \end{pmatrix} \otimes_{N_{GL_2K}H_1} 1.$$

If $X \in N_{GL_2K}(H_1 \cap H_2) \subset N_{GL_2K}H_1$ then

$$Xd(1) = X \otimes_{N_{GL_2K}H_1} 1 - X \begin{pmatrix} 0 & 1 \\ \pi_K^{-1} & 0 \end{pmatrix} \otimes_{N_{GL_2K}H_1} 1 = d(1) = d(X \cdot 1)$$

while

$$\begin{pmatrix} 0 & 1 \\ \pi_K^{-1} & 0 \end{pmatrix} d(1) = \begin{pmatrix} 0 & 1 \\ \pi_K^{-1} & 0 \end{pmatrix} \otimes_{N_{GL_2K}H_1} 1 - 1 \otimes_{N_{GL_2K}H_1} 1$$

$$= -d(1)$$

$$= d\left(\begin{pmatrix} 0 & 1 \\ \pi_K^{-1} & 0 \end{pmatrix} \cdot 1 \right).$$

Since a tree is contractible we have an exact sequence of $k[GL_2K]$-modules

$$0 \longrightarrow C_1 \xrightarrow{d} C_0 \xrightarrow{\epsilon} k \longrightarrow 0$$

where, for $Z \in GL_2K$,

$$d(Z \otimes_{N_{GL_2K}(H_1 \cap H_2)} 1) = Z \otimes_{N_{GL_2K}H_1} 1 - Z \begin{pmatrix} 0 & 1 \\ \pi_K^{-1} & 0 \end{pmatrix} \otimes_{N_{GL_2K}H_1} 1$$

and $\epsilon(Y \otimes_{N_{GL_2K}H_1} v) = v$.

The above action is not simplicial because the subgroup preserving a given 1-simplex does not act on it by the identity. For example, $\{L, L'\}$ is inverted by

$$\begin{pmatrix} 0 & 1 \\ \pi_K^{-1} & 0 \end{pmatrix}.$$

However, it is easy barycentrically to subdivide the simplicial tree by adding L'', the midpoint of $\{L, L'\}$, and all its GL_2K-translates. The stabiliser of L'' is $N_{GL_2K}(H_1 \cap H_2)$. The result is a one-dimensional simplicial complex with 1-simplices given by $\{L, L''\}$ and its GL_2K-translates. The stabiliser of $\{L, L''\}$ is $(H_1 \cap H_2)K^*$ and the resulting GL_2K-action is simplicial. The 0-cells are given by

$$\tilde{C}_0 = c - \text{Ind}_{N_{GL_2K}(H_1)}^{GL_2K}(k) \oplus c - \text{Ind}_{N_{GL_2K}(H_1 \cap H_2)}^{GL_2K}(k)$$

while the 1-cells are

$$\tilde{C}_1 = c - \text{Ind}_{(H_1 \cap H_2)K^*}^{GL_2K}(k).$$

Therefore we have a short exact sequence of $k[GL_2K]$-modules of the form

$$0 \longrightarrow \tilde{C}_1 \xrightarrow{d} \tilde{C}_0 \xrightarrow{\epsilon} k \longrightarrow 0$$

in which

$$d(g \otimes_{(H_1 \cap H_2)K^*} v) = (g \otimes_{N_{GL_2K}(H_1)} v, -g \otimes_{N_{GL_2K}(H_1 \cap H_2)} v)$$

and
$$\epsilon(g_1 \otimes_{N_{GL_2K}(H_1)} v_1, g_2 \otimes_{N_{GL_2K}(H_1 \cap H_2)} v_2) = v_1 + v_2.$$

4.3. If V is an admissible representation of GL_2K as in §3.1 we have (continuing to use as in §4.2 the "crude" algebraic notation described in Example 1.8 for c-induction) an isomorphism of GL_2K-representations

$$\phi : c - \mathrm{Ind}_H^{GL_2K}(W) \otimes V \xrightarrow{\cong} c - \mathrm{Ind}_H^{GL_2K}(W \otimes \mathrm{Res}_H^{GL_2K}(V))$$

given by $\phi((g \otimes_H w) \otimes v) = g \otimes_H (w \otimes g^{-1}v)$, if W is finite-dimensional. This is well-defined because

$$\phi((gh \otimes_H h^{-1}w) \otimes v) = gh \otimes_H (h^{-1}w) \otimes h^{-1}g^{-1}v) = g \otimes_H (w \otimes g^{-1}v)$$

and is a GL_2K-map because

$$g'\phi((g \otimes_H w) \otimes v) = g'g \otimes_H (w \otimes g^{-1}v) = \phi(g'(g \otimes_H w) \otimes g'v).$$

We have $N_{GL_2K}H_1 \bigcap H_2 = \langle H_1 \bigcap H_2, u \rangle$ where

$$u = \begin{pmatrix} 0 & 1 \\ \pi_K^{-1} & 0 \end{pmatrix}, \quad u^2 = \begin{pmatrix} \pi_K^{-1} & 0 \\ 0 & \pi_K^{-1} \end{pmatrix} \in Z(GL_2K) = K^*$$

and $N_{GL_2K}H_1 = \langle H_1, u^2 \rangle = H_1 \cdot K^*$.

The homomorphism

$$c - \mathrm{Ind}_{(H_1 \cap H_2)K^*}^{GL_2K}(k) \otimes V$$

$$d \otimes 1 \downarrow$$

$$c - \mathrm{Ind}_{N_{GL_2K}(H_1)}^{GL_2K}(k) \otimes V \oplus c - \mathrm{Ind}_{N_{GL_2K}(H_1 \cap H_2)}^{GL_2K}(k) \otimes V$$

transforms under ϕ to

$$c - \mathrm{Ind}_{(H_1 \cap H_2)K^*}^{GL_2K}(V)$$

$$\psi \downarrow$$

$$c - \mathrm{Ind}_{N_{GL_2K}(H_1)}^{GL_2K}(V) \oplus c - \mathrm{Ind}_{N_{GL_2K}(H_1 \cap H_2)}^{GL_2K}(V)$$

given by

$$\psi(g \otimes_{(H_1 \cap H_2)K^*} v) = (g \otimes_{N_{GL_2K}(H_1)} v, -g \otimes_{N_{GL_2K}(H_1 \cap H_2)} v)$$

because

$$\psi(\phi((g \otimes_{(H_1 \cap H_2)K^*} 1) \otimes gv))$$

$$= \psi(g \otimes_{(H_1 \cap H_2)K^*} v)$$

$$= (g \otimes_{N_{GL_2K}(H_1)} v, -g \otimes_{N_{GL_2K}(H_1 \cap H_2)} v)$$

while

$$\phi(d \otimes 1((g \otimes_{(H_1 \cap H_2)K^*} 1) \otimes gv))$$

$$= \phi((g \otimes_{N_{GL_2K}(H_1)} 1) \otimes gv, -g \otimes_{N_{GL_2K}(H_1 \cap H_2)} 1) \otimes gv)$$

$$= (g \otimes_{N_{GL_2K}(H_1)} v, -g \otimes_{N_{GL_2K}(H_1 \cap H_2)} v).$$

4.4. For Theorem 2.4 we have bar-monomial resolutions

$$W_{V,*,(H_1 \cap H_2)K^*)} \xrightarrow{\epsilon_1} V$$

in $_{k[(H_1 \cap H_2)K^*],\underline{\phi}}\mathbf{mon}$,

$$W_{V,*,N_{GL_2K}(H_1)} \xrightarrow{\epsilon_0} V$$

in $_{k[(N_{GL_2K}(H_1)],\underline{\phi}}\mathbf{mon}$ and

$$W_{V,*,N_{GL_2K}(H_1 \cap H_2)} \xrightarrow{\epsilon_0'} V$$

in $_{k[N_{GL_2K}(H_1 \cap H_2)],\underline{\phi}}\mathbf{mon}$.

Note that $u^2 \in Z(GL_2K) = K^*$ so that all characters we shall meet are given by $\phi(u^2)$ on u^2.

Suppose that $K^* \cdot J \subseteq GL_2K$ is one of the three above compact open modulo the centre subgroups and suppose that M is a Line Bundle in $_{k[K^* \cdot J],\underline{\phi}}\mathbf{mon}$. As described in Example 1.11, if M is a Line Bundle then $c - \underline{\mathrm{Ind}}_{K^* \cdot J}^{GL_2K}(M)$ is a Line Bundle in $_{k[GL_2K],\underline{\phi}}\mathbf{mon}$.

4.5. *Covering ψ by a monomial-morphism*

The first objective is to produce a $k[GL_2K]$-module homomorphism

$$c - \underline{\mathrm{Ind}}_{(H_1 \cap H_2)K^*}^{GL_2K}(W_{V,0,(H_1 \cap H_2)K^*)})$$

$$\psi_0 \downarrow$$

$$c - \underline{\mathrm{Ind}}_{N_{GL_2K}(H_1)}^{GL_2K}(W_{V,0,N_{GL_2K}(H_1)})$$
$$\oplus c - \underline{\mathrm{Ind}}_{N_{GL_2K}(H_1 \cap H_2)}^{GL_2K}(W_{V,0,N_{GL_2K}(H_1 \cap H_2)})$$

to commute with the augmentations. That is,

$$(\underline{\mathrm{Ind}}_{N_{GL_2K}(H_1)}^{GL_2K}(\epsilon_0) \oplus \underline{\mathrm{Ind}}_{N_{GL_2K}(H_1 \cap H_2)}^{GL_2K}(\epsilon_0'))\psi_0 = \psi\underline{\mathrm{Ind}}_{(H_1 \cap H_2)K^*}^{GL_2K}(\epsilon_1).$$

We begin by constructing a $k[GL_2K]$-module homomorphism and then we sort out the behaviour of Lines under the map.

Start with a Line from $c - \underline{\mathrm{Ind}}_{(H_1 \cap H_2)K^*}^{GL_2K}(W_{V,0,(H_1 \cap H_2)K^*)})$ with stabiliser pair (J, ϕ) where ϕ is a character of J so that $(Z(GL_2K), \underline{\phi}) \leq (J, \phi)$.

This Line will be of the form $g \otimes_{(H_1 \cap H_2)K^*} L$ where L is a Line in $W_{V,0,(H_1 \cap H_2)K^*)}$ and $g \in GL_2K$. If $z \in L$ the action of $j \in J$ satisfies

$$jg \otimes_{(H_1 \cap H_2)K^*} z = \phi(j)g \otimes_{(H_1 \cap H_2)K^*} z.$$

Hence $g^{-1}Jg \subseteq (H_1 \cap H_2)K^*$ and acts on L via $g^*(\phi)$. The $(H_1 \cap H_2)K^*$-orbit of L spans the Line Bundle isomorphic to $c - \underline{\mathrm{Ind}}_{g^{-1}Jg}^{(H_1 \cap H_2)K^*}(g^*(k_\phi))$ and, by conjugation in $(H_1 \cap H_2)K^*$, we may assume that

$$L = \langle 1 \otimes_{g^{-1}Jg} 1 \rangle \subseteq c - \underline{\mathrm{Ind}}_{g^{-1}Jg}^{(H_1 \cap H_2)K^*}(g^*(k_\phi)).$$

Since we need only a representative from the GL_2K-orbit of the Line we may as well assume that $J \subseteq (H_1 \cap H_2)K^*$ and the Line is generated by $1 \otimes_{(H_1 \cap H_2)K^*} (1 \otimes_J 1)$ so that $j \in J$ acts on this line via

$$j(1 \otimes_{(H_1 \cap H_2)K^*} (1 \otimes_J 1)) = \phi(j) \otimes_{(H_1 \cap H_2)K^*} (1 \otimes_J 1)$$

and, by monomial exactness, $v = \epsilon_1(1 \otimes_J 1) \in V^{(J,\phi)}$.

Now consider the two terms in

$$\psi(\epsilon_1(1 \otimes_{(H_1 \cap H_2)K^*} (1 \otimes_J 1))) = \psi(1 \otimes_{(H_1 \cap H_2)K^*} v)$$

$$= 1 \otimes_{N_{GL_2K}(H_1)} v - 1 \otimes_{N_{GL_2K}(H_1 \cap H_2)} v.$$

The action of $j \in J \cap H_1 \cap H_2$ on each of these terms is by multiplication by $\phi(j)$. Therefore, by naturality of the bar-monomial resolution with respect to inclusions of subgroups, there exists $w \in W_{V,0,N_{GL_2K}(H_1)}^{((J,\phi))}$ and $w' \in W_{V,0,N_{GL_2K}(H_1 \cap H_2)}^{((J,\phi))}$ such that $\epsilon_0(w) = v, \epsilon_0'(w') = v$ so that

$$1 \otimes_{N_{GL_2K}(H_1)} \epsilon_0(w) - 1 \otimes_{N_{GL_2K}(H_1 \cap H_2)} \epsilon_0'(w') = \psi(\epsilon_1(1 \otimes_{(H_1 \cap H_2)K^*}(1 \otimes_J 1))).$$

Set

$$\psi_0(1 \otimes_{(H_1 \cap H_2)K^*} (1 \otimes_J 1))) = 1 \otimes_{N_{GL_2K}(H_1)} w - 1 \otimes_{N_{GL_2K}(H_1 \cap H_2)} w'.$$

This defines a $k[GL_2K]$-module homomorphism ψ_0 which commutes with augmentations. In addition, for all $g \in GL_2K$ and $J \subseteq (H_1 \cap H_2)K^*$,

$$\psi_0(g \otimes_{(H_1 \cap H_2)K^*} W_{V,0,(H_1 \cap H_2)K^*)}^{((J,\phi))})$$

lies in

$$g \otimes_{N_{GL_2K}(H_1)} W_{V,0,N_{GL_2K}(H_1)}^{((J,\phi))} \oplus g \otimes_{N_{GL_2K}(H_1 \cap H_2)} W_{V,0,N_{GL_2K}(H_1 \cap H_2)}^{((J,\phi))},$$

which guarantees that ψ_0 is a morphism in $_{k[GL_2K],\phi}\underline{\mathbf{mon}}$.

Now, by induction, we construct similar chain $k[GL_2K]$-module homomorphisms for all $i \geq 0$

$$c - \underline{\mathrm{Ind}}_{(H_1 \cap H_2)K^*}^{GL_2K} (W_{V,i,(H_1 \cap H_2)K^*)})$$

$$\psi_i \downarrow$$

$$c - \underline{\mathrm{Ind}}_{N_{GL_2K}(H_1)}^{GL_2K} (W_{V,i,N_{GL_2K}(H_1)})$$
$$\oplus c - \underline{\mathrm{Ind}}_{N_{GL_2K}(H_1 \cap H_2)}^{GL_2K} (W_{V,i,N_{GL_2K}(H_1 \cap H_2)})$$

which commute with the differentials and satisfy the condition that

$$\psi_i(g \otimes_{(H_1 \cap H_2)K^*} W_{V,i,(H_1 \cap H_2)K^*)}^{((J,\phi))})$$

lies in

$$g \otimes_{N_{GL_2K}(H_1)} W_{V,i,N_{GL_2K}(H_1)}^{((J,\phi))} \oplus g \otimes_{N_{GL_2K}(H_1 \cap H_2)} W_{V,i,N_{GL_2K}(H_1 \cap H_2)}^{((J,\phi))}.$$

Therefore the ψ_i's give a chain map in $_{k[GL_2K],\phi}\mathbf{mon}$.

We start with a Line in $c - \underline{\mathrm{Ind}}_{(H_1 \cap H_2)K^*}^{GL_2K}(M_{1,i})$ with $i \geq 1$ and, as in the case of ψ_0, we may assume that this Line has stabiliser pair (J, ϕ) with $J \subseteq (H_1 \cap H_2)K^*$. In a notation analogous to that of the ψ_0 case, we may assume that the Line is generated by $1 \otimes_{(H_1 \cap H_2)K^*} (1 \otimes_J 1)$. The differential in $W_{V,*,(H_1 \cap H_2)K^*)}$ induces a differential in $c - \underline{\mathrm{Ind}}_{(H_1 \cap H_2)K^*}^{GL_2K}(W_{V,*,(H_1 \cap H_2)K^*)})$ given by

$$d(1 \otimes_{(H_1 \cap H_2)K^*} \otimes(1 \otimes_J 1)) = 1 \otimes_{(H_1 \cap H_2)K^*} \otimes d(1 \otimes_J 1)$$

where $d(1 \otimes_J 1) \in W_{V,i-1,(H_1 \cap H_2)K^*)}^{((J,\phi))}$.

Also $d(1 \otimes_J 1)$ lies in the kernel of the differential if $i \geq 2$ and of the augmentation if $i = 1$. Therefore, by induction,

$$\psi_{i-1}(d(1 \otimes_{(H_1 \cap H_2)K^*} \otimes(1 \otimes_J 1)))$$

lies in

$$1 \otimes_{N_{GL_2K}(H_1)} W_{V,i-1,N_{GL_2K}(H_1)}^{((J,\phi))} \oplus 1 \otimes_{N_{GL_2K}(H_1 \cap H_2)} W_{V,i-1,N_{GL_2K}(H_1 \cap H_2)}^{((J,\phi))}.$$

By the discussion of Lines in $c - \underline{\mathrm{Ind}}_{(H}^{G}(M)$ given in Example 1.11 we have an isomorphism of k-vector spaces

$$W_{V,i-1,N_{GL_2K}(H_1)}^{((J,\phi))} \xrightarrow{\cong} 1 \otimes_{N_{GL_2K}(H_1)} W_{V,i-1,N_{GL_2K}(H_1)}^{((J,\phi))}$$

given by $x \mapsto 1 \otimes_{N_{GL_2K}(H_1)} x$ which transforms the differential d into $1 \otimes_{N_{GL_2K}(H_1)} d$. There is an analogous isomorphism for $W_{V,i-1,N_{GL_2K}(H_1 \cap H_2)}^{((J,\phi))}$ and also for $V^{(J,\phi)}$. Therefore, from the monomial exactness of the monomial resolutions, there exists $w \in W_{V,i,N_{GL_2K}(H_1)}^{((J,\phi))}$ and

$w' \in W_{V,i,N_{GL_2 K}(H_1 \cap H_2)}^{((J,\phi))}$ such that

$$1 \otimes_{N_{GL_2 K}(H_1)} d(w) \oplus 1 \otimes_{N_{GL_2 K}(H_1 \cap H_2)} d(w') = \psi_{i-1}(d(1 \otimes_{(H_1 \cap H_2)K^*} \otimes (1 \otimes_J 1))).$$

Set

$$\psi_i(1 \otimes_{(H_1 \cap H_2)K^*} (1 \otimes_J 1))) = 1 \otimes_{N_{GL_2 K}(H_1)} w + 1 \otimes_{N_{GL_2 K}(H_1 \cap H_2)} w'.$$

This defines a $k[GL_2 K]$-module homomorphism ψ_i which commutes with differentials. In addition, as in the case of ψ_0, for all $g \in GL_2 K$ and $J \subseteq (H_1 \bigcap H_2)K^*$,

$$\psi_i(g \otimes_{(H_1 \cap H_2)K^*} W_{V,i,(H_1 \cap H_2)K^*}^{((J,\phi))})$$

lies in

$$g \otimes_{N_{GL_2 K}(H_1)} W_{V,i,N_{GL_2 K}(H_1)}^{((J,\phi))} \oplus g \otimes_{N_{GL_2 K}(H_1 \cap H_2)} W_{V,i,N_{GL_2 K}(H_1 \cap H_2)}^{((J,\phi))},$$

which guarantees that ψ_i is a morphism in $_{k[GL_2 K],\underline{\phi}}\mathbf{mon}$.

REMARK 4.6. (i) Any two constructions of ψ_* in §4.5 will be chain homotopic as monomial-morphisms because the monomial resolutions of §4.4 are each unique up to chain homotopy (cf. Proposition 2.4).

(ii) By the discussion of §4.2 and §4.3 we have a short exact sequence in $_{k[GL_2 K],\underline{\phi}}\mathbf{mod}$ of the form

$$0 \longrightarrow c - \mathrm{Ind}_{(H_1 \cap H_2)K^*}^{GL_2 K}(V) \xrightarrow{\psi}$$

$$c - \mathrm{Ind}_{N_{GL_2 K}(H_1)}^{GL_2 K}(V) \oplus c - \mathrm{Ind}_{N_{GL_2 K}(H_1 \cap H_2)}^{GL_2 K}(V) \xrightarrow{\epsilon} V \longrightarrow 0.$$

4.7. *The monomial resolution of V in* $_{k[GL_2 K],\underline{\phi}}\mathbf{mon}$

We now consider the chain complex $\underline{M}_* \longrightarrow \overline{V}$ in which, for $i \geq 0$, $\underline{M}_i$ is given by

$$c - \underline{\mathrm{Ind}}_{(H_1 \cap H_2)K^*}^{GL_2 K}(W_{V,i-1,(H_1 \cap H_2)K^*}) \oplus c - \underline{\mathrm{Ind}}_{N_{GL_2 K}(H_1)}^{GL_2 K}(W_{V,i,N_{GL_2 K}(H_1)})$$

$$\oplus c - \underline{\mathrm{Ind}}_{N_{GL_2 K}(H_1 \cap H_2)}^{GL_2 K}(W_{V,i,N_{GL_2 K}(H_1 \cap H_2)})$$

with differential given by

$$d(w_{1,i-1}, w_{0,i}, w'_{0,i}) = (d(w_{1,i-1}), d(w_{0,i}, w'_{0,i}) + (-1)^i \psi_{i-1}(w_{1,i-1})).$$

This is a chain complex because

$$dd(w_{1,i-1}, w_{0,i})$$

$$= (dd(w_{1,i-1}), dd(w_{0,i}, w'_{0,i}) + (-1)^i d\psi_i(w_{1,i-1}) + (-1)^{i-1} \psi_{i-2}(dw_{1,i-1}))$$

$$= (0, (-1)^i d\psi_{i-1}(w_{1,i-1}) + (-1)^{i-1} \psi_{i-2} d(w_{1,i-1}))$$

which is zero because $d\psi_{i-1} = \psi_{i-2}d$, by construction.

The chain complex of $\underline{M}_i$'s is augmented by the map induced by the short exact sequence of Remark 4.6(ii).

PROPOSITION 4.8.

Let $J \subseteq GL_2K$ be a compact open modulo the centre subgroup containing the centre K^*. Then J is conjugate to a subgroup of H_1K^* or of $\langle (H_1 \cap H_2)K^*, u \rangle$ or both.

Proof

Since $J \cap SL_2K \cap K^* = \mathcal{O}_K^*$ we see that $J \cap SL_2K$ is a compact open subgroup of SL_2K. Hence, by well-known properties on the BN-pair for SL_2K with K local (see Chapter Four concerning BN pairs and GL_3K; [4]; [35], [60], [61], [139]), we may assume that $J \cap SL_2K \subseteq SL_2\mathcal{O}_K$. We have a homomorphism

$$v_K \cdot \det : J/(J \cap SL_2K)K^* \longrightarrow \mathbb{Z}/2$$

where v_K is the valuation on K. If this homomorphism is trivial then $J \subseteq H_1K = \text{Ker}(v_K \cdot \det)$. If this homomorphism is non-trivial we have a group extension

$$(J \cap SL_2K)K^* \longrightarrow J \overset{v_K \cdot \det}{\longrightarrow} \mathbb{Z}/2.$$

Suppose that $(J \cap SL_2K)K^* \subseteq (H_1 \cap H_2)K^*$. In this case the extension which pushes out along this inclusion to give an extension

$$(H_1 \cap H_2)K^* \longrightarrow X \overset{v_K \cdot \det}{\longrightarrow} \mathbb{Z}/2.$$

From the simplicial action we see that $X = \langle (H_1 \cap H_2)K^*, u \rangle$. If $(J \cap SL_2K)K^* \nsubseteq (H_1 \cap H_2)K^*$ then pushing out along the inclusion $(J \cap SL_2K)K^* \subseteq H_1K^*$ yields an extension of the form

$$H_1K^* \longrightarrow X \overset{v_K \cdot \det}{\longrightarrow} \mathbb{Z}/2.$$

However the action on the tree shows that there is no such X. $\square$

THEOREM 4.9. *Monomial exactness for GL_2K*

For K local and $G = GL_2K$ the chain complex of §4.7

$$\cdots \longrightarrow \underline{M}_i \overset{d}{\longrightarrow} \underline{M}_{i-1} \overset{d}{\longrightarrow} \cdots \overset{d}{\longrightarrow} \underline{M}_0 \overset{\epsilon}{\longrightarrow} V \longrightarrow 0$$

is a monomial resolution in $_{k[GL_2K],\underline{\phi}}\mathbf{mon}$. That is, for each $(J,\phi) \in \mathcal{M}_{G,\underline{\phi}}$

$$\cdots \longrightarrow \underline{M}_i^{((J,\phi))} \overset{d}{\longrightarrow} \underline{M}_{i-1}^{((J,\phi))} \overset{d}{\longrightarrow} \cdots \overset{d}{\longrightarrow} \underline{M}_0^{((J,\phi))} \overset{\epsilon}{\longrightarrow} V^{(J,\phi)} \longrightarrow 0$$

is an exact sequence of k-vector spaces.

Verification of monomial exactness in the very explicit complex of §4.7 will occupy the rest of this section. However, we pause to record the fact that Theorem 4.9 implies the validity of Conjecture 3.3 for GL_2K.

COROLLARY 4.10. *(Conjecture 3.3 for $GL_2 K$)*
Conjecture 3.3, asserting the existence of a canonical monomial resolution in $_{k[GL_n K], \underline{\phi}}\mathbf{mon}$, is valid when $n = 2$.

Proof

The monomial complex explicitly constructed in §4.7 is isomorphic to that of §3.1 if one uses the simplicial structure on the Bruhat-Tits building for $GL_2 K$, corresponding to that given in §4.1. This is because there is one orbit of 1-cells which is isomorphic to $\tilde{C}_1$ in §4.2 and two orbits of 0-cells whose sum is isomorphic to $\tilde{C}_0$. $\square$

4.11. *Some well-known elementary homological algebra*
If we have two chain complexes

$$\cdots \longrightarrow A_i \longrightarrow A_{i-1} \longrightarrow \cdots \longrightarrow A_{-1} \longrightarrow 0$$

and

$$\cdots \longrightarrow B_i \longrightarrow B_{i-1} \longrightarrow \cdots \longrightarrow B_{-1} \longrightarrow 0$$

with a chain map f_* between them such that

$$0 \longrightarrow A_{-1} \overset{f_{-1}}{\longrightarrow} B_{-1} \longrightarrow V \longrightarrow 0$$

is a short exact sequence, consider the mapping cone chain complex $N_i = A_{i-1} \oplus B_i$ with differential $d(a_{i-1}, b_i) = (d(a_{i-1}), d(b_i) + (-1)^i f_{i-1}(a_{i-1}))$. We have a short exact sequence of chain complexes

$$0 \longrightarrow B_* \longrightarrow N_* \longrightarrow N_*/B_* \longrightarrow 0$$

for $* \geq 0$. Since $N_i/B_i \cong A_{i-1}$ for $i \geq 1$ we have a long exact homology sequence of the form

$$\cdots \longrightarrow H_i(B) \longrightarrow H_i(N) \longrightarrow H_{i-1}(A) \overset{\partial}{\longrightarrow} H_{i-1}(B) \longrightarrow \cdots$$

where $\partial = (-1)^i f_{i-1}$ on $H_{i-1}(A)$. If A_*, B_* are exact (not just in positive dimensions) then we have $H_i(N_*) = 0$ for $i > 0$ while

$$0 \longrightarrow A_{-1} \overset{\partial}{\longrightarrow} B_{-1} \longrightarrow H_0(N_*) \longrightarrow 0$$

yields an isomorphism $H_0(N_*) \cong V$ induced by

$$N_0 \longrightarrow B_0 \longrightarrow B/d(B_1) \cong B_{-1} \longrightarrow V.$$

4.12. *Proof of Theorem 4.9*
Consider the chain complex

$$\cdots \longrightarrow \underline{M}_i \longrightarrow \underline{M}_{i-1} \longrightarrow \cdots \longrightarrow \underline{M}_0 \longrightarrow V \longrightarrow 0.$$

Each of the $\underline{M}_i$'s is a Line-bundle with Lines generated by $g \otimes_{(H_1 \bigcap H_2) K^*} L_1$, $g \otimes_{N_{GL_2 K}(H_1)} L_0$ or $g \otimes_{N_{GL_2 K}(H_1 \cap H_2)} L_0'$ with L_1, L_0, L_0' being Lines in $W_{V, i-1, (H_1 \cap H_2) K^*)}$, $W_{V, i, N_{GL_2 K}(H_1)}$ or $W_{V, i, N_{GL_2 K}(H_1 \cap H_2)}$, respectively.

A Line of the form $g \otimes_{(H_1 \bigcap H_2)K^*} L_1$, $g \otimes_{N_{GL_2K}(H_1)} L_0$ or $g \otimes_{N_{GL_2K}(H_1 \cap H_2)} L_0'$ has stabiliser of the form $g(J', \phi')g^{-1}$ where $J' \subseteq (H_1 \bigcap H_2)K^*$, $N_{GL_2K}(H_1)$ or $N_{GL_2K}(H_1 \cap H_2)$, respectively.

Let $(J, \phi) \in \mathcal{M}_{GL_2K,\underline{\phi}}$ with $(K^*, \underline{\phi}) \leq (J, \phi)$ and J being compact open modulo the centre K^*. This implies that the J-orbit of any 0-simplex or 1-simplex of the (subdivided) tree is finite. For example, if

$$J = N_{GL_2K}H_1 \bigcap H_2 = \langle (H_1 \bigcap H_2)K^*, u \rangle$$

then the J-orbit of an end-point of the fundamental 1-simplex (prior to subdivision) consists of the two end-points.

We wish to examine exactness in the middle of

$$\underline{M}_{i+1}^{((J,\phi))} \longrightarrow \underline{M}_i^{((J,\phi))} \longrightarrow \underline{M}_{i-1}^{((J,\phi))}$$

for $i \geq 1$.

Consider the inclusions of compact open modulo the centre subgroups:

$$H_1 K^* = N_{GL_2K}(H_1) \geq (H_1 \cap H_2)K^* \leq N_{GL_2K}(H_1 \cap H_2)$$

where

$$N_{GL_2K}(H_1 \cap H_2) = \langle H_1 \bigcap H_2, K^*, u \rangle.$$

Since the GL_2K-action is transitive on the subdivided tree and since each the above groups form the set of stabilisers of simplices in the fundamental domain, up to GL_2K-conjugation, we must have one of the following three cases:

Case A: $J \subseteq (H_1 \bigcap H_2)K^*$.

Case B: $J \subseteq H_1 K^*$, but J is not conjugate to a subgroup of $\langle H_1 \bigcap H_2, K^*, u \rangle$.

Case C: $J \subseteq \langle H_1 \bigcap H_2, K^*, u \rangle$, but J is not conjugate to a subgroup of $H_1 K^*$.

Proposition 4.8 together with the following result shows that Cases A–C exhaust the possibilities.

PROPOSITION 4.13.

If $J \subseteq H_1 K^*$ and J is conjugate to a subgroup of $\langle H_1 \bigcap H_2, K^*, u \rangle$ then J is conjugate to a subgroup of $(H_1 \bigcap H_2)K^*$.

Proof

Observe that $H_1 K^* \bigcap \langle H_1 \bigcap H_2, K^*, u \rangle$ stabilises the two ends of the 1-simplex whose stabiliser is $(H_1 \bigcap H_2)K^*$. Pro tem, call this 1-simplex the canonical fundamental domain. Hence

$$H_1 K^* \bigcap \langle H_1 \bigcap H_2, K^*, u \rangle = (H_1 \bigcap H_2)K^*.$$

Now we may assume, by conjugation if necessary, that $J \subseteq \langle H_1 \bigcap H_2, K^*, u \rangle$ and that there exists $g \in GL_2 K$ such that $gJg^{-1} \subseteq H_1 K^*$. Hence J stabilises the end-point, β, of the canonical fundamental domain which was introduced during the barycentric subdivision and also stablises $g\alpha$ where α is the other end of the canonical fundamental domain. Since the tree contains no closed loops J stabilises all the 1-simplices between β and $g\alpha$. In particular J stabilises the canonical fundamental domain or its neighbour. In the first case

$$J \subseteq H_1 K^* \bigcap \langle H_1 \bigcap H_2, K^*, u \rangle = (H_1 \bigcap H_2) K^*$$

and in the second case

$$u^{-1} Ju \subseteq H_1 K^* \bigcap \langle H_1 \bigcap H_2, K^*, u \rangle = (H_1 \bigcap H_2) K^*.$$

$\square$

4.14. *Proof of Theorem 4.9 continued*
Now let us examine $\underline{M}_i^{((J,\phi))}$ in Case A.
We have a short exact sequence of chain complexes

$$0 \longrightarrow c - \underline{\mathrm{Ind}}_{N_{GL_2 K}(H_1)}^{GL_2 K}(W_{V,*,N_{GL_2 K}(H_1)})$$

$$\oplus c - \underline{\mathrm{Ind}}_{N_{GL_2 K}(H_1 \cap H_2)}^{GL_2 K}(W_{V,*,N_{GL_2 K}(H_1 \cap H_2)})$$

$$\longrightarrow \underline{M}_* \longrightarrow c - \underline{\mathrm{Ind}}_{(H_1 \cap H_2) K^*}^{GL_2 K}(W_{V,*-1,(H_1 \cap H_2) K^*})) \longrightarrow 0$$

and taking the $((J, \phi))$-part yields a short exact sequence (because the sum of all the Lines with a fixed stabiliser pair is a direct summand) of the form

$$0 \longrightarrow c - \underline{\mathrm{Ind}}_{N_{GL_2 K}(H_1)}^{GL_2 K}(W_{V,*,N_{GL_2 K}(H_1)})^{((J,\phi))}$$

$$\oplus c - \underline{\mathrm{Ind}}_{N_{GL_2 K}(H_1 \cap H_2)}^{GL_2 K}(W_{V,*,N_{GL_2 K}(H_1 \cap H_2)})^{((J,\phi))}$$

$$\longrightarrow \underline{M}_*^{((J,\phi))} \longrightarrow c - \underline{\mathrm{Ind}}_{(H_1 \cap H_2) K^*}^{GL_2 K}(W_{V,*-1,(H_1 \cap H_2) K^*}))^{((J,\phi))} \longrightarrow 0.$$

Let L be a Line in M so that $g \otimes_H L$ generates a Line in $c - \underline{\mathrm{Ind}}_H^{GL_2 K}(M)$. This Line lies in $c - \underline{\mathrm{Ind}}_H^{GL_2 K}(M)^{((J,\phi))}$ if and only if $g^{-1} Jg \subseteq H$ and $g^{-1} Jg$ acts on L via $g^*(\phi)$. That is, $g^{-1} jg(v) = \phi(j)v$ for $v \in L$.
Therefore the left-hand group in the short exact sequence is equal to

$$\oplus_{g^{-1} Jg \subseteq N_{GL_2 K}(H_1)} \, g \otimes_{N_{GL_2 K}(H_1)} W_{V,*,N_{GL_2 K}(H_1)}^{((g^{-1} Jg, g^*(\phi)))}$$

$$\oplus$$

$$\oplus_{g^{-1} Jg \subseteq N_{GL_2 K}(H_1 \cap H_2)} \, g \otimes_{N_{GL_2 K}(H_1 \cap H_2)} W_{V,*,N_{GL_2 K}(H_1 \cap H_2)}^{((g^{-1} Jg, g^*(\phi)))}$$

while the right-hand group is equal to

$$\oplus_{g^{-1}Jg\subseteq(H_1\cap H_2)K^*} g \otimes_{(H_1\cap H_2)K^*} W^{((g^{-1}Jg,g^*(\phi)))}_{V,*-1,(H_1\cap H_2)K^*}.$$

These direct sums have to be interpreted with care. For example, that for the right-hand group means that we choose coset representatives $\{g_\alpha, \alpha \in \mathcal{A}\}$ then we form the direct sum over the g_α's such that $g_\alpha^{-1}Jg_\alpha \subseteq (H_1 \cap H_2)K^*$ of $g_\alpha \otimes_{(H_1\cap H_2)K^*} L$ where L runs through the Lines of $M^{((g_\alpha^{-1}Jg_\alpha,g_\alpha^*(\phi)))}_{1,*-1}$. The differential on such a Line maps it by $1 \otimes d$ to $g_\alpha \otimes_{(H_1\cap H_2)K^*} d(L)$. Hence the complex is the direct sum of subcomplexes, one for each g_α. As we noted in the discussion of §4.5, by the discussion of Lines in $c - \underline{\mathrm{Ind}}^G_{(H}(M)$ given in Example 1.11 we have an isomorphism of k-vector spaces such as

$$W^{((J,\phi))}_{V,i-1,N_{GL_2K}(H_1)} \xrightarrow{\cong} g \otimes_{N_{GL_2K}(H_1)} W^{((g^{-1}Jg,g^*(\phi)))}_{V,i-1,N_{GL_2K}(H_1)}$$

and similarly for the other two monomial resolutions.

Therefore, by monomial exactness of $W_{V,*,N_{GL_2K}(H_1)}$, $W_{V,*,N_{GL_2K}(H_1\cap H_2)}$ and $W_{V,*-1,(H_1\cap H_2)K^*}$, we have $H_i(\underline{M}^{((J,\phi))}_*) = 0$ for $i \geq 2$ and there is an exact homology sequence of the form

$$0 \longrightarrow H_1(\underline{M}^{((J,\phi))}_*) \longrightarrow \oplus_{g^{-1}Jg\subseteq(H_1\cap H_2)K^*} g \otimes_{(H_1\cap H_2)K^*} V^{(g^{-1}Jg,g^*(\phi))}$$

$$\longrightarrow \oplus_{g^{-1}Jg\subseteq N_{GL_2K}(H_1)} g \otimes_{N_{GL_2K}(H_1)} V^{(g^{-1}Jg,g^*(\phi))}$$

$$\oplus \oplus_{g^{-1}Jg\subseteq N_{GL_2K}(H_1\cap H_2)} g \otimes_{N_{GL_2K}(H_1\cap H_2)} V^{(g^{-1}Jg,g^*(\phi))}$$

$$\longrightarrow H_0(\underline{M}^{((J,\phi))}_*) \longrightarrow 0.$$

Suppose that $g^{-1}Jg \subseteq (H_1 \cap H_2)K^*$ but that $g \notin (H_1 \cap H_2)K^*$. Then $gL \neq L$ where L denotes the canonical fundamental domain. On the other hand J fixes both L and gL. Since the tree has no closed loops this happens only if $J = \{1\}$. A similar argument applies to the other direct sums in the exact sequence, replacing the 1-simplex L by a vertex. Therefore if $J \neq \{1\}$ then the exact sequence takes the form

$$0 \longrightarrow H_1(\underline{M}^{((J,\phi))}_*) \longrightarrow V^{(J,\phi)} \longrightarrow V^{(J,\phi)} \oplus V^{(J,\phi)} \longrightarrow H_0(\underline{M}^{((J,\phi))}_*) \longrightarrow 0.$$

In addition the map in the centre is given by $(v \mapsto (v, -v))$ so that

$$H_i(\underline{M}^{((J,\phi))}_*) = \begin{cases} V^{(J,\phi)} & \text{if } i = 0, \\ \\ 0 & \text{otherwise.} \end{cases}$$

When $J = \{1\}$ the exact sequence becomes

$$0 \longrightarrow H_1(\underline{M}_*^{((\{1\},1))}) \longrightarrow c - \operatorname{Ind}_{(H_1 \cap H_2)K^*}^{GL_2 K} V$$

$$\xrightarrow{\psi} c - \operatorname{Ind}_{N_{GL_2 K}(H_1)}^{GL_2 K} V \oplus c - \operatorname{Ind}_{N_{GL_2 K}(H_1 \cap H_2)}^{GL_2 K} V$$

$$\longrightarrow H_0(\underline{M}_*^{((J,\phi))}) \longrightarrow 0.$$

Therefore when J is trivial we also have

$$H_i(\underline{M}_*^{((\{1\},1))}) = \begin{cases} V & \text{if } i = 0, \\ \\ 0 & \text{otherwise.} \end{cases}$$

In Case B, by a similar argument we find that $H_i(\underline{M}_*^{((J,\phi))}) = 0$ for $i \neq 0$ and

$$H_0(\underline{M}_*^{((J,\phi))}) \cong \oplus_{g^{-1}Jg \subseteq N_{GL_2 K}(H_1)} g \otimes_{N_{GL_2 K}(H_1)} V^{(g^{-1}Jg, g^*(\phi))}.$$

However, if there exists $g \notin N_{GL_2 K}(H_1)$ such that $gJg^{-1} \subseteq N_{GL_2 K}(H_1)$ then J fixes β and $g\beta$. Therefore J fixes all simplices between β and $g\beta$ which include a translate of the canonical fundamental domain L so that J is subconjugate to $(H_1 \cap H_2)K^*$. Therefore there is only one coset in the above direct sum and $H_0(\underline{M}_*^{((J,\phi))}) \cong V^{(J,\phi)}$.

Therefore in all cases we have

$$H_i(\underline{M}_*^{((J,\phi))}) = \begin{cases} V^{(J,\phi)} & \text{if } i = 0, \\ \\ 0 & \text{otherwise.} \end{cases}$$

The proof of Case C is similar but simpler. Arguing as in Case A we have an isomorphism

$$\oplus_{g^{-1}Jg \subseteq N_{GL_2 K}(H_1 \cap H_2)} g \otimes_{N_{GL_2 K}(H_1 \cap H_2)}^{GL_2 K} W_{V,*,N_{GL_2 K}(H_1 \cap H_2)}^{((g^{-1}Jg, g^*(\phi)))}$$

$$\xrightarrow{\cong} M_*^{((J,\phi))}.$$

Now J must contain an element of the coset $(H_1 \cap H_2)K^*u$ denoted by zu, say. If $g^{-1}zug$ lies in $N_{GL_2 K}(H_1 \cap H_2)$ it sends the (pre-subdivision) fundamental simplex of the into itself (switching endpoints) and does the same to the image on the original fundamental simplex under g^{-1}. It is easy to see, either algebraically or from the self-normalising properties of the simplex-stablisers in the simplicially subdivided tree, that this can happen if and only if $g \in N_{GL_2 K}(H_1 \cap H_2)$. Therefore

$$W_{V,*,N_{GL_2 K}(H_1 \cap H_2)}^{((J,\phi))} \xrightarrow{\cong} M_*^{((J,\phi))}$$

which implies monomial exactness in Case C and completes the proof of Theorem 4.9. $\square$

REMARK 4.15. In the construction of the monomial resolution for GL_2K we were able to use any covering of ψ because they are all chain homotopic in the monomial category. However, this would be insufficient for GL_nK when $n \geq 3$ since the result is unlikely to be a chain complex. In fact, the obvious construction of a "differential" d like that of §4.7 would merely result in a composition dd which was chain homotopic to zero. The problem arises because the Bruhat-Tits building is no longer 1-dimensional. This obstacle is what necessitated the construction of the natural bar-monomial resolution, in order to enable the construction of a double complex in §3.1.

4.16. Some subgroups of GL_2K

Let K be a p-adic local field with valuation $v_K : K^* \longrightarrow \mathbb{Z}$. We have homomorphisms

$$\det : GL_2K \longrightarrow K^* \text{ and } v_K \cdot \det : GL_2K \longrightarrow \mathbb{Z}.$$

Following ([115] p. 75) we may define subgroups of GL_2K denoted by SL_2K, GL_2K^0 and GL_2K^+ by

$$SL_2K = \text{Ker}(\det), \quad GL_2K^0 = \text{Ker}(v_K \cdot \det),$$

$$GL_2K^+ = \text{Ker}(v_K \cdot \det \text{ modulo } 2)$$

so that

$$SL_2K \subset GL_2K^0 \subset GL_2K^+ \subset GL_2K.$$

As explained in ([115] pp. 78/79) and in terms of Bruhat-Tits buildings in ([115] p. 91) (i.e. BN-pairs [35] p. 107) each of the first three groups acts transitively on the vertices of the tree and act on a 1-simplex between adjacent vertices simplicially (i.e. any element sending the 1-simplex to itself does so point-wise). Therefore these subgroups act simplicially on the tree and one may perform the constructions of §3.1 and §4.7 without having to perform a barycentric subdivision.

In fact the analogues of Theorem 4.9 and Corollary 4.10 are true for admissible representations of these subgroups.

5. Monomial resolution and π_K-adic levels

5.1. As in §2.1 let K be a p-adic local field with valuation ring $\mathcal{O}_K$ and π_K a generator of the maximal ideal of $\mathcal{O}_K$. Let V be a (left) admissible k-representation of GL_2K with central character ϕ. For $1 \leq m$ let U_K^m denote the subgroup of GL_2K given by

$$U_K^m = \{X \in GL_2\mathcal{O}_K \mid X \equiv I \text{ (modulo } \pi_K^m)\}.$$

Assume that m_0 is the least integer such that the central character $\underline{\phi}$ is trivial on $U_K^{m_0}$ so that

$$V = \bigcup_{m \geq m_0} V^{(K^* U_K^m, \underline{\phi})}.$$

For each $m \geq m_0$ the k-vector space $V^{(K^* U_K^m, \underline{\phi})}$ is a finite-dimensional representation of the finite modulo the centre quotient group $K^* U_K^m / U_K^m$.

THEOREM 5.2.
In the notation of §5.1 and Theorem 4.9
(i)
$$\underline{M}_*^{((K^* \cdot U_n, \underline{\phi}))} \longrightarrow V^{(K^* \cdot U_n, \underline{\phi})}$$
is a monomial resolution in $_{k[K^* U_K^m / U_K^m], \underline{\phi}}\mathbf{mon}$.

(ii) When k is an algebraically closed field of characteristic zero the monomial resolution of part (i), which is unique up to chain homotopy in $_{k[K^* U_K^m / U_K^m], \underline{\phi}}\mathbf{mon}$, contains in its chain homotopy class a monomial resolution which is finitely generated and of finite length.

Proof

Part (i) follows from the fact that $\underline{M}_* \longrightarrow V \longrightarrow 0$ is a monomial resolution of V. The proof of part (ii) is given in Chapter One §6. □

5.3. ϵ-factors and L-functions

If V is an admissible representation of $GL_2 K$ and $\underline{M}_* \longrightarrow V$ is a monomial resolution as in Theorem 4.9 one may possibly construct ϵ-factors for V by some sort of Euler characteristic obtained by applying to each Line an "integral", made from character values, which in the finite case specialises to the Kondo-Gauss sums. These integrals respect induction from one compact, open modulo the centre subgroup to another.

I have not pursued this topic very deeply in this monograph. In the case of $GL_2 K$ the Kondo-style Gauss sum is described in Chapter Six §1. In Chapter Six §2 and §3 I give briefly a number of constructions and questions concerning the local L-function of V and the Tate-style local function equation.

I am assuming an analogue of the result concerning wild ϵ-factors modulo p-power roots of unity [73] holds for all but a finite set of Lines with the result that a well-defined ϵ-factor modulo p-power roots of unity is defined by a finite product of Kondo-style Gauss sums. Here I ought to mention that I slightly disagree with a fundamental result in [73] (see [129] or Appendix II) so the epsilon factor I propose may only be well defined up to ± 1 times a p-power root of unity.

I have yet to develop fully the approach of Tate's thesis to each Line, properly developing Chapter Six §2, in an attempt to get the L-functions of [66].

These methods should apply to GL_nK, since Conjecture 3.3 holds.

6. Galois invariant admissibles for GL_2K

6.1. Let k be an algebraically closed field. Suppose that K is a p-adic local field and $\rho : GL_2K \longrightarrow GL(V)$ is a irreducible admissible k-representation. Let K/F be a finite Galois extension and suppose that $z^*(\rho)$ is equivalent to ρ for each $z \in \mathrm{Gal}(K/F)$. Therefore for $z \in \mathrm{Gal}(K/F)$ there exists $X_z \in GL(V)$ such that

$$X_z \rho(g) X_z^{-1} = \rho(z(g))$$

for all $g \in GL_2K$. Therefore if $z, z_1 \in \mathrm{Gal}(K/F)$ replacing g by $z_1(g)$ gives

$$X_z \rho(z_1(g)) X_z^{-1} = \rho(zz_1(g))$$

and so

$$X_z \rho(z_1(g)) X_z^{-1} = X_z X_{z_1} \rho(g) X_{z_1}^{-1} X_z^{-1} = X_{zz_1} \rho(g) X_{zz_1}^{-1}.$$

By Schur's Lemma $X_{z_1}^{-1} X_z^{-1} X_{zz_1}$ is a k^*-valued scalar matrix and so

$$f(z, z_1) = X_{z_1}^{-1} X_z^{-1} X_{zz_1}$$

is a function from $\mathrm{Gal}(K/F) \times \mathrm{Gal}(K/F)$ to k^*. In fact, f is a 2-cocycle. That is, using commutativity of k,

$df(z, z_1, z_2)$

$$= f(z_1, z_2) f(zz_1, z_2)^{-1} f(z, z_1 z_2) f(z, z_1)^{-1}$$

$$= X_{z_2}^{-1} X_{z_1}^{-1} X_{z_1 z_2} (X_{z_2}^{-1} X_{zz_1}^{-1} X_{zz_1 z_2})^{-1} X_{z_1 z_2}^{-1} X_z^{-1} X_{zz_1 z_2} (X_{z_1}^{-1} X_z^{-1} X_{zz_1})^{-1}$$

$$= (X_{z_2}^{-1} X_{zz_1}^{-1} X_{zz_1 z_2})^{-1} X_{z_2}^{-1} X_{z_1}^{-1} X_{z_1 z_2} X_{z_1 z_2}^{-1} X_z^{-1} X_{zz_1 z_2} (X_{z_1}^{-1} X_z^{-1} X_{zz_1})^{-1}$$

$$= X_{zz_1 z_2}^{-1} X_{zz_1} X_{z_2} X_{z_2}^{-1} X_{z_1}^{-1} X_{z_1 z_2} X_{z_1 z_2}^{-1} X_z^{-1} X_{zz_1 z_2} (X_{z_1}^{-1} X_z^{-1} X_{zz_1})^{-1}$$

$$= X_{zz_1 z_2}^{-1} X_{zz_1} X_{z_1}^{-1} X_z^{-1} X_{zz_1 z_2} (X_{z_1}^{-1} X_z^{-1} X_{zz_1})^{-1}$$

$$= X_{zz_1 z_2}^{-1} X_{zz_1} (X_{z_1}^{-1} X_z^{-1} X_{zz_1})^{-1} X_{z_1}^{-1} X_z^{-1} X_{zz_1 z_2}$$

$$= X_{zz_1 z_2}^{-1} X_{zz_1} X_{zz_1}^{-1} X_z X_{z_1} X_{z_1}^{-1} X_z^{-1} X_{zz_1 z_2}$$

$$= 1.$$

The 2-cocycle f defined a cohomology class in $[f] \in H^2(\mathrm{Gal}(K/F); k^*)$, where $\mathrm{Gal}(K/F)$ acts trivially on k^*. In Proposition 6.2 we shall show that there exists a finite Galois extension E/F containing K such that

$$[f] \in \mathrm{Ker}(H^2(\mathrm{Gal}(K/F); k^*) \longrightarrow H^2(\mathrm{Gal}(E/F); k^*)).$$

Assuming Proposition 6.2 for the moment, this implies that the function

$$f' : \text{Gal}(E/F) \times \text{Gal}(E/F) \longrightarrow \text{Gal}(K/F) \times \text{Gal}(K/F) \xrightarrow{f} k^*$$

is a coboundary $f' = dF$, where F is a function from $\text{Gal}(E/F)$ to k^*. In other words, for $z, z_1 \in \text{Gal}(E/F)$,

$$X_{z_1}^{-1} X_z^{-1} X_{zz_1} = f'(z, z_1) = dF(z, z_1) = F(z_1) F(zz_1)^{-1} F(z_1).$$

Therefore $z \mapsto X_z F(z)$ is a homomorphism from $\text{Gal}(E/F)$ to $GL(V)$, since the image of F is central.

Recall that the semi-direct product $\text{Gal}(E/F) \propto GL_2 K$ is given by the set $\text{Gal}(E/F) \times GL_2 K$ with the product defined by

$$(h_1, g_1) \cdot (h_2, g_2) = (h_1 h_2, g_1 h_1(g_2)).$$

Define a map

$$\tilde{\rho} : \text{Gal}(E/F) \propto GL_2 K \longrightarrow GL(V)$$

by $(z, g) \mapsto \rho(g) X_z F(z)$.

Therefore

$$\tilde{\rho}((h_1 h_2, g_1 h_1(g_2)))$$

$$= \rho(g_1 h_1(g_2)) X_{h_1 h_2} F(h_1 h_2)$$

$$= \rho(g_1) \rho(h_1(g_2)) X_{h_1 h_2} F(h_1 h_2)$$

$$= \rho(g_1) X_{h_1} \rho(g_2) X_{h_1}^{-1} X_{h_1} F(h_1) X_{h_2} F(h_2)$$

$$= \rho(g_1) X_{h_1} F(h_1) \rho(g_2) X_{h_2} F(h_2)$$

$$= \tilde{\rho}((h_1, g_1)) \tilde{\rho}((h_2, g_2))$$

so that

$$\tilde{\rho} : \text{Gal}(E/F) \propto GL_2 K \longrightarrow GL(V)$$

is a k-representation of the semi-direct product, which is irreducible and admissible since it extends ρ[1].

Any two such extensions differ by twisting via a homomorphism $\text{Gal}(E/F) \longrightarrow k^*$.

PROPOSITION 6.2.
In §6.1 for any cohomology class $[f] \in H^2(\text{Gal}(K/F); k^*)$ there exists a finite Galois extension containing K such that the image of $[f]$ in $H^2(\text{Gal}(E/F); k^*)$ is trivial.

---

[1]It would be notationally more satisfying to be able to construct $\tilde{\rho}$ on $\text{Gal}(K/F) \propto GL_2 K$ but, even in the case of finite fields this is not always possible (see Chapter Eight, Theorem 3.11; [117] Theorem 1, p. 406).

Proof

Recall ([**112**] pp. 101–102) that the Galois cohomology group of F with coefficients in k^* is defined as the direct limit

$$H^i(F; k^*) = \varinjlim_{E/F} H^i(\mathrm{Gal}(E/F); (k^*)^{\mathrm{Gal}(E/F)}) = \varinjlim_{E/F} H^i(\mathrm{Gal}(E/F); k^*)$$

where E/F runs through finite Galois extensions of F, since the groups act trivially on k^*. Since k is algebraically closed the quotient $k^*/\mathrm{Tors}(k^*)$ is uniquely divisible and so $H^i(\mathrm{Gal}(E/F); k^*/\mathrm{Tors}(k^*)) = 0$ for $i > 0$, since the Galois group is finite. If p is the characteristic of k then $\mathrm{Tors}(k^*) \cong \mathbb{Q}/\mathbb{Z}[1/p]$ and it is isomorphic to $\mathbb{Q}/\mathbb{Z}$ if k has characteristic zero. The former is a direct summand of the latter so that the vanishing of $H^2(F; \mathbb{Q}/\mathbb{Z})$ implies that of $H^2(F; \mathbb{Q}/\mathbb{Z}[1/p])$. Therefore, from the long exact cohomology sequences, we have isomorphisms

$$H^i(F; \mathbb{Q}/\mathbb{Z}[1/p]) \cong H^i(F; k^*)$$

if $\mathrm{char}(k) = p$ and

$$H^i(F; \mathbb{Q}/\mathbb{Z}) \cong H^i(F; k^*)$$

if $\mathrm{char}(k) = 0$.

To prove that there groups vanish when $i = 2$ it will suffice to choose a prime l and show that the direct limit

$$\varinjlim_t H^2(F; \mathbb{Z}/l^t) = 0.$$

By Tate duality ([**137**] p. 289) there is an isomorphism

$$H^2(F; \mathbb{Z}/l^t) \cong H^0(F; \mu_{l^t}),$$

the Galois invariants of the l^t-th roots of unity. Hence for t large enough $H^0(F; \mu_{l^t})$ is isomorphic to the l-power roots of unity in F, which is independent of t. The inclusion map of $\mathbb{Z}/l^t$ into $\mathbb{Z}/l^{t+1}$ corresponds to the l-th power map, which is nilpotent of the l-power roots of unity in F, which implies the result. □

6.3. *The action of* $\mathrm{Gal}(E/F) \propto GL_2K$ *on the tree*

The Galois action of $\mathrm{Gal}(K/F)$ on $K \oplus K$ preserves the lattices $L = \mathcal{O}_K \oplus \mathcal{O}_K$ and $L' = \mathcal{O}_K \oplus \pi_K \mathcal{O}_K$ and their stabilisers, H_1 and H_2 of §4.1, under the tree-action. Therefore the Galois action of $\mathrm{Gal}(E/F)$, acting via $\mathrm{Gal}(K/F)$, fixes the canonical fundamental domain on the tree and the semi-direct product acts on the tree of GL_2K, extending the action of GL_2K.

The central character $\underline{\phi}$ is fixed by the Galois action.

Recall from §4.2 the cell complex of the simplicially subdivided tree. The GL_2K-normalisers of stabilisers are given by

$$N_{GL_2K}H_1 = H_1K^*, \quad N_{GL_2K}H_2 = H_2K^*,$$

$$N_{GL_2K}H_1 \bigcap H_2 = \langle H_1 \bigcap H_2, K^*, u \rangle.$$

The Galois group $\mathrm{Gal}(E/F)$ preserves each of these normalisers. Setting $G = \mathrm{Gal}(E/F) \propto GL_2K$, the 0-cells are given by

$$\tilde{C}_0 = c - \mathrm{Ind}_{\mathrm{Gal}(E/F) \propto N_{GL_2K}(H_1)}^G(k) \oplus c - \mathrm{Ind}_{\mathrm{Gal}(E/F) \propto N_{GL_2K}(H_1 \cap H_2)}^G(k)$$

while the 1-cells are

$$\tilde{C}_1 = c - \mathrm{Ind}_{\mathrm{Gal}(E/F) \propto (H_1 \cap H_2)K^*}^G(k).$$

Therefore we have a short exact sequence of $k[GL_2K]$-modules of the form

$$0 \longrightarrow \tilde{C}_1 \xrightarrow{d} \tilde{C}_0 \xrightarrow{\epsilon} k \longrightarrow 0$$

in which

$$d(g \otimes_{\mathrm{Gal}(E/F) \propto (H_1 \cap H_2)K^*} v)$$

$$= (g \otimes_{\mathrm{Gal}(E/F) \propto N_{GL_2K}(H_1)} v, -g \otimes_{\mathrm{Gal}(E/F) \propto N_{GL_2K}(H_1 \cap H_2)} v)$$

and

$$\epsilon(g_1 \otimes_{\mathrm{Gal}(E/F) \propto N_{GL_2K}(H_1)} v_1, g_2 \otimes_{\mathrm{Gal}(E/F) \propto N_{GL_2K}(H_1 \cap H_2)} v_2)$$

$$= v_1 + v_2.$$

If $\tilde{V}$ is the admissible representation of G given by $\tilde{\rho}$ we have an isomorphism analogous to that of §4.3.

$$\tilde{\phi} : c - \mathrm{Ind}_{\mathrm{Gal}(E/F) \propto H}^G(W) \otimes V \xrightarrow{\cong} c - \mathrm{Ind}_{\mathrm{Gal}(E/F) \propto H}^G(W \otimes \mathrm{Res}_H^{GL_2K}(V))$$

given by $\phi((g \otimes_H w) \otimes v) = g \otimes_H (w \otimes g^{-1}v)$, if W is finite-dimensional and H is one of $(H_1 \cap H_2)K^*$, $N_{GL_2K}(H_1)$ or $N_{GL_2K}(H_1 \cap H_2)$.

As in §4.3 $\tilde{\phi}$ transforms $d \otimes 1$ to

$$c - \mathrm{Ind}_{\mathrm{Gal}(E/F) \propto (H_1 \cap H_2)K^*}^G(\tilde{V})$$

$$\tilde{\psi} \downarrow$$

$$c - \mathrm{Ind}_{\mathrm{Gal}(E/F) \propto N_{GL_2K}(H_1)}^G(\tilde{V}) \oplus c - \mathrm{Ind}_{\mathrm{Gal}(E/F) \propto N_{GL_2K}(H_1 \cap H_2)}^G(\tilde{V})$$

given by

$$\tilde{\psi}(g \otimes_{\mathrm{Gal}(E/F) \propto (H_1 \cap H_2)K^*} v)$$

$$= (g \otimes_{\mathrm{Gal}(E/F) \propto N_{GL_2K}(H_1)} v, -g \otimes_{\mathrm{Gal}(E/F) \propto N_{GL_2K}(H_1 \cap H_2)} v).$$

Next we define an analogue of the central character

$$\tilde{\phi} : \mathrm{Gal}(E/F) \propto K^* \longrightarrow k^*$$

by $\tilde{\phi}(g, z) = \underline{\phi}(z)$ for $g \in \mathrm{Gal}(E/F), z \in K^*$. This is a well-defined character since ϕ is Galois-invariant.

We define $\mathcal{M}_{G,\tilde{\phi}}$ to be the partially ordered set of pairs (J, ϕ) where $J \subseteq G$ contains the centre $Z(G) = Z(\mathrm{Gal}(E/K)) \times K^*$, is compact open modulo the centre and where ϕ extends $\tilde{\phi}$.

As in §4.4 we have bar-monomial resolutions

$$W_{\tilde{V}, *, \mathrm{Gal}(E/F) \propto (H_1 \cap H_2) K^*)} \xrightarrow{\epsilon_1} \tilde{V}$$

in $_{k[\mathrm{Gal}(E/F) \propto (H_1 \cap H_2) K^*], \tilde{\phi}} \mathbf{mon}$,

$$W_{\tilde{V}, *, \mathrm{Gal}(E/F) \propto N_{GL_2 K}(H_1)} \xrightarrow{\epsilon_0} \tilde{V}$$

in $_{k[\mathrm{Gal}(E/F) \propto (N_{GL_2 K}(H_1))], \tilde{\phi}} \mathbf{mon}$ and

$$W_{\tilde{V}, *, \mathrm{Gal}(E/F) \propto N_{GL_2 K}(H_1 \cap H_2)} \xrightarrow{\epsilon_0'} \tilde{V}$$

in $_{k[\mathrm{Gal}(E/F) \propto N_{GL_2 K}(H_1 \cap H_2)], \tilde{\phi}} \mathbf{mon}$.

Following §4.5 we may construct a $_{k[G], \tilde{\phi}} \mathbf{mon}$ chain map $\{\tilde{\psi}_i \mid i \geq 0\}$ covering $\tilde{\psi}$:

$$c - \underline{\mathrm{Ind}}_{\mathrm{Gal}(E/F) \propto (H_1 \cap H_2) K^*}^{G} \left(W_{\tilde{V}, *, \mathrm{Gal}(E/F) \propto (H_1 \cap H_2) K^*} \right)$$

$$\tilde{\psi}_* \downarrow$$

$$c - \underline{\mathrm{Ind}}_{\mathrm{Gal}(E/F) \propto N_{GL_2 K}(H_1)}^{G} \left(W_{\tilde{V}, *, \mathrm{Gal}(E/F) \propto N_{GL_2 K}(H_1)} \right)$$
$$\oplus c - \underline{\mathrm{Ind}}_{\mathrm{Gal}(E/F) \propto N_{GL_2 K}(H_1 \cap H_2)}^{G} \left(W_{\tilde{V}, *, \mathrm{Gal}(E/F) \propto N_{GL_2 K}(H_1 \cap H_2)} \right).$$

Replacing each of $(H_1 \cap H_2)K^*$, $N_{GL_2 K}(H_1)$ or $N_{GL_2 K}(H_1 \cap H_2)$ by its semi-direct product with $\mathrm{Gal}(E/F)$, the analogue of the construction in §4.7 produces a candidate for a monomial resolution of $\tilde{V}$

$$\underline{\tilde{M}}_* \xrightarrow{\epsilon} \tilde{V}.$$

Explicitly $\underline{\tilde{M}}_i$ is given by

$$c - \underline{\mathrm{Ind}}_{\mathrm{Gal}(E/F) \propto (H_1 \cap H_2) K^*}^{G} \left(W_{\tilde{V}, i-1, \mathrm{Gal}(E/F) \propto (H_1 \cap H_2) K^*} \right)$$

$$\oplus c - \underline{\mathrm{Ind}}_{\mathrm{Gal}(E/F) \propto N_{GL_2 K}(H_1)}^{G} \left(W_{\tilde{V}, i, \mathrm{Gal}(E/F) \propto N_{GL_2 K}(H_1)} \right)$$

$$\oplus c - \underline{\mathrm{Ind}}_{\mathrm{Gal}(E/F) \propto N_{GL_2 K}(H_1 \cap H_2)}^{G} \left(W_{\tilde{V}, i, \mathrm{Gal}(E/F) \propto N_{GL_2 K}(H_1 \cap H_2)} \right)$$

with differential given, as in §4.7, by

$$d(w_{1,i-1}, w_{0,i}, w'_{0,i}) = (d(w_{1,i-1}), d(w_{0,i}, w'_{0,i}) + (-1)^i \tilde{\psi}_{i-1}(w_{1,i-1})).$$

To establish exactness in the middle of

$$\tilde{\underline{M}}_{i+1}^{((J,\phi))} \longrightarrow \tilde{\underline{M}}_i^{((J,\phi))} \longrightarrow \tilde{\underline{M}}_{i-1}^{((J,\phi))}$$

for $i \geq 1$ it suffices, as in §4.12, to consider $J \subset G$ which is a subgroup of $\mathrm{Gal}(E/F) \propto H_1 K^*$, since $\mathrm{Gal}(E/F) \propto H_1$ is a maximal compact open subgroup to which all others are G-conjugate.

Since the Galois group acts trivially on the simplices of the tree the argument of Proposition 4.13 shows that we have just two cases:

Case A: $J \subseteq \mathrm{Gal}(E/F) \propto (H_1 \cap H_2)K^*$.

Case B: $J \subseteq \mathrm{Gal}(E/F) \propto H_1 K^*$, but J is not G-conjugate to a subgroup of either $\mathrm{Gal}(E/F) \propto \langle H_1 \cap H_2, K^*, u \rangle$ or $\mathrm{Gal}(E/F) \propto (H_1 \cap H_2)K^*$.

The analogue of the argument of §4.14 establishes the following result.

THEOREM 6.4. *Monomial resolution for $\tilde{V}$*
Let K, G and $\tilde{\underline{M}}_*$ be as in §6.3. Then

$$\ldots \longrightarrow \tilde{\underline{M}}_i \xrightarrow{d} \tilde{\underline{M}}_{i-1} \xrightarrow{d} \ldots \xrightarrow{d} \tilde{\underline{M}}_0 \xrightarrow{\epsilon} \tilde{V} \longrightarrow 0$$

is a monomial resolution in ${}_{k[G],\tilde{\phi}}\mathbf{mon}$. That is, for each $(J,\phi) \in \mathcal{M}_{G,\tilde{\phi}}$

$$\ldots \longrightarrow \tilde{\underline{M}}_i^{((J,\phi))} \xrightarrow{d} \tilde{\underline{M}}_{i-1}^{((J,\phi))} \xrightarrow{d} \ldots \xrightarrow{d} \tilde{\underline{M}}_0^{((J,\phi))} \xrightarrow{\epsilon} \tilde{V}^{(J,\phi)} \longrightarrow 0$$

is an exact sequence of k-vector spaces.

In ${}_{k[G],\tilde{\phi}}\mathbf{mon}$ the monomial resolution of $\tilde{V}$ is unique up to chain homotopy.

6.5. *Some Galois descent yoga*

Take ρ and form the monomial resolution of $\tilde{V}$ as in Theorem 6.4. Quotient out the monomial complex by the Lines whose stabiliser group is not sub-conjugate in the semi-direct product to $\mathrm{Gal}(E/F) \times GL_2F$. This is a monomial complex for the semi-direct product which originates, via induction, with $\mathrm{Gal}(E/F) \times GL_2F$.

In one case of finite general linear groups this yoga is equivalent to Shintani descent. See [130], which is included for completeness as Appendix I.

QUESTION 6.6. How is the Galois base-change yoga of §6.5 (and its analogues for GL_nK with $n \geq 3$) related to Galois base change for admissible representations of GL_n of local fields in the sense of [7] and [91]?

REMARK 6.7. The following section contains some after-thoughts on the Galois base-change yoga (aka the descent Galois construction), which were added later in light of the existence of the bar-monomial resolution.

7. A descent construction - a folly in the monomial landscape

7.1. *The descent construction*[2]

Let G be a finite group acting on the left on a group H. Form the semi-direct product $G \propto H$ (as in Chapter Two §6). Let $\mathcal{B}$ denote the set of subgroups $J \subseteq G \propto H$ such that $zJz^{-1} \not\subseteq G \times H^G$ for all $z \in G \propto H$.

Suppose that ϕ is a character of $G \propto H$ and that M, M' are objects and that $f : M \longrightarrow M'$ is a morphism in the monomial category $_{k[G\propto H],\underline{\phi}}\mathbf{mon}$. Set

$$M(\mathcal{B}) = \sum_{J \in \mathcal{B},\ (J,\lambda) \in \mathcal{M}_{\underline{\phi}}(G\propto H)} M^{((J,\lambda))} \subseteq M.$$

Then we have $f(M(\mathcal{B})) \subseteq M'(\mathcal{B})$. Hence $M_{\mathcal{B}} = M/M(\mathcal{B})$ is also an object in $_{k[G\propto H],\underline{\phi}}\mathbf{mon}$ such that $M_{\mathcal{B}}^{((J,\lambda))} = 0$ for all $J \in \mathcal{B}$.

There is an isomorphism in $_{k[G\propto H],\underline{\phi}}\mathbf{mon}$ between $M_{\mathcal{B}}$ and the sum of Lines in M whose stabiliser is (H', ϕ') with H' subconjugate to $G \times H^G$. A morphism f induces a morphism

$$f_{\mathcal{B}} : M_{\mathcal{B}} \longrightarrow M'_{\mathcal{B}}$$

which is functorial in the sense that $(f.f')_{\mathcal{B}} = f_{\mathcal{B}} f'_{\mathcal{B}}$ and $1_{\mathcal{B}} = 1_{M_{\mathcal{B}}}$.

Hence we have a functorial ring homomorphism

$$\text{End}_{k[G\propto H,\underline{\phi}]mon}(M) \longrightarrow \text{End}_{k[G\propto H,\underline{\phi}]mon}(M_{\mathcal{B}}).$$

EXAMPLE 7.2. The construction of §7.1 applies, for example, to the case when G is a local Galois group and $H = GL_n K$ as mentioned in §6.5 and §6.6. I first considered it in connection with the Shintani descent example which occupies Appendix I. That appendix was written several years before I discovered the bar-monomial resolution. This section arose since the bar-monomial resolution sheds a little more light — resulting in a some slightly more specific questions and problems, which will be described later in this section.

---

[2]To an 18th century English aristocrat a folly meant some extravagant, pointless construction typically tucked away somewhere on his estate amid the rolling country-side of his Lancelot "Capability" Brown (1716–1783) or Humphry Repton (1752–1818) designed horticulture. Every chap had to have one — a grotto, a tower, a fake lake or bridge and so on. The reader who has noticed this footnote will immediately get the gist of the usage of the term in relation to this section!

7.3. *The descent construction applied to a monomial resolution*

For simplicity, suppose in §7.1 that H is finite modulo the centre and that V is a finite-dimensional k-representation which extends to a representation $\tilde{V}$ of $G \propto H$ with central character $\underline{\phi}$.

Suppose that

$$\cdots \longrightarrow M_n \overset{\partial_{n-1}}{\longrightarrow} M_{n-1} \overset{\partial_{n-2}}{\longrightarrow} \cdots \overset{\partial_0}{\longrightarrow} M_0 \overset{\epsilon}{\longrightarrow} \tilde{V} \longrightarrow 0$$

is a $_{k[G\propto H],\underline{\phi}}$**mon**-resolution of $\tilde{V}$.

Since $\tilde{V}$, up to twists by one dimensional characters of G, is determined by V and since monomial resolutions commute with twists by one-dimensional characters we have a chain complex

$$\cdots \longrightarrow M_{n,\mathcal{B}} \overset{\partial_{n-1}}{\longrightarrow} M_{n-1,\mathcal{B}} \overset{\partial_{n-2}}{\longrightarrow} \cdots \overset{\partial_0}{\longrightarrow} M_{0,\mathcal{B}} \longrightarrow 0$$

in $_{k[G\propto H],\underline{\phi}}$**mon** which depends, up to twists by one-dimensional characters of G and up to chain homotopy in $_{k[G\propto H],\underline{\phi}}$**mon**, only on V.

Recall from Chapter One §3.5 that we have a functor $\mathcal{J}$ giving a full embedding

$$\mathcal{J} :_{k[G\propto H],\underline{\phi}} \textbf{mon} \longrightarrow funct^o_k(_{k[G\propto H],\underline{\phi}}\textbf{mon},_k \textbf{mod})$$

defined by $\mathcal{J}(M) = \text{Hom}_{k[G\propto H],\underline{\phi}\textbf{mon}}(-, M)$. In addition, let $S \in_{k[G\propto H],\underline{\phi}}$ **mon** be the finite $(G \propto H, \underline{\phi})$-Line Bundle over k given by

$$S = \oplus_{(J,\phi)\in \mathcal{M}_{\underline{\phi}}(G\propto H)} \underline{\text{Ind}}_J^{G\propto H}(k_\phi),$$

which was introduced in Chapter One §4.2. As in Chapter One §4.1 we define $\mathcal{A}_S = \text{Hom}_{k[G],\underline{\phi}\textbf{mon}}(S, S)$, the ring of endomorphisms on S under composition.

Then, in the notation of Chapter One §4.1, we have functors

$$\Phi_S : funct^o_k(_{k[G\propto H],\underline{\phi}}\textbf{mon},_k \textbf{mod}) \longrightarrow \textbf{mod}_{\mathcal{A}_S}$$

and

$$\Psi_S : \textbf{mod}_{\mathcal{A}_S} \longrightarrow funct^o_k(_{k[G\propto H],\underline{\phi}}\textbf{mon},_k \textbf{mod}),$$

which are inverse equivalences of categories. In fact, the natural transformations η and ϵ of Chapter One §4.1 are isomorphisms of functors when $M = S$.

Applying $\Phi_S \cdot \mathcal{J}$ to the monomial complex

$$\cdots \longrightarrow M_{n,\mathcal{B}} \overset{\partial_{n-1}}{\longrightarrow} M_{n-1,\mathcal{B}} \overset{\partial_{n-2}}{\longrightarrow} \cdots \overset{\partial_0}{\longrightarrow} M_{0,\mathcal{B}} \longrightarrow 0$$

yields a chain complex in $\textbf{mod}_{\mathcal{A}_S}$ of the form

$$\cdots \to \Phi_S(\mathcal{J}(M_{n,\mathcal{B}})) \overset{\partial_{n-1}}{\longrightarrow} \Phi_S(\mathcal{J}(M_{n-1,\mathcal{B}})) \overset{\partial_{n-2}}{\longrightarrow} \cdots \overset{\partial_0}{\longrightarrow} \Phi_S(\mathcal{J}(M_{0,\mathcal{B}})) \to 0.$$

Up to chain homotopy in the module category $\textbf{mod}_{\mathcal{A}_S}$ this complex depends only on $\tilde{V}$. Therefore, up to chain homotopy in the module category

$\mathbf{mod}_{\mathcal{A}_S}$, we may compute this chain complex from the bar-monomial resolution of $\tilde{V}$.

7.4. $\Phi_S(\mathcal{J}(M_{*,\mathcal{B}}))$ for the bar-monomial resolution

Recall from Chapter One §5.5 that in degree i the bar-monomial resolution of $\tilde{V}$ has the form

$$\tilde{M}_{i,S} \otimes_k S = \mathrm{Hom}_{k[G\alpha H],\underline{\phi}}\mathbf{mod}(\mathcal{V}(S),\tilde{V}) \otimes_k \mathcal{A}_S^{\otimes i} \otimes_k S$$

so that

$$(\tilde{M}_{i,S} \otimes_k S)_{\mathcal{B}} = \mathrm{Hom}_{k[G\alpha H],\underline{\phi}}\mathbf{mod}(\mathcal{V}(S),\tilde{V}) \otimes_k \mathcal{A}_S^{\otimes i} \otimes_k S_{\mathcal{B}}.$$

Therefore in degree i we have

$$\mathcal{J}((\tilde{M}_{i,S} \otimes_k S)_{\mathcal{B}})$$

$$= \mathrm{Hom}_{k[G\alpha H],\underline{\phi}}\mathbf{mod}(\mathcal{V}(S),\tilde{V}) \otimes_k \mathcal{A}_S^{\otimes i} \otimes_k \mathrm{Hom}_{k[G\alpha H],\underline{\phi}}\mathbf{mon}(-,S_{\mathcal{B}})$$

and

$$\Phi_S(\mathcal{J}((\tilde{M}_{i,S} \otimes_k S)_{\mathcal{B}}))$$

$$= \mathrm{Hom}_{k[G\alpha H],\underline{\phi}}\mathbf{mod}(\mathcal{V}(S),\tilde{V}) \otimes_k \mathcal{A}_S^{\otimes i} \otimes_k \mathrm{Hom}_{k[G\alpha H],\underline{\phi}}\mathbf{mon}(S,S_{\mathcal{B}}).$$

Since monomial morphisms only increase Line stabilisers we have an isomorphism of $\mathcal{A}_S$-modules of the form

$$\mathrm{Hom}_{k[G\alpha H],\underline{\phi}}\mathbf{mon}(S,S_{\mathcal{B}}) \cong \mathrm{Hom}_{k[G\alpha H],\underline{\phi}}\mathbf{mon}(S_{\mathcal{B}},S_{\mathcal{B}}) = \mathcal{A}_{S_{\mathcal{B}}}.$$

Therefore, by Chapter One, Theorem 5.4 we have established the following result.

THEOREM 7.5.

In the situation of §7.1, §7.3 and §7.4 the homology of the chain complex $\Phi_S(\mathcal{J}(M_{*,\mathcal{B}}))$, which depends only on $\tilde{V}$ is given by

$$H_i(\Phi_S(\mathcal{J}(M_{*,\mathcal{B}})))$$

$$\cong \mathrm{Tor}^i_{\mathcal{A}_S}(\mathrm{Hom}_{k[G\alpha H],\underline{\phi}}\mathbf{mod}(\mathcal{V}(S),\tilde{V}),\mathcal{A}_{S_{\mathcal{B}}}).$$

EXAMPLE 7.6. In Appendix I we study Shintani descent from Galois invariant complex irreducible representations of $GL_2\mathbb{F}_4$ to irreducibles of $GL_2\mathbb{F}_2 \cong D_6$, the dihedral group of order six. In this situation there are two interesting (that is, of dimension larger than one) which are $GL_2\mathbb{F}_4$-invariant. These are ν_4 and ν_5 of dimensions 4 and 5 respectively. Let $\tilde{\nu}_4$ and $\tilde{\nu}_5$ denote the extensions of these representations to the semi-direct product $GL_2\mathbb{F}_4 \propto GL_2\mathbb{F}_4$. They factor through $GL_2\mathbb{F}_4 \propto PGL_2\mathbb{F}_4$.

By Chapter One, Theorem 6.3 there is a finite length monomial resolution in $_{k[C_2\propto GL_2\mathbb{F}_4],\underline{\phi}}\mathbf{mon}$ for each $\tilde{\nu}_i$. Therefore such a monomial resolution has a well-defined Euler characteristic in the free abelian group of isomorphism classes of objects in $_{k[C_2 \propto GL_2 \mathbb{F}_4], \underline{\phi}}\mathbf{mon}$. This Euler characteristic may be computed without constructing a monomial resolution, using the Explicit Brauer Induction formula, and this was done for $\tilde{\nu}_4$ and $\tilde{\nu}_5$ in Appendix I §6.

The notation of the formulae is explained in the tables of Appendix I §6.

Explicitly we have a $_{k[C_2 \propto GL_2 \mathbb{F}_4], \underline{\phi}}\mathbf{mon}$ resolution of the form

$$0 \longrightarrow M_{i,t} \oplus N_{i,t} \longrightarrow M_{i,t-1} \oplus N_{i,t-1} \longrightarrow \cdots \longrightarrow M_{i,0} \oplus N_{i,0} \longrightarrow \tilde{\nu}_i \longrightarrow 0$$

in which the $M_{i,j}$'s and $N_{i,j}$'s are objects in $_{k[C_2 \propto GL_2 \mathbb{F}_4], \underline{\phi}}\mathbf{mon}$. The calculations of Appendix I §6 imply that for $i = 4, 5$ there are isomorphisms in $_{k[C_2 \propto GL_2 \mathbb{F}_4], \underline{\phi}}\mathbf{mon}$ of the forms

$$\oplus_{0 \leq 2n \leq t} M_{i,2n} \cong \oplus_{0 \leq 2n+1 \leq t} M_{i,2n+1}$$

and

$$\oplus_{0 \leq 2n \leq t} N_{5,2n} \oplus \underline{\mathrm{Ind}}_{V_4}^{C_2 \propto GL_2 \mathbb{F}_4}(1) \oplus \underline{\mathrm{Ind}}_{C_3}^{C_2 \propto GL_2 \mathbb{F}_4}(k_\phi)$$

$$\oplus \underline{\mathrm{Ind}}_{C_2}^{C_2 \propto GL_2 \mathbb{F}_4}(k_\phi) \oplus \underline{\mathrm{Ind}}_{\langle (\sigma,1) \rangle}^{C_2 \propto GL_2 \mathbb{F}_4}(1) \oplus \underline{\mathrm{Ind}}_{\langle (\sigma,1) \rangle}^{C_2 \propto GL_2 \mathbb{F}_4}(k_\tau)$$

$$\oplus \underline{\mathrm{Ind}}_{\langle (\sigma,1),A \rangle}^{C_2 \propto GL_2 \mathbb{F}_4}(1) \oplus \underline{\mathrm{Ind}}_{C_4}^{C_2 \propto GL_2 \mathbb{F}_4}(k_{\phi^2})$$

$$\cong \oplus_{0 \leq 2n+1 \leq t} N_{5,2n+1} \oplus \underline{\mathrm{Ind}}_{A_4}^{C_2 \propto GL_2 \mathbb{F}_4}(k_\phi) \oplus \underline{\mathrm{Ind}}_{\langle (\sigma,B), X_\xi \rangle}^{C_2 \propto GL_2 \mathbb{F}_4}(k_{\tau^2})$$

$$\oplus \underline{\mathrm{Ind}}_{\langle (\sigma,1),A,C \rangle}^{C_2 \propto GL_2 \mathbb{F}_4}(1) \oplus \underline{\mathrm{Ind}}_{\langle (\sigma,1),V_4 \rangle}^{C_2 \propto GL_2 \mathbb{F}_4}(1) \oplus \underline{\mathrm{Ind}}_{\langle (\sigma,1),V_4 \rangle}^{C_2 \propto GL_2 \mathbb{F}_4}(k_\tau)$$

$$\oplus \underline{\mathrm{Ind}}_{\langle (\sigma,1),V_4 \rangle}^{C_2 \propto GL_2 \mathbb{F}_4}(k_\mu) \oplus \underline{\mathrm{Ind}}_{\langle (\sigma,1),C \rangle}^{C_2 \propto GL_2 \mathbb{F}_4}(k_\phi) \oplus \underline{\mathrm{Ind}}_{\langle (\sigma,1),C \rangle}^{C_2 \propto GL_2 \mathbb{F}_4}(k_{\tau\phi})$$

$$\oplus \underline{\mathrm{Ind}}_{C_5}^{C_2 \propto GL_2 \mathbb{F}_4}(k_\phi) \oplus \underline{\mathrm{Ind}}_{\{1\}}^{C_2 \propto GL_2 \mathbb{F}_4}(1)$$

$$\oplus \underline{\mathrm{Ind}}_{\langle (\sigma,1),A \rangle}^{C_2 \propto GL_2 \mathbb{F}_4}(k_\phi) \oplus \underline{\mathrm{Ind}}_{C_4}^{C_2 \propto GL_2 \mathbb{F}_4}(k_\phi)$$

and

$$\oplus_{0 \leq 2n \leq t} N_{4,2n} \oplus \underline{\mathrm{Ind}}_{C_2}^{C_2 \propto GL_2 \mathbb{F}_4}(k_\phi)$$

$$\oplus \underline{\mathrm{Ind}}_{\langle (\sigma,1),A \rangle}^{C_2 \propto GL_2 \mathbb{F}_4}(k_\tau) \oplus \underline{\mathrm{Ind}}_{\langle (\sigma,1),C \rangle}^{C_2 \propto GL_2 \mathbb{F}_4}(k_\tau)$$

$$\cong \oplus_{0 \leq 2n+1 \leq t} N_{5,2n+1} \oplus \underline{\mathrm{Ind}}_{\langle (\sigma,1),A_4 \rangle}^{C_2 \propto GL_2 \mathbb{F}_4}(k_\tau)$$

$$\oplus \underline{\mathrm{Ind}}_{C_2 \times D_6,}^{C_2 \propto GL_2 \mathbb{F}_4}(k_\tau) \oplus \underline{\mathrm{Ind}}_{C_2 \times D_6,}^{C_2 \propto GL_2 \mathbb{F}_4}(k_\phi) \oplus \underline{\mathrm{Ind}}_{C_5}^{C_2 \propto GL_2 \mathbb{F}_4}(k_\phi)$$

$$\oplus \underline{\mathrm{Ind}}_{\langle (\sigma,1),V_4 \rangle}^{C_2 \propto GL_2 \mathbb{F}_4}(k_{\tau\mu}) \oplus \underline{\mathrm{Ind}}_{\langle (\sigma,1),C \rangle}^{C_2 \propto GL_2 \mathbb{F}_4}(k_{\tau\phi}) \oplus \underline{\mathrm{Ind}}_{C_4}^{C_2 \propto GL_2 \mathbb{F}_4}(k_\phi).$$

Applying the descent construction of §7.3 we obtain a chain complexes for $i = 4, 5$ in $_{k[C_2 \propto GL_2\mathbb{F}_4], \phi}\mathbf{mon}$ of the form

$$0 \longrightarrow M_{i,t\mathcal{B}} \oplus N_{i,t\mathcal{B}} \longrightarrow M_{i,t-1\mathcal{B}} \oplus N_{i,t-1\mathcal{B}} \longrightarrow \cdots \longrightarrow M_{i,0\mathcal{B}} \oplus N_{i,0\mathcal{B}} \longrightarrow 0$$

where

$$\oplus_{0 \leq 2n \leq t} N_{5,2n\mathcal{B}} \oplus \underline{\mathrm{Ind}}_{C_3}^{C_2 \propto GL_2\mathbb{F}_4}(k_\phi) \oplus \underline{\mathrm{Ind}}_{C_2}^{C_2 \propto GL_2\mathbb{F}_4}(k_\phi)$$

$$\oplus \underline{\mathrm{Ind}}_{\langle (\sigma,1) \rangle}^{C_2 \propto GL_2\mathbb{F}_4}(1) \oplus \underline{\mathrm{Ind}}_{\langle (\sigma,1) \rangle}^{C_2 \propto GL_2\mathbb{F}_4}(k_\tau) \oplus \underline{\mathrm{Ind}}_{\langle (\sigma,1),A \rangle}^{C_2 \propto GL_2\mathbb{F}_4}(1)$$

$$\cong \oplus_{0 \leq 2n+1 \leq t} N_{5,2n+1\mathcal{B}} \oplus \underline{\mathrm{Ind}}_{\langle (\sigma,1),A,C \rangle}^{C_2 \propto GL_2\mathbb{F}_4}(1) \oplus \underline{\mathrm{Ind}}_{\langle (\sigma,1),C \rangle}^{C_2 \propto GL_2\mathbb{F}_4}(k_\phi)$$

$$\oplus \underline{\mathrm{Ind}}_{\langle (\sigma,1),C \rangle}^{C_2 \propto GL_2\mathbb{F}_4}(k_{\tau\phi}) \oplus \underline{\mathrm{Ind}}_{\{1\}}^{C_2 \propto GL_2\mathbb{F}_4}(1) \oplus \underline{\mathrm{Ind}}_{\langle (\sigma,1),A \rangle}^{C_2 \propto GL_2\mathbb{F}_4}(k_\phi)$$

and

$$\oplus_{0 \leq 2n \leq t} N_{4,2n\mathcal{B}} \oplus \underline{\mathrm{Ind}}_{C_2}^{C_2 \propto GL_2\mathbb{F}_4}(k_\phi)$$

$$\oplus \underline{\mathrm{Ind}}_{\langle (\sigma,1),A \rangle}^{C_2 \propto GL_2\mathbb{F}_4}(k_\tau) \oplus \underline{\mathrm{Ind}}_{\langle (\sigma,1),C \rangle}^{C_2 \propto GL_2\mathbb{F}_4}(k_\tau)$$

$$\cong \oplus_{0 \leq 2n+1 \leq t} N_{5,2n+1\mathcal{B}} \oplus \underline{\mathrm{Ind}}_{C_2 \times D_6,}^{C_2 \propto GL_2\mathbb{F}_4}(k_\tau)$$

$$\oplus \underline{\mathrm{Ind}}_{C_2 \times D_6,}^{C_2 \propto GL_2\mathbb{F}_4}(k_\phi) \oplus \underline{\mathrm{Ind}}_{\langle (\sigma,1),C \rangle}^{C_2 \propto GL_2\mathbb{F}_4}(k_{\tau\phi}).$$

From Euler characteristic equations such as these one can sometimes deduce a little about the homology groups of Theorem 7.5. In this example I believe that one can deduce that some of the odd degree homology groups

$$\mathrm{Tor}_{\mathcal{A}_S}^{2i+1}(\mathrm{Hom}_{k[G \propto H], \phi}\mathbf{mod}(\mathcal{V}(S), \tilde{V}), \mathcal{A}_{S_{\mathcal{B}}})$$

are non-trivial for both $V = \nu_4$ and $V = \nu_5$.

The following result is immediate.

LEMMA 7.7.

Let ψ be a character of the form $G \propto H \longrightarrow G \overset{\psi}{\longrightarrow} k^*$ in which the first map is the canonical surjection. Then the construction of §7.3 commutes with twisting by ψ

$$k_\psi \otimes_k M_{*,\mathcal{B}} \cong k_\psi \otimes_k (M_{*,\mathcal{B}})$$

in $_{k[G \propto H], \psi\phi}\mathbf{mon}$.

In particular, applied to a monomial resolution $M_* \longrightarrow \tilde{V}$ the monomial complex $M_{*,\mathcal{B}}$ depends only on V, up to twists by one dimensional characters of G and up to monomial chain homotopy.

I shall close this section with some (rather pointless[3] related questions.)

---

[3]I give my sincere apologies for these questions to the readers, should there be any. These are the sort of out-of-touch ramblings which one might expect from a mathematically isolated, dilettante retiree!

QUESTION 7.8. Let K be a local field and suppose that V is an irreducible admissible representation of $GL_n K$ with central character $\underline{\phi}$. Let F/K be a finite Galois extension and suppose that $\tilde{V}_F$ is an irreducible admissible representation of the semi-direct product $\mathrm{Gal}(F/K) \propto GL_n F$ whose Galois base-change of $\mathrm{Res}_{GL_n F}^{\mathrm{Gal}(F/K) \propto GL_n F}(\tilde{V}_F)$, in the analogous sense to that of Shintani descent (in Appendix I §4), is V.

Therefore, by Chapter Two §6, there is a monomial resolution

$$M_{*,F} \longrightarrow \tilde{V}_F \longrightarrow 0.$$

Can the monomial complexes $\{M_{*,F}\}$ be chosen coherently? That is, in what sense can they be chosen to form an inverse system? Can the monomial complexes of §7.3 $\{(M_{*,F})_\mathcal{B}\}$ be chosen to form an inverse system?

QUESTION 7.9. The monomial complexes in the family $\{(M_{*,F})_\mathcal{B}\}$ are all "induced from" $\mathrm{Gal}(F/K) \times GL_n K \subseteq \mathrm{Gal}(F/K) \propto GL_n F$. Supposing a fairly strongly affirmative answer to Question 7.8 we would have an inverse system of monomial complexes induced from $\mathrm{Gal}(F/K) \times GL_n K$ and depending only on V, up to one-dimensional Galois twists.

Is it possible to use this structure to associate to V a Galois representation of $\mathrm{Gal}(\overline{K}/K)$, modulo one-dimensional twists Galois twists, in some sort of "dual pair" [75] relation?

REMARK 7.10. *Galois descent, Functoriality and functoriality*

Let L/K be a Galois extension of local fields. Suppose that V is an admissible representation of $GL_n L$ with central character $\underline{\phi}$.

Suppose that V is irreducible then Galois base change (aka Galois descent) may be characterised in terms of character values ([91] Chapter Two) which is analogous to the finite field case of Shintani descent described in Appendix I, §4. It may also be characterised in a manner which extends immediately to the global case in terms of the L-group and the Principle of Functoriality ([91] Chapter One). However, although it sounds like it, the Principle of Functoriality is not functorial. It is inter alia a bijection between sets of irreducible admissible representation with a functorial-like behaviour. For the local $GL_n L$ it was established for cyclic extensions L/K (and hence for nilpotent extensions, presumably) in [7].

It seems to me that Galois descent should ideally aim to feature a sheaf of representations on the poset of local Galois groups. In the spirit of this monograph, an equivalent aim would be a sheaf of monomial complexes on the poset of local Galois groups similar to the one I constructed in Chapter Four on the Bruhat-Tits building.

Allow me to illustrate what I have in mind by an example. Suppose that we have Galois base change admissible representations of all nilpotent subgroups of $\mathrm{Gal}(L/K)$. Let us take the example of the case when $\mathrm{Gal}(L/K)$ is isomorphic to one of the icosahedral, tetrahedral or octahedral groups.

Each of these has a 2-dimensional irreducible complex representation which is not a monomial representation. Using this representation the group $\mathrm{Gal}(L/K)$ acts simplicially on $U(2,\mathbb{C})/N_{U(2,\mathbb{C})}T^2$, the coset space of the normaliser of the maximal torus in the 2×2 unitary group. The homology group $H^*(U(2,\mathbb{C})/N_{U(2,\mathbb{C})}T^2;\mathbb{Q})$ is isomorphic to the rational homology of a point and the stabiliser of each simplex is a proper, nilpotent subgroup of $\mathrm{Gal}(L/K)$ [**122**]. If the base-change data for the stabiliser of each simplex were functorial for inclusions we would have a sheaf of representations on $U(2,\mathbb{C})/N_{U(2,\mathbb{C})}T^2$ and would be able to form a double complex whose terms were admissible representations of $GL_n K$. The total complex of this double complex probably would not have homology concentrated in degree zero, but the same construction with $U(2,\mathbb{C})/N_{U(2,\mathbb{C})}T^2$ replaced by the tom Dieck-Baum-Connes space (see Appendix IV) with respect to the family of nilpotent subgroups of $\mathrm{Gal}(L/K)$ definitely would. This non-zero homology group should be the candidate for the Galois descent of V.

One can see just where functoriality is missing in the "descent" correspondences involving finite groups. If S is a finite group acting on the finite group G then there is a canonical correspondence due to Glauberman [**65**] between complex irreducible representations of G fixed by the action of S and complex irreducibles of G^S, the subgroup of S-fixed elements in G. Let this correspondence be denoted by $V \mapsto Gl(V)$. Extend this map to set of isomorphism classes of representations. How is this to extend functorially to maps between representations? For example, when S is cyclic of order p, which is prime, then $Gl(V)$ is the unique irreducible irreducible of G^S in $\mathrm{Res}_{G^S}^G(V)$ whose multiplicity is congruent to ± 1 modulo p ([**5**] Lemma 3.3). Given a map between S-invariant representations of G it is by no means clear how to map Glauberman correspondents because, despite having copies of $Gl(V)$ contained in V, because the multiplicities prevent a characterisation of $Gl(V)$ as a subspace of V.

Consider the finite group example of Appendix I. Here the Shintani correspondence [**117**] is denoted by $V \mapsto Sh(V)$. There are three $\mathrm{Gal}(\mathbb{F}_4/\mathbb{F}_2)$-invariant irreducible complex representations of $GL_2\mathbb{F}_4$ denoted by $1, \nu_4, \nu_5$ where ν_i is i-dimensional. Perhaps we can characterise $Sh(\nu_i)$ as a subspace of ν_i and thereby extend the correspondence to morphisms? Each ν_i extends, uniquely up to one-dimensional twists by Galois characters, to an irreducible representation $\tilde{\nu}_i$ of the semi-direct product $\mathrm{Gal}(\mathbb{F}_4/\mathbb{F}_2) \propto GL_2\mathbb{F}_4$. We have $Sh(\nu_4) = \nu$, the unique two-dimensional irreducible of $GL_2\mathbb{F}_2 \cong D_6$. The results of Appendix I, §2 and §3 imply that

$$\mathrm{Res}_{\mathrm{Gal}(\mathbb{F}_4/\mathbb{F}_2)\times GL_2\mathbb{F}_2}^{\mathrm{Gal}(\mathbb{F}_4/\mathbb{F}_2)\propto GL_2\mathbb{F}_4}(\tilde{\nu}_4) = (1+\tau) \otimes \chi + \tau \otimes \nu$$

where τ and χ are the non-trivial one-dimensional characters of $\mathrm{Gal}(\mathbb{F}_4/\mathbb{F}_2)$ and D_6 respectively. This leads one to hope that $Sh(\nu_i)$ can be given by a quotient of the restriction of $\tilde{\nu}_i$ by a subrepresentation of the form $\mathrm{Ind}_{GL_2\mathbb{F}_2}^{\mathrm{Gal}(\mathbb{F}_4/\mathbb{F}_2)\times GL_2\mathbb{F}_2}(W)$. However, if the formulae of Appendix I are correct, we have $Sh(\nu_5) = \chi$ but, rather disappointingly,

$$\mathrm{Res}_{\mathrm{Gal}(\mathbb{F}_4/\mathbb{F}_2)\times GL_2\mathbb{F}_2}^{\mathrm{Gal}(\mathbb{F}_4/\mathbb{F}_2)\ltimes GL_2\mathbb{F}_4}(\tilde{\nu}_5) = (1+\tau)\otimes\nu + 1\otimes 1.$$

Too bad!

8. A curiosity - or dihedral voodoo

8.1. In this section, merely out of curiosity, I am going to apply the descent construction of §7.3 to the case of a cyclic group of order two acting on a dihedral 2-group. To my knowledge there is nothing known in this case which might be considered an analogue of the Shintani descent correspondence of [117] or the Glauberman correspondence of [65]. Of course, I am not going to get anything interesting when the generator is merely acting via an inner automorphism. Fortunately, a theorem of Gaschutz states that every p-group has an outer automorphism. Therefore I propose to take C_2 acting via an involutory outer automorphism of a dihedral 2-group.

The formulae become quite complicated so I shall restrict to the example of C_2 acting on D_8 by an outer automorphism which becomes inner in D_{16}. I strongly believe that the descent construction is rather interesting in each of the other dihedral cases, too.

In this example the representations will be defined over an algebraically closed field of characteristic different from two. In this case D_8 has a unique 2-dimensional irreducible ν which is fixed by the involution. The subgroup of fixed points is the central C_2 so that any interesting descent construction should send ν to the non-trivial character of the centre. On the other hand there are two extension $\tilde{\nu}$ of ν to the semi-direct product of C_2 with D_8 differing by a one-dimensional twist which commutes with the descent construction. Therefore the descent construction should give us — by some sort of yoga — a representation of the product of C_2 with the centre of D_8. If $\tilde{\chi}_2$ is the non-trivial character of the form and $\tilde{\chi}_1$ of the latter then the outcome

$$\tilde{\chi}_1 \oplus \tilde{\chi}_1\tilde{\chi}_2 = (1 \oplus \tilde{\chi}_2) \otimes \tilde{\chi}_1$$

would be quite satisfactory!

8.2. *The group* $\langle x, y, t \rangle$

Write the dihedral group D_8 in the form

$$D_8 = \langle x, y \mid x^4 = 1 = y^2, \ yxy = x^3 \rangle.$$

An involutory outer automorphism of D_8 denoted by

$$\lambda : D_8 \xrightarrow{\cong} D_8$$

is defined by the formula $\lambda(y) = xy, \lambda(x) = x^3$.

Define a group G to be the semi-direct product of order sixteen in which the outer automorphism λ has become the inner automorphism of conjugation by t

$$G = \langle x, y, t \mid x^4 = 1 = y^2 = t^2, yxy = x^3, tyt = xy, txt = x^3 \rangle.$$

The two-dimensional irreducible of D_8 is $\nu = \mathrm{Ind}_{\langle x \rangle}^{D_8}(\phi)$ where ϕ is the character defined by $\phi(x) = \sqrt{-1}$.

The action of D_8 on ν extends to a representation $\tilde{\nu}$ of G by the action

$$t(1 \otimes_{\langle x \rangle} 1) = \overline{\xi}_8 y \otimes_{\langle x \rangle} 1, \; t(y \otimes_{\langle x \rangle} 1) = \xi_8 \otimes_{\langle x \rangle} 1$$

where $\xi_8 = \frac{1 + \sqrt{-1}}{\sqrt{2}}$.

In terms of 2×2 matrices $\tilde{\nu}$ is given by

$$t = \begin{pmatrix} 0 & \xi_8 \\ \overline{\xi}_8 & 0 \end{pmatrix}, \; x = \begin{pmatrix} \sqrt{-1} & 0 \\ 0 & -\sqrt{-1} \end{pmatrix}, \; y = \begin{pmatrix} 0 & 1 \\ 1 & 0 \end{pmatrix}.$$

In G we have $tyty = xyy = x$ so that $\langle ty \rangle$ is cyclic of order eight containing x so that y and ty generate G and from the matrices one sees that

$$G = \langle ty, y \mid (ty)^8 = 1 = y^2, y(ty)y = (ty)^7 \rangle \cong D_{16}.$$

Write

$$D_{16} = \langle X, Y \mid X^8 = 1 = Y^2, YXY = X^7 \rangle$$

where $Y = y, X = ty$ in the previous notation.

8.3. *Subgroups of D_{16} and characters on them*

We have eight elements of order two $X^i Y \in D_{16}$ for $0 \le i \le 7$ which fall into two conjugacy classes represented by Y and XY since $XX^i YX^{-1} = X^{i+2}Y$.

Up to conjugation the subgroups of D_{16} are given by the following table.

H	Order	Generators	Number in conjugacy class
D_{16}	16	X, Y	1
D_8	8	X^2, Y	1
C_8	8	X	1
C_4	4	X^2	1
V_4	4	$\langle X^4, Y \rangle$	4
C_2	2	X^4	1
C_2'	2	Y	4
$\{1\}$	1	1	1

The following table gives the one-dimensional characters of the subgroups up to conjugation (denoted by $\sim$).

H	$\hat{H}$	formulae
D_{16}	$1, \lambda_1, \lambda_2, \lambda_1\lambda_2$	$\lambda_1(X) = -1$ $\lambda_1(Y) = \lambda_2(X), \lambda_2(Y) = -1$
D_8	$1, \chi_1, \chi_2, \chi_1\chi_2$	$\chi_1(X^2) = -1, \chi_2(Y) = -1$ $\chi_1(Y) = 1 = \chi_2(X^2)$
C_8	$1, \phi \sim \phi^7, \phi^2 \sim \phi^6, \phi^3 \sim \phi^5, \phi^4$	$\phi(X) = \xi_8$
C_4	$1, \alpha \sim \alpha^3, \alpha^2$	$\alpha(X^2) = \sqrt{-1}$
V_4	$1, \tilde{\chi}_1, \tilde{\chi}_2, \tilde{\chi}_1\tilde{\chi}_2$	$\tilde{\chi}_1(X^4) = -1, \tilde{\chi}_2(Y) = -1$ $\tilde{\chi}_1(Y) = 1 = \tilde{\chi}_2(X^4)$
C_2	$1, \tau$	$\tau(X^4) = -1$
C_2'	$1, \tau'$	$\tau'(Y) = -1$
$\{1\}$	1	$-$

If H is a subgroup of D_{16} the H-abelian part of $\tilde{\nu}$ is the sum of the subspaces $\tilde{\nu}^{(H,\mu)}$ as μ runs through $\hat{H}$.

In the notation of the above table, the representation of D_8 given by $\nu = \text{Ind}_{\langle X^2 \rangle}^{D_8}(\alpha)$ is equal to the restriction of $\tilde{\nu} = \text{Ind}_{\langle X \rangle}^{D_{16}}(\phi)$ so we may calculate a monomial resolution of $\tilde{\nu}$ and apply the descent yoga to it, out of curiosity.

The following table gives, up to conjugation, the H-abelian parts of $\tilde{\nu}$.

H	H − abelian part of $\tilde{\nu}$
D_{16}	0
D_8	0
C_8	$\phi + \phi^7$
C_4	$\alpha + \alpha^3$
V_4	$\tilde{\chi}_1 + \tilde{\chi}_1\tilde{\chi}_2$
C_2	$2 \cdot \tau$
C_2'	$1 + \tau$
$\{1\}$	$2 \cdot 1$

8.4. *A monomial resolution of $\tilde{\nu}$*

The Line bundle $\underline{\mathrm{Ind}}_{\langle X \rangle}^{D_{16}}(\phi)$ has two Lines one with stabiliser pair (C_8, ϕ) and one with stabiliser pair (C_8, ϕ^7). Hence the isomorphism of representations $\iota : \underline{\mathrm{Ind}}_{\langle X \rangle}^{D_{16}}(\phi) \longrightarrow \tilde{\nu}$ yields isomorphisms

$$\underline{\mathrm{Ind}}_{\langle X \rangle}^{D_{16}}(\phi)^{((C_8, \phi))} \xrightarrow{\cong} \tilde{\nu}^{(C_8, \phi)}$$

$$\underline{\mathrm{Ind}}_{\langle X \rangle}^{D_{16}}(\phi)^{((C_8, \phi^7))} \xrightarrow{\cong} \tilde{\nu}^{(C_8, \phi^7)}$$

$$\underline{\mathrm{Ind}}_{\langle X \rangle}^{D_{16}}(\phi)^{((C_4, \alpha))} \xrightarrow{\cong} \tilde{\nu}^{(C_4, \alpha)}$$

$$\underline{\mathrm{Ind}}_{\langle X \rangle}^{D_{16}}(\phi)^{((C_4, \alpha^3))} \xrightarrow{\cong} \tilde{\nu}^{(C_4, \alpha^3)}$$

$$\underline{\mathrm{Ind}}_{\langle X \rangle}^{D_{16}}(\phi)^{((C_2, \tau))} \xrightarrow{\cong} \tilde{\nu}^{(C_2, \tau)}$$

$$\underline{\mathrm{Ind}}_{\langle X \rangle}^{D_{16}}(\phi)^{((\{1\}, 1))} \xrightarrow{\cong} \tilde{\nu}^{(\{1\}, 1)}.$$

However, for the other non-zero cases of abelian parts have zero image from $\underline{\mathrm{Ind}}_{\langle X \rangle}^{D_{16}}(\phi)^{((H, \psi))}$'s.

Next consider the maps of representations

$$f_1 : \underline{\mathrm{Ind}}_{V_4}^{D_{16}}(\tilde{\chi}_1) \longrightarrow \tilde{\nu}$$

and

$$f_2 : \underline{\mathrm{Ind}}_{V_4}^{D_{16}}(\tilde{\chi}_1 \tilde{\chi}_2) \longrightarrow \tilde{\nu}$$

given by

$$f_1(1 \otimes_{V_4} 1) = 1 \otimes_{C_8} 1 + Y \otimes_{C_8} 1, \ \ f_2(1 \otimes_{V_4} 1) = 1 \otimes_{C_8} 1 - Y \otimes_{C_8} 1.$$

These formulae define linear maps because $X^4(1 \otimes_{C_8} 1 \pm Y \otimes_{C_8} 1) = -(1 \otimes_{C_8} 1 \pm Y \otimes_{C_8} 1)$ and $Y(1 \otimes_{C_8} 1 \pm Y \otimes_{C_8} 1) = (Y \otimes_{C_8} 1 \pm 1 \otimes_{C_8} 1)$.

Consider the map of representations

$$\epsilon = \iota + f_1 + f_2 : M_0 = \underline{\mathrm{Ind}}_{\langle X \rangle}^{D_{16}}(\phi) \oplus \underline{\mathrm{Ind}}_{V_4}^{D_{16}}(\tilde{\chi}_1) \oplus \underline{\mathrm{Ind}}_{V_4}^{D_{16}}(\tilde{\chi}_1 \tilde{\chi}_2) \longrightarrow \tilde{\nu}$$

which satisfies

$$M_0^{((H, \psi))} \xrightarrow{\cong} \tilde{\nu}^{(H, \psi)}$$

for $H = D_{16}, D_8, D_8', C_8, C_4$ and is surjective for all other (H, ψ)'s.

Next observe that

$$M_0^{((V_4, \tilde{\chi}_1))} \longrightarrow \tilde{\nu}^{(V_4, \tilde{\chi}_1)}$$

and

$$M_0^{((V_4, \tilde{\chi}_1 \tilde{\chi}_2))} \longrightarrow \tilde{\nu}^{(V_4, \tilde{\chi}_1 \tilde{\chi}_2)}$$

are also isomorphisms.

Consider
$$M_0 = M_0^{((C_2,\tau))} \longrightarrow \tilde{\nu}^{(C_2,\tau)} = \tilde{\nu}$$
whose kernel is a $\mathbb{C}[D_{16}]$ of complex dimension eight and containing

$(1 \otimes_{C_8} 1 + Y \otimes_{C_8} 1, -1 \otimes_{V_4} 1, 0)$ and $(1 \otimes_{C_8} 1 - Y \otimes_{C_8} 1, 0, -1 \otimes_{V_4} 1)$.

Notice that
$\epsilon(1 \otimes_{C_8} 1, (-1/2)1 \otimes_{V_4} 1, (-1/2)1 \otimes_{V_4} 1)$

$= 1 \otimes_{C_8} 1 + (-1/2)(1 \otimes_{C_8} 1 + Y \otimes_{C_8} 1) + (-1/2)(1 \otimes_{C_8} 1 - Y \otimes_{C_8} 1)$

$= 0$

and

$(1+Y)(1 \otimes_{C_8} 1, (-1/2)1 \otimes_{V_4} 1, (-1/2)1 \otimes_{V_4} 1) = (1 \otimes_{C_8} 1 + Y \otimes_{C_8} 1, -1 \otimes_{V_4} 1, 0)$

and

$(1-Y)(1 \otimes_{C_8} 1, (-1/2)1 \otimes_{V_4} 1, (-1/2)1 \otimes_{V_4} 1) = (1 \otimes_{C_8} 1 - Y \otimes_{C_8} 1, 0, -1 \otimes_{V_4} 1)$.

Define
$$d : M_1 = \mathrm{Ind}_{C_2}^{D_{16}}(\tau) \longrightarrow \mathrm{Ker}(M_0^{((C_2,\tau))} \longrightarrow \tilde{\nu}^{(C_2,\tau)})$$
by
$$d(1 \otimes_{C_2} 1) = (1 \otimes_{C_8} 1, (-1/2)1 \otimes_{V_4} 1, (-1/2)1 \otimes_{V_4} 1).$$
Hence we have a candidate for a monomial resolution
$$0 \longrightarrow M_1 \longrightarrow M_0 \longrightarrow \tilde{\nu} \longrightarrow 0.$$
We must verify the exactness of each of the sequences of vector spaces
$$0 \longrightarrow M_1^{((H,\psi))} \longrightarrow M_0^{((H,\psi))} \longrightarrow \tilde{\nu}^{(H,\psi)} \longrightarrow 0.$$
When $H = D_{16}, D_8$ this is a sequence of zeroes. The right-hand map is always surjective. When $H = C_8, C_4, V_4$ we have $M_1^{((H,\psi))} = 0$ and the right-hand map is an isomorphism. When $H = C_2, \{1\}$ the complex is equal to the entire candidate monomial resolution which is exact by surjectivity of the right-hand map and a dimension count. When $(H, \psi) = (C_2', 1)$ or $(H, \psi) = (C_2', \tau)$ the left-hand vector space is trivial and the right-hand map is a surjection of one-dimensional spaces and hence an isomorphism.

8.5. *Applying the descent construction*

Now let us apply the descent yoga. The subgroup $D_8^{\langle t \rangle} = \langle X^4, XY \rangle$ which is conjugate to V_4 therefore, equivalently, I shall apply the yoga to V_4 to receive
$$0 \longrightarrow \mathrm{Ind}_{C_2}^{D_{16}}(\tau) \xrightarrow{\partial} \mathrm{Ind}_{V_4}^{D_{16}}(\tilde{\chi}_1) \oplus \mathrm{Ind}_{V_4}^{D_{16}}(\tilde{\chi}_1\tilde{\chi}_2) \longrightarrow 0$$
where ∂ is given by
$$\partial(1 \otimes_{C_2} 1) = ((-1/2)1 \otimes_{V_4} 1, (-1/2)1 \otimes_{V_4} 1).$$

Therefore $\underline{\mathrm{Ind}}_{C_2}^{D_{16}}(\tau)^{((H,\psi))}$ is trivial unless H contains X^4 and $\psi(X^4) = -1$ but in that case the other term is zero, too. For (C_2, τ) we find that $\partial^{((C_2,\tau))}$ is an isomorphism and so is $\partial^{((\{1\},1))}$. For (V_4, α) with $\alpha = \tilde{\chi}_1, \tilde{\chi}_1\tilde{\chi}_2$ we obtain a chain complex

$$0 \longrightarrow 0 \longrightarrow V_\alpha \longrightarrow 0$$

where V_α is one-dimensional spanned by $1 \otimes_{V_4} 1$ in the appropriate summand.

Hence the complexes

$$0 \longrightarrow \underline{\mathrm{Ind}}_{C_2}^{D_{16}}(\tau)^{((H,\mu))} \xrightarrow{\partial} \underline{\mathrm{Ind}}_{V_4}^{D_{16}}(\tilde{\chi}_1)^{((H,\mu))} \oplus \underline{\mathrm{Ind}}_{V_4}^{D_{16}}(\tilde{\chi}_1\tilde{\chi}_2)^{((H,\mu))} \longrightarrow 0$$

are all exact except when $H = V_4$ and $\mu = \tilde{\chi}_1, \tilde{\chi}_1\tilde{\chi}_2$ in which case the homology is V_μ in dimension zero.

Hence if $P^{((H))} = \sum_{\mu \in \hat{H}} P^{((H,\mu))}$ and P_* is the monomial complex produced by the descent construction then

$$H_i(P_*^{((H))}) = \begin{cases} \tilde{\chi}_1 \oplus \tilde{\chi}_1\tilde{\chi}_2 & \text{if } H = V_4, i = 0, \\ \\ 0 & \text{otherwise.} \end{cases}$$

QUESTION 8.6. Is the outcome of §8.5 the result of something systematic or just a black magical coincidence?

REMARK 8.7. (i) The descent construction complex P_* of §8.5 is *not* a $k[V_4]$-monomial resolution of $\tilde{\chi}_1 \oplus \tilde{\chi}_1\tilde{\chi}_2$ because

$$0 \longrightarrow P_1^{((C_2,\tau))} \xrightarrow{\cong} P_0^{((C_2,\tau))} \longrightarrow (\tilde{\chi}_1 \oplus \tilde{\chi}_1\tilde{\chi}_2)^{((C_2,\tau))} = 2 \cdot \tau \longrightarrow 0$$

is not exact.

However a monomial resolution is easily found and takes the form

$$0 \longrightarrow P_1 \longrightarrow P_0 \oplus \underline{\mathrm{Ind}}_{C_2}^{V_4}(\tau) \longrightarrow (\tilde{\chi}_1 \oplus \tilde{\chi}_1\tilde{\chi}_2) \longrightarrow 0.$$

(ii) The descent construction applied to $G \times H^G \subseteq G \propto H$ yields a monomial complex $M_{\mathcal{B}}$ and an abelian representation

$$M_{\mathcal{B}}^{((G \times H^G))} = \sum_{\lambda \in G \times \hat{H}^G} M_{\mathcal{B}}^{((G \times H^G, \lambda))}$$

which is naturally a complex of representations of the normaliser $N_{G \propto H}G \times H^G$.

This happens in the example of §8.5 where the normaliser of V_4 in D_{16} is D_8 and $\tilde{\chi}_1 + \tilde{\chi}_1\tilde{\chi}_2$ is the restriction of the irreducible representation ν of D_8.

8.8. *The descent construction for ν_i of Appendix I revisited*

The subgroups of D_6 and their characters up to conjugation (denoted on characters by $\sim$) are given in the following table.

H	generators	$\hat{H}$
D_6	A, C	$1, \psi$
C_3	C	$1 \; \phi \sim \phi^2$
C_2	A	$1, \mu$
$\{1\}$	1	1

A monomial resolution for $\nu = \mathrm{Ind}_{C_3}^{D_6}(\phi)$ over an algebraically closed field of characteristic different from 2 is

$$M_* : \quad 0 \longrightarrow \mathrm{Ind}_{\{1\}}^{D_6}(1) \xrightarrow{\ \partial\ } \mathrm{Ind}_{C_3}^{D_6}(\phi) \oplus \mathrm{Ind}_{C_2}^{D_6}(1) \oplus \mathrm{Ind}_{C_2}^{D_6}(\mu)) \xrightarrow{\ \epsilon\ } \nu \longrightarrow 0$$

where μ is the non-trivial character and the differentials are given by

$$\partial(1 \otimes_{\{1\}} 1) = (1 \otimes_{C_3} 1, -(1/2) \otimes_{C_2} 1, -(1/2) \otimes_{C_2} 1),$$

$$\epsilon(1 \otimes_{C_3} 1, 0, 0)) = 1 \otimes_{C_3} 1,$$

$$\epsilon(0, 1 \otimes_{C_2} 1, 0) = 1 \otimes_{C_3} 1 + A \otimes_{C_3} 1,$$

$$\epsilon(0, 0, 1 \otimes_{C_2} 1) = 1 \otimes_{C_3} 1 - A \otimes_{C_3} 1.$$

The Euler characteristics of $M_*^{((H))}$ in $R_+(D_6)$ are given by

$\chi(M_*^{((D_6))})$	0
$\chi(M_*^{((C_3))})$	$\phi + \phi^2$
$\chi(M_*^{((C_2))})$	$1 + \mu$
$\chi(M_*^{((\{1\}))})$	2

From the calculations of Appendix I §6 we have $a_G(\tilde{\nu}_4)$ and $a_G(\tilde{\nu}_5)$ in $R_+(\mathrm{Gal}(\mathbb{F}_4/\mathbb{F}_2) \propto PGL_2\mathbb{F}_4)$, which are the Euler characteristics of the monomial resolutions of $\tilde{\nu}_4$ and $\tilde{\nu}_5$ respectively. Applying the descent construction relative to $\mathrm{Gal}(\mathbb{F}_4/\mathbb{F}_2) \times GL_2\mathbb{F}_2 = C_2 \times D_6$ we obtain the $M(\mathcal{B})$ monomial complexes whose Euler characteristics in $R_+(\mathrm{Gal}(\mathbb{F}_4/\mathbb{F}_2) \propto PGL_2\mathbb{F}_4)$ are obtained by applying the descent construction term-by-term to the $a_G(\tilde{\nu}_i)$'s.

Write $\mathrm{Des}_{C_2 \times D_6}(\tilde{\nu}_i)$ for the descent construction monomial complexes and write $\chi(\mathrm{Des}_{C_2 \times D_6}(\tilde{\nu}_i)) \in R_+(\mathrm{Gal}(\mathbb{F}_4/\mathbb{F}_2) \propto PGL_2\mathbb{F}_4)$ for their Euler characteristics.

From Appendix I §6 we have the formulae:

$$\mathrm{Des}_{C_2 \times D_6}(a_G(\tilde{\nu}_4))$$

$$= (C_2 \times D_6, \tau)^G + (C_2 \times D_6, \phi)^G + (\langle(\sigma,1),C\rangle,\tau\phi)^G$$

$$-(C_2,\phi)^G - (\langle(\sigma,1),A\rangle,\tau)^G - (\langle(\sigma,1),C\rangle,\tau)^G$$

and

$$\chi(\mathrm{Des}_{C_2 \times D_6}(a_G(\tilde{\nu}_5)))$$

$$= (\langle(\sigma,1),A,C\rangle,1)^G + (\langle(\sigma,1),C\rangle,\phi)^G + (\langle(\sigma,1),C\rangle,\tau\phi)^G$$

$$-(C_3,\phi)^G - (C_2,\phi)^G - (\langle(\sigma,1)\rangle,1)^G - (\langle(\sigma,1)\rangle,\tau)^G$$

$$+(\{1\},1)^G - (\langle(\sigma,1),A\rangle,1)^G + (\langle(\sigma,1),A\rangle,\phi)^G.$$

From the tables of Appendix I §11 we obtain the following tables of Euler characteristics of $(-)^{((J,\lambda))}$ data. The notation for subgroups is that of Appendix I §11 but the notation for characters of D_6 is that of the table beginning this subsection and τ (resp. ϕ') is the non-trivial character of $\mathrm{Gal}(\mathbb{F}_4/\mathbb{F}_2)$ (resp. C_2') as in Appendix I.

J	$\mathrm{Des}_{C_2 \times D_6}(a_G(\tilde{\nu}_4))^{((J))}$
D_6	$1 + \psi$
$C_2 \times C_3$	$1 - \tau + \tau(\phi + \phi^2)$
$C_2 \times C_2$	$\tau + \tau\mu$
C_3	$\phi + \phi^2$
C_2	-2μ
C_2'	$1 - 5\phi'$
$\{1\}$	-70

J	$\mathrm{Des}_{C_2 \times D_6}(a_G(\tilde{\nu}_5))^{((J))}$
D_6	1
$C_2 \times C_3$	$1 + (1 + \tau)(\phi + \phi^2)$
$C_2 \times C_2$	$\tau + \tau\mu$
C_3	1
C_2	-2μ
C_2'	$-12 - \phi'$
$\{1\}$	-50

QUESTION 8.9. Is the outcome of §8.8 the result of something system-atic or just another black magical coincidence? More precisely[4], does the comparison of the C_3-row of the $(-)^{((J))}$ data suggest the Shintani cor-respondence $Sh(\nu_4) = \nu$ and the combination of the C_3/C_2-rows suggest $Sh(\nu_5) = \mu$?

---

[4]This question is reminiscent of the punch-line of the joke which starts "What is the definition of an optimist?"

CHAPTER 3

Automorphic representations

One encounters the profound relation between automorphic representations and modular forms in [[51], [62], [67], [80]], for example. The topic is a breathtaking mathematical story of local-global flavour which has proved so important in number theory and arithmetic-algebraic geometry. Having already introduced monomial resolutions in the admissible local case, in this chapter I shall give a brief sketch of their introduction for global automorphic representations via the Tensor Product Theorem.

In §1 we recapitulate automorphic representations of GL_2 as manifested in terms of the $(\mathcal{U}(gl_2\mathbb{C}), K_\infty) \times GL_2\mathbb{A}_{fin}$-modules of ([67] Vol. I). Most importantly §1 describes the tensor product theorem which enables one to construct automorphic representations from local admissible representations together with some Archimedean data. In §2 the tensor product theorem for local monomial resolutions is proved. This guarantees, in Theorem 2.5, the existence of a monomial resolution for any $(\mathcal{U}(gl_2\mathbb{C}), K_\infty) \times GL_2\mathbb{A}_{fin}$-module. §3 recalls how modular forms and their Hecke operators enter into the theory of $(\mathcal{U}(gl_2\mathbb{C}), K_\infty) \times GL_2\mathbb{A}_{fin}$-modules. Monomial resolutions (local or global) of V give important "resolutions" of the subspaces $V^{(H,\phi)}$. In §4 we recall from [62] how, in the case of automorphic representations, the $V^{(H,\phi)}$'s include, inter alia, the all-important spaces of classical modular forms.

1. Automorphic representations of $GL_2\mathbb{A}_\mathbb{Q}$

1.1. In this section I am going to recall from ([67] Vol. I) what an irreducible automorphic representation is and how they are constructed by the tensor product theorem. I am only going to do this for $GL_2\mathbb{A}_\mathbb{Q}$ since I can then get all the technical details from ([67] Vol. I) with the minimum of technical elaborations which are needed for the case of a general number field (see [51], [62], [80]). In this chapter representations will be defined over the complex numbers.

My objective is to describe how the analogue of the tensor product theorem works in terms of monomial resolutions.

1.2. Adèles and idèles for GL_1 and GL_2

The ring of adèles of $\mathbb{Q}$ is given by

$$\mathbb{A}_{\mathbb{Q}} = \{(x_\infty, x_2, x_3, \dots, x_p, \dots) \mid x_\infty \in \mathbb{R}, x_p \in \mathbb{Q}_p, \; p \text{ prime}, \; x_p \in \mathbb{Z}_p \text{ p.p.}\}$$

with ring operations performed coordinatewise ([67] Vol. I p. 7). The multiplicative group of idèles is given by

$$\mathbb{A}_{\mathbb{Q}}^* = \{(x_\infty, x_2, x_3, \dots, x_p, \dots) \mid x_\infty \in \mathbb{R}^*, x_p \in \mathbb{Q}_p^*, \; p \text{ prime}, \; x_p \in \mathbb{Z}_p^* \text{ p.p.}\}.$$

Set

$$\mathbb{A}_{fin} = \{(0, x_2, \dots, x_p, \dots) \in \mathbb{A}_{\mathbb{Q}}\}$$

and

$$\mathbb{A}_{fin}^* = \{(1, x_2, \dots, x_p, \dots) \in \mathbb{A}_{\mathbb{Q}}^*\}.$$

The topology on $\mathbb{A}_{\mathbb{Q}}$, which makes it into a locally compact topological ring, has a basis of open sets given by taking any finite set of places containing ∞ and taking U as any open set in the product topology of $\mathbb{R} \times \prod_{p \in S} \mathbb{Q}_p$ and forming

$$O = U \times \prod_{p \notin S} \mathbb{Z}_p.$$

The topology on $\mathbb{A}_{\mathbb{Q}}^*$, which makes it into a locally compact topological group under coordinatewise multiplication, is given by taking U' as any open set in $\mathbb{R}^* \times \prod_{p \in S} \mathbb{Q}_p^*$ and forming

$$O' = U' \times \prod_{p \notin S} \mathbb{Z}_p^*.$$

Note that the topology on the idèles is *not* the subspace topology induced from the adèles.

The rationals embed diagonally into the adèles and the non-zero rationals embeds diagonally into the idèles.

PROPOSITION 1.3. *Adèlic fundamental domain ([67] Vol. I p. 10)*
A fundamental domain for $\mathbb{Q} \backslash \mathbb{A}_{\mathbb{Q}}$ is

$$D = [0, 1) \times \prod_p \mathbb{Z}_p$$

so that

$$\mathbb{A}_{\mathbb{Q}} = \bigcup_{\beta \in \mathbb{Q}} \beta + D \quad \text{(disjoint union)}.$$

PROPOSITION 1.4. *Idèlic fundamental domain ([67] Vol. I p. 11)*
A fundamental domain for $\mathbb{Q}^* \backslash \mathbb{A}_{\mathbb{Q}}^*$ is

$$D' = (0, \infty) \times \prod_p \mathbb{Z}_p^*$$

so that

$$\mathbb{A}_\mathbb{Q}^* = \bigcup_{\alpha \in \mathbb{Q}^*} \alpha D' \quad \text{(disjoint union)}.$$

1.5. The adèlic GL_2 - $GL_2\mathbb{A}_\mathbb{Q}$

The adèlic GL_2 for the rationals consists of

$$\{(g_\infty, g_2, \ldots, g_p, \ldots) \mid g_\infty \in GL_2\mathbb{R}, g_p \in GL_2\mathbb{Q}_p, \ p \text{ prime}, \ g_p \in GL_2\mathbb{Z}_p \text{ p.p.}\}$$

with coordinatewise multiplication. There is a diagonal embedding of $GL_2\mathbb{Q}$ into $GL_2\mathbb{A}_\mathbb{Q}$. We also have

$$GL_2\mathbb{A}_{fin} = \{(I, g_2, \ldots, g_p, \ldots) \in GL_2\mathbb{A}_\mathbb{Q}\}.$$

PROPOSITION 1.6. Adèlic GL_2 fundamental domain ([67] Vol. I p. 109 and p. 111)

Let D_∞ be a fundamental domain for $GL_2\mathbb{Z}\backslash GL_2\mathbb{R}$. Then a fundamental domain for $GL_2\mathbb{Q}\backslash GL_2\mathbb{A}_\mathbb{Q}$ is

$$D_\infty \times \prod_{p \text{ prime}} GL_2\mathbb{Z}_p.$$

Every element of $GL_2\mathbb{A}_\mathbb{Q}$ may be uniquely written in the form

$$\gamma \cdot \left(\left(\begin{array}{cc} y_\infty & x_\infty \\ 0 & 1 \end{array}\right) \cdot \left(\begin{array}{cc} r_\infty & 0 \\ 0 & r_\infty \end{array}\right), I, \ldots, I, \ldots\right) \cdot k$$

with $\gamma \in GL_2\mathbb{Q}$, $-1/2 \leq x_\infty \leq 0$, $y_\infty > 0$, $x_\infty^2 + y_\infty^2 \geq 1$, $r_\infty > 0$ and $k \in O_2\mathbb{R} \cdot \prod_{p \text{ prime}} GL_2\mathbb{Z}_p$.

DEFINITION 1.7. Unitary Hecke character of $\mathbb{A}_\mathbb{Q}^*$ ([67] Vol. I p. 40)

A Hecke character of $\mathbb{A}_\mathbb{Q}^*$ is a continuous homomorphism

$$\omega : \mathbb{Q}^*\backslash \mathbb{A}_\mathbb{Q}^* \longrightarrow \mathbb{C}^*.$$

A Hecke character is unitary if all its values have absolute value 1. The following four properties characterise a unitary Hecke operator:

(i) $\omega(gg') = \omega(g)\omega(g')$ for all $g, g' \in \mathbb{A}_\mathbb{Q}^*$,
(ii) $\omega(\gamma g) = \omega(g)$ for all $\gamma \in \mathbb{Q}^*$, $g \in \mathbb{A}_\mathbb{Q}^*$,
(iii) ω is continuous at $(1, 1, 1, \ldots, 1, \ldots)$ and
(iv) $|\omega|_\mathbb{C} = 1$.

DEFINITION 1.8. Automorphic forms on GL_1 and GL_2 ([67] Vol. I p. 40 and pp. 117–119)

Fix a unitary Hecke character ω as in Definition 1.7. An automorphic form on $GL_1\mathbb{A}_\mathbb{Q} = \mathbb{A}_\mathbb{Q}^*$ is a function

$$\phi : GL_1\mathbb{A}_\mathbb{Q} \longrightarrow \mathbb{C}$$

such that

(i) $\phi(\gamma g) = \phi(g)$ for all $\gamma \in \mathbb{Q}^*$, $g \in \mathbb{A}_{\mathbb{Q}}^*$,

(ii) $\phi(zg) = \omega(z)\phi(g)$ for all $g, z \in \mathbb{A}_{\mathbb{Q}}^*$,

(iii) ϕ is of moderate growth. In other words, for each $(g_\infty, g_2, \ldots, g_p, \ldots) \in \mathbb{A}_{\mathbb{Q}}^*$ there exists positive constants C and M such that

$$|\phi(tg_\infty, g_2, \ldots, g_p, \ldots)|_{\mathbb{C}} < C(1 + |t|_\infty)^M.$$

The space of automorphic forms on $GL_1\mathbb{A}_{\mathbb{Q}}$ is a one-dimensional complex vector space. The condition (ii) just means that ϕ is a Hecke character of moderate growth but when we come to GL_2 the analogue of (ii) will have some more significance.

An automorphic form for GL_2 is a function

$$\phi : GL_2\mathbb{Q} \backslash GL_2\mathbb{A}_{\mathbb{Q}} \longrightarrow \mathbb{C}$$

which is smooth, of moderate growth, right-K-finite and $Z(\mathcal{U}(gl_2))$-finite.

Here K is the maximal compact adèlic subgroup of $GL_2\mathbb{A}_{\mathbb{Q}}$, both it and K-finiteness are defined below.

The action on the ϕ's of the universal enveloping algebras (see Definition 1.9) of the Lie algebras $gl_2\mathbb{R}$ and $gl_2\mathbb{C}$ is given in terms of differential operators D. If $Z(\mathcal{U}(gl_2))$ is the centre of the universal enveloping algebra then a smooth ϕ is $Z(\mathcal{U}(gl_2))$-finite if the set

$$\{D\phi(g) \mid D \in Z(\mathcal{U}(gl_2))\}$$

spans a finite-dimensional vector space.

A function ϕ is smooth if for every $g_0 \in GL_2\mathbb{A}_{\mathbb{Q}}$ there exists an open set $U \subseteq GL_2\mathbb{A}_{\mathbb{Q}}$ containing g_0 and a smooth function

$$\phi_\infty^U : GL_2\mathbb{R} \longrightarrow \mathbb{C}$$

such that $\phi(g) = \phi_\infty^U(g_\infty)$ for all $g \in U$.

Let $g = \begin{pmatrix} a & b \\ c & d \end{pmatrix} \in GL_2\mathbb{A}_{\mathbb{Q}}$ be given by adèles $a = (a_\infty, a_2, \ldots, a_p, \ldots)$, $b = (b_\infty, b_2, \ldots, b_p, \ldots)$, $c = (c_\infty, c_2, \ldots, c_p, \ldots)$ and $d = (d_\infty, d_2, \ldots, d_p, \ldots)$. Define a norm function by

$$||g|| = \prod_{v \leq \infty} \max\left\{|a_v|_v, |b_v|_v, |c_v|_v, |d_v|_v, |a_vd_v - b_vc_v|_v^{-1}\right\}.$$

Then ϕ has moderate growth if there exist constants $C, B > 0$ such that $|\phi(g)|_{\mathbb{C}} < C||g||^B$ for all $g \in GL_2\mathbb{A}_{\mathbb{Q}}$.

Let $K = O_2\mathbb{R} \cdot \prod_{p \text{ prime}} GL_2\mathbb{Z}_p$ be, as above, the maximal compact subgroup of $GL_2\mathbb{A}_{\mathbb{Q}}$. Then ϕ is right-K-finite if the set of right K-translates of ϕ given by the functions

$$\{g \mapsto \phi(gk), \ k \in K\}$$

generates a finite-dimensional subspace of the space of all functions

$$GL_2\mathbb{Q}\backslash GL_2\mathbb{A}_\mathbb{Q} \longrightarrow \mathbb{C}.$$

DEFINITION 1.9. *The universal enveloping algebra of gl_2 ([67] Vol. I pp. 112–117)*

The real Lie algebra $gl_2\mathbb{R}$ is the real vector space of 2×2 matrices with real entries. The Lie bracket is given by the commutator $[\alpha, \beta] = \alpha\beta - \beta\alpha$. The universal enveloping algebra $\mathcal{U}(gl_2\mathbb{R})$ is an associative $\mathbb{R}$-algebra which contains $gl_2\mathbb{R}$ and in which the Lie bracket and the algebra product "$\circ$" are compatible in the sense that $[\alpha, \beta] = \alpha \circ \beta - \beta \circ \alpha$. The universal algebra is constructed as a quotient of the tensor algebra

$$\mathcal{U}(gl_2\mathbb{R}) = T(gl_2\mathbb{R})/\{[\alpha, \beta] - \alpha \otimes \beta - \beta \otimes \alpha \text{ for all } \alpha, \beta \in gl_2\mathbb{R}\}.$$

If A is an $\mathbb{R}$-algebra and $\phi : gl_2\mathbb{R} \longrightarrow A$ is a linear map such that $\phi([\alpha, \beta]) = \phi(\alpha) \cdot \phi(\beta) - \phi(\beta) \cdot \phi(\alpha) \in A$ for all $\alpha, \beta \in gl_2\mathbb{R}$ then there exists a unique $\mathbb{R}$-algebra homomorphism

$$\Phi : \mathcal{U}(gl_2\mathbb{R}) \longrightarrow A$$

extending ϕ.

One of the most important applications of this universal property is to give an isomorphism between the enveloping algebra and an algebra of differential operators.

Let $\alpha \in gl_2\mathbb{R}$ be a 2×2 real matrix and let $F : GL_2\mathbb{R} \longrightarrow \mathbb{C}$ be a smooth function. The differential operator D_α acts on functions such as F by the formula

$$(D_\alpha F)(g) = \frac{\partial}{\partial t}F(g \cdot \exp(t\alpha))|_{t=0} = \frac{\partial}{\partial t}F(g + tg \cdot \alpha)|_{t=0}$$

where $\exp(t\alpha) = I + \sum_{k \geq 1} \frac{t^k \alpha^k}{k!}$.

The differential operators satisfy, as usual,

$$(D_\alpha(c_1 F_1 + c_2 F_2))(g) = c_1 D_\alpha(F_1)(g) + c_2 D_\alpha(F_2)(g)$$

$$(D_\alpha(F_1 \cdot F_2))(g) = (D_\alpha(F_1))(g) \cdot F_2(g) + F_1(g) \cdot (D_\alpha(F_2))(g)$$

for all smooth functions F_1, F_2, constants $c_1, c_2 \in \mathbb{C}$ and matrices $g \in GL_2\mathbb{R}$.

Under the product given by composition, written $D_\alpha \circ D_\beta$, the differential operators generate an associative $\mathbb{R}$-algebra denoted by $\mathcal{D}_\mathbb{R}^2$ consisting of $\mathbb{R}$-linear combinations of finitely iterated compositions $D_{\alpha_1} \circ D_{\alpha_2} \circ \ldots \circ D_{\alpha_k}$.

PROPOSITION 1.10. *([67] Vol. I Proposition 4.5.2 p. 113)*

Let $\alpha_1, \alpha_2 \in gl_2\mathbb{R}$ and $r_1, r_2 \in \mathbb{R}$. Then

(i) $D_{r_1\alpha_1 + r_2\alpha_2} = r_1 D_{\alpha_1} + r_2 D_{\alpha_2}$ and

(ii) $D_{\alpha_1} \circ D_{\alpha_2} - D_{\alpha_2} \circ D_{\alpha_1} = D_{[\alpha_1, \alpha_2]}.$

1.11. ([**67**] *Vol. I Proposition 4.5.2 p. 113*)

Let $V = C^\infty(GL_2\mathbb{R})$, the space of smooth complex-valued functions on $GL_2\mathbb{R}$. Then the linear map $\alpha \mapsto D_\alpha$ extends, via the universal property of Definition 1.9 and Proposition 1.10, to an $\mathbb{R}$-algebra homomorphism

$$\delta : \mathcal{U}(gl_2\mathbb{R}) \longrightarrow \text{End}(V)$$

such that $\delta(\alpha) = D_\alpha$ for each 2×2 matrix α.

The algebra homomorphism δ is injective ([**67**] Vol. I Lemma 4.5.4 p. 114) and therefore yields an isomorphism

$$\delta : \mathcal{U}(gl_2\mathbb{R}) \xrightarrow{\cong} \mathcal{D}_\mathbb{R}^2.$$

If $\alpha, \beta \in gl_2\mathbb{R}$ then $\alpha + \sqrt{-1}\beta \in gl_2\mathbb{C}$ is a general element and defining $D_{\alpha+\sqrt{-1}\beta} = D_\alpha + \sqrt{-1}D_\beta$ we obtain a $\mathbb{C}$-algebra of differential operators on $C^\infty(GL_2\mathbb{R})$ and an isomorphism

$$\delta : \mathcal{U}(gl_2\mathbb{C}) \xrightarrow{\cong} \mathcal{D}_\mathbb{C}^2.$$

DEFINITION 1.12. $(\mathcal{U}(gl_2\mathbb{C}), K_\infty)$-*modules*

Following ([**67**] p. 102) let $K_\infty = SO_2(\mathbb{R})$, the special orthogonal group of 2×2 orthogonal matrices with determinant equal to one. A $(\mathcal{U}(gl_2\mathbb{C}), K_\infty)$-module is a complex vector space V together with actions

$$\pi_g : \mathcal{U}(gl_2\mathbb{C}) \longrightarrow \text{End}(V)$$

$$\pi_{K_\infty} : K_\infty \longrightarrow GL(V)$$

such that for all $v \in V$ the subspace spanned by $\{\pi_{K_\infty}(k)(v) \mid k \in K_\infty\}$ is finite-dimensional and

$$\pi_g(D_\alpha) \cdot \pi_{K_\infty}(k) = \pi_{K_\infty}(k) \cdot \pi_g(D_{k^{-1}\alpha k}).$$

We also require that

$$\pi_g(D_\alpha)(v) = \lim_{t \to 0} \frac{1}{t}(\pi_{K_\infty}(\exp)(t\alpha)) \cdot v - v)$$

for all $v \in V$ and α in the Lie algebra of K_∞ is contained in $gl_2\mathbb{C}$. Note that the limit is defined, without the topology, because $\pi_{K_\infty}(\exp)(t\alpha)) \cdot v$ remains within a finite-dimensional subspace.

We shall denote the pair (π_g, π_{K_∞}) by π and call (π, V) a $(\mathcal{U}(gl_2\mathbb{C}), K_\infty)$-module.

The condition that the subspace spanned by $\{\pi_{K_\infty}(k)(v) \mid k \in K_\infty\}$ is finite-dimensional for each $v \in V$ can be replaced by the equivalent condition, which is more explicit, that for all $v \in V$ there exist integers $M < N$ and complex numbers c_l and vectors $v_l \in V$ with $M \leq l \leq N$ such that $v = \sum_M^N c_l v_l$ and

$$\pi_{K_\infty}\left(\begin{pmatrix} cos(\theta) & sin(\theta) \\ -sin(\theta) & cos(\theta) \end{pmatrix} \right)(v_l) = e^{l\theta\sqrt{-1}} v_l$$

for all $M \le l \le N$ and $\theta \in \mathbb{R}$.

1.13. *Equivalent formulation of* $(\mathcal{U}(gl_2\mathbb{C}), K_\infty)$*-modules*

We have an inclusion $K_\infty = SO_2(\mathbb{R}) \subset O_2(\mathbb{R})$ into the orthogonal group of 2×2 real matrices. The latter acts on $gl_2\mathbb{C}$ by conjugation, $k \cdot z = kzk^{-1}$, on 2×2 matrices with complex entries. This gives, by universality, an algebra automorphism

$$\Phi_k : \mathcal{U}(gl_2\mathbb{C}) \xrightarrow{\cong} \mathcal{U}(gl_2\mathbb{C})$$

for each $k \in K_\infty$.

We may form the twisted tensor algebra

$$\mathcal{U}(gl_2\mathbb{C}) \tilde{\otimes}_\mathbb{R} \mathbb{R}[K_\infty]$$

whose multiplication is given, for $k, k_1 \in K_\infty, X, X_1 \in \mathcal{U}(gl_2\mathbb{C})$, by

$$(X \otimes k) \cdot (X_1 \otimes k_1) = \Phi_{k_1^{-1}}(X) X_1 \otimes kk_1.$$

From Definition 1.12 we have the composition identity

$$\pi_\infty(k^{-1}) \cdot \pi_g(D_\alpha) \cdot \pi_\infty(k) = \pi_g(D_{k^{-1}\alpha k}) = \pi_g(\Phi_{k^{-1}}(D_\alpha)).$$

Since the D_α's generate the enveloping algebra we have

$$\pi_\infty(k^{-1}) \cdot \pi_g(X) \cdot \pi_\infty(k) = \pi_g(\Phi_{k^{-1}}(X))$$

for all $X \in \mathcal{U}(gl_2\mathbb{C})$.

Let $X \otimes k$ act on V by the map $\pi_\infty(k) \cdot \pi_g(X)$. This action makes V into a left module over the twisted tensor algebra $\mathcal{U}(gl_2\mathbb{C}) \tilde{\otimes}_\mathbb{R} \mathbb{R}[K_\infty]$ via

$$(X \otimes k) \cdot v = \pi_\infty(k)(\pi_g(X)(v)).$$

This action makes sense because

$$\pi_\infty(k) \cdot \pi_g(X) \cdot \pi_\infty(k_1) \cdot \pi_g(X_1)$$

$$= \pi_\infty(k) \cdot \pi_\infty(k_1) \cdot \pi_\infty(k_1^{-1}) \cdot \pi_g(X) \cdot \pi_\infty(k_1) \cdot \pi_g(X_1)$$

$$= \pi_\infty(kk_1) \cdot \pi_g(\Phi_{k_1^{-1}}(X)) \cdot \pi_g(X_1)$$

$$= \pi_\infty(kk_1) \cdot \pi_g(\Phi_{k_1^{-1}}(X)X_1)$$

so that

$$(X \otimes k) \cdot ((X_1 \otimes k_1) \cdot v)$$

$$= \pi_\infty(kk_1)(\pi_g(\Phi_{k_1^{-1}}(X)X_1)(v))$$

$$= (\Phi_{k_1^{-1}}(X)X_1 \otimes kk_1) \cdot v,$$

as required.

Therefore a $(\mathcal{U}(gl_2\mathbb{C}), K_\infty)$-module is equivalent to a left $\mathcal{U}(gl_2\mathbb{C})\tilde{\otimes}_\mathbb{R}\mathbb{R}[K_\infty]$-module, which satisfies the additional conditions of Definition 1.12.

DEFINITION 1.14. $(\mathcal{U}(gl_2\mathbb{C}), K_\infty) \times GL_2\mathbb{A}_{fin}$-modules

Let $GL_2\mathbb{A}_{fin}$ denote the finite adèlic GL_2 as defined in §1.4. Define a $(\mathcal{U}(gl_2\mathbb{C}), K_\infty) \times GL_2\mathbb{A}_\mathbb{Q}$-module to be a complex vector space V with actions:

$$\pi_g : \mathcal{U}(gl_2\mathbb{C}) \longrightarrow \mathrm{End}(V)$$

$$\pi_{K_\infty} : K_\infty \longrightarrow GL(V)$$

$$\pi_{fin} : GL_2\mathbb{A}_{fin} \longrightarrow GL(V)$$

satisfying the relations

$$\pi_{fin}(a_{fin})\pi_g(D_\alpha) = \pi_g(D_\alpha)\pi_{fin}(a_{fin})$$

$$\pi_{fin}(a_{fin})\pi_{K_\infty}(k) = \pi_{K_\infty}(k)\pi_{fin}(a_{fin}).$$

If we let $\pi = (\pi_g, \pi_{K_\infty}), \pi_{fin})$ then we refer to the pair (π, V) as a $(\mathcal{U}(gl_2\mathbb{C}), K_\infty) \times GL_2\mathbb{A}_{fin}$-module.

DEFINITION 1.15. Smooth $(\mathcal{U}(gl_2\mathbb{C}), K_\infty) \times GL_2\mathbb{A}_{fin}$-modules

Let V be a $(\mathcal{U}(gl_2\mathbb{C}), K_\infty) \times GL_2\mathbb{A}_{fin}$-module as in Definition 1.14. We say that V is smooth if every $v \in V$ is fixed by some compact, open subgroup of $GL_2\mathbb{A}_{fin}$. The $(\mathcal{U}(gl_2\mathbb{C}), K_\infty) \times GL_2\mathbb{A}_{fin}$-module V is said to be irreducible if it is non-zero and has no proper non-zero subspace which is preserved by the actions $\pi_g, \pi_{K_\infty}, \pi_{fin}$.

1.16. The space of adèlic automorphic forms $\mathcal{A}_\omega(GL_2\mathbb{A}_\mathbb{Q})$

Fix a unitary Hecke character as in Definition 1.7. Let $\mathcal{A}_\omega(GL_2\mathbb{A}_\mathbb{Q})$ denote the complex vector space of all adèlic automorphic forms for $GL_2\mathbb{A}_\mathbb{Q}$, as defined in Definition 1.8.

We shall now examine three linear actions which make sense on the space of all functions on $GL_2\mathbb{A}_\mathbb{Q}$. In fact, these three actions preserve all the conditions of Definition 1.8 and therefore give well-defined actions on $\mathcal{A}_\omega(GL_2\mathbb{A}_\mathbb{Q})$.

Note that there is also a natural action of $GL_2\mathbb{R}$ by right translation on the vector space of all functions on $GL_2\mathbb{R}$. This action does *not* preserve the space $\mathcal{A}_\omega(GL_2\mathbb{A}_\mathbb{Q})$!

Right translation by the finite adèles

Define an action

$$\pi_{fin} : GL_2\mathbb{A}_{fin} \longrightarrow GL(\mathcal{A}_\omega(GL_2\mathbb{A}_\mathbb{Q}))$$

by

$$(\pi_{fin}(a_{fin})(\phi))(g) = \phi(g \cdot a_{fin})$$

where $\phi \in \mathcal{A}_\omega(GL_2\mathbb{A}_\mathbb{Q})$, $g \in GL_2\mathbb{A}_\mathbb{Q}$ and $a_{fin} \in GL_2\mathbb{A}_{fin}$.

Right translation by $O_2(\mathbb{R})$

Consider $k \in K_\infty = O_2(\mathbb{R})$ as embedded in $GL_2\mathbb{A}_\mathbb{Q}$ by inclusion into the Archimedean factor. There we define

$$(\pi_{K_\infty}(k)(\phi))(g) = \phi(gk).$$

Action of $gl_2\mathbb{C}$ by differential operators

If D is a differential operator as in §1.11 then we define an action by

$$(\pi_{gl_2\mathbb{C}}(D)(\phi))(g) = D\phi(g).$$

With these actions the vector space $\mathcal{A}_\omega(GL_2\mathbb{A}_\mathbb{Q})$ is a smooth $(\mathcal{U}(gl_2\mathbb{C}), K_\infty) \times GL_2\mathbb{A}_{fin}$-module in the sense of Definition 1.15 ([**67**] Vol. I Lemma 5.1.7 p. 157).

The notion of an intertwining map of $(\mathcal{U}(gl_2\mathbb{C}), K_\infty) \times GL_2\mathbb{A}_{fin}$-modules is defined in ([**67**] Vol. I Lemma 5.1.7 p. 159) in such a manner that quotients of these are again $(\mathcal{U}(gl_2\mathbb{C}), K_\infty) \times GL_2\mathbb{A}_{fin}$-modules.

One can then define an automorphic representation with central character ω as a smooth $(\mathcal{U}(gl_2\mathbb{C}), K_\infty) \times GL_2\mathbb{A}_{fin}$-module which is isomorphic to a subquotient of $\mathcal{A}_\omega(GL_2\mathbb{A}_\mathbb{Q})$.

1.17. *Infinite tensor products of local representations*

Let $\{V_v\}_{v \le \infty}$ be a family of vector spaces indexed by the rational primes and ∞. Let S be a finite set of primes including ∞. For $v \notin S$ let $\xi_v^0 \in V_v$ be a choice of non-zero vector. The restricted tensor product of the V_v's with respect to the ξ_v^0's is the space of all finite linear combinations of vectors

$$\xi = \bigotimes_{v \le \infty}' \xi_v$$

where $\xi \in V_v$ and $\xi_v = \xi_v^0$ for all but finitely many v's.

Consider a $(\mathcal{U}(gl_2\mathbb{C}), K_\infty) \times GL_2\mathbb{A}_{fin}$-module as in Definition 1.14. It is a complex vector space together with actions

$$\pi_g : \mathcal{U}(gl_2\mathbb{C}) \longrightarrow \text{End}(V)$$

$$\pi_{K_\infty} : K_\infty \longrightarrow GL(V)$$

$$\pi_{fin} : GL_2\mathbb{A}_{fin} \longrightarrow GL(V).$$

The tensor product theorem ([67] Vol. I §10.8 pp. 406–413) yields an isomorphism with a $(\mathcal{U}(gl_2\mathbb{C}), K_\infty) \times GL_2\mathbb{A}_{fin}$-module constructed on a restricted infinite tensor product. In order to define such a $(\mathcal{U}(gl_2\mathbb{C}), K_\infty) \times GL_2\mathbb{A}_{fin}$-module we require the following data:

(i) a $(\mathcal{U}(gl_2\mathbb{C}), K_\infty)$-module (π_∞, V_∞),

(ii) a local representation (π_p, V_p) of $GL_2\mathbb{Q}_p$ for each prime p,

(iii) a finite set of primes S containing ∞,

(iv) a distinguished, non-zero vector $\xi_p^0 \in V_p$ for $p \notin S$ which is fixed by $GL_2\mathbb{Z}_p$.

DEFINITION 1.18. *Unramified local representations*

Fix a prime p. A representation (π, V) of $GL_2\mathbb{Q}_p$ is called unramified if the subspace of $GL_2\mathbb{Z}_p$-fixed points is non-zero ([67] Vol. I Definition 6.2.1 p. 192).

The local components of a $(\mathcal{U}(gl_2\mathbb{C}), K_\infty) \times GL_2\mathbb{A}_{fin}$-module are unramified at all but a finite set of primes. This is related to condition (iv) of §1.17 by the following crucial result.

THEOREM 1.19. *([67] Vol. I Theorem 10.6.12 pp. 400–402)*

Let (π, V) be an unramified admissible irreducible representation of $GL_2\mathbb{Q}_p$ then

$$\dim_\mathbb{C}(V^{GL_2\mathbb{Z}_p}) = 1.$$

In the notation of §2.2 the $GL_2\mathbb{Z}_p$-fixed points of V would be denoted by $V^{((GL_2\mathbb{Z}_p,1))}$.

1.20. *Infinite tensor products of local representations (continued)*

Let (π_∞, V_∞) be a $(\mathcal{U}(gl_2\mathbb{C}), K_\infty)$-module. Let S be a finite set of primes not containing ∞. For each $p \notin S$ let (π_p, V_p) be a representation of $GL_2\mathbb{Q}_p$ such that $V_p^{GL_2\mathbb{Z}_p} \neq 0$. Choose a non-zero $\xi_p \in V_p^{GL_2\mathbb{Z}_p}$ for each $p \notin S$. Set V equal to the restricted tensor product

$$V = \bigotimes_{v \leq \infty}' V_v.$$

Define actions:

$$\pi'_g : \mathcal{U}(gl_2\mathbb{C}) \longrightarrow \text{End}(V)$$

$$\pi'_{K_\infty} : K_\infty \longrightarrow GL(V)$$

$$\pi'_{fin} : GL_2\mathbb{A}_{fin} \longrightarrow GL(V)$$

by

$$\pi'_g(D)(\bigotimes \xi_v) = (\pi_g(D)(v_\infty) \otimes (\bigotimes_{v<\infty} \xi_v),$$

$$\pi'_{K_\infty}(k)(\bigotimes \xi_v) = (\pi_\infty(k)(v_\infty) \otimes (\bigotimes_{v<\infty} \xi_v),$$

$$\pi'_{fin}(a_{fin})(\bigotimes \xi_v) = \xi_\infty \otimes (\bigotimes_{v<\infty} (\pi_v(a_v)(\xi_v)).$$

Observe that if T is the set consisting of ∞ and all the primes such that $a_p \notin GL_2\mathbb{Z}_p$ then T is finite and for all $p \notin S \bigcup T$ we have

$$\pi_p(a_p)(\xi_p) = \pi_p(a_p)(\xi_p^0) = \xi_p^0$$

so that the action π'_{fin} preserves the restricted product V.

THEOREM 1.21. *Tensor product theorem ([67] Vol. I Theorem 10.8.2 p. 407)*

Let (π, V) denote an irreducible admissible $(\mathcal{U}(gl_2\mathbb{C}), K_\infty) \times GL_2\mathbb{A}_{fin}$-module. Let $\{q_1, \ldots, q_m\}$ be the finite set of primes where π is ramified. Let $S = \{\infty, q_1, \ldots, q_m\}$. Then there exists
(i) an irreducible admissible $(\mathcal{U}(gl_2\mathbb{C}), K_\infty)$-module (π_∞, V_∞),
(ii) an irreducible admissible representation (π_p, V_p) of $GL_2\mathbb{Q}_p$ for each finite prime p,
(iii) a non-zero vector $v_p^0 \in V_p^{GL_2\mathbb{Z}_p}$ for each prime $p \notin S$ such that

$$\pi \cong \bigotimes_{v \leq \infty}^{'} \pi_v.$$

The factors are unique ([67] Vol. I Theorem 10.8.12 p. 412).

2. Tensor products of monomial resolutions

2.1. Let p be a rational prime and let (π_p, V_p) be an irreducible admissible complex representations of $GL_2\mathbb{Q}_p$. Denote the central character of (π_p, V_p) by $\underline{\phi}_p : \mathbb{Q}_p^* \longrightarrow \mathbb{C}^*$. Let $\{(\pi_p, V_p) \mid p \text{ prime}\}$ be a family of such representations such that $\{\underline{\phi}_p \mid p \text{ prime}\}$ induces a Hecke character $\omega : \mathbb{Q}^* \backslash \mathbb{A}_\mathbb{Q}^* \longrightarrow \mathbb{C}^*$, as in Definition 1.7.

Let $\mathcal{S} = \{q_1, \ldots, q_m\}$ be a finite set of primes. Assume, in addition, that for each $p \notin \mathcal{S}$ the central character $\underline{\phi}_p$ is trivial when restricted to $\mathbb{Q}_p^* \bigcap GL_2\mathbb{Z}_p$ and

$$V_p^{(GL_2\mathbb{Z}_p, 1)} = V_p^{(\mathbb{Q}_p^* GL_2\mathbb{Z}_p, \underline{\phi}_p)}$$

is non-zero and hence one-dimensional by Theorem 1.19.

As in Theorem 4.9 let

$$\cdots \longrightarrow \underline{M}(p)_i \xrightarrow{d} \underline{M}(p)_{i-1} \xrightarrow{d} \cdots \xrightarrow{d} \underline{M}(p)_0 \xrightarrow{\epsilon} V_p \longrightarrow 0$$

be a monomial resolution in $_{\mathbb{C}[GL_2\mathbb{Q}_p],\phi_p}$**mon**.

That is, for each $(J, \phi) \in \mathcal{M}_{GL_2\mathbb{Q}_p, \underline{\phi}_p}$

$$\cdots \rightarrow \underline{M}(p)_i^{((J,\phi))} \xrightarrow{d} \underline{M}(p)_{i-1}^{((J,\phi))} \xrightarrow{d} \cdots \xrightarrow{d} \underline{M}(p)_0^{((J,\phi))} \xrightarrow{\epsilon} V_p^{(J,\phi)} \rightarrow 0$$

is an exact sequence of complex vector spaces.

2.2. The unramified vectors $\underline{\xi}_p^0$

If (π_p, V_p) is an unramified representation in §2.1 for $p \notin \mathcal{S}$ then

$$\dim_{\mathbb{C}}(V_p^{(\mathbb{Q}_p^* GL_2\mathbb{Z}_p, \underline{\phi}_p)}) = 1$$

generated by a vector denoted by ξ_p^0, say. In the notation of §2.1, if m is large enough,

$$V_p^{(\mathbb{Q}_p^* GL_2\mathbb{Z}_p, \underline{\phi}_p)} = V_p(m)^{(\mathbb{Q}_p^* GL_2\mathbb{Z}_p/U_{\mathbb{Q}_p}^m, \underline{\phi}_p)}.$$

In the bar-monomial resolution for $V_p(m)$ we have $M_{V_p(m),0,G(m+r)}$ we find a canonical summand $V_p^{(\mathbb{Q}_p^* GL_2\mathbb{Z}_p, \underline{\phi}_p)} \otimes S_{G(m+r)}$ where $S_{G(m+r)}$ is the associated monomial module endomorphism ring used in the construction of the bar-monomial resolution (see Theorem 5.6 and §2.1). Therefore we have a canonical vector $\underline{\xi}_p^0 = \xi_p^0 \otimes 1$ where 1 is the identity monomial endomorphism of $\underline{\text{Ind}}_{G(m+r)}^{G(m+r)}(\mathbb{C}_{\underline{\phi}_p}) \subset S_{G(m+r)}$. When r is large the vector $\underline{\xi}_p^0$ is independent of r and defines $\underline{\xi}_p^0 \in W_{V_p, 0, \mathbb{Q}_p^* GL_2\mathbb{Z}_p}$.

Hence we obtain

$$\underline{\xi}_p^0 = \xi_p^0 \otimes_{\mathbb{Q}_p^* GL_2\mathbb{Z}_p} 1 \in \underline{\text{Ind}}_{\mathbb{Q}_p^* GL_2\mathbb{Z}_p}^{GL_2\mathbb{Q}_p}(W_{V_p, 0, \mathbb{Q}_p^* GL_2\mathbb{Z}_p}),$$

which is a summand of $\underline{M}_0$ in the monomial resolution of V_p as in §4.7.

Note that the subspace $\langle \underline{\xi}_p^0 \rangle$ is one of the Lines in $\underline{M}_0^{((\mathbb{Q}_p^* GL_2\mathbb{Z}_p, \underline{\phi}_p))}$ and that the augmentation sends $\underline{\xi}_p^0$ to ξ_p^0.

DEFINITION 2.3. *Infinite monomial tensor products*

For $i \geq 0$ define a $_{\mathbb{C}[GL_2\mathbb{A}_{fin}],\omega}$**mon**-Line Bundle $\underline{M}_i$ by the degree i component of the graded restricted tensor product of the family $\{\underline{M}(p)_* \mid p \text{ prime}\}$ with respect to the $\underline{\xi}_p^0$'s. That is, following §1.17, $\underline{M}_i$ is the space of all finite linear combinations of vectors

$$\xi = \bigotimes_{p \text{ prime}}{}' \underline{\xi}_p$$

where $\underset{=p}{\xi} \in \underline{M}(p)_{i_p}$ and $\sum_p i_p = i$. When ξ is non-zero then all but finitely many of the i_p's must be zero and of these all but finitely many must satisfy $\underset{=p}{\xi} = \underset{=p}{\xi^0}$.

The $_{\mathbb{C}[GL_2\mathbb{A}_{fin}],\omega}$**mon**-Lines of $\underline{M}_i$ are defined to be the one dimensional subspaces generated by the ξ's where each $\underset{=p}{\xi}$ generated a Line in $\underline{M}(p)_{i_p}$.

Finally we have to define the adèlic poset $\mathcal{M}(GL_2\mathbb{A}_{fin}, \omega)$. This is the set of pairs (H, ϕ) where ϕ is a continuous complex-valued character and

$$H = \prod_p H_p$$

is a compact, open modulo the centre subgroup of $GL_2\mathbb{A}_{fin}$ given by the product over p of compact, open modulo the centre subgroups of $H_p \subseteq GL_2\mathbb{Q}_p$.

Restricting the subgroups in $\mathcal{M}(GL_2\mathbb{A}_{fin}, \omega)$ to adèlic products ensures that the stabiliser pair of the Line

$$\langle \xi \rangle = \langle \bigotimes_{p \text{ prime}}' \underset{=p}{\xi} \rangle$$

is in $\mathcal{M}(GL_2\mathbb{A}_{fin}, \omega)$. It also ensures that the graded tensor product of the differentials in the local monomial resolutions gives a $_{\mathbb{C}[GL_2\mathbb{A}_{fin}],\omega}$**mon**-morphism for all $i \geq 1$

$$\underline{M}_i \xrightarrow{d} \underline{M}_{i-1}.$$

The augmentations induce a $_{\mathbb{C}[GL_2\mathbb{A}_{fin}],\omega}$**mod**-morphism of the form

$$\underline{M}_0 \xrightarrow{\epsilon} \bigotimes_{p \text{ prime}}' V_p.$$

THEOREM 2.4. *Monomial resolution for* $GL_2\mathbb{A}_{fin}$

In the situation of §2.1, §2.2 and Definition 2.3 let $V = \bigotimes_{p \text{ prime}}' V_p$. Then the chain complex

$$\cdots \longrightarrow \underline{M}_i \xrightarrow{d} \underline{M}_{i-1} \xrightarrow{d} \cdots \xrightarrow{d} \underline{M}_0 \xrightarrow{\epsilon} V \longrightarrow 0$$

is a monomial resolution in $_{\mathbb{C}[GL_2\mathbb{A}_{fin}],\omega}$**mon**. That is, for each $(J, \phi) \in \mathcal{M}(GL_2\mathbb{A}_{fin}, \omega)$

$$\cdots \longrightarrow \underline{M}_i^{((J,\phi))} \xrightarrow{d} \underline{M}_{i-1}^{((J,\phi))} \xrightarrow{d} \cdots \xrightarrow{d} \underline{M}_0^{((J,\phi))} \xrightarrow{\epsilon} V^{(J,\phi)} \longrightarrow 0$$

is an exact sequence of complex vector spaces.

Proof

Take $(J, \phi) \in \mathcal{M}(GL_2\mathbb{A}_{fin}, \omega)$ where $J = \prod_p J_p$. Then $\phi = \prod_p \phi_p$ and $\underline{M}_i^{((J,\phi))}$ is the degree i component of the restricted product of the

$\underline{\underline{M}}(p)_*^{((J_p, \phi_p))}$. In the case of a finite tensor product of chain complexes over a field the Künneth formula [132] states that the homology of the tensor product chain complex is the (graded) tensor product of the homology of the constituent complexes. Since any vector in the restricted tensor product chain complex lies in a finite tensor product the Künneth formula applies to show that the complex $\underline{\underline{M}}_i^{((J,\phi))}$ is exact in every degree except zero where its homology is the restricted tensor product of the $V_p^{(J_p,\phi_p)}$'s, which is $V^{(J,\phi)}$ because J is a product of the J_p's. $\square$

THEOREM 2.5. *Monomial resolution for* $(\mathcal{U}(gl_2\mathbb{C}), K_\infty) \times GL_2\mathbb{A}_{fin}$
Let (π, V) denote an irreducible admissible $(\mathcal{U}(gl_2\mathbb{C}), K_\infty) \times GL_2\mathbb{A}_{fin}$-module. Let $\{q_1, \ldots, q_m\}$ be the finite set of primes where π is ramified. Let $S = \{\infty, q_1, \ldots, q_m\}$. Then there exists
   (i)   a unique irreducible admissible $(\mathcal{U}(gl_2\mathbb{C}), K_\infty)$-module (π_∞, V_∞),
   (ii)   a unique irreducible admissible representation (π_p, V_p) of $GL_2\mathbb{Q}_p$ for each finite prime p,
   (iii)   a monomial resolution in $_{\mathbb{C}[GL_2\mathbb{A}_{fin}], \omega}\mathbf{mon}$

$$\underline{\underline{M}}_* \xrightarrow{\epsilon} \bigotimes_{p \text{ prime}}' V_p,$$

such that there is an isomorphism of $(\mathcal{U}(gl_2\mathbb{C}), K_\infty) \times GL_2\mathbb{A}_{fin}$-modules

$$V \cong V_\infty \otimes \bigotimes_{p \text{ prime}}' V_p$$

and a $(\mathcal{U}(gl_2\mathbb{C}), K_\infty) \times GL_2\mathbb{A}_{fin}$-monomial resolution

$$V_\infty \otimes \underline{\underline{M}}_* \xrightarrow{1 \otimes \epsilon} V_\infty \otimes \bigotimes_{p \text{ prime}}' V_p \cong V.$$

That is, for each $(J, \phi) \in \mathcal{M}(GL_2\mathbb{A}_{fin}, \omega)$

$$\cdots \longrightarrow V_\infty \otimes \underline{\underline{M}}_i^{((J,\phi))} \xrightarrow{1 \otimes d} V_\infty \otimes \underline{\underline{M}}_{i-1}^{((J,\phi))} \xrightarrow{1 \otimes d} \cdots$$

$$\xrightarrow{1 \otimes d} V_\infty \otimes \underline{\underline{M}}_0^{((J,\phi))} \xrightarrow{1 \otimes \epsilon} V_\infty \otimes V^{(J,\phi)} \longrightarrow 0$$

is an exact sequence of complex vector spaces. The $(\mathcal{U}(gl_2\mathbb{C}), K_\infty) \times GL_2\mathbb{A}_{fin}$-module structures are given by the formulae of §1.20.

Proof
   This follows from the Künneth formula [132] together with Theorem 2.4 and Theorem 1.21. $\square$

3. Maass forms and their adèlic lifts

DEFINITION 3.1. *Automorphic functions of integral weight k*

Let Γ be a subgroup of finite index in $SL_2\mathbb{Z}$. For a non-negative integer k an automorphic function of weight k and character $\psi : \Gamma \longrightarrow \mathbb{C}$ is a smooth, complex-valued function f of moderate growth ([**67**] Definition 3.3.3; see also [**51**] §11) on the upper half-plane $\mathcal{H}$ which satisfies

$$f(\gamma z) = \psi(\gamma)(cz + d)^k f(z)$$

where $\gamma = \begin{pmatrix} a & b \\ c & d \end{pmatrix}$. The vector space of all automorphic functions of weight k and character ψ is denoted by $\mathcal{A}_{k,\psi}(\Gamma)$.

Fixing an integer $N \geq 1$ define the subgroup $\Gamma_0(N)$ by

$$\Gamma_0(N) = \{ \begin{pmatrix} a & b \\ c & d \end{pmatrix} \in SL_2\mathbb{Z} \mid c \equiv 0 \text{ (modulo } N)\}.$$

If χ is a Dirichlet character of conductor N then

$$\tilde{\chi}\begin{pmatrix} a & b \\ c & d \end{pmatrix} = \chi(d)$$

gives a well-defined character. Hence we have $\mathcal{A}_{k,\chi}(\Gamma_0(N))$.

A Maass form is said to have level N if it is a Maass form for $\Gamma_0(M)$ but not for any $\Gamma_0(N)$ with $M < N$.

DEFINITION 3.2. *Adèlic lifts of even weight zero, level one Maass forms*

For an integer k the weight k Laplace operator is given by

$$\Delta_k = -y^2(\frac{\partial^2}{\partial x^2} + \frac{\partial^2}{\partial y^2}) + \sqrt{-1}ky\frac{\partial}{\partial x}.$$

Then Δ_k maps $\mathcal{A}_{k,\chi}(\Gamma_0(N))$ to itself ([**67**] Lemma 3.5.4).

Let $\mathcal{H}$ denote the upper half-plane envisioned as

$$\mathcal{H} = GL_2\mathbb{R}/O_2(\mathbb{R}) \cdot \mathbb{R}^*$$

by sending $X = \begin{pmatrix} a & b \\ c & d \end{pmatrix}$ to $X \cdot \sqrt{-1} = \frac{a\sqrt{-1}+b}{c\sqrt{-1}+d}$. Every point in $\mathcal{H}$ has a unique representative of the form

$$g = \begin{pmatrix} 1 & x \\ 0 & 1 \end{pmatrix}\begin{pmatrix} y & 0 \\ 0 & 1 \end{pmatrix}$$

with $x, y \in \mathbb{R}$ and $y > 0$.

Let ν be a complex number then an even weight zero Maass form of type ν for $SL_2\mathbb{Z}$ is a non-zero smooth function $f \in \mathcal{L}^2(SL_2\mathbb{Z}\backslash\mathcal{H})$ such that

(i) $f(\gamma g) = f(g)$ for all $\gamma \in SL_2\mathbb{Z}$, $g = \begin{pmatrix} y & x \\ 0 & 1 \end{pmatrix} \in \mathcal{H}$,

(ii) $\Delta(f) = \Delta_0(f) = \nu(1-\nu)f$,

(iii) $\int_0^1 f(\begin{pmatrix} 1 & x \\ 0 & 1 \end{pmatrix} g)dx = 0$ for all $g \in GL_2(\mathbb{R})$,

(iv) $f(\begin{pmatrix} y & -x \\ 0 & 1 \end{pmatrix}) = f(\begin{pmatrix} y & x \\ 0 & 1 \end{pmatrix})$ for all $\begin{pmatrix} y & x \\ 0 & 1 \end{pmatrix} \in \mathcal{H}$.

By the coset space description of $\mathcal{H}$ we may consider a Maass form f as a function

$$f : GL_2\mathbb{R} \longrightarrow \mathbb{C}$$

such that

$$f(\gamma gkz) = f(g)$$

for $\gamma \in SL_2\mathbb{Z}, g \in GL_2\mathbb{R}, k \in O_2(\mathbb{R}) = K_\infty, z = \begin{pmatrix} r & 0 \\ 0 & r \end{pmatrix}$.

From Proposition 1.6 we know that every element $g \in GL_2\mathbb{A}_\mathbb{Q}$ may be uniquely written in the form

$$\gamma \cdot \begin{pmatrix} y_\infty & x_\infty \\ 0 & 1 \end{pmatrix} \cdot (\begin{pmatrix} r_\infty & 0 \\ 0 & r_\infty \end{pmatrix}, I, \ldots, I, \ldots) \cdot k$$

with $\gamma \in GL_2\mathbb{Q}$, $-1/2 \le x_\infty \le 0$, $y_\infty > 0$, $x_\infty^2 + y_\infty^2 \ge 1$, $r_\infty > 0$ and $k \in O_2\mathbb{R} \cdot \prod_{p \text{ prime}} GL_2\mathbb{Z}_p$.

The adèlic lift of an even weight zero Maass form f is the function

$$f_{adèlic} : GL_2\mathbb{A}_\mathbb{Q} \longrightarrow \mathbb{C}$$

given by

$$f_{adèlic}(g) = f(\begin{pmatrix} y_\infty & x_\infty \\ 0 & 1 \end{pmatrix}).$$

DEFINITION 3.3. *Adèlic lifts of odd weight zero, level one Maass forms*

There is a similar construction for odd forms to that of Definition 3.2.

Let ν be a complex number then an odd weight zero Maass form of type ν for $SL_2\mathbb{Z}$ is a non-zero smooth function $f \in \mathcal{L}^2(SL_2\mathbb{Z}\backslash\mathcal{H})$ such that

(i) $f(\gamma g) = f(g)$ for all $\gamma \in SL_2\mathbb{Z}$, $g = \begin{pmatrix} y & x \\ 0 & 1 \end{pmatrix} \in \mathcal{H}$,

(ii) $\Delta(f) = \Delta_0(f) = \nu(1-\nu)f$,

(iii) $\int_0^1 f(\begin{pmatrix} 1 & x \\ 0 & 1 \end{pmatrix} g)dx = 0$ for all $g \in GL_2(\mathbb{R})$,

(iv) $f(\begin{pmatrix} y & -x \\ 0 & 1 \end{pmatrix}) = -f(\begin{pmatrix} y & x \\ 0 & 1 \end{pmatrix})$ for all $\begin{pmatrix} y & x \\ 0 & 1 \end{pmatrix} \in \mathcal{H}$.

By the coset space description of $\mathcal{H}$ we may consider a Maass form f as a function

$$f : GL_2\mathbb{R} \longrightarrow \mathbb{C}$$

such that

$$f(\gamma g k z) = f(g)$$

for $\gamma \in SL_2\mathbb{Z}, g \in GL_2\mathbb{R}, k \in O_2(\mathbb{R}) = K_\infty, z = \begin{pmatrix} r & 0 \\ 0 & r \end{pmatrix}$.

From Proposition 1.6 we know that every element $g \in GL_2\mathbb{A}_\mathbb{Q}$ may be uniquely written in the form

$$\gamma \cdot \begin{pmatrix} y_\infty & x_\infty \\ 0 & 1 \end{pmatrix} \cdot (\begin{pmatrix} r_\infty & 0 \\ 0 & r_\infty \end{pmatrix}, I, \ldots, I, \ldots) \cdot k$$

with $\gamma \in GL_2\mathbb{Q}, -1/2 \leq x_\infty \leq 0, y_\infty > 0, x_\infty^2 + y_\infty^2 \geq 1, r_\infty > 0$ and $k \in O_2\mathbb{R} \cdot \prod_{p \text{ prime}} GL_2\mathbb{Z}_p$.

The adèlic lift of an odd weight zero Maass form f is the function

$$f_{adèlic} : GL_2\mathbb{A}_\mathbb{Q} \longrightarrow \mathbb{C}$$

given by

$$f_{adèlic}(g) = f(\begin{pmatrix} y_\infty & x_\infty \\ 0 & 1 \end{pmatrix})\det(k_\infty).$$

In both the even and odd case the adèlic lift is an adèlic cusp form in the sense of ([**67**] Definition 4.7.7).

DEFINITION 3.4. *Adèlic lifts of Maass forms with arbitrary weight, level and character*

This is similar but quite involved, dealing with prime power level first. Details are given in ([**67**] §4.12 pp. 136–140).

3.5. *Explicit realisation of a $(\mathcal{U}(gl_2\mathbb{C}), K_\infty)$-module*

This material comes from ([**67**] pp. 161–166). A convenient basis for $gl_2\mathbb{C}$ is

$$Z = \begin{pmatrix} 1 & 0 \\ 0 & 1 \end{pmatrix}, X = \begin{pmatrix} 1 & 0 \\ 0 & -1 \end{pmatrix}, Y = \begin{pmatrix} 0 & 1 \\ 1 & 0 \end{pmatrix}, H = \begin{pmatrix} 0 & 1 \\ -1 & 0 \end{pmatrix}.$$

As in Definition 1.9 we have associated differential operators D_Z, D_X, D_Y, D_H and in terms of the coordinates

$$g = \begin{pmatrix} 1 & x \\ 0 & 1 \end{pmatrix} \begin{pmatrix} y & 0 \\ 0 & 1 \end{pmatrix} \begin{pmatrix} r & 0 \\ 0 & r \end{pmatrix} \begin{pmatrix} cos(\theta) & sin(\theta) \\ -sin(\theta) & cos(\theta) \end{pmatrix}$$

we have

$$D_Z = r\frac{\partial}{\partial z}, \quad D_H = \frac{\partial}{\partial \theta}.$$

In ([67] Lemma 5.2.4) the relations for a $(\mathcal{U}(gl_2\mathbb{C}), K_\infty)$-module are verified. The operators

$$R = (D_X + \sqrt{-1}D_Y)/2, \quad L = (D_X - \sqrt{-1}D_Y)/2$$

are called the raising and lowering operators because they correspond with the classical raising and lowering operations on Maass forms which raise or lower the weight by 2.

EXAMPLE 3.6. *The $(\mathcal{U}(gl_2\mathbb{C}), K_\infty)$-module associated to a Maass form*

Let $N, k \in \mathbb{Z}$ with $N \geq 1$ and let χ (modulo N) be a Dirichlet character. Fix a Maass form f of type ν, weight k and character χ for $\Gamma_0(N)$. Consider $f_{adèlic}$ as in Definitions 3.2 and 3.3 and define a vector space

$$V_f = \{\sum_{l=1}^{M} c_l R^{m_l} f_{adèlic}(g \cdot k_l) \mid M = 0, 1, 2, 3, \dots \; m_l \in \mathbb{Z}, c_l \in \mathbb{C}\}.$$

Here $g \in GL_2(\mathbb{A}_\mathbb{Q})$ and

$$k_l = (k_{\infty,l}, \begin{pmatrix} 1 & 0 \\ 0 & 1 \end{pmatrix}, \begin{pmatrix} 1 & 0 \\ 0 & 1 \end{pmatrix}, \dots)$$

with $k_{\infty,l} \in O_2(\mathbb{R}) = K_\infty$. The formulae referred to in §3.5 show that V_f is a $(\mathcal{U}(gl_2\mathbb{C}), K_\infty)$-module.

3.7. *Central characters and Hecke operators*

(i) If f is an automorphic form of weight k, level N and character χ (modulo N) then the adèlic lift $f_{adèlic}$ is an adèlic automorphic form whose central character is the idèlic lift of the Dirichlet character χ, which is denoted by $\chi_{idèlic}$ and is given by the formula of ([67] Definition 2.1.7).

(ii) Hecke operators may be defined on spaces of adèlic automorphic forms, by means of sums over suitable adèlic double cosets ([51] §11), which correspond to the classical Hecke operators under the adèlic lift.

4. $V^{(H,\psi)}$ and spaces of modular forms

4.1. The following consists of extracts from [62]. Let Γ denote a congruence subgroup of $SL_2\mathbb{Z}$. That is, Γ contains a subgroup of the form

$$\Gamma(N) = \{\begin{pmatrix} a & b \\ c & d \end{pmatrix} \in SL_2\mathbb{Z} \mid a - 1 \equiv d - 1 \equiv b \equiv c \equiv 0 \text{ (modulo } N)\}$$

for some positive integer N. An important example is given by Hecke's group

$$\Gamma_0(N) = \{\begin{pmatrix} a & b \\ c & d \end{pmatrix} \in SL_2\mathbb{Z} \mid c \equiv 0 \text{ (modulo } N)\}.$$

Let $GL_2^+\mathbb{R}$ denote the group of 2×2 real matrices with positive determinant. The matrix

$$g = \begin{pmatrix} a & b \\ c & d \end{pmatrix} \in GL_2^+\mathbb{R}$$

acts on the complex upper half-plane by $g(z) = \frac{az+b}{cz+d}$.
If k is a positive integer define

$$j(g, z) = (cz + d)\det(g)^{-1/2} \text{ and } f|_{[g]_k}(z) = f(g(z))j(g, z)^{-k}.$$

The map $f \mapsto f|_{[g]_k}$ defines an operator on the complex-valued functions on the upper half-plane, $\{z \in \mathbb{C} \mid \text{Im}(z) > 0\}$. In fact, it defines a right Γ-action on such f's, as the following calculation shows. Let

$$g = \begin{pmatrix} a & b \\ c & d \end{pmatrix} \text{ and } g' = \begin{pmatrix} a' & b' \\ c' & d' \end{pmatrix}$$

so that

$$gg' = \begin{pmatrix} aa' + bc' & ab' + bd' \\ ca' + dc' & cb' + dd' \end{pmatrix}.$$

Therefore

$$f|_{[gg']_k}(z) = f(gg'(z))((ca' + dc')z + cb' + dd')^{-k}\det(gg')^{k/2}.$$

On the other hand

$$(f|_{[g]_k})|_{[g']_k}(z)$$

$$= f|_{[g]_k}(g'(z))j(g', z)^{-k}$$

$$= f(gg'(z))j(g, g'(z))^{-k}j(g', z)^{-k}$$

$$= f(gg'(z))\det(g)^{k/2}\det(g')^{k/2}(c\tfrac{a'z+b'}{c'z+d'} + d)^{-k}(c'z + d')^{-k}$$

$$= f|_{[gg']_k}(z).$$

Two points z_1, z_2 are called Γ-equivalent if there exists $g \in \Gamma$ such that $g(z_1) = z_2$; i.e. they belong to the same Γ-orbit. A fundamental domain F for the Γ-action on the upper half-plane is a connected open subset which intersects each Γ-orbit at most once and every z is Γ-equivalent to a point in the closure of F. A point $s \in \mathbb{R} \bigcup \infty$ is called a cusp of Γ if there exists an element of the parabolic form

$$g = \begin{pmatrix} a & b \\ 0 & d \end{pmatrix} \in \Gamma$$

such that $g(s) = s$. If $\mathcal{H}^*$ denotes the union of the upper half-plane and the cusps then Γ also acts on $\mathcal{H}^*$ and the resulting quotient space possesses a natural Hausdorff topology and a complex structure such that $\Gamma \backslash \mathcal{H}^*$ is a compact Riemann surface. The cusps we shall consider are various rational points on the real line together with $\infty = \sqrt{-1}\infty$.

DEFINITION 4.2. A complex-valued function $f(z)$ is called a Γ-automorphic form of weight k if it is defined on the upper half-plane, $\{z \in \mathbb{C} \mid \mathrm{Im}(z) > 0\}$ and satisfies:
  (i) automorphy: $f|_{[g]_k} = f$ for all $\gamma \in \Gamma$.
  (ii) f is holomorphic in $\{z \in \mathbb{C} \mid \mathrm{Im}(z) > 0\}$.
  (iii) f is holomorphic at every cusp of Γ.
  The space of such functions is denoted by $M_k(\Gamma)$. If $\Gamma = \Gamma(N)$ then $M_k(\Gamma(N))$ is often called modular forms of level N.
  If ψ is a character of $(\mathbb{Z}/N)^*$ then $M_k(N, \psi)$ is the space of functions which satisfy conditions (ii) and (iii) as well as

$$f(\frac{az + b}{cz + d}) = \psi(a)^{-1}(cz + d)^k f(z),$$

which is condition (i) if $\psi = 1$.

  4.3. *Regularity at a cusp*
  Suppose that s is a cusp of Γ. Therefore exists $\sigma \in SL_2\mathbb{Z}$ such that $\sigma(\infty) = s$. The matrix σ exists because $z \mapsto 1/z$ maps points near $\sqrt{-1}\infty$

to close to the origin and this maps close to b/d so all rational points on the real axis are in the $SL_2\mathbb{Z}$-orbit of ∞. Therefore, by §4.2(i), $f|_{[\sigma]_k}$ is invariant under $\rho = \sigma^{-1}\gamma\,\sigma$ if $\gamma(s) = s$. Let Γ_s denote the group of γ's fixing s. We have the translation $z \mapsto z + 1$ and each ρ in $\sigma^{-1}\Gamma_s\sigma$ is translation by some h_1 since $\sigma^{-1}\Gamma_s\sigma$ fixes ∞. Let h be the smallest such translation and suppose k is even then

$$f|_{[\sigma]_k}(z + h) = f|_{[\sigma]_k}(z)$$

from which it follows that

$$\hat{f}_s(\zeta) = f|_{[\sigma]_k}(z)$$

where $\zeta = e^{2\pi\sqrt{-1}z/h}$, which is called the local uniformising variable at s. Then $\hat{f}_s(\zeta)$ is well-defined in $|\zeta| < 1$ and is holomorphic in the punctured unit disc, by condition (ii). The meaning of condition (iii) is that $\hat{f}_s(\zeta)$ is holomorphic at $\zeta = 0$ for every cusp s.

When k is odd and

$$\begin{pmatrix} -1 & 0 \\ 0 & -1 \end{pmatrix} \in \Gamma$$

then $M_k(\Gamma) = 0$. If we assume that $-1 \notin \Gamma$ then if $\sigma^{-1}\Gamma_s\sigma$ is generated by

$$\begin{pmatrix} -1 & -h \\ 0 & -1 \end{pmatrix}$$

then

$$f|_{[\sigma]_k}(z + h) = -f|_{[\sigma]_k}(z)$$

so that the variable for regularity at s should be $\zeta = e^{\pi\sqrt{-1}z/h}$ rather than $\zeta = e^{2\pi\sqrt{-1}z/h}$.

4.4. Fourier expansion at a cusp

Suppose that f lies in $M_k(\Gamma)$. Since f is regular at the cusp s implies that $\hat{f}_s(\zeta) = f|_{[\sigma]_k}(z)$ has a Taylor series at $\zeta = 0$. This series yields an expansion

$$f|_{[\sigma]_k}(z) = \sum_{n=0}^{\infty} a_n e^{2\pi\sqrt{-1}nz/h}.$$

The series converges absolutely and uniformly on compact subsets and is called the Fourier expansion of f at the cusp s.

If Γ is Hecke's group $\Gamma_0(N)$ then the Fourier expansion at ∞ of any $f \in M_k(\Gamma_0(N))$ will have the form

$$f(z) = \sum_{n=0}^{\infty} a_n e^{2\pi\sqrt{-1}nz}.$$

DEFINITION 4.5. A Γ-automorphic form is a cusp form if it vanishes at every cusp of Γ. In other words, the Fourier coefficient $a_0 = 0$ at each cusp. The cusp forms of weight k will be denoted by $S_k(\Gamma)$ and we set $S_k(\Gamma, \psi) = S_k(\Gamma) \bigcap M_k(\Gamma, \psi)$.

4.6. The following is an adelic explicit description of the classical modular forms in terms of automorphic functions on $GL_2 A_{\mathbb{Q}}$. It is taken from ([**51**] §11.1) which is, in turn, a version of the description in ([**62**] Prop. 3.1). After we have recalled this then, in §4.7 et seq we shall recall the more general setting involving Maass forms as in §3.1–§3.3.

Temporarily we shall write $G_{\mathbb{Q}} = GL_2 \mathbb{Q}$, $G_A = GL_2 A_{\mathbb{Q}}$, $G_\infty = GL_2 \mathbb{R}$ and $G_f = GL_2 A_{fin}$. Put $\mathcal{H}^\pm = \mathbb{C} - \mathbb{R}$ and let

$$U_\infty = SO_2 \mathbb{R} \cdot \left\{ \begin{pmatrix} a & 0 \\ 0 & a \end{pmatrix} \mid a \in \mathbb{R}^* \right\}.$$

U_∞ is the stabiliser of $i = \sqrt{-1}$ in the action on the upper half-plane since

$$\begin{pmatrix} cos(\theta) & sin(\theta) \\ -sin(\theta) & cos(\theta) \end{pmatrix} (i) = \frac{cos(\theta)i + sin(\theta)}{-sin(\theta)i + cos(\theta)} = (-i)^{-1} = i.$$

Identify G_∞ / U_∞ with $\mathcal{H}^\pm$ via the map $gU_\infty \mapsto g(i)$. Define j' (closely related to the function j which was introduced earlier)

$$j' : G_\infty \times \mathcal{H}^\pm \longrightarrow \mathbb{C}$$

by the formula

$$j'\left(\begin{pmatrix} a & b \\ c & d \end{pmatrix}, z \right) = cz + d.$$

Let S_k de the space of functions $\phi : G_{\mathbb{Q}} \backslash G_A \longrightarrow \mathbb{C}$ such that:
   (i) $\phi(gu) = \phi(g)$ for all u lying in some compact open subgroup U.
   (ii) $\phi(gu_\infty) = j(u_\infty, i)^{-k} det(u_\infty) \phi(g)$ for all $u_\infty \in U_\infty, g \in G_A$.
   (iii) For all $g \in G_f$ the map $\mathcal{H}^\pm \longrightarrow \mathbb{C}$ given, for $h \in G_\infty$, by

$$hi \mapsto \phi(gh)j'(h, i)^k (det(h))^{-1}$$

is holomorphic.
   (iv) ϕ is slowly increasing in the sense that for every $c > 0$ and every compact subset $K \subset G_A$ there exist constants A, B satisfying

$$\phi\left(\begin{pmatrix} a & 0 \\ 0 & 1 \end{pmatrix} h \right) \leq A ||a||^B$$

for all $h \in K, a \in A^*$ with $||a|| > c$.

(v) ϕ is cuspidal in the sense that for all $g \in G_{\mathbb{A}}$

$$\int_{\mathbb{Q}\backslash\mathbb{A}} \phi\left(\begin{pmatrix} 1 & x \\ 0 & 1 \end{pmatrix} g\right) dx = 0,$$

where dx is a non-trivial Haar measure.

The space S_k is a G_f-module via the action given by right translation. For every compact open subgroup U let $S_k(U) = S_k^{(U,1)}$, the fixed points of U. Hence S_k is the union of the $S_k(U)$'s.

For $N > 0$ define subgroups

$$U_0(N) = \left\{ \begin{pmatrix} a & b \\ c & d \end{pmatrix} \in G_f \mid c \in N \cdot \mathbb{A}_{fin} \right\}$$

and

$$U_1(N) = \left\{ \begin{pmatrix} a & b \\ c & d \end{pmatrix} \in G_f \mid c, d - 1 \in N \cdot \mathbb{A}_{fin} \right\}.$$

For $\phi \in S_k(U_1(N))$ define a function f_ϕ by the formula

$$f_\phi(hi) = \phi(h)j'(h,i)^k(\det(h))^{-1}$$

for $h \in GL_2^+\mathbb{R}$. Then $\phi \mapsto f_\phi$ defined an isomorphism

$$S_k(U_1(N)) \cong S_k(\Gamma_1(N)).$$

Moreover, if ϵ is a mod N Dirichlet character then

$$S_k^{(U_0(N),\epsilon)} \cong S_k(\Gamma_0(N), \epsilon).$$

4.7. Now we shall consider some results from ([**67**] p. 170 and pp. 176–177). Given an automorphic representation of $GL_2\mathbb{A}_{\mathbb{Q}}$ we may restrict it to $GL_2\mathbb{A}_{fin}$. Suppose we have an integer $M = \prod_p p^{f_p}$ define compact open subgroups of $GL_2\mathbb{A}_{fin}$ as follows:

$$K_0(M) = \left\{ k = (1, k_2, \ldots, k_p, \ldots) \mid k_p \equiv \begin{pmatrix} * & * \\ 0 & * \end{pmatrix} \pmod{p^{f_p}} \right\}$$

and

$$K_1(M) = \left\{ k = (1, k_2, \ldots, k_p, \ldots) \mid k_p \equiv \begin{pmatrix} 1 & * \\ 0 & 1 \end{pmatrix} \pmod{p^{f_p}} \right\}$$

so that $K_1(M) \subset K_0(M)$. Hence every continuous character of $K_0(M)/K_1(M)$ inflates to give a continuous character of $K_0(M)$ which is trivial on $K_1(M)$.

PROPOSITION 4.8.

Given a smooth representation (π, V) of $GL_2 \mathbb{A}_{fin}$ there exists an integer $M = \prod_p p^{f_p}$ and a continuous character λ of $K_0(M)$, which is trivial on $K_1(M)$, such that $V^{(K_0(M),\lambda)} \neq \{0\}$.

REMARK 4.9. In Proposition 4.8 the character λ need not necessarily be an idélic lift of a classical Dirichlet character as in ([**67**] §4.12). However, when λ is such a lift and when V is the restriction of an automorphic representation, then the space $V^{(K_0(M),\lambda)}$ would be isomorphic to a space of newforms, as in §4.6.

4.10. *Proof of Proposition 4.8*

Let $v_1 \in V$ be non-zero. Then v_1 is fixed by a compact open subgroup K' and for some N we must have $K(N) \subseteq K'$ where

$$K_1(M) = \{k = (1, k_2, \dots, k_p, \dots) \mid k_p \equiv \begin{pmatrix} 1 & 0 \\ 0 & 1 \end{pmatrix} \ (\mathrm{mod}\ p^{f_p})\}$$

for $N = \prod_p p^{f_p}$. Set

$$v_2 = \pi(i_{fin}(\begin{pmatrix} N & 0 \\ 0 & 1 \end{pmatrix}^{-1}))v_1.$$

Now

$$K_1(N^2) \subset \begin{pmatrix} N & 0 \\ 0 & 1 \end{pmatrix}^{-1} K(N) \begin{pmatrix} N & 0 \\ 0 & 1 \end{pmatrix}$$

so that $\pi(k)v_2 = v_2$ for all $k \in K_1(N^2)$.

Now there is an isomorphism $K_0(M)/K_1(M) \cong \mathbb{Z}/M)^* \times \mathbb{Z}/M)^*$ ([**67**] p. 171). Now let λ run through the characters of this finite abelian group. Define

$$v_\lambda = \frac{1}{|(\mathbb{Z}/N^2)^*|^2} \sum_g \lambda(g)^{-1} \pi(g)(v_2) \in V^{(K_0(M),\lambda)}$$

where g runs through $K_0(N^2)/K_1(N^2)$. Well-known properties of characters of finite groups ([**126**]) yield the relation

$$v_2 = \sum_\lambda v_\lambda$$

so that at least one v_λ is non-zero. □

REMARK 4.11. Let V be an irreducible cuspidal automorphic representation, as defined in ([**67**] §5.1.14), which is a subspace of the space of all cuspidal automorphic forms with central character ω. Set

$$V_k = \{v \in V \mid \pi_{K_\infty}(\begin{pmatrix} \cos(\theta) & \sin(\theta) \\ -\sin(\theta) & \cos(\theta) \end{pmatrix})v = e^{\sqrt{-1}k\theta}, \text{ all } \theta \in \mathbb{R}\}.$$

Then V is an admissible $(\mathcal{U}(gl_2\mathbb{C}), K_\infty) \times GL_2\mathbb{A}_{fin}$-module.

In particular, the spaces of newforms $V_k^{(K(N),1)}$ (and hence the subspaces $V_k^{(K(N),\lambda)}$) are all finite-dimensional.

CHAPTER 4

GL_nK in general

In this chapter I shall verify Conjecture 3.3 for GL_nK for all $n \geq 2$ where K is a p-adic local field. For GL_2K this was accomplished (in Chapter Two, Theorem 4.9 and Corollary 4.10) by means of explicit formulae, in order to introduce the ideas of the general proof gradually. In this chapter I shall adopt a similar gradual approach, going into considerable detail in the GL_3K case before giving the general case.

For GL_2K the proof of Chapter Two Conjecture 3.3 was accomplished by constructing a double complex in $_{k[GL_2K],\phi}\mathbf{mon}$ using several bar-monomial resolutions together with a simplicial action on the tree for GL_2K. For GL_2K, by some low-dimensional good fortune, the construction of the differential in the double complex was made particularly easy (see the introduction to Chapter Two). For GL_nK with $n \geq 3$ we have to use in a crucial way the naturality of the bar-monomial resolutions in order to apply the construction of the monomial complex given in Chapter Two §3. This requires a simplicial action on a space Y which, for GL_nK with $n \geq 2$, we take to be the Bruhat-Tits building. Such buildings are constructed from BN-pairs.

In §1 we recall the definition and properties of BN-pairs. In §2 we recall the association of a building to a BN-pair with particular emphasis on SL_2K, GL_2K, SL_3K and GL_3K when K is a p-adic local field. In §3 we verify Chapter Two Conjecture 3.3 for all GL_nK using the contractibility properties of the Bruhat-Tits building, which are explained in detail in several GL_3K examples in §2.

As explained in §2.11 the Bruhat-Tits building for GL_nK is a factor of the Baum-Connes space $\underline{E}GL_nK$ and from this the crucial contractibility properties follow. Here I should point out that Appendix IV explains the construction of $\underline{E}(G,\mathcal{C})$ for any locally p-adic Lie group and the family $\mathcal{C}$ of compact open modulo the centre subgroups. This simplicial G-space has all the contractibility properties required for the verification of the analogue of Chapter Two, Conjecture 3.3 for all admissible representations of G with a fixed choice of central character ϕ. I leave to the reader the mustering of all the details for that verification!

1. BN-pairs

1.1. We shall start by recalling the theory of BN-pairs and their buildings in order to examine closely the cases of SL_nK and GL_nK for $n = 2, 3$. More complete accounts of this topic are to be found in [4] and [35]. See also [11], [60] and [61].

DEFINITION 1.2. ([35] p. 107)
Let G be a group with subgroups B and N. This is a BN-pair if
(i) $G = \langle B, N \rangle$,
(ii) $T = B \bigcap N \lhd N$,
(iii) $W = N/T = \langle S \rangle$ for some set S such that the following conditions hold:

(BN1): $C(s)C(w) \subseteq C(w) \bigcup C(sw)$ for all $s \in S, w \in W$, where $C(w) = BwB$ which depends only on the coset of $w \in N/T$ since $T \subseteq B$ and T is normal in N.

(BN2): $sBs^{-1} \nsubseteq B$ for all $s \in S$.

The terminology is: (G, B, N, S) is a Tits system and W is the Weyl group. A special subgroup of W, $W' \subseteq W$, is one of the form $W' = \langle S' \rangle$ with S' a subset of S.

PROPOSITION 1.3. ([35] p. 107)
Assume that S consists of elements of order two and that **(BN1)** holds. Then
(a) $B \bigcup C(s)$ is a subgroup of G for every $s \in S$.
(b) $BW'B$ is a subgroup of G for every special subgroup $W' \subseteq W$.
(c) As a set G is the disjoint union of the $C(w)$ as w runs through W.
(d) $C(s)C(w) = C(sw)$ if $l(sw) \geq l(w)$ where $l(w)$, the length of $w \in W$ ([35] p. 34), is the minimal d such that $w = s_1 s_2 \ldots s_d$ with $s_i \in S$.

Proof

Clearly, taking $W' = \{1, s\}$ we have (b) implies (a). To prove (b) we shall show that $C(w)C(w') \subseteq BW'B$ for $w, w' \in W''$. Write $w = s_1 \ldots s_d$ with $s_i \in S'$ and $W' = \langle S' \rangle$. When $d = 1$ the axiom **(BN1)** implies that $C(w)C(w') \subseteq BW'B$. By induction on d we shall show that

$$C(w)C(w') \subseteq \bigcup_{\epsilon_i = 0,1} C(s_1^{\epsilon_1} s_2^{\epsilon_2} \ldots s_d^{\epsilon_d} w')$$

which implies (b) since $s_1^{\epsilon_1} s_2^{\epsilon_2} \ldots s_d^{\epsilon_d} w' \in W'$. Equivalently we may show that

$$wBw' \subseteq \bigcup_{\epsilon_i = 0,1} Bs_1^{\epsilon_1} s_2^{\epsilon_2} \ldots s_d^{\epsilon_d} w'B.$$

By induction

$$wBw' = s_1(s_2 \ldots s_d Bw')$$

$$\subseteq s_1(\bigcup_{\epsilon_i=0,1} Bs_2^{\epsilon_2} \ldots s_d^{\epsilon_d} w'B)$$

$$\subseteq \bigcup_{\epsilon_i=0,1} Bs_2^{\epsilon_2} \ldots s_d^{\epsilon_d} w'B \cup (\bigcup_{\epsilon_i=0,1} Bs_1 s_2^{\epsilon_2} \ldots s_d^{\epsilon_d} w'B)$$

as required.

To prove (c) we first observe that BWB is a subgroup of G which contains both B and N so that it must be equal to G. To complete the proof of (c) we must show that $C(w) = C(w')$ implies that $w = w' \in W$. Assume that $d = l(w') \leq l(w)$ then we proof (c) by induction on d. If $d = 0$ then $w' = 1$ and so $C(w) = B$. Therefore the image of $w \in W$ in $W = N/T = N/(B \cap N)$ is trivial, as required. Now suppose that $d > 0$ and write $w' = sw''$ with $s \in S$ and $l(w'') = d - 1$. The condition $BwB = Bw'B$ implies that $w'B \subseteq BwB$ and so $w''B \subseteq sBwB$ and therefore, by **(BN1)**, $C(w'') \subseteq C(w) \bigcup C(sw)$. Therefore, since $C(w'')$ is the double B-coset of a single element either $C(w'') = C(w)$ or $C(w'') = C(sw)$. By induction this implies that either $w'' = w$ or $w'' = sw$. If $w'' = w$ then $l(w) = l(w'') < d \leq l(w)$ which is a contradiction. Hence $w'' = sw$ and $w' = sw'' = ssw = w$, as required.

Finally we shall prove (d) by induction on $l(w)$. If $l(w) = 0$ then $w = 1$ and the result is obvious. Suppose that $l(w) > 0$ and write $w = w't$ with $t \in S$ and $l(w') = l(w) - 1$. If $C(s)C(w) \neq C(sw)$ then **(BN1)** implies that sBw intersects BwB and so sBw' intersects $BwBt$. From **(BN1)** we have $tBw^{-1} \subseteq Bw^{-1}B \bigcup Btw^{-1}B$ and taking inverses we obtain an inclusion $BwBt \subseteq BwB \bigcup BwtB$. Therefore sBw' meets $C(w) \bigcup C(wt) = C(w) \bigcup C(w')$.

Assume for the moment that $l(sw) \geq l(w)$ then we must have $l(sw') \geq l(w')$ for if not

$$l(sw) = l(sw't) \leq l(sw') + 1 < l(w') + 1 = l(w).$$

Therefore, by induction, $C(s)C(w') = C(sw')$ and the proof of (c) shows that either $C(sw') = C(w)$ or $C(sw') = C(w')$. Hence, by (c), either $sw' = w$ or $sw' = w'$. The latter is impossible since s has order two and so is non-trivial. However, $sw' = w$ implies that $l(sw) = l(w') < l(w)$, which is also impossible. The only possibility remaining is that $C(s)C(w) = C(sw)$, which proves (d). $\square$

PROPOSITION 1.4. *([35] p. 108)*

Assume that S consists of elements of order two and that both **(BN1)** and **(BN2)** hold. If Proposition 1.3(a) and (d) hold then, for all $s \in S, w \in W$,

$$C(s)C(w) = C(w) \bigcup C(sw)$$

if $l(sw) \leq l(w)$.

Proof

By **(BN1)** $C(s)C(s) \subseteq B \cup C(s)$. Since $C(s)C(s)$ is closed under left and right B-multiplication and each of B and $C(s)$ are generated by any of their elements under left and right B-multiplication we must have $C(s)C(s) = B$ or $C(s)C(s) = B \cup C(s)$, because $C(s)C(s)$ contains B. Since $C(s)C(s) = B$ implies $sBs^{-1} = sBs \subseteq B$, contradicting **(BN2)**, we have $C(s)C(s) = B \cup C(s)$. If $l(sw) \leq l(w)$ then $l(s \cdot sw) = l(w) \geq l(sw)$ so, replacing w by sw in §1.3(d), we obtain

$$C(s)C(sw) = C(ssw) = C(w).$$

Multiplying by $C(s)$ we obtain

$$
\begin{aligned}
C(s)Cw) \quad &= C(s)C(s)C(sw) \\[6pt]
&= (B \cup C(s))C(sw) \\[6pt]
&= C(sw) \cup C(s)C(sw) \\[6pt]
&= C(sw) \cup C(w),
\end{aligned}
$$

by Proposition 1.3(d). □

DEFINITION 1.5. *Coxeter systems*

Suppose that W is a group and that S is a subset of W whose elements each have order two and which satisfy $W = \langle S \rangle$. The pair (W, S) is called a Coxeter system ([35] pp. 46–53) if, for all $w \in W, s, t \in S$ satisfying $l(sw) = l(w) + 1 = l(wt)$, either $l(swt) = l(w) + 2$ or $swt = w$.

PROPOSITION 1.6. *([35] p. 109)*

Assume that S consists of elements of order two and that Proposition 1.3(c),(d) and Proposition 1.4 hold then (W, S) is a Coxeter system.

Proof

Suppose that $l(sw) = l(w) + 1 = l(wt)$ and that $l(swt) < l(w) + 2$. Then we have, by Proposition 1.4,

$$C(s)C(wt) = C(wt) \cup C(swt)$$

which must be a disjoint union, by Proposition 1.3(c). By Proposition 1.3(d) we have $C(wt) = C(w)C(t)$ so that $C(s)C(w)C(t)$ is the disjoint union of $C(wt)$ and $C(swt)$. By Proposition 1.3(d) we have $C(s)C(w) = C(sw)$.

By Proposition 1.4, if $l(sw) \leq l(w)$, we have

$$C(s)C(w) = C(w) \cup C(sw)$$

and taking inverses we find

$$C(w^{-1})C(s) = C(w^{-1}) \bigcup C(w^{-1}s).$$

Replacing w^{-1} by sw and s by t we obtain, since length is preserved under taking inverses,

$$C(sw)C(t) = C(sw) \bigcup C(swt)$$

since $l(swt) \leq l(w) + 1 = l(sw)$. Combining all this we have that the disjoint union of $C(sw)$ and $C(swt)$ equals the disjoint union of $C(wt)$ and $C(swt)$. Hence $C(sw) = C(wt)$. Therefore $sw = wt$, by Proposition 1.3(c).

$\square$

PROPOSITION 1.7. ([35] p. 109)

If **(BN1)** and **(BN2)** hold then every $s \in S$ has order two.

Proof

By **(BN1)** we have $sBs^{-1} \subseteq C(s)C(s^{-1}) \subseteq B \bigcup C(s^{-1})$. Hence **(BN2)** implies that $C(s)C(s^{-1})$ meets $C(s^{-1})$ and so, by left and right multiplication by B, we must have $C(s^{-1}) \subseteq C(s)C(s^{-1})$. Also $B \subseteq C(s)C(s^{-1})$ so we must have $C(s)C(s^{-1}) = B \bigcup C(s^{-1})$ and the union is a disjoint union, since each of B and $C(s^{-1})$ is generated by any of its elements by left and right multiplication by B.

Taking inverses shows that $C(s)C(s^{-1})$ is the disjoint union of B and $C(s)$. Therefore $C(s^{-1}) = C(s)$ and so $C(s)C(s)$ is the disjoint union of $C(s)$ and B. By **(BN1)** with $w = s$ we have $C(s)C(s) \subseteq C(s) \bigcup C(s^2)$. Since $C(s)C(s)$ is the disjoint union of two double cosets we must have $C(s)C(s)$ equals the disjoint union of $C(s)$ and $C(s^2)$. Therefore $C(s) \neq B$ and $C(s^2) = B$ so that $s \notin B$ but $s^2 \in B$ which implies that s has order two in $W = N/B \cap N$. $\square$

DEFINITION 1.8. ([35] p. 110)

Let (G, B, N, S) be a Tits system as in §1.2. For $S' \subseteq S$ let W' denote the special subgroup of W given by $W' = \langle S' \rangle$. Then a special subgroup of G is a subgroup of the form $BW'B$, which is a subgroup by Proposition 1.3(b). When $S' = S$ then $W' = G$, which for the moment will be allowed as a special subgroup. However, when we come to the building of a BN-pair in Definition 2.2 we shall only use the proper special subgroups.

It is shown in ([35] §1D and §2B) that the map $S' \mapsto BW'B$ is a bijection of posets from the poset of subsets of S to that of special subgroups of G.

PROPOSITION 1.9. ([35] p. 110)

Let $w \in W$ with $l(w) = d$ and $w = s_1 \ldots s_d$ for $s_i \in S$. Then the subgroup of G generated by $C(w)$ contains the $C(s_i)$ for $i = 1, \ldots, d$. Moreover this subgroup is generated by B and wBw^{-1}.

Proof

Since the subgroup generated by $C(w)$ contains w and B we have inclusions

$$\langle B, wBw^{-1} \rangle \subseteq \langle C(w) \rangle \subseteq \langle C(s_1), \dots, C(s_d) \rangle.$$

Therefore the result will follow if we establish the inclusions $C(s_i) \subseteq P = \langle B, wBw^{-1} \rangle$ for each i. Since $l(s_1 w) < l(w)$ we know, by Proposition 1.4 (proof), that $s_1 Bw$ meets BwB so $s_1 B$ meets $BwBw^{-1}$ which implies that $C(s_1) \subseteq P$. Hence P also contains $s_1 w Bw^{-1} s_1$ and applying the induction hypothesis to $s_1 w$ shows that P contains each of $C(s_2), \dots, C(s_d)$. $\square$

THEOREM 1.10. *([35] p. 110)*
The special subgroups of G are precisely the subgroups containing B.

Proof

Clearly each special subgroup contains B. Conversely suppose that P is a subgroup containing B. Therefore P is the union of double cosets and so $P = BW'B$ where W' is the subset of W defined by

$$W' = \{w \in W \mid C(w) \subseteq P\}.$$

Since $C(w^{-1}) = C(w)^{-1}$ and $C(ww') \subset C(w)C(w')$ we see that W' is a subgroup of W. By Proposition 1.9 W' contains, for each of its elements w, the generators $s \in S$ which occur in any minimal decomposition of w. Hence W' is a special subgroup of W generated by $S' = W' \cap S$ and therefore P is a special subgroup of G. $\square$

PROPOSITION 1.11. *([35] p. 111)*
The set S consists of all non-trivial elements $w \in W$ such that $B \cup C(w)$ is a subgroup of G.

Proof

Any $s \in S$ satisfies the condition that $B \cup C(s)$ is a subgroup of G, by Proposition 1.3(a). Conversely, if $w \in W$ and $B \cup C(w) = P$ is a subgroup then it is a special subgroup, by Theorem 1.10. The proof of Theorem 1.10 shows that $W' = W \cap S$ and

$$W' = \{w' \in W \mid C(w') \subseteq P\}.$$

Since $C(w) \subseteq P$ we have $w \in W' \subseteq S$, as required. $\square$

EXAMPLE 1.12. *SL_n and $GL_n K$ for a p-adic local field*
Let $G = GL_n K$ with $n \geq 2$ and K a p-adic local field. Let B denote the inverse image of the upper triangular subgroup under the homomorphism $GL_n \mathcal{O}_K \longrightarrow GL_n \mathcal{O}_K / (\pi_K)$. Let N denote the subgroup of monomial matrices in $GL_n K$; that is, the matrices which have precisely one non-zero entry in each row and column.

When $G = SL_n K$ we set $B = B \cap SL_n K$ and $N = N \cap SL_n K$. We are going to construct a BN-pair from this example, following ([35] pp. 128–138).

PROPOSITION 1.13. ([35] p. 129)
Let $G = SL_n K, GL_n K$ as in Example 1.12. Then G is generated by N and the elementary matrices in $SL_n \mathcal{O}_K, GL_n \mathcal{O}_K$, respectively.

Proof

Let $X = (x_{i,j})$ be a matrix in G. Choose a matrix entry $x_{i,j}$ such that the valuation $v_K(x_{i,j})$ is minimal. Then pivot to clear out every other non-zero entry in the i-th row and j-th column. This can be done using elementary matrices in $SL_n \mathcal{O}_K$. Now ignore the i-th row and j-th column and repeat the process to eventually produce a monomial matrix, which will be in $SL_n K$ if X was. $\square$

COROLLARY 1.14. ([35] p. 130)
Let $G = SL_n K, GL_n K$ as in Example 1.12. Then G is generated by N and B.

Proof

The subgroup B contains all the upper triangular matrices in $G \cap GL_n \mathcal{O}_K$ and N contains all the permutation matrices. Therefore the group generated by B and N contains N and all the elementary matrices in $G \cap GL_n \mathcal{O}_K$, respectively. $\square$

1.15. *The BN-pair of $SL_n K$ when K is a p-adic local field*
Continuing with $G = SL_n K, GL_n K$ as in Example 1.12 we set $T = B \cap N$ and $W = N/T$. Therefore T is the subgroup of diagonal matrices in $G \cap GL_n \mathcal{O}_K$. If T_K is the subgroup of all the diagonal matrices in G then the quotient group N/T_K is isomorphic to the symmetric group, Σ_n, in both cases. Therefore there is a split surjection from W/T to Σ_n whose kernel consists of $(K^*/\mathcal{O}_K^*)^m \cong \mathbb{Z}^m$ where $m = n - 1$ if $G = SL_n K$ and $m = n$ if $G = GL_n K$.

Therefore W/T is given isomorphic to the semi-direct product $W/T \cong \Sigma_m \propto \mathbb{Z}^m$.

When $n = 2$ and $G = SL_2 K$ we have $W \cong \Sigma_2 \propto \mathbb{Z}$ which is generated by

$$s_1 = \begin{pmatrix} 0 & -1 \\ 1 & 0 \end{pmatrix} \text{ and } s_2 = \begin{pmatrix} 0 & -\pi_K^{-1} \\ \pi_K & 0 \end{pmatrix}$$

so in this case we shall set $S = \{s_1, s_2\}$.

With this definition $(SL_2 K, B, N, S)$ is a Tits system ([35] pp. 131–132).

When $n \geq 3$ take $S = \{s_1, s_2, \dots, s_n\}$ where s_i for $1 \leq i \leq n - 1$ are the involutions given by matrices made using the first and the $(i + 1)$-th elements in the standard basis for K^n in the same manner as the first and second standard basis elements were used to construct s_1 for $SL_2 K$. Define s_n to be the matrix

$$
s_n = \begin{pmatrix} 1 & 0 & 0 & \ldots & -\pi_K^{-1} \\ 0 & 1 & 0 & \ldots & 0 \\ \vdots & \vdots & \vdots & \vdots & \vdots \\ \pi_K & 0 & 0 & \ldots & 0 \end{pmatrix}.
$$

Then $S = \{s_1, s_2, \ldots, s_n\}$ generates W and $(SL_n K, B, N, S)$ is a Tits system ([35] pp. 135–137).

2. Buildings and BN-pairs

2.1. *The building associated to a BN-pair*

A Coxeter complex is a simplicial complex associated to a pair (W, S) where S is a set of generators for a group W, each of order 2. The cosets of the form $w\langle S' \rangle$ with $w \in W$ and $S' \subseteq S$ form a poset under inclusion. The poset with the same objects but the opposite ordering is a simplicial complex which is called the Coxeter complex associated to (W, S). More generally a Coxeter complex will mean any simplicial complex which is simplicially isomorphic to the Coxeter complex associated to (W, S).

A building ([35] p. 76) is a simplicial complex Δ which is the union of subcomplexes Σ called apartments which satisfy the following axioms:

(**B0**) Each apartment Σ is a Coxeter complex.

(**B1**) For any two simplices $A, B \in \Delta$ there is an apartment containing both of them.

(**B2**) If Σ and Σ' are two apartments containing simplices $A, B \in \Delta$ then there is a simplicial isomorphism $\Sigma \xrightarrow{\cong} \Sigma'$ fixing A and B pointwise. In particular, any two apartments are isomorphic.

Let (G, B, N, S) be a Tits system. Consider the poset of left cosets gP for $g \in G$ and P a proper special subgroup as in Definition 1.8 endowed with the opposite partial ordering to that given by inclusion. The building associated to this BN-pair consists of the simplicial complex given by this oppositely ordered poset of cosets of proper special subgroups[1]. It is denoted by $\Delta(G, B)$.

---

[1] The example in §2.2 of how the building for $SL_2 K$ gives rise to the tree for $GL_2 K$, when K is a local field, illustrates the fact that only proper special subgroups are used to construct the building.

The subgroup N is used to define the apartments of $\Delta(G, B)$. The fundamental apartment $\Sigma \subseteq \Delta(G, B)$ is defined to be the subcomplex whose vertices are the special cosets of the form wP with $w \in W$. Since every special subgroup contains B this is the set of vertices wP with a coset representative in N. Define the set $\mathcal{A}$ of apartments to consist of all G-translates of the fundamental apartment.

In a BN-pair every special subgroup is its own normaliser and no two special subgroups are conjugate ([35] p. 111). Furthermore $\Delta(G, B)$ is a (thick) building with apartment system $\mathcal{A}$ on which the G-action is (type-preserving[2] and) strongly transitive ([35] p. 112).

A maximal element of an apartment is called a chamber.

Since there is a bijection between cosets of proper special subgroups gP and their conjugates gPg^{-1} the building is also describeable as the simplicial complex given by the oppositely ordered poset of conjugates of proper special subgroups of G.

2.2. The building associated to SL_nK when K is a p-adic local field

The building Δ associated to SL_nK is the tree of classes of lattices when $n = 2$. Similarly ([35] p. 137) the building of SL_nK is made from lattices in K^n. The fundamental chamber is the simplex with vertices $(e_1, \dots, e_i, \pi_K e_{i+1}, \dots, \pi_K e_n)$ for $1 \leq i \leq n$ and $\{e_i\}$ the standard basis. The resulting building is not spherical and therefore is contractible ([35] p. 94).

The action of SL_nK extends to an action of GL_nK which, as in the case of $n = 2$, does not preserve type but a mild barycentric subdivision renders the GL_nK-action simplicial.

Let us look in detail at the cases of SL_2K and SL_3K.

EXAMPLE 2.3. SL_2K and GL_2K when K is a local field

We have

$$B = \left\{ \begin{pmatrix} \alpha & \beta \\ \pi_K\gamma & \delta \end{pmatrix} \mid \alpha, \beta, \gamma, \delta \in \mathcal{O}_K,\ \alpha\delta - \pi_K\beta\gamma = 1 \right\} \subseteq SL_2\mathcal{O}_K$$

and

$$W \cong C_2 \propto \mathbb{Z} = \langle s_1 = \begin{pmatrix} 0 & -1 \\ 1 & 0 \end{pmatrix}, s_2 = \begin{pmatrix} 0 & -\pi_K^{-1} \\ \pi_K & 0 \end{pmatrix} \rangle.$$

Consider the group Bs_1B. Since B contains the upper triangular subgroup of $SL_2\mathcal{O}_K$ and $Bs_1Bs_1B \subseteq Bs_1B$ we must have the lower triangular

---

[2]By way of example the type of a vertex in the tree for GL_2 of a local field is the distance (mod 2) from the vertex to the vertex represented by the lattice $\mathcal{O}_K \oplus \mathcal{O}_K$. In general we shall not need the notion of type or of "thickness" here.

subgroup of $SL_2 \mathcal{O}_K$ also contained in $Bs_1 B$. Suppose that

$$\begin{pmatrix} u & v \\ w & z \end{pmatrix} \in SL_2 \mathcal{O}_K$$

and consider the equation

$$\begin{pmatrix} a\alpha + c\beta & \beta a^{-1} \\ c\alpha^{-1} & \alpha^{-1} a^{-1} \end{pmatrix}$$

$$= \begin{pmatrix} \alpha & \beta \\ 0 & \alpha^{-1} \end{pmatrix} \begin{pmatrix} a & 0 \\ c & a^{-1} \end{pmatrix}$$

$$= \begin{pmatrix} u & v \\ w & z \end{pmatrix}.$$

If $z = 0$ then

$$\begin{pmatrix} 0 & -1 \\ 1 & 0 \end{pmatrix} \begin{pmatrix} u & v \\ v^{-1} & 0 \end{pmatrix} = \begin{pmatrix} v^{-1} & 0 \\ u & v \end{pmatrix}$$

so that

$$\begin{pmatrix} u & v \\ v^{-1} & 0 \end{pmatrix} \in Bs_1 B.$$

Otherwise we have

$$\begin{pmatrix} z^{-1}\pi_K^m(1 + vw) & v\pi_K^{-m} \\ w\pi_K^m & z\pi_K^{-m} \end{pmatrix}$$

$$= \begin{pmatrix} \alpha & av\pi_K^{-m} \\ 0 & \alpha^{-1} \end{pmatrix} \begin{pmatrix} a & 0 \\ \alpha w\pi_K^m & a^{-1} \end{pmatrix}$$

and

$$\begin{pmatrix} z^{-1}\pi_K^m(1 + vw) & v\pi_K^{-m} \\ w\pi_K^m & z\pi_K^{-m} \end{pmatrix} \begin{pmatrix} \pi_K^{-m} & 0 \\ 0 & \pi_K^m \end{pmatrix}$$

$$= \begin{pmatrix} u & v \\ w & z \end{pmatrix}$$

since $1 = uz - vw$ so that $u = z^{-1}(1 + vw)$.

So far then any $X \in SL_2\mathcal{O}_K$ lies in

$$Bs_1B \begin{pmatrix} \pi_K^{-1} & 0 \\ 0 & \pi_K \end{pmatrix}^m$$

for some $m \geq 0$. If $m = 0$ then $X \in Bs_1B$. If $m > 0$ then

$$X = \begin{pmatrix} u & v \\ w & z \end{pmatrix}$$

and $z \in \mathcal{O}_K\pi_K^m$. Since $uz - 1 = vw$ we have $v, w \in \mathcal{O}_K^*$ so that

$$\begin{pmatrix} 1 & 0 \\ -z/v & 1 \end{pmatrix} \begin{pmatrix} u & v \\ w & z \end{pmatrix} = \begin{pmatrix} u & v \\ w - zu/v & 0 \end{pmatrix}$$

which lies in Bs_1B as does the left-hand matrix so $X \in Bs_1B$ in this remaining case, too. Hence we have $SL_2\mathcal{O}_K \subseteq Bs_1B \subseteq SL_2\mathcal{O}_K$.

Hence we have verified that $Bs_1B = SL_2\mathcal{O}_K$. Next we shall determine the identity of Bs_2B.

We begin by observing that if $a, b, c, d \in \mathcal{O}_K$ and $1 = ad - bc$ then

$$X = \begin{pmatrix} a & b\pi_K^{-1} \\ c\pi_K & d \end{pmatrix} \in SL_2K$$

and

$$X^{-1} = \begin{pmatrix} d & -b\pi_K^{-1} \\ -c\pi_K & a \end{pmatrix} \in SL_2K$$

is a matrix of the same form so that these matrices form a subgroup of SL_2K because, if $a', b', c', d' \in \mathcal{O}_K$ and $1 = a'd' - b'c'$,

$$\begin{pmatrix} a & b\pi_K^{-1} \\ c\pi_K & d \end{pmatrix} \begin{pmatrix} a' & b'\pi_K^{-1} \\ c'\pi_K & d' \end{pmatrix}$$

$$= \begin{pmatrix} aa' + bc' & ab'\pi_K^{-1} + bd'\pi_K^{-1} \\ ac'\pi_K + dc'\pi_K & cb' + dd' \end{pmatrix}.$$

Pro tem let us denote this subgroup by H so that $B \subseteq H, s_2 \in H$ and therefore $Bs_2B \subseteq H$. We shall establish the reverse inclusion.

Suppose that $a, b, c, d \in \mathcal{O}_K$ and $1 = ad - bc$ and that

$$X = \begin{pmatrix} a & b\pi_K^{-1} \\ c\pi_K & d \end{pmatrix} \in H.$$

If $b = 0$ then $X \in B$ and B lies inside every special subgroup so $X \in Bs_2B$. If $d = d'\pi_K$ with $d' \in \mathcal{O}_K$ then

$$
\begin{pmatrix} 0 & -\pi_K^{-1} \\ \pi_K & 0 \end{pmatrix}
\begin{pmatrix} c & d' \\ -a\pi_K & -b \end{pmatrix}
= \begin{pmatrix} a & b\pi_K^{-1} \\ c\pi_K & d \end{pmatrix} = X.
$$

It remains to consider X when $d \in \mathcal{O}_K^*$. Applying the same computation to X^{-1} shows that $X \in Bs_2B$ unless $a \in \mathcal{O}_K^*$, too.

In $s_2 B s_2 \subseteq Bs_2B$ we have

$$
\begin{pmatrix} 0 & -\pi_K^{-1} \\ \pi_K & 0 \end{pmatrix}
\begin{pmatrix} -d & c' \\ b\pi_K & -a \end{pmatrix}
\begin{pmatrix} 0 & -\pi_K^{-1} \\ \pi_K & 0 \end{pmatrix}
$$

$$
= \begin{pmatrix} -b & a\pi_K^{-1} \\ -d\pi_K & c'\pi_K \end{pmatrix}
\begin{pmatrix} 0 & -\pi_K^{-1} \\ \pi_K & 0 \end{pmatrix}
$$

$$
= \begin{pmatrix} a & b\pi_K^{-1} \\ c'\pi_K^2 & d \end{pmatrix}.
$$

Then

$$
\begin{pmatrix} a & b\pi_K^{-1} \\ c'\pi_K^2 & d \end{pmatrix}
\begin{pmatrix} d & -b\pi_K^{-1} \\ -c\pi_K & a \end{pmatrix}
$$

$$
= \begin{pmatrix} 1 & 0 \\ d(c'\pi_K^2 - c\pi_K) & 1 + b(c - c'\pi_K) \end{pmatrix}
$$

which is in B and hence in Bs_2B. Therefore we have $Y, Y' \in Bs_2B$ such that $YX^{-1} = Y'$ and therefore $X \in Bs_2B$, as required.

The only other special subgroup is B itself and we have $B = SL_2\mathcal{O}_K \bigcap H$. The building of SL_2K is the opposite poset of the set of SL_2K-conjugates of $B, H, SL_2\mathcal{O}_K$.

From §2.1 we know that each of $B, H, SL_2\mathcal{O}_K$ is its own normaliser and since $SL_2K = BN$ the conjugates of these groups are contained in the sets gBg^{-1}, $gSL_2\mathcal{O}_K g^{-1}$, gHg^{-1} as g varies through coset representatives of $W = N/B \cap N$. The elements of W are represented by the matrices

$$
(s_1^3 s_2)^m = \begin{pmatrix} \pi_K^m & 0 \\ 0 & \pi_K^{-m} \end{pmatrix}, \quad
s_1(s_1^3 s_2)^m = \begin{pmatrix} 0 & -\pi_K^{-m} \\ \pi_K^m & 0 \end{pmatrix}
$$

for $m \in \mathbb{Z}$. It is clear that the distinct conjugates of B are precisely

$$\begin{pmatrix} \pi_K^m & 0 \\ 0 & \pi_K^{-m} \end{pmatrix} B \begin{pmatrix} \pi_K^m & 0 \\ 0 & \pi_K^{-m} \end{pmatrix}^{-1},$$

$$\begin{pmatrix} 0 & -\pi_K^{-m} \\ \pi_K^m & 0 \end{pmatrix} B \begin{pmatrix} 0 & -\pi_K^{-m} \\ \pi_K^m & 0 \end{pmatrix}^{-1}$$

as m runs through the integers. Explicitly these subgroups have the following forms.

For $\alpha, \beta, \gamma, \delta \in \mathcal{O}_K$ and $1 = \alpha\delta - \beta\gamma\pi_K$ we have

$$\begin{pmatrix} 0 & -\pi_K^{-m} \\ \pi_K^m & 0 \end{pmatrix} \begin{pmatrix} \alpha & \beta \\ \gamma\pi_K & \delta \end{pmatrix} \begin{pmatrix} 0 & \pi_K^{-m} \\ -\pi_K^m & 0 \end{pmatrix}$$

$$= \begin{pmatrix} -\gamma\pi_K^{1-m} & -\delta\pi_K^{-m} \\ \alpha\pi_K^m & \beta\pi_K^m \end{pmatrix} \begin{pmatrix} 0 & \pi_K^{-m} \\ -\pi_K^m & 0 \end{pmatrix}$$

$$= \begin{pmatrix} \delta & -\gamma\pi_K^{1-2m} \\ -\beta\pi_K^{2m} & \alpha \end{pmatrix}$$

and

$$\begin{pmatrix} \pi_K^m & 0 \\ 0 & \pi_K^{-m} \end{pmatrix} \begin{pmatrix} \alpha & \beta \\ \gamma\pi_K & \delta \end{pmatrix} \begin{pmatrix} \pi_K^{-m} & 0 \\ 0 & \pi_K^m \end{pmatrix}$$

$$= \begin{pmatrix} \alpha\pi_K^m & \beta\pi_K^m \\ \gamma\pi_K^{1-m} & \delta\pi_K^{-m} \end{pmatrix} \begin{pmatrix} \pi_K^{-m} & 0 \\ 0 & \pi_K^m \end{pmatrix}$$

$$= \begin{pmatrix} \alpha & \beta\pi_K^{2m} \\ \gamma\pi_K^{1-2m} & \delta \end{pmatrix}.$$

Therefore we have

$$(s_1^3 s_2)^m B (s_1^3 s_2)^{-m}$$

$$= \left\{ \begin{pmatrix} \alpha & \beta\pi_K^{2m} \\ \gamma\pi_K^{1-2m} & \delta \end{pmatrix} \mid \alpha, \beta, \gamma, \delta \in \mathcal{O}_K, \ 1 = \alpha\delta - \beta\gamma\pi_K \right\}$$

and

$$s_1(s_1^3 s_2)^m B(s_1^3 s_2)^{-m} s_1^{-1}$$

$$= \left\{ \begin{pmatrix} \delta & -\gamma \pi_K^{1-2m} \\ -\beta \pi_K^{2m} & \alpha \end{pmatrix} \mid \alpha, \beta, \gamma, \delta \in \mathcal{O}_K, \ 1 = \alpha\delta - \beta\gamma\pi_K \right\}.$$

Similarly the distinct conjugates of $SL_2\mathcal{O}_K$ are contained (with some repetition) in the set

$$(s_1^3 s_2)^m SL_2\mathcal{O}_K (s_1^3 s_2)^{-m} \bigcup s_1(s_1^3 s_2)^m SL_2\mathcal{O}_K (s_1^3 s_2)^{-m} s_1^{-1}$$

which are explicitly given by

$$(s_1^3 s_2)^m SL_2\mathcal{O}_K (s_1^3 s_2)^{-m}$$

$$= \left\{ \begin{pmatrix} \alpha & \beta \pi_K^{2m} \\ \gamma \pi_K^{-2m} & \delta \end{pmatrix} \mid \alpha, \beta, \gamma, \delta \in \mathcal{O}_K, \ 1 = \alpha\delta - \beta\gamma \right\}$$

and

$$s_1(s_1^3 s_2)^m SL_2\mathcal{O}_K (s_1^3 s_2)^{-m} s_1^{-1}$$

$$= \left\{ \begin{pmatrix} \delta & -\gamma \pi_K^{-2m} \\ -\beta \pi_K^{2m} & \alpha \end{pmatrix} \mid \alpha, \beta, \gamma, \delta \in \mathcal{O}_K, \ 1 = \alpha\delta - \beta\gamma \right\}.$$

However when $m = 0$ we have $s_1 SL_2\mathcal{O}_K s_1^{-1} = SL_2\mathcal{O}_K$ and also for ξ a representative of an element in the residue field $\mathcal{O}_K/(\pi_K)$ then

$$\begin{pmatrix} 1 & 0 \\ \xi & 1 \end{pmatrix}$$

lies in $SL_2\mathcal{O}_K - B$ so that $SL_2\mathcal{O}_K$ contains $|\mathcal{O}_K/(\pi_K)| + 1$ conjugates of B, which agrees with the number of edges out of a vertex in the tree (see Chapter Two §4.1).

For H we have a similar assertion (with some repetition) and

$$(s_1^3 s_2)^m H (s_1^3 s_2)^{-m}$$

$$= \left\{ \begin{pmatrix} \alpha & \beta \pi_K^{2m-1} \\ \gamma \pi_K^{1-2m} & \delta \end{pmatrix} \mid \alpha, \beta, \gamma, \delta \in \mathcal{O}_K, \ 1 = \alpha\delta - \beta\gamma \right\}$$

and

$$s_1(s_1^3 s_2)^m H(s_1^3 s_2)^{-m} s_1^{-1}$$

$$= \left\{ \begin{pmatrix} \delta & -\gamma\pi_K^{1-2m} \\ -\beta\pi_K^{2m-1} & \alpha \end{pmatrix} \mid \alpha,\beta,\gamma,\delta \in \mathcal{O}_K,\ 1 = \alpha\delta - \beta\gamma \right\}.$$

Since, in GL_2K, we have

$$H = \begin{pmatrix} \pi_K^{-1} & 0 \\ 0 & 1 \end{pmatrix} SL_2\mathcal{O}_K \begin{pmatrix} \pi_K & 0 \\ 0 & 1 \end{pmatrix}$$

one finds that H also contains $|\mathcal{O}_K/(\pi_K)| + 1$ conjugates of B.

We see that $SL_2\mathcal{O}_K$ is the stabiliser of the lattice $\mathcal{O}_K \oplus \mathcal{O}_K$ (as column vectors with left multiplication by G. Also H is the stabiliser of $\mathcal{O}_K \oplus \pi_K\mathcal{O}_K$ since

$$\begin{pmatrix} \alpha & \beta\pi_K^{-1} \\ \gamma\pi_K & \delta \end{pmatrix}\begin{pmatrix} u \\ v\pi_K \end{pmatrix} = \begin{pmatrix} \alpha u + \beta v \\ u\gamma\pi_K + \delta v\pi_K \end{pmatrix}.$$

The inclusion of lattices is opposite to the inclusion of stabilisers so the building may be equivalently described in terms of SL_2K-translates of lattices. The action extends to GL_2K but only gives exactly the same simplicial complex if we use lattices up to homothety (i.e. multiplication by K^*-scalars) in which case we obtain the GL_2K-action on the barycentric subdivision of the tree which we used to make the monomial resolution.

EXAMPLE 2.4. SL_3K and GL_3K when K is a local field

$$B = \left\{ X = \begin{pmatrix} a & b & c \\ d\pi_K & e & f \\ g\pi_K & h\pi_K & i \end{pmatrix} \mid a,b,c,d,e,f,g,h,i \in \mathcal{O}_K \right\} \subseteq SL_3\mathcal{O}_K$$

and

$$W \cong \Sigma_3 \propto (\mathbb{Z} \oplus \mathbb{Z}).$$

The generators of the symmetric group Σ_3 in W are represented by

$$s_1 = \begin{pmatrix} 0 & -1 & 0 \\ 1 & 0 & 0 \\ 0 & 0 & 1 \end{pmatrix} \text{ and } s_2 = \begin{pmatrix} 1 & 0 & 0 \\ 0 & 0 & -1 \\ 0 & 1 & 0 \end{pmatrix}$$

and the third member of the set S is

$$s_3 = \begin{pmatrix} 0 & 0 & -\pi_K^{-1} \\ 0 & 1 & 0 \\ \pi_K & 0 & 0 \end{pmatrix}.$$

The special subgroups are

$$B, B\langle s_1 \rangle B, B\langle s_2 \rangle B, B\langle s_3 \rangle B, B\langle s_1, s_2 \rangle B, B\langle s_1, s_3 \rangle B, B\langle s_2, s_3 \rangle B.$$

Arguing as in the SL_2K case we find that $B\langle s_1, s_2 \rangle B = SL_3\mathcal{O}_K$. These groups are related to the stabilisers of lattices in K^3. For example the lattice of column vectors

$$L_1 = \left\{ \begin{pmatrix} \alpha \\ \beta \\ \gamma \end{pmatrix} \mid \alpha, \beta, \gamma \in \mathcal{O}_K \right\}$$

is stabilised under left multiplication by $B\langle s_1, s_2 \rangle B$. The lattice

$$L_2 = \left\{ \begin{pmatrix} \alpha \\ \beta \\ \pi_K \gamma \end{pmatrix} \mid \alpha, \beta, \gamma \in \mathcal{O}_K \right\}$$

is stabilised by $B\langle s_1 \rangle B$ and $B\langle s_3 \rangle B$ because

$$\begin{pmatrix} a & b & c \\ d\pi_K & e & f \\ \pi_K g & \pi_K h & j \end{pmatrix} \begin{pmatrix} 0 & -1 & 0 \\ 1 & 0 & 0 \\ 0 & 0 & 1 \end{pmatrix} \begin{pmatrix} a' & b' & c' \\ d'\pi_K & e' & f' \\ \pi_K g' & \pi_K h' & j' \end{pmatrix} \begin{pmatrix} \alpha \\ \beta \\ \pi_K \gamma \end{pmatrix}$$

$$= \begin{pmatrix} a & b & c \\ d\pi_K & e & f \\ \pi_K g & \pi_K h & j \end{pmatrix} \begin{pmatrix} 0 & -1 & 0 \\ 1 & 0 & 0 \\ 0 & 0 & 1 \end{pmatrix} \begin{pmatrix} a'\alpha + b'\beta + c'\pi_K\gamma \\ d'\pi_K\alpha + e'\beta + f'\pi_K\gamma \\ \pi_K g'\alpha + \pi_K h'\beta + j'\pi_K\gamma \end{pmatrix}$$

$$= \begin{pmatrix} a & b & c \\ d\pi_K & e & f \\ \pi_K g & \pi_K h & j \end{pmatrix} \begin{pmatrix} \alpha'' \\ \beta'' \\ \pi_K\gamma'' \end{pmatrix} = \begin{pmatrix} a\alpha'' + b\beta'' + c\pi_K\gamma'' \\ d\pi_K\alpha'' + e\beta'' + f\pi_K\gamma'' \\ \pi_K g\alpha'' + \pi_K h\beta'' + j\pi_K\gamma'' \end{pmatrix}$$

and

$$
\begin{pmatrix} 0 & 0 & -\pi_K^{-1} \\ 0 & 1 & 0 \\ \pi_K & 0 & 0 \end{pmatrix} \begin{pmatrix} \alpha \\ \beta \\ \pi_K\gamma \end{pmatrix} = \begin{pmatrix} -\gamma \\ \beta \\ \pi_K\alpha \end{pmatrix}.
$$

Therefore $B\langle s_1, s_3\rangle B$ is the stabiliser of the above lattice. Similarly $B\langle s_2, s_3\rangle B$ is the stabiliser of the lattice

$$
L_3 = \left\{ \begin{pmatrix} \alpha \\ \pi_K\beta \\ \pi_K\gamma \end{pmatrix} \mid \alpha, \beta, \gamma \in \mathcal{O}_K \right\}.
$$

These facts are sketchily mentioned in ([**35**] p. 137).

The incidence condition on two lattices L, L' which implies they define a 1-simplex in the building is $\pi_K L \subseteq L' \subseteq L$. We have incidence relations

$$
\pi_K L_1 \subseteq L_2 \subseteq L_1
$$

$$
\pi_K L_1 \subseteq L_3 \subseteq L_1
$$

$$
\pi_K L_2 \subseteq L_3 \subseteq L_2
$$

and the fundamental simplex in the building for $SL_3 K$ is as shown below, with stabilisers adjacent to the simplex which they stabilise.

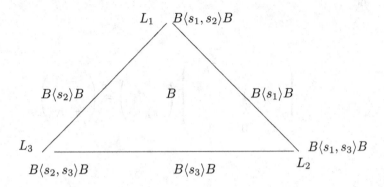

Consider the matrix

$$W = \begin{pmatrix} 0 & -1 & 0 \\ 1 & 0 & 0 \\ 0 & 0 & 1 \end{pmatrix} \begin{pmatrix} 1 & 0 & 0 \\ 0 & \pi_K^{-1} & 0 \\ 0 & 0 & 1 \end{pmatrix} = \begin{pmatrix} 0 & -\pi_K^{-1} & 0 \\ 1 & 0 & 0 \\ 0 & 0 & 1 \end{pmatrix}.$$

Therefore

$$W \begin{pmatrix} \alpha \\ \beta\pi_K \\ \gamma\pi_K \end{pmatrix} = \begin{pmatrix} 0 & -\pi_K^{-1} & 0 \\ 1 & 0 & 0 \\ 0 & 0 & 1 \end{pmatrix} \begin{pmatrix} \alpha \\ \beta\pi_K \\ \gamma\pi_K \end{pmatrix} = \begin{pmatrix} -\beta \\ \alpha \\ \gamma\pi_K \end{pmatrix}$$

so that $W(L_3) = L_2$. Similarly

$$W \begin{pmatrix} \alpha \\ \beta \\ \gamma\pi_K \end{pmatrix} = \begin{pmatrix} -\beta\pi_K^{-1} \\ \alpha \\ \gamma\pi_K \end{pmatrix}$$

and

$$W \begin{pmatrix} \alpha \\ \beta \\ \gamma \end{pmatrix} = \begin{pmatrix} -\beta\pi_K^{-1} \\ \alpha \\ \gamma \end{pmatrix}.$$

Therefore

$$W(L_2) = \{ \begin{pmatrix} -\beta\pi_K^{-1} \\ \alpha \\ \gamma\pi_K \end{pmatrix} \mid \alpha, \beta, \gamma \in \mathcal{O}_K \} = L_4$$

and

$$W(L_1) = \{ \begin{pmatrix} -\beta\pi_K^{-1} \\ \alpha \\ \gamma \end{pmatrix} \mid \alpha, \beta, \gamma \in \mathcal{O}_K \} = \pi_K^{-1} L_3.$$

We have incidence relations

$$\pi_K(\pi_K^{-1} L_3) = L_3 \subseteq L_1 \subseteq \pi_K^{-1} L_3$$

$$\pi_K(\pi_K^{-1} L_3) = L_3 \subseteq L_2 \subseteq \pi_K^{-1} L_3$$

$$\pi_K(\pi_K^{-1} L_3) = L_3 \subseteq L_4 \subseteq \pi_K^{-1} L_3$$

so that $\{L_2, \pi_K^{-1} L_3, L_4\}$ is a 2-simplex in the building and W maps the fundamental simplex to it simplicially.

The action by $SL_3 K$ preserves the type of a lattice. This is the valuation (modulo 3) of the determinant whose columns are an $\mathcal{O}_K$-basis for the lattice. Hence we find that

lattice	type mod 3
L_1	0
L_2	1
L_3	2
L_4	0
$\pi_K^{-1} L_3$	2

This means that $W(L_1, L_2, L_3) = (\pi_K^{-1} L_3, L_4, L_2)$ acts like $(0, 1, 2) \mapsto (2, 0, 1)$ on types.

The action by $SL_3 K$ on the building, whose vertices are represented by lattices (or by their stabilisers) is simplicial. This action extends to $GL_3 K$ if we represent vertices by the homothety classes of lattices but the action is not simplicial, because U rotates the fundamental simplex. However the $GL_3 K$-action becomes simplicial if we barycentrically subdivide the fundamental simplex and all its translates by adding the centroid as a 0-simplex and the three 1-simplices given by the lines from the vertices to the centroid. All these added simplices are stabilised only by $K^* \cdot B$.

The calculation

$$
U \begin{pmatrix} \alpha \\ \beta \\ \gamma \end{pmatrix} = \begin{pmatrix} 0 & \pi_K^{-1} & 0 \\ 0 & 0 & \pi_K^{-1} \\ 1 & 0 & 0 \end{pmatrix} \begin{pmatrix} \alpha \\ \beta \\ \gamma \end{pmatrix} = \begin{pmatrix} \beta \pi_K^{-1} \\ \gamma \pi_K^{-1} \\ \alpha \end{pmatrix}
$$

shows that

$$ U(L_1) = \pi_K^{-1} L_2, \quad U(L_2) = \pi_K^{-1} L_3, \quad U(L_3) = L_1. $$

Therefore $U^3 = \pi_K^{-2}$ and $(uU)^3 = \pi_K$.

In ([**140**] p. 48) the building of $GL_3 K$ is described as a plane triangulated by equilateral triangles[3]. This description agrees with the above analysis. The action by U rotates the entire plane through $2\pi/3$ fixing only the barycentre of the fundamental simplex.

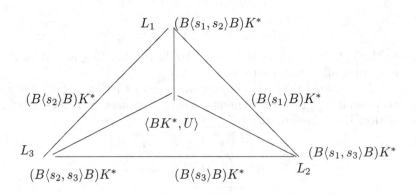

---

[3]This is actually only one apartment. Tits says that the building itself is obtained by "ramifying along every edge" of the triangulated plane. My thanks to Gerry Cliff for correcting me on this point.

Now we examine how U conjugates these stabilisers. We have

$$UXU^{-1} = \begin{pmatrix} 0 & \pi_K^{-1} & 0 \\ 0 & 0 & \pi_K^{-1} \\ 1 & 0 & 0 \end{pmatrix} \begin{pmatrix} a & b & c \\ d\pi_K & e & f \\ g\pi_K & h\pi_K & i \end{pmatrix} \begin{pmatrix} 0 & 0 & 1 \\ \pi_K & 0 & 0 \\ 0 & \pi_K & 0 \end{pmatrix}$$

$$= \begin{pmatrix} d & e\pi_K^{-1} & f\pi_K^{-1} \\ g & h & i\pi_K^{-1} \\ a & b & c \end{pmatrix} \begin{pmatrix} 0 & 0 & 1 \\ \pi_K & 0 & 0 \\ 0 & \pi_K & 0 \end{pmatrix}$$

$$= \begin{pmatrix} e & f & d \\ h\pi_K & i & g \\ b\pi_K & c\pi_K & a \end{pmatrix}$$

so that $UBU^{-1} = B$.

Also

$$Us_1U^{-1} = \begin{pmatrix} 0 & \pi_K^{-1} & 0 \\ 0 & 0 & \pi_K^{-1} \\ 1 & 0 & 0 \end{pmatrix} \begin{pmatrix} 0 & -1 & 0 \\ 1 & 0 & 0 \\ 0 & 0 & 1 \end{pmatrix} \begin{pmatrix} 0 & 0 & 1 \\ \pi_K & 0 & 0 \\ 0 & \pi_K & 0 \end{pmatrix}$$

$$= \begin{pmatrix} \pi_K^{-1} & 0 & 0 \\ 0 & 0 & \pi_K^{-1} \\ 0 & -1 & 0 \end{pmatrix} \begin{pmatrix} 0 & 0 & 1 \\ \pi_K & 0 & 0 \\ 0 & \pi_K & 0 \end{pmatrix}$$

$$= \begin{pmatrix} 0 & 0 & \pi_K^{-1} \\ 0 & 1 & 0 \\ -\pi_K & 0 & 0 \end{pmatrix}$$

so that $Us_1U^{-1} \in s_3 B \cap Bs_3$. In addition

$$
Us_3U^{-1} = \begin{pmatrix} 0 & \pi_K^{-1} & 0 \\ 0 & 0 & \pi_K^{-1} \\ 1 & 0 & 0 \end{pmatrix} \begin{pmatrix} 0 & 0 & -\pi_K^{-1} \\ 0 & 1 & 0 \\ \pi_K & 0 & 0 \end{pmatrix} \begin{pmatrix} 0 & 0 & 1 \\ \pi_K & 0 & 0 \\ 0 & \pi_K & 0 \end{pmatrix}
$$

$$
= \begin{pmatrix} 0 & \pi_K^{-1} & 0 \\ 1 & 0 & 0 \\ 0 & 0 & -\pi_K^{-1} \end{pmatrix} \begin{pmatrix} 0 & 0 & 1 \\ \pi_K & 0 & 0 \\ 0 & \pi_K & 0 \end{pmatrix}
$$

$$
= \begin{pmatrix} 1 & 0 & 0 \\ 0 & 0 & 1 \\ 0 & -1 & 0 \end{pmatrix}
$$

so that $Us_3U^{-1} \in s_2 B \cap Bs_2$.

Now let me describe what one gets from the action of $GL_3 K$ on homothety classes of lattices. The normaliser in $N_{GL_3 K}BK^* = \langle BK^*, U \rangle$ and $N_{GL_3 K}BK^*/BK^*$ is a cyclic group of order three generated by the image of U. Let $\lambda : N_{GL_3 K}BK^* \longrightarrow k^*$ be the resulting character of order three.

The non-simplicially subdivided building gives an exact sequence of admissibles

$$
0 \longrightarrow c - \text{Ind}_{\langle BK^*, U \rangle}^{GL_3 K}(k_\lambda) \longrightarrow c - \text{Ind}_{B\langle s_1 \rangle BK^*}^{GL_3 K}(k)
$$

$$
\longrightarrow c - \text{Ind}_{B\langle s_1, s_2 \rangle BK^*}^{GL_3 K}(k) \longrightarrow k \longrightarrow 0.
$$

When we make the simplicial subdivision we obtain

$$
0 \longrightarrow c - \text{Ind}_{BK^*}^{GL_3 K}(k) \longrightarrow c - \text{Ind}_{BK^*}^{GL_3 K}(k) \oplus c - \text{Ind}_{B\langle s_1 \rangle BK^*}^{GL_3 K}(k)
$$

$$
\longrightarrow c - \text{Ind}_{B\langle s_1, s_2 \rangle BK^*}^{GL_3 K}(k) \oplus c - \text{Ind}_{\langle BK^*, U \rangle}^{GL_3 K}(k) \longrightarrow k \longrightarrow 0.
$$

Next we take four bar-monomial resolutions of restrictions of V:

$$W_{V,*,BK^*} \xrightarrow{\epsilon} V$$

$$W_{V,*,B\langle s_1 \rangle BK^*} \xrightarrow{\epsilon_1} V$$

$$W_{V,*,B\langle s_1,s_2 \rangle BK^*} \xrightarrow{\epsilon_0} V$$

$$W_{V,*,\langle BK^*,U \rangle} \xrightarrow{\epsilon_0'} V.$$

By analogy with the $GL_2 K$ case we construct monomial morphisms of chain complexes $\tilde{\psi}$ which cover, via the augmentations of the monomial resolutions onto V, the 3-term complex obtained from the simplicial subdivision complex by replacing the k's by V's.

$$c - \operatorname{Ind}_{BK^*}^{GL_3 K}(W_{V,*,BK^*})$$

$$\tilde{\psi} \downarrow$$

$$c - \operatorname{Ind}_{BK^*}^{GL_3 K}(W_{V,*,BK^*}) \oplus c - \operatorname{Ind}_{B\langle s_1 \rangle BK^*}^{GL_3 K}(W_{V,*,B\langle s_1 \rangle BK^*})$$

$$\tilde{\psi} \downarrow$$

$$c - \operatorname{Ind}_{B\langle s_1,s_2 \rangle BK^*}^{GL_3 K}(W_{V,*,B\langle s_1,s_2 \rangle BK^*}) \oplus c - \operatorname{Ind}_{\langle BK^*,U \rangle}^{GL_3 K}(W_{V,*,\langle BK^*,U \rangle}).$$

In a manner similar to the $GL_2 K$ case we obtain the total complex of this double complex which gives a candidate for a $GL_3 K$-monomial resolution of V

$$\underline{M}_* \xrightarrow{\epsilon} V \longrightarrow 0$$

in which, for $i \geq 0$, $\underline{M}_i$ is given by

$$c - \operatorname{Ind}_{BK^*}^{GL_3 K}(W_{V,i-2,BK^*}) \oplus$$

$$c - \operatorname{Ind}_{BK^*}^{GL_3 K}(W_{V,i-1,BK^*}) \oplus c - \operatorname{Ind}_{B\langle s_1 \rangle BK^*}^{GL_3 K}(W_{V,i-1,B\langle s_1 \rangle BK^*}) \oplus$$

$$c - \operatorname{Ind}_{B\langle s_1,s_2 \rangle BK^*}^{GL_3 K}(W_{V,i,B\langle s_1,s_2 \rangle BK^*}) \oplus c - \operatorname{Ind}_{\langle BK^*,U \rangle}^{GL_3 K}(W_{V,i,\langle BK^*,U \rangle}).$$

Now we work towards monomial exactness for $GL_3 K$.

Let $(J,\phi) \in \mathcal{M}_{GL_3 K,\phi}$ satisfy $(K^*,\underline{\phi}) \leq (J,\phi)$ with J being compact open modulo the centre $\bar{K}^*$. We wish to examine exactness in the middle of

$$\underline{M}_{i+1}^{((J,\phi))} \longrightarrow \underline{M}_i^{((J,\phi))} \longrightarrow \underline{M}_{i-1}^{((J,\phi))}$$

for $i \geq 1$. A Line of $\underline{M}_i$ is given by one of the following types $g \otimes_{BK^*} L_2$, $g \otimes_{BK^*} L_1$, $g \otimes_{B\langle s_1 \rangle BK^*} L_1'$, $g \otimes_{B\langle s_1,s_2 \rangle BK^*} L_0$ or $g \otimes_{\langle BK^*,U \rangle} L_0'$. Here

$L_2, L_1, L_1', L_0, L_0'$ are Lines in $W_{V,i-2,BK^*}$, $W_{V,i-1,BK^*}$, $W_{V,i-1,B\langle s_1 \rangle BK^*}$, $W_{V,i,B\langle s_1,s_2 \rangle BK^*}$ or $W_{V,i,\langle BK^*,U \rangle}$ respectively.

The stabiliser pair of the above five types of Lines has the form $g(J', \phi')g^{-1}$ where $J' \subseteq BK^*$, $J' \subseteq BK^*$, $J' \subseteq B\langle s_1 \rangle BK^*$, $J' \subseteq B\langle s_1, s_2 \rangle BK^*$ or $J' \subseteq \langle BK^*, U \rangle$ respectively.

Consider the inclusions

$$\langle BK^*, U \rangle \geq BK^* \leq B\langle s_1 \rangle BK^* \leq B\langle s_1, s_2 \rangle BK^*.$$

PROPOSITION 2.5.

If $J \subseteq B\langle s_1, s_2 \rangle BK^*$ is conjugate to a subgroup of $\langle BK^*, U \rangle$ then J is conjugate to a subgroup of BK^*.

Proof

The following works for all $n \geq 2$.

$\langle BK^*, U \rangle$ is not subconjugate to $B\langle s_1, s_2 \rangle BK^* = GL_3 \mathcal{O}_K K^*$ because the valuation of the determinant of any matrix in the latter is congruent to zero modulo 3 but for any conjugate of U this is 1 (modulo 3). $\square$

2.6. *Example 2.4 continued*

Up to $GL_3 K$-conjugation we have one of the following four mutually exclusive (by Proposition 2.5) cases:

Case A: $J \subseteq BK^*$.

Case B: $J \subseteq B\langle s_1 \rangle BK^*$ but J is not conjugate to a subgroup of $\langle BK^*, U \rangle$.

Case C: $J \subseteq B\langle s_1, s_2 \rangle BK^*$ but J is not conjugate to a subgroup of either $\langle BK^*, U \rangle$ or $B\langle s_1 \rangle BK^*$.

Case D: $J \subseteq \langle BK^*, U \rangle$ but J is not conjugate to a subgroup of $B\langle s_1, s_2 \rangle BK^*$.

REMARK 2.7. The four cases in §2.6 exhaust all possibilities provided that every compact open modulo the centre subgroup of $GL_3 K$ is conjugate to a subgroup of at least one simplex stabiliser in the Bruhat-Tits building. For now we shall assume that this condition holds.

In fact, I am going to illustrate the verification only in Case A. To make matters even more tedious, I am going to do the verification by computing the spectral sequence of the double-complex $\underline{M}_*$ in "slow motion". I hope that proceeding in this manner will gradually introduce the technical homotopy theoretic properties of the building which make the verification work.

The final general proof will then be a sort of short "reprise" which experts would understand without the preparatory examination of the $GL_2 K$ and $GL_3 K$ situations.

2.8. *Monomial exactness in Case A*

If $J \subseteq BK^*$ then $\underline{M}_i^{((J,\phi))}$ for $i \geq 0$ is equal to

$$c - \operatorname{Ind}_{BK^*}^{GL_3K}(W_{V,i-2,BK^*})^{((J,\phi))}$$

$$\oplus c - \operatorname{Ind}_{BK^*}^{GL_3K}(W_{V,i-1,BK^*})^{((J,\phi))}$$

$$\oplus c - \operatorname{Ind}_{B\langle s_1 \rangle BK^*}^{GL_3K}(W_{V,i-1,B\langle s_1 \rangle BK^*})^{((J,\phi))}$$

$$\oplus c - \operatorname{Ind}_{B\langle s_1,s_2 \rangle BK^*}^{GL_3K}(W_{V,i,B\langle s_1,s_2 \rangle BK^*})^{((J,\phi))}$$

$$\oplus c - \operatorname{Ind}_{\langle BK^*,U \rangle}^{GL_3K}(W_{V,i,\langle BK^*,U \rangle})^{((J,\phi))}.$$

As in the GL_2K-case this is equal to

$$\oplus_{g^{-1}Jg \subseteq BK^*} g \otimes_{BK^*} W_{V,i-2,BK^*}^{((g^{-1}Jg,g^*(\phi)))}$$

$$\oplus_{g^{-1}Jg \subseteq BK^*} g \otimes_{BK^*} W_{V,i-1,BK^*}^{((g^{-1}Jg,g^*(\phi)))}$$

$$\oplus \oplus_{g^{-1}Jg \subseteq B\langle s_1 \rangle BK^*} g \otimes_{B\langle s_1 \rangle BK^*} W_{V,i-1,B\langle s_1 \rangle BK^*}^{((g^{-1}Jg,g^*(\phi)))}$$

$$\oplus_{g^{-1}Jg \subseteq B\langle s_1,s_2 \rangle BK^*} g \otimes_{B\langle s_1,s_2 \rangle BK^*} W_{V,i,B\langle s_1,s_2 \rangle BK^*}^{((g^{-1}Jg,g^*(\phi)))}$$

$$\oplus_{g^{-1}Jg \subseteq \langle BK^*,U \rangle} g \otimes_{\langle BK^*,U \rangle} W_{V,i,\langle BK^*,U \rangle}^{((g^{-1}Jg,g^*(\phi)))}.$$

Recall the chain complex $(\underline{M}_*, \underline{d})$ is the admissible monomial double complex and $(\underline{M}_*^{((J,\phi))}, \underline{d})$ is a sub-double complex. The differential is given by the formula $\underline{d} = d_Y \pm d$ where d_Y comes from the simplicial structure of the building Y and d complex from the differentials in the various bar-monomial resolutions which were used in Chapter Two §3.1 to construct $(\underline{M}_*, \underline{d})$.

We want to evaluate the homology groups $H_m(\underline{M}_*^{((J,\phi))}, \underline{d})$ for $m \geq 0$, expecting to discover that this is zero unless $m = 0$ in which case it is $V^{(J,\phi)}$. The homology of a double-complex may be computed using a spectral sequence. This spectral sequence is derived from a filtration $F^n\underline{M}_*^{((J,\phi))}$. Explicitly the filtration is defined by

$$F^2\underline{M}_i^{((J,\phi))} = C_{2.i-2} \oplus C_{1.i-1} \oplus C_{0.i}$$

$$F^1\underline{M}_i^{((J,\phi))} = C_{1.i-1} \oplus C_{0.i}$$

$$F^0\underline{M}_i^{((J,\phi))} = C_{0.i}$$

where

$$C_{2,i-2} = \oplus_{g^{-1}Jg\subseteq BK^*} \; g \otimes_{BK^*} W_{V,i-2,BK^*}^{((g^{-1}Jg,g^*(\phi)))}$$

$$C_{1,i-1} = \oplus_{g^{-1}Jg\subseteq BK^*} \; g \otimes_{BK^*} W_{V,i-1,BK^*}^{((g^{-1}Jg,g^*(\phi)))}$$

$$\oplus \oplus_{g^{-1}Jg\subseteq B\langle s_1\rangle BK^*} \; g \otimes_{B\langle s_1\rangle BK^*} W_{V,i-1,B\langle s_1\rangle BK^*}^{((g^{-1}Jg,g^*(\phi)))}$$

$$C_{0,i} = \oplus_{g^{-1}Jg\subseteq B\langle s_1,s_2\rangle BK^*} \; g \otimes_{B\langle s_1,s_2\rangle BK^*} W_{V,i,B\langle s_1,s_2\rangle BK^*}^{((g^{-1}Jg,g^*(\phi)))}$$

$$\oplus \oplus_{g^{-1}Jg\subseteq\langle BK^*,U\rangle} \; g \otimes_{\langle BK^*,U\rangle} W_{V,i,\langle BK^*,U\rangle}^{((g^{-1}Jg,g^*(\phi)))}.$$

This filtration is increasing with n so that

$$0 = F^{-1}\underline{M}_i^{((J,\phi))} \subseteq F^0\underline{M}_i^{((J,\phi))} \subseteq F^1\underline{M}_i^{((J,\phi))}$$

$$\subseteq F^2\underline{M}_i^{((J,\phi))} = F^3\underline{M}_i^{((J,\phi))} \cdots$$

and $\underline{d}(F^n\underline{M}_*^{((J,\phi))}) \subseteq F^n\underline{M}_{*-1}^{((J,\phi))}$.

The first step in the spectral sequence computation is to define

$$E_{n,i-n}^0 = F^n\underline{M}_i^{((J,\phi))}/F^{n-1}\underline{M}_i^{((J,\phi))}$$

and to compute the differential induced by $\underline{d}$

$$d^0 : E_{n,*-n}^0 \longrightarrow E_{n,*-n-1}^0$$

and the homology groups

$$E_{n,i-n}^1 = \frac{\text{Ker}(E_{n,i-n}^0 \longrightarrow E_{n,i-n-1}^0)}{\text{Im}(E_{n,i+1-n}^0 \longrightarrow E_{n,i-n}^0)}.$$

Since $d_Y(F^n\underline{M}_i^{((J,\phi))}) \subseteq F^{n-1}\underline{M}_{i-1}^{((J,\phi))}$ the differential d^0 is equal to that induced by d, the differential in the various bar-monomial resolutions. Therefore the only possibly non-zero d^0-differentials which we must calculate are

$$E_{2,i-2}^0 \cong \oplus_{g^{-1}Jg\subseteq BK^*} \; g \otimes_{BK^*} W_{V,i-2,BK^*}^{((g^{-1}Jg,g^*(\phi)))}$$

$$\downarrow d^0$$

$$E_{2,i-3}^0 \cong \oplus_{g^{-1}Jg\subseteq BK^*} \; g \otimes_{BK^*} W_{V,i-3,BK^*}^{((g^{-1}Jg,g^*(\phi)))},$$

$$E^0_{1,i-1} \cong \oplus_{g^{-1}Jg \subseteq BK^*} \ g \otimes_{BK^*} \ W^{((g^{-1}Jg,g^*(\phi)))}_{V,i-1,BK^*}$$

$$\oplus \oplus_{g^{-1}Jg \subseteq B\langle s_1 \rangle BK^*} \ g \otimes_{B\langle s_1 \rangle BK^*} \ W^{((g^{-1}Jg,g^*(\phi)))}_{V,i-1,B\langle s_1 \rangle BK^*}$$

$$\downarrow d^0$$

$$E^0_{1,i-2} \cong \oplus_{g^{-1}Jg \subseteq BK^*} \ g \otimes_{BK^*} \ W^{((g^{-1}Jg,g^*(\phi)))}_{V,i-2,BK^*}$$

$$\oplus \oplus_{g^{-1}Jg \subseteq B\langle s_1 \rangle BK^*} \ g \otimes_{B\langle s_1 \rangle BK^*} \ W^{((g^{-1}Jg,g^*(\phi)))}_{V,i-2,B\langle s_1 \rangle BK^*}$$

and

$$E^0_{0,i} \cong \oplus_{g^{-1}Jg \subseteq B\langle s_1,s_2 \rangle BK^*} \ g \otimes_{B\langle s_1,s_2 \rangle BK^*} \ W^{((g^{-1}Jg,g^*(\phi)))}_{V,i,B\langle s_1,s_2 \rangle BK^*}$$

$$\oplus_{g^{-1}Jg \subseteq \langle BK^*,U \rangle} \ g \otimes_{\langle BK^*,U \rangle} \ W^{((g^{-1}Jg,g^*(\phi)))}_{V,i,\langle BK^*,U \rangle}$$

$$\downarrow d^0$$

$$E^0_{0,i-1} \cong \oplus_{g^{-1}Jg \subseteq B\langle s_1,s_2 \rangle BK^*} \ g \otimes_{B\langle s_1,s_2 \rangle BK^*} \ W^{((g^{-1}Jg,g^*(\phi)))}_{V,i-1,B\langle s_1,s_2 \rangle BK^*}$$

$$\oplus_{g^{-1}Jg \subseteq \langle BK^*,U \rangle} \ g \otimes_{\langle BK^*,U \rangle} \ W^{((g^{-1}Jg,g^*(\phi)))}_{V,i-1,\langle BK^*,U \rangle} \ .$$

Since d^0 is induced by d the argument used in the GL_2K cases shows that the only possibly non-zero $E^1_{n,i-n}$'s are

$$E^1_{2,0} \cong \oplus_{g^{-1}Jg \subseteq BK^*} \ g \otimes_{BK^*} \ V^{(g^{-1}Jg,g^*(\phi))},$$

$$E^1_{1,0} \cong \oplus_{g^{-1}Jg \subseteq BK^*} \ g \otimes_{BK^*} \ V^{(g^{-1}Jg,g^*(\phi))}$$

$$\oplus \oplus_{g^{-1}Jg \subseteq B\langle s_1 \rangle BK^*} \ g \otimes_{B\langle s_1 \rangle BK^*} \ V^{(g^{-1}Jg,g^*(\phi))}$$

and

$$E^1_{0,0} \cong \oplus_{g^{-1}Jg \subseteq B\langle s_1,s_2 \rangle BK^*} \ g \otimes_{B\langle s_1,s_2 \rangle BK^*} \ V^{(g^{-1}Jg,g^*(\phi))}$$

$$\oplus_{g^{-1}Jg \subseteq \langle BK^*,U \rangle} \ g \otimes_{\langle BK^*,U \rangle} \ V^{(g^{-1}Jg,g^*(\phi))}.$$

The next step in the spectral sequence computation is to compute the differentials induced by d_Y in the chain complex of $E^1_{p,q}$'s

$$0 \longrightarrow E^1_{2,0} \xrightarrow{d_Y} E^1_{1,0} \xrightarrow{d_Y} E^1_{0,0} \longrightarrow 0.$$

The homology group at $E^1_{p,q}$ is denoted by $E^2_{p,q}$.

Consider the group $E^1_{2,0}$. In the picture of the subdivided fundamental simplex in Example 2.4 the upper right 2-simplex has stabiliser BK^* and every other 2-simplex in the building Y is a translate of this one. Hence the summands in $E^1_{2,0}$ are in bijective correspondence with the 2-simplices of Y which are fixed by J. On the summand corresponding to the coset representative g the map $g \otimes_{BK^*} w \mapsto g \otimes_{BK^*} gw$ combine to give an isomorphism

$$E^1_{2,0} \cong C_2(Y^J; k) \otimes_k V^{(J,\phi)}$$

where $C_i(Z; k)$ denotes the i-dimensional simplicial chains with coefficients in k of a simplicial complex Z. Similar remarks hold for $E^1_{1,0}$ and $E^1_{0,0}$ and the complex $(E^1_{n,0}, d_Y)$ is identified with the complex $(C_*(Y^J; k) \otimes_k V^{(J,\phi)}, d_Y \otimes 1)$.

By Proposition 2.12 the subcomplex Y^J is non-empty and contractible for each compact open modulo K^* subgroup J. Therefore the groups $E^2_{p,q}$ are all zero except for $E^2_{0,0}$.

In a general spectral sequence calculation one inductively has differentials of the form $d^r : E^r_{p,q} \longrightarrow E_{p-r,q+r-1}$ for $r \geq 2$ satisfying $d^r d^r = 0$ from which one calculates the homology at $E^r_{p,q}$ which gives $E^{r+1}_{p,q}$. For a given pair (p, q) the groups $E^r_{p,q}$ stabilise as r increases to give $E^\infty_{p,q}$. The family $E^\infty_{p,m-p}$ form the associated graded group to a filtration on the homology, in this case, $H_m(\underline{M}^{((J,\phi))}_*, \underline{d})$, which the spectral sequence "calculates".

In our case the spectral sequence stabilises at $E^2_{*,*}$ and the filtration in the homology has only one step so that

$$H_m(\underline{M}^{((J,\phi))}_*, \underline{d}) = \begin{cases} V^{(J,\phi)} & \text{if } m = 0, \\ \\ 0 & \text{otherwise} \end{cases}$$

which establishes monomial exactness for $\underline{M}_* \longrightarrow V$.

Before proceeding to Proposition 2.12 we shall pause for two of the simplest examples of the pair (J, ϕ).

EXAMPLE 2.9. $(J, \phi) = (K^*, \underline{\phi})$

The group $M_i^{((K^*, \underline{\phi}))}$ is equal to

$$\bigoplus_{g^{-1} K^* g \subseteq BK^*} g \otimes_{BK^*} W^{((K^*, \underline{\phi}))}_{V, i-2, BK^*}$$

$$\bigoplus_{g^{-1} K^* g \subseteq BK^*} g \otimes_{BK^*} W^{((K^*, \underline{\phi}))}_{V, i-1, BK^*}$$

$$\oplus \bigoplus_{g^{-1} K^* g \subseteq B\langle s_1 \rangle BK^*} g \otimes_{B\langle s_1 \rangle BK^*} W^{((K^*, \underline{\phi}))}_{V, i-1, B\langle s_1 \rangle BK^*}$$

$$\bigoplus_{g^{-1} K^* g \subseteq B\langle s_1, s_2 \rangle BK^*} g \otimes_{B\langle s_1, s_2 \rangle BK^*} W^{((K^*, \underline{\phi}))}_{V, i, B\langle s_1, s_2 \rangle BK^*}$$

$$\bigoplus_{g^{-1} K^* g \subseteq \langle BK^*, U \rangle} g \otimes_{\langle BK^*, U \rangle} W^{((K^*, \underline{\phi}))}_{V, i, \langle BK^*, U \rangle}.$$

Taking the homology using the internal d-differentials first we see that the $E^1_{*,*}$-term of the spectral sequence of the double-complex is isomorphic to the chain complex

$$0 \longrightarrow c - \mathrm{Ind}_{BK^*}^{GL_3K}(V) \longrightarrow c - \mathrm{Ind}_{BK^*}^{GL_3K}(V) \oplus c - \mathrm{Ind}_{B\langle s_1 \rangle BK^*}^{GL_3K}(V)$$

$$\longrightarrow c - \mathrm{Ind}_{B\langle s_1, s_2 \rangle BK^*}^{GL_3K}(V) \oplus c - \mathrm{Ind}_{\langle BK^*, U \rangle}^{GL_3K}(V) \longrightarrow 0.$$

In turn this is equal to

$$0 \longrightarrow c - \mathrm{Ind}_{BK^*}^{GL_3K}(k) \otimes V$$

$$\longrightarrow c - \mathrm{Ind}_{BK^*}^{GL_3K}(k) \otimes V \oplus c - \mathrm{Ind}_{B\langle s_1 \rangle BK^*}^{GL_3K}(k) \otimes V$$

$$\longrightarrow c - \mathrm{Ind}_{B\langle s_1, s_2 \rangle BK^*}^{GL_3K}(k) \otimes V \oplus c - \mathrm{Ind}_{\langle BK^*, U \rangle}^{GL_3K}(k) \otimes V \longrightarrow 0.$$

The homology of this is just $V = V^{(K^*, \phi)}$ in dimension zero because

$$0 \longrightarrow c - \mathrm{Ind}_{BK^*}^{GL_3K}(k) \longrightarrow c - \mathrm{Ind}_{BK^*}^{GL_3K}(k) \oplus c - \mathrm{Ind}_{B\langle s_1 \rangle BK^*}^{GL_3K}(k)$$

$$\longrightarrow c - \mathrm{Ind}_{B\langle s_1, s_2 \rangle BK^*}^{GL_3K}(k) \oplus c - \mathrm{Ind}_{\langle BK^*, U \rangle}^{GL_3K}(k) \longrightarrow 0.$$

is the simplicial chain complex of the building Y, which equals the fixed points Y^{K^*} and which is contractible.

EXAMPLE 2.10. $(J, \phi) = (\langle K^*, U \rangle, \phi)$
Here we must have $\phi(U)^3 = \phi(\pi_K)^{-2}$.

Write $\hat{\sigma}$ for the barycentre of the original fundamental simplex and observe that U fixes $\hat{\sigma}$ and no other vertex. This is because there are precisely two vertex orbits, that of $\hat{\sigma}$ and that of any one of the three vertices of the original fundamental simplex (a triangle). Therefore $Y^{\hat{\sigma}}$ is a point in this example.
The group $M_i^{((\langle K^*, U \rangle, \phi))}$ is equal to

$$\bigoplus_{g^{-1}\langle K^*, U \rangle g \subseteq BK^*} g \otimes_{BK^*} W_{V, i-2, BK^*}^{((g^{-1}\langle K^*, U \rangle g, g^*(\phi)))}$$

$$\bigoplus_{g^{-1}(\langle K^*, U \rangle g \subseteq BK^*} g \otimes_{BK^*} W_{V, i-1, BK^*}^{((g^{-1}\langle K^*, U \rangle g, g^*(\phi)))}$$

$$\oplus \bigoplus_{g^{-1}\langle K^*, U \rangle g \subseteq B\langle s_1 \rangle BK^*} g \otimes_{B\langle s_1 \rangle BK^*} W_{V, i-1, B\langle s_1 \rangle BK^*}^{((g^{-1}\langle K^*, U \rangle g, g^*(\phi)))}$$

$$\bigoplus_{g^{-1}\langle K^*, U \rangle g \subseteq B\langle s_1, s_2 \rangle BK^*} g \otimes_{B\langle s_1, s_2 \rangle BK^*} W_{V, i, B\langle s_1, s_2 \rangle BK^*}^{((g^{-1}\langle K^*, U \rangle g, g^*(\phi)))}$$

$$\bigoplus_{g^{-1}\langle K^*, U \rangle g \subseteq \langle BK^*, U \rangle} g \otimes_{\langle BK^*, U \rangle} W_{V, i, \langle BK^*, U \rangle}^{((g^{-1}\langle K^*, U \rangle g, g^*(\phi)))}$$

which is simply isomorphic to

$$W_{V,i,\langle BK^*,U\rangle}^{((\langle K^*,U\rangle,\phi))},$$

which immediately establishes the monomial exactness for the pair $(\langle K^*,U\rangle,\phi)$.

2.11. *Buildings, extended buildings and $\underline{E}G$*

Let Y denote the building associated to $SL_n K$ when K is a p-adic local field, which was described in §2.2. The building of $SL_n K$ is the simplicial complex whose vertices are lattices in K^n ([35] p. 137). The fundamental chamber is the simplex with vertices $(e_1, \ldots, e_i, \pi_K e_{i+1}, \ldots, \pi_K e_n)$ for $1 \leq i \leq n$ and $\{e_i\}$ the standard basis. The resulting building is not spherical and therefore is contractible ([35] p. 94).

The action of $SL_n K$ on Y is simplicial extends to an action of $GL_n K$ which, as in the case of $n = 2$, where the vertices are now thought of as homothety classes of lattices. This extended action does not preserve type but a mild barycentric subdivision renders the $GL_n K$-action simplicial. Let Y also denote this simplicial $GL_n K$-space. Since the central K^* acts trivially Y is also a building for the projective linear group $PGL_n K$.

If one lets $GL_n K$ act on the real line $\mathbb{R}$ where $X \in GL_n K$ acts via translation by $v_K(\det(X))$ the product $Y \times \mathbb{R}$ is denoted by $\underline{E}GL_n K$, a space which is central to the classification of spaces with proper $GL_n K$-actions ([13], [100], [11]).

PROPOSITION 2.12.

In the notation of §2.11 let $J \subseteq GL_n K$ be a compact open modulo the centre subgroup. Then, after a suitable simplicial subdivision if necessary, the fixed point subcomplex Y^J is non-empty and contractible.

Proof

In order to show that Y^J is non-empty subcomplex it suffices to consider the J's which are maximal in the poset of conjugacy classes of compact open modulo the centre subgroups. This is a finite set of "ends". For each such J we have only to show that there is a J-fixed point in Y which, by subdivision, we may assume is a vertex. Then Y^J will be a non-empty subcomplex. Since the set of "ends" is finite only a finite number of simplicial subdivisions is necessary.

The existence of a J-fixed point is an immediate consequence of the Bruhat-Tits fixed point theorem for groups acting on CAT(0) spaces ([4], [139], [140]). Alternatively, from the proper $GL_n K$-actions point of view, if Y^J were empty then one easily finds that $\underline{E}GL_n K$ does not classify all $GL_n K$-spaces with proper actions.

The groups $GL_n K$, $SL_n K$, $PGL_n K$ are locally compact topological groups. Suppose that G is a locally compact topological group and that $\mathcal{F}$

is a family of subgroups which is closed under conjugation and passage to subgroups. The space $\underline{E}G$ is the universal space in the sense of [100] (see also [13] p. 7 Remark 2.5) for G-CW-complexes having stabilisers which lie in $\mathcal{F}$. This universal space always exists and for $J \in \mathcal{F}$ the fixed point subspace $\underline{E}G^J$ is contractible. When $G = SL_nK$ and $\mathcal{F}$ is the family of compact subgroups then $\underline{E}SL_nK = Y$. When $G = GL_nK$ and $\mathcal{F}$ is the family of compact subgroups then $\underline{E}GL_nK = Y \times \mathbb{R}$ with the action described in §2.11. If $J \subseteq GL_nK^*$ contains K^* and is compact open modulo K then we may write $J = K^*H$ where H is compact open. Then $\underline{E}GL_nK^H$ is contractible. Therefore the image of $\underline{E}GL_nK^H$ under projection onto Y is also contractible but this is $Y^H = Y^J$. $\square$

REMARK 2.13. In §2.6 I delineated four cases of stabiliser group and in §2.8 showed how to prove monomial exactness in Case A. In fact, by the contractibility part of Proposition 2.12, it is clear that the argument applies also to Cases B–D. There remained the question whether for all the pairs (J, ϕ) under consideration the subcomplex Y^J was non-empty. The fixed-point part of Proposition 2.12 takes care of this problem. By a suitable simplicial subdivision we may assume that every compact open modulo the centre stabilises a simplex of Y. In the GL_3K example the subdivision may introduce some new simplex stabilisers which do not occur in Cases A–D (actually this does not happen) but the argument illustrated in Case A works in all cases for all GL_nK. I shall give the complete verification of Chapter Two, Conjecture 3.3 in the next section.

3. Verification of Chapter Two, Conjecture 3.3

3.1. We begin this section by recapitulating the situation of Chapter Two, §3. We are studying (left) admissible k-representations of GL_nK with central character ϕ. Here K continues to be a p-adic local field. As usual k is an algebraically closed field of arbitrary characteristic. If V is such an admissible k-representation.

Let Y be the simplicial complex upon which GL_nK acts simplicially given by a simplicial subdivision of the Bruhat-Tits building of GL_nK (§2.2; for more details [4], [11], [35], [36], [37], [60], [61], [139], [140]).

We shall assume that Y has been chosen according to Proposition 2.12, namely such that for every compact open modulo the centre subgroup $J \subseteq GL_nK$ the fixed point subcomplex Y^J is non-empty and contractible.

For each simplex σ of Y, by Chapter Two, Theorem 2.4, we have a $k[H_\sigma],_\phi$**mon**-bar monomial resolution of V

$$W_{V,*,H_\sigma} \longrightarrow V \longrightarrow 0.$$

Form the graded k-vector space which in degree m is equal to

$$\underline{M}_m = \oplus_{\alpha+n=m} W_{V,\alpha,H_{\sigma^n}}$$

which is a double complex with two commuting differentials d_Y and d. The differential d_Y comes from the simplicial structure of Y together with the natural chain maps between bar-monomial resolutions. Explicitly, for $x \in W_{V,\alpha,H_{\sigma^n}}$, it is given by

$$d_Y(x) = \sum_{\sigma^{n-1} \text{ face of } \sigma^n} d(\sigma^{n-1}, \sigma^n)\, i_{H_{\sigma^n}, H_{\sigma^{n-1}}}(x).$$

If d denotes any of the differentials in the bar-monomial resolutions the total differential $\underline{d} : \underline{M}_m \longrightarrow \underline{M}_{m-1}$ when $m = \alpha + n$ is given by

$$\underline{d}(x) = d_Y(x) + (-1)^n d_{\sigma^n}(x)$$

and $\underline{dd} = 0$.

Finally $\underline{M}_*$ has an obvious structure of a $_{k[GL_n K],\underline{\phi}}\mathbf{mon}$-Line Bundle since the $GL_n K$-action permutes the summands $W_{V,*,H_{\sigma^n}}$, each of which is a $_{k[H_{\sigma^n}],\underline{\phi}}\mathbf{mon}$-Line Bundle.

We are now ready to complete the verification of Chapter Two, Conjecture 3.3 in general having, by way of illustration and introduction, verified the case $n = 2$ in Chapter Two §§4.1–4.15 and sketched the verification for $n = 3$ in Example 2.4 and §2.8.

THEOREM 3.2. *(Verification of Chapter Two, Conjecture 3.3)*
(i) If Y is the Bruhat-Tits building for $GL_n K$, suitably subdivided to make the $GL_n K$-action simplicial as in §3.1, then $(\underline{M}_*, \underline{d})$ is a chain complex in $_{k[GL_n K],\underline{\phi}}\mathbf{mon}$ equipped with a canonical augmentation homomorphism in $_{k[GL_n K],\underline{\phi}}\mathbf{mod}$ of the form $\underline{M}_0 \xrightarrow{\epsilon} V$.
(ii) For $n \geq 2$

$$\cdots \longrightarrow \underline{M}_i \xrightarrow{d} \underline{M}_{i-1} \xrightarrow{d} \cdots \xrightarrow{d} \underline{M}_0 \xrightarrow{\epsilon} V \longrightarrow 0$$

is a monomial resolution in $_{k[GL_n K],\underline{\phi}}\mathbf{mon}$. That is, for each $(H, \phi) \in \mathcal{M}_{GL_n K,\underline{\phi}}$

$$\cdots \longrightarrow \underline{M}_i^{((H,\phi))} \xrightarrow{d} \underline{M}_{i-1}^{((H,\phi))} \xrightarrow{d} \cdots \xrightarrow{d} \underline{M}_0^{((H,\phi))} \xrightarrow{\epsilon} V^{(H,\phi)} \longrightarrow 0$$

is an exact sequence of k-vector spaces.

Proof

Part (i) is established in Chapter Two, Theorem 3.2.

For part (ii) we begin by choosing $GL_n K$-orbit representatives of the q-dimensional simplices $\{\sigma_\alpha^q \mid \alpha \in \mathcal{A}(q)\}$ for each $q \geq 0$. Recall that Y is finite-dimensional. Let $H_{\sigma_\alpha^q} = \text{stab}_{GL_n K}(\sigma_\alpha^q)$, which is compact open modulo the centre K^*.

By naturality of the bar-monomial resolutions the monomial complex given by the direct sum of the $W_{V,*,H_\sigma}$ as σ varies through the $GL_n K$-orbit of σ_α^q is isomorphic to

$$c - \underline{\operatorname{Ind}}_{H_{\sigma_\alpha^q}}^{GL_n K}(W_{V,*,H_{\sigma_\alpha^q}}).$$

Therefore

$$\underline{M}_* \cong \oplus_q \; \oplus_{\alpha \in \mathcal{A}(q)} \; c - \underline{\operatorname{Ind}}_{H_{\sigma_\alpha^q}}^{GL_n K}(W_{V,*-q,H_{\sigma_\alpha^q}}).$$

As in the case $n = 3$ in Example 2.4 and §2.8, for $(H,\phi) \in \mathcal{M}_{GL_n K,\underline{\phi}}$ we have

$$\underline{M}_*^{((H,\phi))}$$

$$\cong \oplus_q \; \oplus_{\alpha \in \mathcal{A}(q)} \; c - \underline{\operatorname{Ind}}_{H_{\sigma_\alpha^q}}^{GL_n K}(W_{V,*-q,H_{\sigma_\alpha^q}})^{((H,\phi))}$$

$$\cong \oplus_q \; \oplus_{\alpha \in \mathcal{A}(q)} \; \oplus_{g^{-1}Hg \leq H_{\sigma_\alpha^q}} \; g \otimes_{H_{\sigma_\alpha^q}} W_{V,*-q,H_{\sigma_\alpha^q}}^{((g^{-1}Hg,g^*(\phi)))}.$$

Define a decreasing filtration on $\underline{M}_*^{((H,\phi))}$ by

$$F^p \underline{M}_i^{((H,\phi))} = \oplus_{j \leq p} C_{j,i-j}$$

where

$$C_{j,i-j} = \oplus_{\alpha \in \mathcal{A}(j)} \; \oplus_{g^{-1}Hg \leq H_{\sigma_\alpha^j}} \; g \otimes_{H_{\sigma_\alpha^j}} W_{V,i-j,H_{\sigma_\alpha^j}}^{((g^{-1}Hg,g^*(\phi)))}.$$

In the spectral sequence associated to this filtration we have

$$E_{p,i-p}^0 = F^p \underline{M}_i^{((H,\phi))} / F^{p-1} \underline{M}_i^{((H,\phi))} \cong C_{p,i-p}.$$

The differential $d^0 : E_{p,i-p}^0 \longrightarrow E_{p,i-p-1}^0$ is induced by the internal differential of the $W_{V,*-q,H_{\sigma_\alpha^q}}$'s and by the monomial exactness of these resolutions we find that, as in the $GL_2 K$ and $GL_3 K$ examples, the homology $E_{p,s}^1$ vanishes unless $s = 0$. Furthermore, as explained in the $GL_3 K$ example,

$$E_{p,0}^1 \cong \oplus_{\alpha \in \mathcal{A}(p)} \; \oplus_{g^{-1}Hg \leq H_{\sigma_\alpha^p}} \; g \otimes_{H_{\sigma_\alpha^p}} V^{(g^{-1}Hg,g^*(\phi))} \cong C_p(Y^H;k) \otimes_k V^{(H,\phi)}$$

and $d^1 : E_{p,0}^1 \longrightarrow E_{p-1,0}^1$ may be identified with $d_Y \otimes 1$. The contractibility of Y^H implies that, in the spectral sequence,

$$E_{p,r}^\infty = \begin{cases} V^{(H,\phi)} & \text{if } (p,r) = (0,0) \\ 0 & \text{otherwise.} \end{cases}$$

This established part (ii) and completes the proof. $\square$

REMARK 3.3. Results analogous to Theorem 3.2 are true for the groups

$$SL_n K \subset GL_n K^0 \subset GL_n K^+,$$

which were defined in Chapter Two, §4.16, and for $PGL_n K$. The details are left to the reader.

CHAPTER 5

Monomial resolutions and Deligne representations

The Langlands correspondence for $GL_n K$ when K is a p-adic local field concerns a canonical bijection between n-dimensional semisimple Deligne representations and irreducible smooth complex representations of $GL_n K$. This correspondence is characterised in terms of local L-functions and ϵ-factors of these two types of representations (see, for example, [40] Chapter Eight). Deligne representations are finite-dimensional representations of the Weil group $\mathcal{W}_K$ together with a nilpotent operator satisfying certain properties. The importance of Deligne representations lies in the result that, if l is different from p, the category of finite-dimensional representations of $\mathcal{W}_K$ over $\overline{\mathbb{Q}}_l$ is isomorphic to a category of Deligne representations of $\mathcal{W}_K$ over the complex numbers (see Theorem 1.8).

In §1 we recall the definition and properties of Deligne representations of the Weil group. In §2 we define what is meant by a Deligne representation. In Conjecture 2.4 we describe the bar-monomial resolution resolution for a finite-dimensional Deligne representation $(\rho, V, \mathbf{n})$. The verification of Conjecture 2.4 should be straightforward but for the time being, out of laziness, I shall leave it unproved.

1. Weil groups and representations

1.1. Galois and Weil groups
The material of this section is a sketch of ([40] Chapter 7).

Let K be a p-adic local field in characteristic zero with residue field $\underline{k} = \mathcal{O}_K/\mathcal{P}_K$. Choose an algebraic closure $\overline{K}$. Then

$$\mathrm{Gal}(\overline{K}/K) = \varprojlim \mathrm{Gal}(E/K)$$

where the limit runs over finite Galois subextensions E/K of $\overline{K}/K$. We have an extension of groups

$$\{1\} \longrightarrow \mathcal{I}_K \longrightarrow \mathrm{Gal}(\overline{K}/K) \longrightarrow \hat{\mathbb{Z}} \longrightarrow \{1\}$$

where the inertia group $\mathcal{I}_K = \mathrm{Gal}(\overline{K}/K_\infty)$ where K_∞ is the unique maximal unramified extension of K lying in $\overline{K}$. For each integer n such that $HCF(p, n) = 1$ then there is a unique cyclic extension of K_∞ of degree

n given by $E_n = K_\infty(\pi_K^{1/n})$ so that E_∞ is the maximal tamely ramified extension of K with

$$t_0 : \mathrm{Gal}(E_\infty/K_\infty) \xrightarrow{\cong} \prod_{l \neq p} \mathbb{Z}_l.$$

Let Φ_K denote the geometric Frobenius — that is, the inverse of the lift of $x \mapsto x^{|\underline{k}|}$ on residue fields. Then $t_0(\Phi_K g \Phi_K^{-1}) = |\underline{k}|^{-1} t_0(g)$ for $g \in \mathrm{Gal}(E_\infty/K_\infty)$. For this choice of geometric Frobenius the Weil group is the locally profinite group in the centre of the subextension

$$\{1\} \longrightarrow \mathcal{I}_K \longrightarrow \mathcal{W}_K \longrightarrow \mathbb{Z} = \langle \Phi_K \rangle \longrightarrow \{1\}.$$

Hence the Weil group is the dense subgroup of the Galois group generated by Frobenius elements and the inertia group is an open subgroup. Sending a geometric Frobenius to 1 yields $v_K : \mathcal{W}_K \longrightarrow \mathbb{Z}$ and we define

$$||x|| = q^{-v_K(x)} \text{ for all } x \in \mathcal{W}_K, \ q = |\underline{k}|.$$

For each finite extension E/K we set $\mathcal{W}_E = \mathcal{W}_K \bigcap \mathrm{Gal}(\overline{K}/E)$, it is open and of finite index in $\mathcal{W}_K$. It is isomorphic to the Weil group of E and ([**40**] p. 183) this system of Weil groups enjoys all the well-known properties of absolute Galois groups.

We consider representations over an algebraically closed field k of characteristic zero.

Since $\mathrm{Gal}(\overline{K}/K)$ is profinite any smooth representation is semisimple. This not true for the Weil group which is locally profinite and has $\mathbb{Z}$ as a quotient. However an irreducible smooth representation of $\mathcal{W}_K$ is finite-dimensional. Therefore a smooth irreducible Weil representation is semisimple with finite image when restricted to the inertia subgroup — the subtleties come from the geometric Frobenius elements.

Smooth irreducible Galois representations restrict to smooth irreducible Weil representations and two such are equivalent if and only if they restrict to give equivalent Weil representations.

An irreducible smooth Weil representation has finite image if and only if it is the restriction of a Galois representation if and only if its determinantal character has finite order ([**40**] p. 184).

If E/K is a finite separable extension then a smooth representation ρ of $\mathcal{W}_K$ is semisimple if and only if $\mathrm{Res}_{E/K}(\rho)$ is a semisimple smooth representation of $\mathcal{W}_E$. Conversely a smooth representation ρ of $\mathcal{W}_E$ is semisimple if and only if $\mathrm{Ind}_{E/K}(\rho)$ is a semisimple smooth representation of $\mathcal{W}_K$.

Let $\mathcal{G}_n^{ss}(K)$ denote the set of isomorphism classes of semisimple smooth representations of $\mathcal{W}_K$ of dimension n. We have induction and restriction maps between these sets. A smooth finite-dimensional representation ρ of $\mathcal{W}_K$ is semisimple if and only if $\Phi_K(x)$ is semisimple for every element x ([**40**] pp. 185–186). $\mathcal{G}_n^0(K)$ is the set isomorphism classes of irreducible smooth Weil representations of dimension n. The L-function extends to $\mathcal{G}_n^{ss}(K)$ via Artin's Euler factor definition and the ϵ-factors extend also ([40] Chapter 7, §30). The L-function is therefore invariant under induction ([40] p. 189).

Local class field theory yields the reciprocity map $\alpha_K : \mathcal{W}_K \longrightarrow K^*$ with commutative diagrams featuring restriction, the norm and the Verlagerung.

1.2. Deligne representations

A Deligne representation of $\mathcal{W}_K$ is a triple $(\rho, V, \mathbf{n})$ in which (ρ, V) is a finite-dimensional smooth Weil representation and $\mathbf{n} \in \mathrm{End}_k(V)$ is a nilpotent endomorphism satisfying $\rho(x)\mathbf{n}\rho(x)^{-1} = ||x||\mathbf{n}$. We call $(\rho, V, \mathbf{n})$ semisimple if and only if (ρ, V) is semisimple. Write $\mathcal{G}_n(K)$ for the equivalence classes of n-dimensional semisimple Deligne representation of the Weil group so that we have

$$\mathcal{G}_n^0(K) \subset \mathcal{G}_n^{ss}(K) \subset \mathcal{G}_n(K).$$

We have analogues of the usual constructions of operations on representations:

$(\rho, V, \mathbf{n})^\vee = (\rho^\vee, V^\vee, -\mathbf{n}^\vee)$ (contragredients),

$(\rho_1, V_1, \mathbf{n}_1) \otimes (\rho_2, V_2, \mathbf{n}_2) = (\rho_1 \otimes \rho_2, V_1 \otimes V_2, 1 \otimes \mathbf{n}_2 + \mathbf{n}_1 \otimes 1)$,

$(\rho_1, V_1, \mathbf{n}_1) \oplus (\rho_2, V_2, \mathbf{n}_2) = (\rho_1 \oplus \rho_2, V_1 \oplus V_2, \mathbf{n}_1 \oplus \mathbf{n}_2)$.

$\mathrm{Ker}(\mathbf{n})$ carries a Weil representation and the L-functions and ϵ-factors are extended to Deligne representations via this.

EXAMPLE 1.3. $\mathrm{Sp}(n)$

Let $V = k^n$ and definite $\mathbf{n}(v_0, v_1, \ldots, v_{n-1}) = (0, v_0, v_1, \ldots, v_{n-2})$ and $\rho_0(x)(v_0, v_1, \ldots, v_{n-1}) = (v_0, ||x||v_1, ||x||^2 v_2, \ldots, ||x||^{n-1}v_{n-1})$. Then we set $\rho(x) = ||x||^{(1-n)/2}\rho_0(x)$. The triple $(\rho, k^n, \mathbf{n})$ is a semisimple Deligne representation denoted by $\mathrm{Sp}(n)$.

A semisimple Deligne representation of $\mathcal{W}_K$ is indecomposable if it cannot be written as the direct sum of two non-zero Deligne representations. The indecomposable semisimple Deligne representations are precisely those of the form $\rho \otimes \mathrm{Sp}(n)$ for some $\rho \in \mathcal{G}_n^0(K)$.

1.4. l-adic representations

Let l be a prime different from p. Let G be a locally profinite group and let C be a field of characteristic zero. A C-representation $\pi : G \longrightarrow \mathrm{Aut}_C(V)$ is defined to be smooth if $\mathrm{Stab}_G(v)$ is an open subgroup of G for every $v \in V$. Smooth representations form a category $\mathrm{Rep}_C(G)$. An isomorphism of fields gives, by extension of scalars, an equivalence of smooth representation categories. For example an isomorphism of the form $\overline{\mathbb{Q}}_l \cong \mathbb{C}$. Similarly there is an equivalence of categories of Deligne representations.

If $\mathbf{n}$ is a nilpotent endomorphism then $\exp(\mathbf{n}) = 1 + \sum_{j=1}^{\infty} \frac{\mathbf{n}^j}{j!}$ is a unipotent automorphism and if $\mathbf{u}$ is such then $\log(\mathbf{u}) = \sum_{j=1}^{\infty} (-1)^{j-1} \frac{\mathbf{u}^j}{j!}$ is a nilpotent endomorphism.

Consider V a d-dimensional $\underline{\mathbb{Q}}_l$-vector space. The valuation $\underline{\mathbb{Q}}_l \longrightarrow \mathbb{Q} \bigcup \{\infty\}$ gives a metric on $\underline{\mathbb{Q}}_l$ — which is *not* complete. Hence $GL_d(\underline{\mathbb{Q}}_l)$ has an entry-by-entry topology. A representation $\pi : G \longrightarrow \mathrm{Aut}(V) \cong GL_d(\underline{\mathbb{Q}}_l)$ is continuous if, viewed as a homomorphism to invertible matrices, it is continuous in this topology. A smooth representation of G on V is always continuous but not conversely.

We have $t : \mathcal{I}_K \longrightarrow \mathbb{Z}_l$ given by mapping to $\mathrm{Gal}(E_\infty/K_\infty)$ then composing with

$$t_0 : \mathrm{Gal}(E_\infty/K_\infty) \overset{\cong}{\longrightarrow} \prod_{l \neq p} \mathbb{Z}_l$$

and finally projecting to the l-adic coordinate. If $\underline{P}_K$ is the wild ramification group then we have a short exact sequence

$$0 \longrightarrow \underline{P}_K \longrightarrow \mathrm{Ker}(t) \longrightarrow \prod_{m \text{ prime } \neq l,p} \mathbb{Z}_m \longrightarrow 0.$$

We have $t(gxg^{-1}) = ||g|| t(x)$ for $x \in \mathcal{I}_K, g \in \mathcal{W}_K$. The kernel of t contains no open subgroup of $\mathcal{I}_K$.

The following result is important in classifying l-adic Weil representations.

THEOREM 1.5.

Let (σ, V) be a finite-dimensional continuous representation of $\mathcal{W}_K$ over $\overline{\mathbb{Q}}_l$ with $l \neq p$. Then there exists a unique nilpotent $\mathbf{n}_\sigma \in \mathrm{End}_{\overline{\mathbb{Q}}_l}(V)$ such that

$$\sigma(y) = \exp(t(y)\mathbf{n}_\sigma)$$

for all y in some open subgroup of $\mathcal{I}_K$.

Proof

Uniqueness follows from $\mathbf{n}_\sigma = t(y)^{-1}\log(\sigma(y))$, which is independent of y providing $t(y) \neq 0$. For existence assume that σ takes values in $GL_d\overline{\mathbb{Q}}_l$ and let $\overline{\mathbb{Z}}_l$ denote the integral closure of $\mathbb{Z}_l$ in $\overline{\mathbb{Q}}_l$. Then $1 + l^m M_d \overline{\mathbb{Z}}_l$ for $m \geq 1$ is an open subgroup of $GL_d\overline{\mathbb{Q}}_l$ normalising $1 + l^{m+1} M_d \overline{\mathbb{Z}}_l$ with abelian quotient of exponent l.

Viewing σ as a continuous homomorphism $\mathcal{W}_K \longrightarrow GL_d\overline{\mathbb{Q}}_l$, let J denote the set of $g \in \mathrm{Ker}(t)$ such that $\sigma(g) \in 1 + l^2 M_d \overline{\mathbb{Z}}_l$. Thus J is an open subgroup of $\mathrm{Ker}(t)$ with $\sigma(J) \subseteq 1 + l^2 M_d \overline{\mathbb{Z}}_l$. Its image in $\frac{1 + l^2 M_d \overline{\mathbb{Z}}_l}{1 + l^3 M_d \overline{\mathbb{Z}}_l}$ is trivial so that $\sigma(J) \subseteq 1 + l^3 M_d \overline{\mathbb{Z}}_l$. By induction $\sigma(J) = \{1\}$. Since J is open there is an open subgroup H_0 of $\mathcal{I}_K$ such that $H_0 \bigcap \mathrm{Ker}(t) \subseteq J$. Shrinking H_0

if necessary we may assume that $\sigma(H_0) \subseteq 1 + l^2 M_d \overline{\mathbb{Z}}_l$. There is an open, normal subgroup of finite index in $\mathcal{W}_K$ such that $H \cap \mathcal{I}_K \subset H_0$.

The restriction of σ to $H \cap \mathcal{I}_K$ therefore factors through a continuous homomorphism $\phi : t(H \cap \mathcal{I}_K) \longrightarrow 1 + l^2 M_d \overline{\mathbb{Z}}_l$; that is $\sigma(h) = \phi(t(h))$ for all $h \in H \cap \mathcal{I}_K$.

Therefore we have $\sigma(\Phi h \Phi^{-1})^q = \sigma(h)$ for all $h \in H \cap \mathcal{I}_K$ and every Frobenius element of $\mathcal{W}_K$. Suppose that $\sigma(h)(v) = \alpha v$ then $\sigma(\Phi)v$ is an eigenvector for $\sigma(\Phi h \Phi^{-1})$ with eigenvalue α. Hence α^q is also an eigenvalue for $\sigma(h)$. As $\sigma(h)$ is invertible this implies that α is a root of unity. Since $\sigma(h) \in 1 + l^2 M_d \overline{\mathbb{Z}}_l$ then $(\sigma(h) - 1)/l^2$ is integral over $\mathbb{Z}_l$.

However ([**40**] Lemma p. 205) if α is a root of unity such that $(\alpha - 1)/l^2$ is integral over $\mathbb{Z}_l$ then $\alpha = 1$.

Next choose $h_0 \in H \cap \mathcal{I}_K$ with $t(h_0) \neq 0$ and set $\mathbf{n}_\sigma = t(h_0)^{-1} \log(\sigma(h_0))$ which is nilpotent. Now put $A = \mathbb{Z}_l t(h_0)$. We have two continuous homomorphisms — $x \mapsto \phi(x)$ and $x \mapsto \exp(x \mathbf{n}_\sigma)$ — which coincide on h_0 and hence on the closure A of $\mathbb{Z}_l h_0$. Putting $H' = t^{-1}(A)$, which is open, since A is open in $\mathbb{Z}_l$, yields the result. $\square$

REMARK 1.6. (i) In Theorem 1.5 (σ, V) is smooth if and only if $\mathbf{n}_\sigma = 0$. In particular, (σ, V) is smooth if V is one-dimensional.

Also, since $\|g\| = 1$ for $g \in \mathcal{I}_K$ we see that $\mathbf{n}_\sigma$ commutes with $\sigma(\mathcal{I}_K)$.

(ii) If $x \in \mathcal{I}_K, g \in \mathcal{W}_K$ we have — provided we are in H in some sense —

$$\sigma(gxg^{-1}) = \exp(t(gxg^{-1})\mathbf{n}) = \exp(\|g\| t(x)\mathbf{n})$$

and

$$\sigma(gxg^{-1}) = \sigma(g) \exp(t(x)\mathbf{n}) \sigma(g)^{-1} = \exp(t(x)\sigma(g)\mathbf{n}\sigma(g)^{-1})$$

so that

$$\|g\| \mathbf{n} = \sigma(g) \mathbf{n} \sigma(g)^{-1}.$$

1.7. *The equivalence of representation categories*
Fix a Frobenius $\Phi \in \mathcal{W}_K$ and define

$$\sigma_\Phi(\Phi^a x) = \sigma(\Phi^a x) \exp(-t(x)\mathbf{n}_\sigma) \quad a \in \mathbb{Z}, \ x \in \mathcal{I}_K.$$

Therefore, by Theorem 1.5, the homomorphism

$$\sigma_\Phi : \mathcal{W}_K \longrightarrow \mathrm{Aut}_{\overline{\mathbb{Q}}_l}(V)$$

is trivial on some open subgroup of $\mathcal{I}_K$. It therefore yields a smooth representation of $\mathcal{W}_K$. By Theorem 1.5, the triple $(\sigma_\Phi, V, \mathbf{n}_\sigma)$ is a Deligne representation of $\mathcal{W}_K$ on V.

THEOREM 1.8. *([**40**] p. 206)*
Let $\mathrm{Rep}^f_{\overline{\mathbb{Q}}_l}(\mathcal{W}_K)$ denote the category of finite-dimensional continuous representations of $\mathcal{W}_K$ over $\overline{\mathbb{Q}}_l$. Let $\Phi \in \mathcal{W}_K$ be a Frobenius element and

$t : \mathcal{I}_K \longrightarrow \mathbb{Z}_l$ a continuous surjection. Then the map

$$(\sigma, V) \mapsto (\sigma_\Phi, V, \mathbf{n}_\sigma)$$

is functorial and induces an equivalence of categories

$$\mathrm{Rep}_{\overline{\mathbb{Q}}_l}^f (\mathcal{W}_K) \xrightarrow{\simeq} \mathrm{D} - \mathrm{Rep}_{\overline{\mathbb{Q}}_l}^f (\mathcal{W}_K).$$

The isomorphism of the Deligne representation $(\sigma_\Phi, V, \mathbf{n}_\sigma)$ depends only on the isomorphism class of (σ, V); that is, it does not depend on the choice of Φ and t.

1.9. Theorem 1.8 gives a canonical bijection between the set of isomorphism classes of finite-dimensional continuous representations of $\mathcal{W}_K$ over $\overline{\mathbb{Q}}_l$ and the set of isomorphism classes of Deligne representations of $\mathcal{W}_K$ over $\overline{\mathbb{Q}}_l$. The latter can be transported from $\overline{\mathbb{Q}}_l$ to $\mathbb{C}$.

The Langlands programme concerns the Φ-semisimple (σ, V)'s — that is, those for which $(\sigma_\Phi, V, \mathbf{n}_\sigma)$ is semisimple.

PROPOSITION 1.10. ([40] p. 208)
Let (σ, V) be a finite-dimensional continuous representation of $\mathcal{W}_K$ over $\overline{\mathbb{Q}}_l$. The following are equivalent:

(i) (σ, V) is Φ-semisimple.

(ii) There is a Frobenius element $\Psi \in \mathcal{W}_K$ such that $\sigma(\Psi)$ is semisimple.

(iii) For every $g \in \mathcal{W}_K - \mathcal{I}_K$ the automorphism $\sigma(g)$ is semisimple.

THEOREM 1.11. ([40] p. 208)
Let l be a prime not equal to p and let $n \geq 1$ be an integer. There is a canonical bijection between the following sets of isomorphism classes of representations:

(i) n-dimensional Φ-semisimple continuous representations of $\mathcal{W}_K$ over $\overline{\mathbb{Q}}_l$

(ii) n-dimensional semisimple continuous Deligne representations of $\mathcal{W}_K$ over $\overline{\mathbb{Q}}_l$.

The choice of an isomorphism $\overline{\mathbb{Q}}_l \cong \mathbb{C}$ induces a bijection of these sets with isomorpism classes of n-dimensional semisimple Deligne representations of $\mathcal{W}_K$ over $\mathbb{C}$.

REMARK 1.12. $GL_d\mathbb{F}_q$
(i) In [93] and in [40] §25.4 p. 159) we find two approaches to correspondences involving Deligne representations of the Weil group and irreducible complex representations of $GL_d\mathbb{F}_q$. The first is combinatorial and the second is a special case of the Langlands correspondence.

(ii) It is worth pointing out the analogy between the nilpotent operator in a Deligne representation and the differential operators of $(\mathcal{U}(gl_2\mathbb{C}), K_\infty)$-modules associated to automorphic representations (Chapter Three, §§1.14–1.16) via the $(\mathcal{U}(gl_2\mathbb{C}), K_\infty) \times GL_2\mathbb{A}_{fin}$-module formulation.

2. The bar-monomial resolution of a Deligne representation

2.1. We continue with the situation and notation of §1.1.

Let $\mathcal{M}(\mathcal{W}_K)$ denote the poset of pairs (H, ϕ) where H is a subgroup of finite index in $\mathcal{M}(\mathcal{W}_K)$ and $\phi : H \longrightarrow k^*$ is a continuous character. Therefore the image of ϕ restricted to $H \bigcap \mathcal{I}_K$ is finite. Given (H, ϕ) there are infinitely many continuous characters which agree with ϕ on $H \bigcap \mathcal{I}_K$, since we may tensor ϕ with any homomorphism of the form

$$H \subseteq \mathcal{W}_K \longrightarrow \mathcal{W}_K / \mathcal{I}_K \cong \mathbb{Z} \longrightarrow k^*.$$

Let $(\rho, V, \mathbf{n})$ be a Deligne representation in $\mathcal{G}_n(K)$. If $v \in V^{(H,\phi)}$ then

$$\phi(g)^{-1}\rho(g)(\mathbf{n}(v)) = \rho(g)(\mathbf{n}(\rho(g)^{-1}(v))) = ||g||\mathbf{n}(v) = q^{-v_K(g)}\mathbf{n}(v)$$

so that $\mathbf{n}(V^{(H,\phi)}) \subseteq V^{(H,||-||\cdot\phi)}$.

PROPOSITION 2.2.
If the characteristic of k is not equal to p then there are only finitely pairs $(H, \phi) \in \mathcal{M}(\mathcal{W}_K)$ for which $V^{(H,\phi)} \neq 0$.

Proof
The representation ρ factors through a quotient $\mathcal{W}_K/N$ where N is a normal subgroup which lies in the inertia group and where $\mathcal{I}_K/N$ is finite. Therefore, in order that $V^{(H,\phi)}$ may possibly be non-zero it is necessary that ϕ is trivial on $H \bigcap \mathcal{I}_K/H \bigcap N$. Hence there are only a finite number of possibilities for the restriction of ϕ to $\mathcal{I}_K$. Therefore it suffices to choose ϕ and prove that there are only a finite number of characters of the form $\phi_i = || - ||^i \cdot \phi$ such that $V^{(H,\phi_i)} \neq 0$.

If the characteristic of k is non-zero and not equal to p then q is non-zero and of finite order in k^* so that there are only a finite number of ϕ_i's. If the characteristic of k is zero assume that the result is false and choose non-zero vectors $v_{i_1}, v_{i_2}, \ldots, v_{i_t}$ with t strictly greater than the dimension of V and $v_{i_s} \in V^{(H,\phi_{i_s})}$. There is a non-trivial linear dependence relation between the v_{i_s}'s. Choose the shortest possible such linear dependence relation and assume, rearranging the v_{i_s}'s if necessary, that it involves $v_{i_1}, v_{i_2}, \ldots, v_{i_r}$ with $i_j \leq i_r$ for all $1 \leq j \leq r - 1$. That is, we have

$$a_1 v_{i_1} + a_2 v_{i_2} + \ldots + a_r v_{i_r} = 0$$

with each a_j non-zero. Choose any $g \in H$ which does not lie in the inertia group. Hence $||g|| \neq 0, 1$. Applying $\rho(g)$ to the relation yields

$$a_1 \phi(g) ||g||^{i_1} v_{i_1} + a_2 \phi(g) ||g||^{i_2} v_{i_2} + \ldots + a_r \phi(g) ||g||^{i_r} v_{i_r} = 0.$$

Subtracting $\phi(g)||g||^{i_r}$ times the first relation from the second leads to a shorter non-trivial linear dependence relation, which is a contradiction. $\square$

2.3. We may define a monomial category $_{k[\mathcal{W}_K]}\mathbf{mon}$ of Line Bundles and morphism by replacing $\mathcal{M}_{\underline{\phi}}(G)$ by $\mathcal{M}(\mathcal{W}_K)$ (and relinquishing the central character condition).

Let $(\rho, V, \mathbf{n})$ be a Deligne representation in $\mathcal{G}_n(K)$. Set

$$S = \oplus_{(H,\phi) \in \mathcal{M}(\mathcal{W}_K), V^{(H,\phi)} \neq 0} \, \underline{\mathrm{Ind}}_H^{\mathcal{W}_K}(k_\phi)$$

and set $\mathcal{A}_S = \mathrm{End}_{k[\mathcal{W}_K]\mathbf{mon}}(S)$. Following Chapter One §5 set

$$\tilde{M}_{S,i} = \mathrm{Hom}_{k[\mathcal{W}_K]\mathbf{mod}}(\mathcal{V}(S), V) \otimes_k \mathcal{A}_S \otimes_k \cdots \otimes_k \mathcal{A}_S$$

and, by the same formulae as in the bar-monomial resolution define a complex

$$\cdots \xrightarrow{d} \tilde{M}_{S,i} \otimes_k S \xrightarrow{d} \cdots \xrightarrow{d} \tilde{M}_{S,1} \otimes_k S \xrightarrow{d} \tilde{M}_{S,0} \otimes_k S \xrightarrow{\epsilon} V \longrightarrow 0.$$

All the differentials and the augmentation commute with the $\mathbf{n}$ on V and post-composition with $\mathbf{n}$ on $\mathrm{Hom}_{k[\mathcal{W}_K]\mathbf{mod}}(\mathcal{V}(S), V)$, because post-composition commutes with pre-composition. Endowed with post-composition with $\mathbf{n}$ each $\tilde{M}_{S,i} \otimes_k S$ becomes a Deligne $k[\mathcal{W}_K]$-monomial Line-Bundle and $(\tilde{M}_{S,*} \otimes_k S, d)$ is a chain complex of such. We define a monomial resolution of a Deligne representation in the obvious manner.

The bar-monomial chain complex given above is a monomial resolution of the representation V restricted to the inertia group. This follows from the properties of the bar-monomial resolution for finite-dimensional representations of finite groups.

The following conjecture should not be too difficult to prove — perhaps by an explicit chain homotopy.

CONJECTURE 2.4. Let $(\rho, V, \mathbf{n})$ be a finite-dimensional Deligne representation over an algebraically closed field of characteristic zero. Then the complex of §2.3

$$\cdots \xrightarrow{d} \tilde{M}_{S,i} \otimes_k S \xrightarrow{d} \cdots \xrightarrow{d} \tilde{M}_{S,1} \otimes_k S \xrightarrow{d} \tilde{M}_{S,0} \otimes_k S \xrightarrow{\epsilon} V \longrightarrow 0,$$

endowed with the nilpotent endomorphism induced by $\mathbf{n}$, is a monomial resolution of the Deligne representation V.

CHAPTER 6

Kondo style invariants

In [85] a Gauss sum is attached to each finite-dimensional complex irreducible representation V of $GL_n\mathbb{F}_q$. The Kondo-Gauss sum is a scalar $d \times d$-matrix where $d = \dim_\mathbb{C}(V)$. In Appendix III, §3 I recapitulate the construction of [85] but giving the formulae in terms of character values, which simultaneously removes the irreducibility condition and reveals the functorial properties (e.g. invariance under induction; see Appendix III, Theorem 3.2).

In this chapter the theme is the association of ϵ-factors, L-functions and Kondo-style invariants to the terms in a monomial resolution of an admissible representation V of $GL_n K$ when K is a p-adic local field. The examples here suggest that eventually one may be able to construct the ϵ-factors and L-functions of [66] by merely applying variations of my constructions to the monomial modules which occur in the monomial resolution of V and taking the Euler characteristic.

In §1 ρ is a finite-dimensional complex representation ρ of a compact modulo the centre subgroup J of $GL_n K$. To this I associate a Kondo-style Gauss sum $\tau_J(\rho)$, defined by the formula used in Appendix III, §3, and show that, at least for $GL_2 K$, that $\tau_J(\rho)$ is given by a "Haar integral" over J when the multiplicity of the trivial representation in ρ is zero (as in Appendix III, Lemma 1.7).

In §2 we recapitulate the properties and construction of the Haar integral on a locally p-adic Lie group G. Then we recall in the case $G = K, K^*$ Weil's approach [142] to Tate's thesis, which derives the local functional equation (Corollary 2.23). Then we study the simple case of $G = \langle K^*, u \rangle$ which is a finite modulo the centre subgroup of $GL_2 K$. From the case $G = K, K^*$ we construct meromorphic extension to the whole complex plane of eigendistributions which are analogous to those of [142] and derive a functional equation in Example 2.28.

In §3 I explain how the case when G is finite modulo the centre extends the local functional equation to the compact open modulo the centre subgroups of $GL_n K$. I conclude the section with several questions related to what conjecturally might happen if one could take the Euler characteristic of the constructions in §2 applied term-by-term to a monomial resolution of an admissible representation V of $GL_n K$.

1. Kondo style epsilon factors

1.1. Let K be a p-adic local field with valuation ring $\mathcal{O}_K$ and prime ideal $\mathcal{P}_K$. Write $U_K^n = 1 + \mathcal{P}_K^n$ for $n \geq 1$ and $U_K^0 = \mathcal{O}_K^*$. The standard additive character on K is

$$\psi_K : K \xrightarrow{\text{trace}_{K/\mathbb{Q}_p}} \mathbb{Q}_p \longrightarrow \mathbb{Q}_p/\mathbb{Z}_p \subset \mathbb{C}^*$$

where the final map is given by $1/n + \mathbb{Z} \mapsto e^{2\pi\sqrt{-1}/n}$.

We have a chain of fractional ideals

$$\cdots \subset \mathcal{P}_K^3 \subset \mathcal{P}_K^2 \subset \mathcal{P}_K \subset \mathcal{O}_K \subset \mathcal{P}_K^{-1}\mathcal{O}_K \subset \mathcal{P}_K^{-2}\mathcal{O}_K \subset \cdots$$

and $\mathcal{D}_K^{-1} = \mathcal{P}_K^e$ — a fractional ideal called the codifferent (or inverse different) — is the biggest fractional ideal on which ψ_K is trivial. That is, $\psi_K(\mathcal{P}_K^e) \subseteq \mathbb{Z}_p$ and $\psi_K(\mathcal{P}_K^{e-1}) \not\subseteq \mathbb{Z}_p$. The different is the fractional ideal $\mathcal{D}_K = \mathcal{P}_K^{-e}$.

1.2. *Kondo-Gauss sums for compact modulo the centre subgroups*

For $n \geq 1$ let $U_K^n = 1 + \pi_K^n M_m \mathcal{O}_K \subseteq U_K^0 = GL_m \mathcal{O}_K$ where $M_m \mathcal{O}_K$ is the ring of $m \times m$ matrices with entries in $\mathcal{O}_K$.

Let J be a compact modulo the centre subgroup of $GL_m K$ which contains the centre K^* and let ρ be a continuous, finite-dimensional complex representation of J. Let $n_J(\rho)$ be the least integer such that ρ factorises through $J/J \bigcap U_K^{n_J(\rho)}$. Set $f_J(\rho) = \mathcal{P}_K^{n_J(\rho)}$, which shall be called the J-conductor of ρ (or sometimes merely the conductor of ρ if the identity of J is clear).

Choose $c \in K$ such that $\mathcal{O}_K \cdot c = f_J(\rho)\mathcal{D}_K$, where $\mathcal{D}_K$ is the different of K so that $\mathcal{D}_K \mathcal{D}_K^{-1} = \mathcal{O}_K$.

Define the Kondo-Gauss sum $\tau_J(\rho)$ by

$$\tau_J(\rho) = \frac{1}{\dim(\rho)} \sum_{X \in J \cap U_K^0 / J \cap U_K^{n_J(\rho)}} \chi_\rho(c^{-1}X)\psi_K(\text{Trace}(c^{-1}X)).$$

Here χ_ρ is the character function given by $X \mapsto \text{Trace}(\rho) \in \mathbb{C}$.

LEMMA 1.3.

Suppose that ρ restricted to the centre $J \bigcap K^* = K^*$ is given by a central character ϕ (for example, if ρ is irreducible). Then $\tau_J(\rho)$ is well-defined in §1.2.

Proof

The $\chi_\rho(c^{-1}X)$ term is well-defined because if $X, X' \in J \bigcap U_K^0$ satisfy $X' = XU$ with $U \in J \bigcap U_K^{n_J(\rho)}$ then we have $\rho(U) = I$, the identity, so that

$$\chi_\rho(c^{-1}X') = \underline{\phi}(c)^{-1}\chi_\rho(XU) = \underline{\phi}(c)^{-1}\chi_\rho(X) = \chi_\rho(c^{-1}X).$$

Also, if $U = I + W$, then

$$\psi_K(\mathrm{Trace}(c^{-1}X')) = \psi_K(\mathrm{Trace}(c^{-1}X + c^{-1}W))$$

$$= \psi_K(\mathrm{Trace}(c^{-1}X))\psi_K(\mathrm{Trace}(c^{-1}W))$$

$$= \psi_K(\mathrm{Trace}(c^{-1}X))$$

because $\mathrm{Trace}(c^{-1}W) \in \mathcal{D}_K^{-1}$. $\square$

EXAMPLE 1.4. If $n_J(\rho) = 0$ then ρ is J-unramified and the formula becomes

$$\tau_J(\rho) = \frac{1}{\dim(\rho)} \dim(\rho)\underline{\phi}(c^{-1})\psi_K(\mathrm{Trace}(c^{-1})) = \underline{\phi}(c^{-1}) = \underline{\phi}(\mathcal{D}_K^{-1}).$$

LEMMA 1.5.
If $n \geq 0$ and $d \in \mathcal{P}_K^{-n}\mathcal{D}_K^{-1}$ then

$$\sum_{x \in \mathcal{O}_K/\mathcal{P}_K^n} \psi_K(xd) = \begin{cases} |N\mathcal{P}_K|^n & \text{if } d \in \mathcal{D}_K^{-1}, \\ \\ 0 & \text{otherwise.} \end{cases}$$

Proof

If $d \in \mathcal{D}_K^{-1}$ then $xd \in \mathcal{D}_K^{-1}$ and $\psi_K(xd) = 1$. Otherwise, if $x \equiv y$ (modulo $\mathcal{P}_K^n$) then $xd \equiv yd$ (modulo $\mathcal{P}_K^n d$) and $\mathcal{P}_K^n d \subseteq \mathcal{D}_K^{-1}$ so that $\psi_K(xd) = \psi_K(yd)$ and the sum is well-defined. But if $d \notin \mathcal{D}_K^{-1}$ there exists $x_1 \in \mathcal{O}_K$ such that $\psi_K(x_1 d) \neq 1$ and so

$$\sum_{x \in \mathcal{O}_K/\mathcal{P}_K^n} \psi_K(xd) = \sum_{x \in \mathcal{O}_K/\mathcal{P}_K^n} \psi_K((x+x_1)d) = \psi_K(x_1 d) \sum_{x \in \mathcal{O}_K/\mathcal{P}_K^n} \psi_K(xd),$$

which shows that the sum is zero. $\square$

COROLLARY 1.6.
If $n \geq 0$ and $d \in \mathcal{P}_K^{-n}\mathcal{D}_K^{-1}$ then

$$\sum_{x \in \mathcal{O}_K^*/U_K^n} \psi_K(xd) = \begin{cases} 1 & \text{if } n = 0, \\ \\ |N\mathcal{P}_K|^{n-1}(|N\mathcal{P}_K| - 1) & \text{if } d \in \mathcal{D}_K^{-1} \text{ and } n \geq 1, \\ \\ -|N\mathcal{P}_K|^{n-1} & \text{if } d \notin \mathcal{D}_K^{-1}, \ \pi_K d \in \mathcal{D}_K^{-1}, \\ & n \geq 1, \\ \\ 0 & \text{if } \pi_K d \notin \mathcal{D}_K^{-1}, \ n \geq 1. \end{cases}$$

Proof

Again the sum is well-defined. Since $\mathcal{O}_K/\mathcal{P}_K^n$ is local, $\mathcal{O}_K^*/U_K^n = (\mathcal{O}_K/\mathcal{P}_K^n)^* = \mathcal{O}_K/\mathcal{P}_K^n - \mathcal{P}_K/\mathcal{P}_K^n$. Therefore the order of $\mathcal{O}_K^*/U_K^n$ is equal to $|N\mathcal{P}_K|^{n-1}(|N\mathcal{P}_K| - 1)$ when $n \geq 1$ and to 1 when $n = 0$. This yields the formulae when $d \in \mathcal{D}_K^{-1}$. The formula

$$\sum_{x \in \mathcal{O}_K^*/U_K^n} \psi_K(xd) = \sum_{x \in \mathcal{O}_K/\mathcal{P}_K^n} \psi_K(xd) - \sum_{x \in \mathcal{O}_K/\mathcal{P}_K^{n-1}} \psi_K(\pi_K xd)$$

yields the other two formulae. $\square$

LEMMA 1.7.

In the situation of §1.2 and Lemma 1.3 suppose that $1 \leq n < n_J(\rho) << t$. Then, if $c_n \in \mathcal{P}_K^{-n}\mathcal{D}_K^{-1}$ and the multiplicity of 1 in ρ is zero (i.e. the Schur inner product satisfies $\langle \rho, 1 \rangle_{J \cap U_K^n / J \cap U_K^t} = 0$),

$$\sum_{X \in J \cap U_K^n / J \cap U_K^t} \chi_\rho(c_n X)\psi_K(\mathrm{Trace}(c_n X)) = 0.$$

In general

$$\sum_{X \in J \cap U_K^n / J \cap U_K^t} \chi_\rho(c_n X)\psi_K(\mathrm{Trace}(c_n X))$$

$$= \underline{\phi}(c_n)\psi_K(c_n)^m [J \bigcap U_K^n : J \bigcap U_K^t]\langle \rho, 1 \rangle_{J \cap U_K^n / J \cap U_K^t}.$$

Proof

This is a well-defined sum. If $X \equiv X'$ (modulo $\mathcal{P}_K^t \cdot M_m \mathcal{O}_K$) then $\mathrm{Trace}(c_n X - c_n X') \in \mathcal{D}_K^{-1}$ so that

$$1 = \psi_K(\mathrm{Trace}(c_n X - c_n X')) = \psi_K(\mathrm{Trace}(c_n X))\psi_K(\mathrm{Trace}(c_n X'))^{-1}.$$

Also

$$\chi_\rho(c_n X') = \underline{\phi}(c_n)\chi_\rho(X') = \underline{\phi}(c_n)\chi_\rho(XU)$$

for some $U \in U_K^t$ so that $\rho(XU) = \rho(X)$ and therefore

$$\chi_\rho(c_n X') = \underline{\phi}(c_n)\chi_\rho(X) = \chi_\rho(c_n X).$$

Then, if $X \in U_K^n$ with $n \geq 1$ we have

$$c_n X \in c_n + c_n \mathcal{P}_K^n \cdot M_m \mathcal{O}_K \subseteq c_n + \mathcal{D}_K^{-1} \cdot M_m \mathcal{O}_K.$$

Therefore the sum is equal to

$$\sum_{X \in J \cap U_K^n / J \cap U_K^t} \chi_\rho(c_n X)\psi_K(mc_n)$$

$$= \psi_K(c_n)^m \underline{\phi}(c_n) \sum_{X \in J \cap U_K^n / J \cap U_K^t} \chi_\rho(X)$$

$$= \underline{\phi}(c_n)\psi_K(c_n)^m [J \bigcap U_K^n : J \bigcap U_K^t]\langle \rho, 1 \rangle_{J \cap U_K^n / J \cap U_K^t},$$

as required. $\square$

The proof of Lemma 1.7 also yields the following result, since

$$\langle \rho, 1 \rangle_{J \cap U_K^{n_J(\rho)}/J \cap U_K^t} = \dim_{\mathbb{C}}(\rho).$$

LEMMA 1.8.

In the situation of §1.2 and Lemma 1.3 suppose that $1 \le n_J(\rho) << t$. Then, if $c_{n_J(\rho)} \in \mathcal{P}_K^{-n_J(\rho)} \mathcal{D}_K^{-1}$,

$$\sum_{X \in J \cap U_K^{n_J(\rho)}/J \cap U_K^t} \chi_\rho(c_{n_J(\rho)}X)\psi_K(\mathrm{Trace}(c_{n_J(\rho)}X))$$

$$= \underline{\phi}(c_{n_J(\rho)})\psi_K(c_{n_J(\rho)})^m [J \cap U_K^{n_J(\rho)} : J \cap U_K^t]\dim_{\mathbb{C}}(\rho).$$

LEMMA 1.9.

In the situation of §1.2 and Lemma 1.3 suppose that $1 \le n < n_J(\rho) <<$ t and $c_n \in \mathcal{P}_K^{-n} \mathcal{D}_K^{-1}$. Suppose that $n = n_J(\rho) - 1$, then

$$\sum_{X \in J \cap U_K^0/J \cap U_K^t} \chi_\rho(c_n X)\psi_K(\mathrm{Trace}(c_n X))$$

$$= [J \cap U_K^n : J \cap U_K^t]\sum_{X \in J \cap U_K^0/J \cap U_K^n} \chi_{V^{J \cap U_K^n}}(c_n Z_i)\psi_K(\mathrm{Trace}(c_n Z_i)),$$

where V is the vector space which affords the representation ρ.

Proof

Let $Z_1, \ldots, Z_s \in J \cap U_K^0$ be a set of coset representatives for $J \cap U_K^0/J \cap U_K^n$ and let $W_1, W_2 \ldots W_v \in J \cap U_K^n$ be a set of coset representatives for $J \cap U_K^n/J \cap U_K^t$. Hence $\{Z_i W_j \mid 1 \le i \le s, 1 \le j \le v\}$ is a set of coset representatives for $J \cap U_K^0/J \cap U_K^t$. In addition

$$c_n Z_i W_j \in c_n Z_i + c_n Z_i \mathcal{P}_K^n \subseteq M_m \mathcal{O}_K c_n Z_i + Z_i \mathcal{D}_K^{-1} M_m \mathcal{O}_K.$$

Therefore

$$\sum_{X \in J \cap U_K^0/J \cap U_K^t} \chi_\rho(c_n X)\psi_K(\mathrm{Trace}(c_n X))$$

$$= \sum_{i=1}^s \sum_{j=1}^v \underline{\phi}(c_n)\chi_\rho(Z_i W_j)\psi_K(\mathrm{Trace}(c_n Z_i)).$$

Now we shall use the fact that $n = n_J(\rho) - 1$ and we consider the matrix representation of

$$\sum_{i=1}^s \sum_{j=1}^v \underline{\phi}(c_n)\rho(Z_i W_j)\psi_K(\mathrm{Trace}(c_n Z_i))$$

$$= \sum_{i=1}^s \sum_{j=1}^v \underline{\phi}(c_n)\rho(Z_i)\rho(W_j)\psi_K(\mathrm{Trace}(c_n Z_i))$$

$$= \sum_{i=1}^s \underline{\phi}(c_n)\rho(Z_i)\psi_K(\mathrm{Trace}(c_n Z_i))(\sum_{j=1}^v \rho(W_j)).$$

The W_j's runs through $J \cap U_K^{n_J(\rho)-1}/J \cap U_K^t$ and ρ on this quotient group factors through the abelian quotient $J \cap U_K^{n_J(\rho)-1}/J \cap U_K^{n_J(\rho)}$. Therefore, in terms of matrices, we may diagonalise ρ on $J \bigcap U_K^{n_J(\rho)-1} / J \bigcap U_K^t$ as

$$W_j \mapsto \begin{pmatrix} \lambda_1(W_j) & 0 & 0 & \cdots & \cdots & 0 & & 0 & 0 \\ 0 & \lambda_2(W_j) & 0 & \cdots & \cdots & 0 & & 0 & 0 \\ \vdots & \vdots & \vdots & \vdots & \vdots & \vdots & \vdots & \vdots & \vdots \\ 0 & 0 & 0 & 0 & 0 & \lambda_k(W_j) & \cdots & \cdots & 0 & 0 \\ 0 & 0 & 0 & 0 & 0 & 0 & 1 & \cdots & 0 & 0 \\ \vdots & \vdots & \vdots & \vdots & \vdots & \vdots & \vdots & \vdots & \vdots \end{pmatrix}$$

where $\lambda_1, \ldots, \lambda_k$ are the only non-trivial characters

$$J \bigcap U_K^{n_J(\rho)-1} / J \bigcap U_K^t \longrightarrow \mathbb{C}^*$$

appearing on the diagonal of this matrix representation of ρ. Therefore the matrix $\sum_{j=1}^v \rho(W_j)$ has the form, summed over $J \cap U_K^{n_J(\rho)-1} / J \cap U_K^t$,

$$\sum_{j=1}^v \rho(W_j)$$

$$= \sum_W \begin{pmatrix} \lambda_1(W) & 0 & 0 & \cdots & \cdots & 0 & & 0 & 0 \\ \vdots & \vdots & \vdots & \vdots & \vdots & \vdots & \vdots & \vdots & \vdots \\ 0 & 0 & 0 & 0 & 0 & \lambda_k(W) & \cdots & \cdots & 0 & 0 \\ 0 & 0 & 0 & 0 & 0 & 0 & 1 & \cdots & 0 & 0 \\ \vdots & \vdots & \vdots & \vdots & \vdots & \vdots & \vdots & \vdots & \vdots \end{pmatrix}$$

$$= [J \cap U_K^{n_J(\rho)-1} : J \cap U_K^t] \begin{pmatrix} 0 & 0 \\ 0 & I_{d-k} \end{pmatrix}$$

where the three 0's are the $k \times k, k \times (d-k), (d-k) \times k$ zero matrices and I_{d-k} is the $(d-k) \times (d-k)$ identity matrix and $d = \dim_{\mathbb{C}}(V)$. The matrix I_{d-k} comes from the trivial action of ρ on the fixed-points $V^{J \cap U_K^{n_J(\rho)-1}}$.

Therefore, for each $1 \leq i \leq s$

$$\underline{\phi}(c_n) \sum_{j=1}^{v} \chi_\rho(Z_i W_j) \psi_K(\text{Trace}(c_n Z_i))$$

$$= [J \cap U_K^{n_J(\rho)-1} : J \cap U_K^t] \underline{\phi}(c_n) \times$$

$$\sum_{j=1}^{v} \chi_{V^{J \cap U_K^{n_J(\rho)-1}}}(Z_i) \psi_K(\text{Trace}(c_n Z_i))$$

$$= [J \cap U_K^{n_J(\rho)-1} : J \cap U_K^t] \sum_{j=1}^{v} \chi_{V^{J \cap U_K^{n_J(\rho)-1}}}(c_n Z_i) \psi_K(\text{Trace}(c_n Z_i))$$

$$= [J \cap U_K^n : J \cap U_K^t] \sum_{X \in J \cap U_K^0 / J \cap U_K^n} \chi_{V^{J \cap U_K^n}}(c_n Z_i) \psi_K(\text{Trace}(c_n Z_i)).$$

Next I shall prove Lemma 1.9 in general, by an induction based on the proof of that special case.

LEMMA 1.10.

In the situation of §1.2 and Lemma 1.3 suppose that $1 \leq n < n_J(\rho) << t$ and $c_n \in \mathcal{P}_K^{-n} \mathcal{D}_K^{-1}$. Then

$$\sum_{X \in J \cap U_K^0 / J \cap U_K^t} \chi_\rho(c_n X) \psi_K(\text{Trace}(c_n X))$$

$$= [J \cap U_K^n : J \cap U_K^t] \sum_{Z \in J \cap U_K^0 / J \cap U_K^n} \chi_{V^{J \cap U_K^n}}(c_n Z) \psi_K(\text{Trace}(c_n Z)),$$

where V is the vector space which affords the representation ρ.

Proof

Let $Z_1, \ldots, Z_s \in J \bigcap U_K^0$ be a set of coset representatives for $J \bigcap U_K^0 / J \bigcap U_K^n$ and let $W_1, \ldots, W_v \in J \bigcap U_K^n$ be a set of coset representatives for $J \bigcap U_K^n / J \bigcap U_K^t$. Hence $\{Z_i W_j\}$ is a set of coset representatives for $J \bigcap U_K^0 / J \bigcap U_K^t$. Let us subdivide the W_i's so that

$W_1, \ldots, W_{v_1}$ is a set of coset representatives for $J \bigcap U_K^{n_J(\rho)-1} / J \bigcap U_K^t$
$W_{v_1+1}, \ldots, W_{v_1+v_2}$ are coset representatives of
$$J \bigcap U_K^{n_J(\rho)-2} / J \bigcap U_K^{n_J(\rho)-1}$$
$W_{v_1+v_2+1}, \ldots, W_{v_1+v_2+v_3}$ is a set $\ldots$ for $J \bigcap U_K^{n_J(\rho)-3} / J \bigcap U_K^{n_J(\rho)-2}$

$$\vdots \quad \vdots \quad \vdots \quad \vdots \quad \vdots \quad \vdots \quad \vdots \quad \vdots$$

$W_{v_1+\ldots+v_{n_J(\rho)-n-1}+1}, \ldots, W_{v_1+\ldots+v_{n_J(\rho)-n}}$ is a set $\ldots$ for
$$J \bigcap U_K^n / J \bigcap U_K^{n+1}.$$

Let $d_u = [J \cap U_K^u : J \cap U_K^t]$. By the inductive step explained in the proof of Lemma 1.9 we have

$$\sum_{X \in J \cap U_K^0 / J \cap U_K^t} \chi_\rho(c_n X) \psi_K(\text{Trace}(c_n X))$$

$$= \sum_{i=1}^{s} \sum_{j=1}^{v_1 + \ldots + v_{n_J(\rho)} - n} \underline{\phi}(c_n) \chi_\rho(Z_i W_j) \psi_K(\text{Trace}(c_n Z_i))$$

$$= d_{n_J(\rho)-1} \sum_{i=1}^{s} \sum_{j=1}^{v_2 + \ldots + v_{n_J(\rho)} - n} \underline{\phi}(c_n) \chi_{V^{J \cap U_K^{n_J(\rho)-1}}}(Z_i W_j) \times$$
$$\psi_K(\text{Trace}(c_n Z_i))$$

$$= d_{n_J(\rho)-2} \sum_{i=1}^{s} \sum_{j=1}^{v_3 + \ldots + v_{n_J(\rho)} - n} \underline{\phi}(c_n) \chi_{V^{J \cap U_K^{n_J(\rho)-2}}}(Z_i W_j) \times$$
$$\psi_K(\text{Trace}(c_n Z_i))$$

$$= d_n \sum_{i=1}^{s} \underline{\phi}(c_n) \chi_{V^{J \cap U_K^n}}(Z_i) \psi_K(\text{Trace}(c_n Z_i)),$$

as required. □

LEMMA 1.11.

In the situation of §1.2 and Lemma 1.3 suppose that $0 = n = n_J(\rho) << t$. Then, if $c_0 \in \mathcal{D}_K^{-1}$,

$$\sum_{X \in J \cap U_K^0 / J \cap U_K^t} \chi_\rho(c_0 X) \psi_K(\text{Trace}(c_0 X))$$

$$= [J \cap U_K^0 : J \cap U_K^t] \dim_{\mathbb{C}}(\rho) \underline{\phi}(c_0).$$

Proof

Since ρ is trivial on $J \bigcap U_K^0$

$$\sum_{X \in J \cap U_K^0 / J \cap U_K^t} \chi_\rho(c_0 X) \psi_K(\text{Trace}(c_0 X))$$

$$= \sum_{X \in J \cap U_K^0 / J \cap U_K^t} \underline{\phi}(c_0) \dim_{\mathbb{C}}(\rho) \psi_K(\text{Trace}(c_0 X))$$

$$= [J \cap U_K^0 : J \cap U_K^t] \dim_{\mathbb{C}}(\rho) \underline{\phi}(c_0),$$

because $\text{Trace}(c_0 X) \in \mathcal{D}_K^{-1}$ so that $\psi_K(\text{Trace}(c_0 X)) = 1$. □

LEMMA 1.12.

In the situation of §1.2 and Lemma 1.3 suppose that

$$\langle c_{n_J(\rho)} \rangle = \mathcal{P}^{-n_J(\rho)} \mathcal{D}_K^{-1},$$

$$\sum_{X \in J \cap U_K^0 / J \cap U_K^{n_J(\rho)}} \chi_\rho(X) \psi_K(\text{Trace}(c_0 X)) = \dim_{\mathbb{C}}(\rho) \frac{\tau_J(\rho)}{\underline{\phi}(c_0)}.$$

Proof

By definition. □

LEMMA 1.13.

In the situation of §1.2 and Lemma 1.3 suppose that $c_n \in \mathcal{P}_K^{-n}\mathcal{D}_K^{-1}$ and $1 \leq n_J(\rho) < n << t$. Then

$$\sum\nolimits_{X \in J \cap U_K^{n_J(\rho)}/J \cap U_K^t} \chi_\rho(c_n X)\psi_K(\text{Trace}(c_n X))$$

$$= \underline{\phi}(c_n)[J \cap U_K^n : J \cap U_K^t]\sum\nolimits_{X \in J \cap U_K^{n_J(\rho)}/J \cap U_K^n} \psi_K(\text{Trace}(c_n X)).$$

Proof

Let $Z_1, \ldots, Z_s \in J \cap U_K^{n_J(\rho)}$ be coset representatives for $J \cap U_K^{n_J(\rho)}/J \cap U_K^n$ and let $W_1, \ldots, W_v \in J \cap U_K^n$ be a set of coset representatives for $J \cap U_K^n/J \cap U_K^t$. Hence $\{Z_i W_j\}$ are coset representatives for $J \cap U_K^{n_J(\rho)}/J \cap U_K^t$. Then

$$\sum\nolimits_{X \in J \cap U_K^{n_J(\rho)}/J \cap U_K^t} \chi_\rho(c_n X)\psi_K(\text{Trace}(c_n X))$$

$$= \underline{\phi}(c_n) \sum\nolimits_{i=1}^s \sum\nolimits_{j=1}^v \psi_K(\text{Trace}(c_n Z_i W_j))$$

$$= \underline{\phi}(c_n) \sum\nolimits_{i=1}^s \sum\nolimits_{j=1}^v \psi_K(\text{Trace}(c_n Z_i))\psi_K(\text{Trace}(c_n Z_i(W_j - 1)))$$

$$= \underline{\phi}(c_n)[J \cap U_K^n : J \cap U_K^t]\sum\nolimits_{X \in J \cap U_K^{n_J(\rho)}/J \cap U_K^n} \psi_K(\text{Trace}(c_n X)),$$

as required. $\square$

The formula of Lemma 1.13 features a constant which depends on the identity of J, which is an issue I shall examine below.

LEMMA 1.14.

In the situation of §1.2 and Lemma 1.3 suppose that $c_n \in \mathcal{P}_K^{-n}\mathcal{D}_K^{-1}$ and $1 \leq n_J(\rho) < n << t$. Then

$$\sum\nolimits_{X \in J \cap U_K^0/J \cap U_K^t} \chi_\rho(c_n X)\psi_K(\text{Trace}(c_n X))$$

$$= \sum\nolimits_{Z \in J \cap U_K^0/J \cap U_K^{n_J(\rho)}} \underline{\phi}(c_n)\chi_\rho(Z)[J \cap U_K^n : J \cap U_K^t]A(Z, W)$$

where $A(Z, W) = \sum\nolimits_{W \in J \cap U_K^{n_J(\rho)}/J \cap U_K^n} \psi_K(\text{Trace}(c_n ZW))$.

Proof

Let $Z_1, \ldots, Z_s \in J \cap U_K^0$ be a set of coset representatives for $J \cap U_K^0/J \cap U_K^{n_J(\rho)}$ and let $W_1, \ldots, W_v \in J \cap U_K^{n_J(\rho)}$ be a set of coset representatives for $J \cap U_K^{n_J(\rho)}/J \cap U_K^t$. Therefore

$$\sum_{X \in J \cap U_K^0/J \cap U_K^t} \chi_\rho(c_n X)\psi_K(\mathrm{Trace}(c_n X))$$

$$= \sum_{i=1}^s \sum_{j=1}^v \underline{\phi}(c_n)\chi_\rho(Z_i W_j)\psi_K(\mathrm{Trace}(c_n Z_i W_j))$$

$$= \sum_{Z \in J \cap U_K^0/J \cap U_K^{n_J(\rho)}} \underline{\phi}(c_n)\chi_\rho(Z)[J \cap U_K^n : J \cap U_K^t]A(Z,W),$$

as required. $\square$

The formula of Lemma 1.14 contains occurrences of the constant from Lemma 1.13, which depends on the identity of J, which is an issue I shall examine below.

LEMMA 1.15.

In the situation of §1.2 and Lemma 1.3 suppose that

$$\langle c_{n_J(\rho)} \rangle = \mathcal{P}_K^{-n_J(\rho)}\mathcal{D}_K^{-1}$$

and $1 \leq n_J(\rho) << t$. Then

$$\sum_{X \in J \cap U_K^0/J \cap U_K^t} \chi_\rho(c_{n_J(\rho)}X)\psi_K(\mathrm{Trace}(c_{n_J(\rho)}X))$$

$$= [J \cap U_K^{n_J(\rho)} : J \cap U_K^t]\tau_J(\rho).$$

Proof

Let $Z_1, \ldots, Z_s \in J \cap U_K^0$ be a set of coset representatives for $J \cap U_K^0/J \cap U_K^{n_J(\rho)}$ and let $W_1, \ldots, W_v \in J \cap U_K^{n_J(\rho)}$ be a set of coset representatives for $J \cap U_K^{n_J(\rho)}/J \cap U_K^t$. Therefore

$$\sum_{X \in J \cap U_K^0/J \cap U_K^t} \chi_\rho(c_{n_J(\rho)}X)\psi_K(\mathrm{Trace}(c_{n_J(\rho)}X))$$

$$= \sum_{i=1}^s \sum_{j=1}^v \underline{\phi}(c_{n_J(\rho)})\chi_\rho(Z_i W_j)\psi_K(\mathrm{Trace}(c_{n_J(\rho)}Z_i W_j))$$

$$= \sum_{i=1}^s \sum_{j=1}^v \underline{\phi}(c_{n_J(\rho)})\chi_\rho(Z_i)\psi_K(\mathrm{Trace}(c_{n_J(\rho)}Z_i)) \times$$
$$\psi_K(\mathrm{Trace}(c_{n_J(\rho)}Z_i(W_j - 1)))$$

$$= \sum_{i=1}^s \chi_\rho(c_{n_J(\rho)}Z_i)\psi_K(\mathrm{Trace}(c_{n_J(\rho)}Z_i))[J \cap U_K^{n_J(\rho)} : J \cap U_K^t]$$

$$= [J \cap U_K^{n_J(\rho)} : J \cap U_K^t]\tau_J(\rho),$$

as required. $\square$

1.16. In the situation of §1.2 and Lemma 1.3 suppose that $1 \leq n < n_J(\rho) << t$. Set ρ_n equal to the representation given on $V^{J \cap U_K^n}$

$$\rho_n : J \bigcap U_K^0 \longrightarrow \mathrm{Aut}(V^{J \cap U_K^n})$$

so that $n_J(\rho_n) \leq n << t$.

If $1 \leq n_J(\rho_n) \leq n \ll t$ we have

$$\sum_{X \in J \cap U_K^0 / J \cap U_K^t} \chi_\rho(c_n X) \psi_K(\text{Trace}(c_n X))$$

$$= [J \cap U_K^n : J \cap U_K^t] \sum_{X \in J \cap U_K^0 / J \cap U_K^{n_J(\rho_n)}} \chi_{\rho_n}(c_n X) \psi_K(\text{Trace}(c_n X))$$

$$+ [J \cap U_K^n : J \cap U_K^t] \underline{\phi}(c_n) \langle \psi_K(c_n \cdot -), 1 \rangle_{J \cap U_K^{n_J(\rho_n)}}$$

$$= [J \cap U_K^{n_J(\rho_n)} : J \cap U_K^t] \tau_J(\rho_n)$$
$$+ [J \cap U_K^n : J \cap U_K^t] \underline{\phi}(c_n) \langle \psi_K(c_n \cdot -), 1 \rangle_{J \cap U_K^{n_J(\rho_n)}}.$$

THEOREM 1.17.

In the situation of §1.2 and Lemma 1.3 suppose that $H \subseteq J$ are two compact modulo the centre subgroups with $[J : H]$ finite and that t is large enough so that $U_t \subseteq H$. Let σ be a finite-dimensional representation of H. Then

$$\frac{1}{\dim_{\mathbb{C}}(\sigma)} \sum_{X \in H \cap U_K^0 / H \cap U_K^t} \chi_\sigma(c_n X) \psi_K(\text{Trace}(c_n X))$$

$$= \frac{1}{\dim_{\mathbb{C}}(\text{Ind}_{H \cap U_K^0}^{J \cap U_K^0}(\sigma))} \sum_{X \in J \cap U_K^0 / J \cap U_K^t} \chi_{\text{Ind}_{H \cap U_K^0}^{J \cap U_K^0}(\sigma)}(c_n X) \psi_K(\text{Trace}(c_n X)).$$

Proof

The proof is going to use the same argument as the proof of Theorem 3.2 in Appendix III. Since $H \cap U_K^t = J \cap U_K^t$ we may consider σ as a representation of $\tilde{H} = H \cap U_K^0 / H \cap U_K^t$. If $\tilde{J} = J \cap U_K^0 / J \cap U_K^t$ we have $\tilde{H} \subseteq \tilde{J}$ and we may set $\rho = \text{Ind}_{\tilde{H}}^{\tilde{J}}(\sigma)$.

By definition

$$\frac{|\tilde{H}|}{|\tilde{J}| \cdot \dim_{\mathbb{C}}(\sigma)} \sum_{X \in \tilde{J}} \chi_\rho(c_n X) \psi_K(\text{Trace}(c_n X))$$

$$= \frac{1}{|\tilde{J}| \cdot \dim_{\mathbb{C}}(\sigma)} \sum_{X \in \tilde{J}} \sum_{Y \in \tilde{J}, \, YXY^{-1} \in \tilde{H}} \chi_\sigma(c_n YXY^{-1}) \psi_K(\text{Trace}(c_n X))$$

$$= \frac{1}{|\tilde{J}| \cdot \dim_{\mathbb{C}}(\sigma)} \sum_{X \in \tilde{J}} \sum_{Y \in \tilde{J}, \, YXY^{-1} \in \tilde{H}} \chi_\sigma(c_n YXY^{-1}) \times$$
$$\psi_K(\text{Trace}(c_n YXY^{-1}))$$

by the character formula for an induced representation ([**126**] Theorem 1.2.43). Consider the free action of $\tilde{J}$ on $\tilde{J} \times \tilde{J}$ given by $(X, Y)Z = (Z^{-1}XZ, YZ)$ for $X, Y, Z \in \tilde{J}$. The map from $\tilde{J} \times \tilde{J}$ to $\tilde{J}$ sending (X, Y) to

YXY^{-1} is constant on each $\tilde{J}$-orbit. Therefore

$$\frac{1}{|\tilde{J}|\cdot\dim_{\mathbb{C}}(\sigma)} \sum_{X\in\tilde{J}} \sum_{Y\in\tilde{J},\, YXY^{-1}\in\tilde{H}} \chi_\sigma(c_n YXY^{-1})\times$$
$$\psi_K(\text{Trace}(c_n YXY^{-1}))$$

$$= \frac{1}{|\tilde{J}|\cdot\dim_{\mathbb{C}}(\sigma)} |\tilde{J}| \sum_{U\in\tilde{H}} \chi_\sigma(c_n U)\psi_K(\text{Trace}(c_n U)),$$

as required. $\square$

1.18. *The constant from Lemma 1.13*
In the situation of §1.2 and Lemma 1.3 suppose that $c_n \in \mathcal{P}_K^{-n}\mathcal{D}_K^{-1}$ and $1 \leq n_J(\rho) < n$. Then, in Lemma 1.13, we encountered the constant

$$\sum_{X\in J\cap U_K^{n_J(\rho)}/J\cap U_K^n} \psi_K(\text{Trace}(c_n X)).$$

I would like to evaluate this in the case when $J \subseteq GL_2 K$ is one of:
Case (i): $GL_2\mathcal{O}_K$ and
Case (ii):

$$J = \left\{ \begin{pmatrix} a & b\pi_K \\ c & d \end{pmatrix} \mid a,b,c,d \in \mathcal{O}_K,\ ad \in \mathcal{O}_K^* \right\}.$$

Case (iii):

$$J = \left\langle \begin{pmatrix} 0 & 1 \\ \pi_K^{-1} & 0 \end{pmatrix}, \begin{pmatrix} a & b\pi_K \\ c & d \end{pmatrix} \mid a,b,c,d \in \mathcal{O}_K,\ ad \in \mathcal{O}_K^* \right\rangle.$$

LEMMA 1.19.
In Case (i), Case (ii) and Case (iii) of §1.18

$$\sum_{X\in J\cap U_K^{n_J(\rho)}/J\cap U_K^n} \psi_K(\text{Trace}(c_n X)) = 0$$

Proof
In Case (i) $GL_2\mathcal{O}_K \cap U_K^{n_J(\rho)} = U_K^{n_J(\rho)}$ and $GL_2\mathcal{O}_K \cap U_K^n = U_K^n$ so that the coset representatives X have the form

$$I_2 + \pi_K^{n_J(\rho)} \begin{pmatrix} a & b \\ c & d \end{pmatrix}$$

with $a,b,c,d \in \mathcal{O}_K/\mathcal{P}_K^{n-n_J(\rho)}$. Therefore

$$\psi_K(c_n X) = \psi_K(c_n)^2 \psi_K(c_n \pi_K^{n_J(\rho)}(a+d))$$

and so

$$\sum_{X \in GL_2\mathcal{O}_K \cap U_K^{n_J(\rho)}/GL_2\mathcal{O}_K \cap U_K^n} \psi_K(\text{Trace}(c_n X))$$

$$= \sum_{X \in U_K^{n_J(\rho)}/U_K^n} \psi_K(\text{Trace}(c_n X))$$

$$= \psi_K(c_n)^2 |N\mathcal{P}_K|^{2n-2n_J(\rho)} \sum_{a,d \in \mathcal{O}_K/\mathcal{P}_K^{n-n_J(\rho)}} \psi_K(c_n \pi_K^{n_J(\rho)}(a+d))$$

$$= \psi_K(c_n)^2 |N\mathcal{P}_K|^{3n-3n_J(\rho)} \sum_{z \in \mathcal{O}_K/\mathcal{P}_K^{n-n_J(\rho)}} \psi_K(c_n \pi_K^{n_J(\rho)} z)$$

$$= 0,$$

by Lemma 1.5 since $c_n \pi_K^{n_J(\rho)}$ generates $\mathcal{P}_K^{n_J(\rho)-n}\mathcal{D}_K^{-1} \not\subseteq \mathcal{D}_K^{-1}$.

In Case (ii) also we have $J \cap U_K^{n_J(\rho)} = U_K^{n_J(\rho)}$ and $J \cap U_K^n = U_K^n$ so that the sum is again zero.

In Case (iii) the sum to be evaluated coincides with that of Case (ii) because the intersections of these two J's with U_K^m are the same. $\square$

1.20. *Results in Cases (i), (ii) and (iii) of §1.18*
Assume that J is as in Cases (i) and (ii) of §1.18 so that

$$\sum_{X \in J \cap U_K^{n_J(\rho)}/J \cap U_K^n} \psi_K(\text{Trace}(c_n X))$$

$$= \sum_{X \in U_K^{n_J(\rho)}/U_K^n} \psi_K(\text{Trace}(c_n X))$$

$$= 0$$

when $1 \leq n_J(\rho) < n$ and $\langle c_n \rangle = \mathcal{P}_K^{-n}\mathcal{D}_K^{-1}$. In addition, if $Z \in GL_2\mathcal{O}_K$, then we also have

$$\sum_{X \in J \cap U_K^{n_J(\rho)}/J \cap U_K^n} \psi_K(\text{Trace}(c_n ZX))$$

$$= \sum_{X \in U_K^{n_J(\rho)}/U_K^n} \psi_K(\text{Trace}(c_n ZX))$$

$$= \sum_{X \in U_K^{n_J(\rho)}/U_K^n} \psi_K(\text{Trace}(c_n Z))\psi_K(\text{Trace}(c_n Z(X - I_2)))$$

$$= \sum_{X \in U_K^{n_J(\rho)}/U_K^n} \psi_K(\text{Trace}(c_n Z))\psi_K(\text{Trace}(c_n(X - I_2)))$$

$$= 0,$$

because during this sum the matrices $Z(X - I_2)$ run over a set which is independent of $Z \in GL_2\mathcal{O}_K$.

If $\rho^{J \cap U_K^n}$ denotes the representation of J given by the $J \cap U_K^n$-fixed points of ρ then for any irreducible ρ we have

$$\rho^{J \cap U_K^n} = \begin{cases} \rho & \text{if } n_J(\rho) \leq n, \\ 0 & \text{otherwise.} \end{cases}$$

Therefore, by Lemma 1.10, $1 \leq n < n_J(\rho) << t$ and $\langle c_n \rangle = \mathcal{P}_K^{-n} \mathcal{D}_K^{-1}$ and ρ is an irreducible J-representation

$$\sum_{X \in J \cap U_K^0 / J \cap U_K^t} \chi_\rho(c_n X) \psi_K(\text{Trace}(c_n X))$$

$$= [J \cap U_K^n : J \cap U_K^t] \sum_{X \in J \cap U_K^0 / J \cap U_K^n} \chi_{\rho^{J \cap U_K^n}}(c_n Z_i) \psi_K(\text{Trace}(c_n Z_i))$$

$$= 0.$$

In addition, if $n = 0$ with ρ non-trivial, irreducible and $1 \leq n_J(\rho) << t$ then $\langle c_0 \rangle = \mathcal{D}_K^{-1}$ so that

$$\sum_{X \in J \cap U_K^0 / J \cap U_K^t} \chi_\rho(c_0 X) \psi_K(\text{Trace}(c_0 X))$$

$$= \underline{\phi}(c_0) \sum_{X \in J \cap U_K^0 / J \cap U_K^t} \chi_\rho(X)$$

$$= \underline{\phi}(c_0) [J \cap U_K^0 : J \cap U_K^t] \langle \rho, 1 \rangle_{J \cap U_K^0}$$

$$= 0.$$

From Lemma 1.15 we have, if $1 \leq n_J(\rho) << t$ and

$$\langle c_{n_J(\rho)} \rangle = \mathcal{P}_K^{-n_J(\rho)} \mathcal{D}_K^{-1},$$

$$\sum_{X \in J \cap U_K^0 / J \cap U_K^t} \chi_\rho(c_{n_J(\rho)} X) \psi_K(\text{Trace}(c_{n_J(\rho)} X))$$

$$= [J \cap U_K^{n_J(\rho)} : J \cap U_K^t] \sum_{X \in J \cap U_K^0 / J \cap U_K^{n_J(\rho)}} \chi_\rho(c_{n_J(\rho)} X) \times$$
$$\psi_K(\text{Trace}(c_{n_J(\rho)} X)).$$

Next we turn to the range $1 \leq n_J(\rho) < n << t$ and $\langle c_n \rangle = \mathcal{P}_K^{-n} \mathcal{D}_K^{-1}$. By Lemma 1.14

$$\sum_{X \in J \cap U_K^0 / J \cap U_K^t} \chi_\rho(c_n X) \psi_K(\text{Trace}(c_n X))$$

$$= \sum_{Z \in J \cap U_K^0 / J \cap U_K^{n_J(\rho)}} \underline{\phi}(c_n) \chi_\rho(Z) [J \cap U_K^n : J \cap U_K^t] A(Z, W)$$

$$= 0,$$

where $A(Z, W) = \sum_{W \in J \cap U_K^{n_J(\rho)} / J \cap U_K^n} \psi_K(\text{Trace}(c_n Z W))$.

Therefore if ρ is non-trivial, irreducible with $n_J(\rho) \geq 1$ and $\langle c_n \rangle = \mathcal{P}_K^{-n}\mathcal{D}_K^{-1}$ then

$$\lim_{N_0,t\to\infty} \sum_{0\leq n\leq N_0 <<t} \sum_{X\in J\cap U_K^0/J\cap U_K^t} \frac{\chi_\rho(c_n X)\psi_K(\text{Trace}(c_n X))}{\dim_{\mathbb{C}}(\rho)[J\cap U_K^n:J\cap U_K^t]}$$

$$= \sum_{X\in J\cap U_K^0/J\cap U_K^{n_J(\rho)}} \frac{\chi_\rho(c_{n_J(\rho)}X)\psi_K(\text{Trace}(c_{n_J(\rho)}X))}{\dim_{\mathbb{C}}(\rho)}$$

$$= \tau_J(\rho).$$

Next we examine the case when $n_J(\rho) = 0$ so that

$$\lim_{N_0,t\to\infty} \sum_{0\leq n\leq N_0<<t} \sum_{X\in J\cap U_K^0/J\cap U_K^t} \frac{\chi_\rho(c_n X)\psi_K(\text{Trace}(c_n X))}{\dim_{\mathbb{C}}(\rho)[J\cap U_K^n:J\cap U_K^t]}$$

$$= \lim_{N_0,t\to\infty} \sum_{0\leq n\leq N_0<<t} \sum_{X\in J\cap U_K^0/J\cap U_K^t} \frac{\phi(c_n)\dim_{\mathbb{C}}(\rho)\psi_K(\text{Trace}(c_n X))}{\dim_{\mathbb{C}}(\rho)[J\cap U_K^n:J\cap U_K^t]}$$

$$= \lim_{N_0,t\to\infty} \sum_{0\leq n\leq N_0<<t} \sum_{X\in J\cap U_K^0/J\cap U_K^t} \frac{\phi(c_n)\psi_K(\text{Trace}(c_n X))}{[J\cap U_K^n:J\cap U_K^t]}$$

$$= \underline{\phi}(c_0) + \lim_{N_0,t\to\infty} \sum_{1\leq n\leq N_0<<t} \sum_{X\in J\cap U_K^0/J\cap U_K^t} \frac{\phi(c_n)\psi_K(\text{Trace}(c_n X))}{[J\cap U_K^n:J\cap U_K^t]}.$$

However, if $n \geq 1$, let $X_1,\ldots,X_s \in J\cap U_K^0$ be a set of coset representatives of $J\cap U_K^0/J\cap U_K^1$ and let $W_1,\ldots,W_v \in U_K^1$ be a set of coset representatives of $J\cap U_K^1/J\cap U_K^t$. Therefore

$$\sum_{X\in J\cap U_K^0/J\cap U_K^t} \psi_K(\text{Trace}(c_n X))$$

$$= \sum_{i=1}^s \sum_{j=1}^v \psi_K(\text{Trace}(c_n Z_i W_j))$$

$$= \sum_{i=1}^s \psi_K(\text{Trace}(c_n Z_i))\sum_{j=1}^v \psi_K(\text{Trace}(c_n Z_i(W_j - I_2)))$$

$$= \sum_{i=1}^s \psi_K(\text{Trace}(c_n Z_i))\sum_{j=1}^v \psi_K(\text{Trace}(c_n(W_j - I_2)))$$

$$= 0,$$

by the argument used in §1.20. Hence we have shown that

$$\lim_{N_0,t\to\infty} \sum_{0\leq n\leq N_0<<t} \sum_{X\in J\cap U_K^0/J\cap U_K^t} \frac{\chi_\rho(c_n X)\psi_K(\text{Trace}(c_n X))}{\dim_{\mathbb{C}}(\rho)[J\cap U_K^n:J\cap U_K^t]}$$

$$= \underline{\phi}(c_0)$$

$$= \tau_J(\rho)$$

in this case also.

2. Tate's thesis in the compact modulo the centre case

2.1. Let K be a p-adic local field and let $\mathcal{O}_K$, π_K, $\mathcal{P}_K$ etc. be as in §1.1. Throughout this section G will be a locally p-adic Lie group containing K^* within its centre and such that G/K^* is compact. In fact we rather favour the case when G/K^* is finite, reducing the more general case to this one. For example, within GL_2K we have the subgroup

$$G = \langle K^*, u \rangle \subset GL_2K$$

where, as usual,

$$u = \begin{pmatrix} 0 & 1 \\ \pi_K^{-1} & 0 \end{pmatrix}$$

so that $u^2 = \pi_K^{-1} \in K^*$.

2.2. Let G be a locally profinite group and $C_c^\infty(G)$ the space of locally constant functions $f : G \longrightarrow \mathbb{C}$ of compact support [40]. For any such function there are compact open subgroups C_1, C_2 such that $f(c_1gc_2) = f(g)$ for all $g \in G$, $c_i \in C_i$. Hence f is a linear combination of characteristic functions of $C\backslash G/C$ with $C = C_1 \bigcap C_2$.

For example, if G is the additive group of a p-adic local field K then any compact open subgroup contains a compact open subgroup of the form $\mathcal{P}_K^n$ of finite index. Hence any $f \in C_c^\infty(K)$ is a finite linear combination of characteristic functions $\Phi_{x+\mathcal{P}_K^n}$ of cosets $x + \mathcal{P}_K^n$. Hence there exists an integer $r \geq 0$ such that $f(x) \neq 0$ implies $x \in \mathcal{P}_K^{-r}$ and f is constant on each coset $x + \mathcal{P}_K^r$. This Schwartz-Bruhat space is denoted by $S(K)$ in ([86] p. 115; see also [143]) (see §2.3).

Another important example is given by $G = K^*$. As usual, set $U_K^0 = \mathcal{O}_K^*$ and $U_K^n = 1 + \mathcal{P}_K^n$ for $n \geq 1$. Any compact open subgroup of K^* contains a compact open subgroup of the form U_K^n of finite index. Hence any $f \in C_c^\infty(K^*)$ is a finite linear combination of characteristic functions $\Phi_{xU_K^n}$ (with $n \geq 0$). Suppose that $f = \sum_{i=1}^t \alpha_i \Phi_{x_iU_K^{n_i}}$ then we may assume that all n_i's are equal so that f is constant on each coset intersection $(x + \mathcal{P}_K^r) \bigcap K^*$ for large enough r. Also $x_iU_K^{n_i} \subset \mathcal{P}_K^{v_K(x_i)} + \mathcal{P}_K^{n_i+v_K(x_i)}$ so that, for large enough $r \geq 0$, $f(x) \neq 0$ implies that $x \in \mathcal{P}_K^r \bigcap K^*$. Hence we have an inclusion $C_c^\infty(K^*) \subset C_c^\infty(K) = S(K)$ given by extending f by $f(0) = 0$.

On $C_c^\infty(G)$ we have left and right translations given respectively by $\lambda_g(f)(x) = f(g^{-1}x)$ and $\rho_g(f)(x) = f(xg)$. A right Haar integral is a non-zero linear functional

$$I_G : C_c^\infty(G) \longrightarrow \mathbb{C}$$

such that $I(\rho_g(f)) = I_G(f)$ and $I_G(f) \geq 0$ if $f \geq 0$. If it exists then I_G is unique up to multiplication by positive scalars. G is called unimodular if any right Haar measure is also a left Haar measure.

We shall start by recalling the explicit formulae for the Haar integral on K and K^*.

EXAMPLE 2.3. *Haar integration on K*

Let r be a positive integer. A measure on $\mathcal{P}_K^{-r}$ is a family of functions $\phi_{K,r,n} : \mathcal{P}_K^{-r}/\mathcal{P}_K^n \longrightarrow \mathbb{C}$ for $n \geq N_0$ which satisfy

$$\phi_{K,r,n}(x + \mathcal{P}_K^n) = \sum_{y+\mathcal{P}_K^{n+1} \mid y \in x+\mathcal{P}_K^n} \phi_{K,r,n+1}(y + \mathcal{P}_K^{n+1}).$$

In this case the sum

$$I_{K,r,n}(f) = \sum_{x \in \mathcal{P}_K^{-r}/\mathcal{P}_K^n} f(x)\phi_{K,r,n}(x + \mathcal{P}_K^n)$$

is well-defined for each $n >> 0$ and independent of n. This is because f is locally constant so that there is an n such that $f(x)$ depends only on the coset $x + \mathcal{P}_K^n$ and in this case

$$I_{K,r,n+1}(f)$$

$$= \sum_{y \in \mathcal{P}_K^{-r}/\mathcal{P}_K^{n+1}} f(y)\phi_{K,r,n+1}(y + \mathcal{P}_K^{n+1})$$

$$= \sum_{x \in \mathcal{P}_K^{-r}/\mathcal{P}_K^n} f(x) \sum_{y+\mathcal{P}_K^{n+1} \mid y \in x+\mathcal{P}_K^n} \phi_{K,r,n+1}(y + \mathcal{P}_K^{n+1})$$

$$= I_{K,r,n}(f).$$

If, for example, we set $\phi_{K,r,n}(x + \mathcal{P}_K^n) = |\mathcal{O}_K/\mathcal{P}_K|^{(1/2)-n}$ for all $n \geq 0$ then

$$\sum_{y+\mathcal{P}_K^{n+1} \mid y \in x+\mathcal{P}_K^n} \phi_{K,n+1}(y + \mathcal{P}_K^{n+1}) = |\mathcal{P}_K^n/\mathcal{P}_K^{n+1}||\mathcal{O}_K/\mathcal{P}_K|^{(1/2)-n-1}$$

$$= |\mathcal{O}_K/\mathcal{P}_K|^{(1/2)-n},$$

as required. Usually the integer $|\mathcal{O}_K/\mathcal{P}_K|$ is denoted by q.

Now let $f \in C_c^\infty(K)$, the set of compactly supported and locally constant functions

$$f : K \longrightarrow \mathbb{C}.$$

This means (see §2.2) that there exists an integer $t \geq 0$, depending on f, such that

(i) $f(x) \neq 0$ implies that $x \in \mathcal{P}_K^{-t}$ and
(ii) if $x, y \in K$ and $x - y \in \mathcal{P}_K^t$ then $f(x) = f(y)$.

Recall that there is a chain of fractional ideals of the form

$$\mathcal{P}_K^{-r} \subset \mathcal{P}_K^{-r-1} \subset \ldots \subset K.$$

Choosing integers $r, n \gg 0$ we define

$$I_K(f) = I_{K,r,n}(f)$$

which will serve as a formula for a Haar integral on K once we have verified invariance under right translation, which is seen as follows.

For $a \in K$ set $f_a(x) = f(a + x)$. Choose r so large that $a \in \mathcal{P}_K^{-r}$ then as $x + \mathcal{P}_K^n$ runs through $\mathcal{P}_K^{-r}/\mathcal{P}_K^n$ so does $a + x + \mathcal{P}_K^n$ and vice versa so that $I_{K,r,n}(f) = I_{K,r,n}(f_a)$, as required.

In the integral notation it is usual to write $I_K(f) = \int_K f(x)dx$, in the spirit of calculus!

EXAMPLE 2.4. *Haar integration on K^**

If $f \in C_c^\infty(K^*)$ it is simple enough to find a multiplicative measure on K^* to define $I_{K^*}(f) = \int f(x)d^*x$. Alternatively, we may extend f by zero at 0 to give a function (denoted still by f) lying in $C_c^\infty(K)$ and set

$$I_{K^*}(f(x)) = I_K(f(x)|x|^{-1}).$$

Recall that the normalised absolute value of $x \in K^*$ is given by

$$|x|_K = |\mathcal{O}_K/\mathcal{P}_K|^{-v_K(x)} = q^{-v_K(x)}.$$

Explicitly, for $r, n \gg 0$

$$I_{K^*}(f(x))$$

$$= I_K(f(x)|x|^{-1})$$

$$= I_{K,r,n}(f(x)|x|^{-1})$$

$$= \sum_{x \in \mathcal{O}_K \pi_K^{-r}/\mathcal{P}_K^n} f(x)q^{v_K(x)}\phi_{K,r,n}(x + \mathcal{P}_K^n).$$

This is well-defined and will serve as a formula for a Haar integral on K^* once we have verified invariance under right multiplication, which is seen as follows. Given $a \in K^*$ we may write $a = b\pi_K^j$ with $b \in \mathcal{O}_K^*$. Therefore as x runs through $\mathcal{P}_K^{-r}/\mathcal{P}_K^n$ the element ax runs through $\mathcal{P}_K \pi_K^{-r+j}/\mathcal{P}_K^{n+j}$

and vice versa when $r, r + j, n, n + j \gg 0$. Therefore

$$I_K(f(x)|x|^{-1})$$

$$= \sum_{ax \in \mathcal{O}_K \pi_K^{-r+j}/\mathcal{P}_K^{n+j}} f(ax) q^{v_K(ax)} \phi_{K,r,n+j}(ax + \mathcal{P}_K^{n+j})$$

$$= \sum_{x \in \mathcal{O}_K \pi_K^{-r}/\mathcal{P}_K^n} f(ax) q^{v_K(x)} q^j \phi_{K,r,n+j}(ax + \mathcal{P}_K^{n+j})$$

$$= \sum_{x \in \mathcal{O}_K \pi_K^{-r}/\mathcal{P}_K^n} f(ax) q^{v_K(x)} q^j q^{(1/2)-n-j}$$

$$= \sum_{x \in \mathcal{O}_K \pi_K^{-r}/\mathcal{P}_K^n} f(ax) q^{v_K(x)} q^{(1/2)-n}$$

$$= I_K(f(ax)|x|^{-1}),$$

as required.

DEFINITION 2.5. *Eigendistributions* [86]

Let $S(K)$ denote the space of Schwartz-Bruhat functions on K and let $S'(K)$ be the space of tempered distributions on K, which is the space of continuous linear functionals $\lambda : S(K) \longrightarrow \mathbb{C}$. A complex-valued function is in $S(K)$ if and only if there exists an integer $r \geq 0$ such that (a) supp$(f) \subseteq \mathcal{P}^{-r}$ (compact support) and f is constant on cosets of $\mathcal{P}^r$ (locally constant) (see [15]).

The multiplicative group K^* acts on $S(K)$ by

$$(r(a)f)(x) = f(xa) \text{ for } x \in K, a \in K^*$$

and on $S'(K)$ by

$$\langle r'(a)(\lambda), f \rangle = \langle \lambda, r(a^{-1})f \rangle \text{ for } \lambda \in S'(K), f \in S(K)$$

where $\langle -, - \rangle : S'(K) \times S(K) \longrightarrow \mathbb{C}$ is the evaluation/integration pairing.

For a character ω of K^* define

$$S'(\omega) = \{\lambda \in S'(K) \mid r'(a)(\lambda) = \omega(a)\lambda \text{ for all } a \in K^*\},$$

the space of ω-eigendistributions.

2.6. The space $S'(\omega)$ can be analysed geometrically. First observe that K^* acting on K has only two orbits $\{0\}$ and K^*. We have an inclusion (extension by zero; see §2.2)

$$\mathcal{C}_c^\infty(K^*) \subset S(K).$$

By duality there is a short exact sequence of distributions

$$0 \longrightarrow S'(K)_0 \longrightarrow S'(K) \longrightarrow \mathcal{C}_c^\infty(K^*)' \longrightarrow 0$$

where $S'(K)_0$ is the space of distributions supported at 0. This sequence is compatible with the K^*-action and, taking ω-eigenspaces, we have the following exact sequence

$$0 \longrightarrow S'(\omega)_0 \longrightarrow S'(\omega) \longrightarrow \mathcal{C}_c^\infty(K^*)'(\omega)$$

where $S'(\omega)_0$ is the space of ω-eigendistributions supported at 0.

LEMMA 2.7. ([86] *Lemma 3.2 p. 116*)

The space $\mathcal{C}_c^\infty(K^*)'(\omega)$ is one-dimensional and is spanned by $\omega(x)d^*x$. In particular, for any $\lambda \in S'(\omega)$ there is a complex number c such that

$$\mathrm{Res}_{\mathcal{C}_c^\infty(K^*)}^{S(K)}(\lambda) = c \cdot \omega(x)d^*x.$$

That is, if $\mathrm{supp}(f) \subset K^*$ is compact then

$$\langle \lambda, f \rangle = c \cdot \int_{K^*} f(x)\omega(x)d^*x.$$

LEMMA 2.8. ([86] *Lemma 3.3 p. 116*)

The delta distribution δ_0 is defined by $\langle \delta_0, f \rangle = f(0)$, which is obviously K^*-invariant and supported at 0. Then

$$S'(K)_0 = \mathbb{C} \cdot \delta_0 \subset S'(\omega_0),$$

where $\omega_0 = 1$ and for $\omega \neq \omega_0$, $S'(\omega)_0 = 0$.

The fundamental local uniqueness result is:

THEOREM 2.9. ([86] *Theorem 3.4 p. 117*)

For any character ω of K^* $\dim_{\mathbb{C}} S'(\omega) = 1$.

DEFINITION 2.10. *Zeta integrals on K^**

For a character ω of K^* (i.e. a unitary quasicharacter) the local zeta integral is

$$\zeta(s,\omega,f) = \int_{K^*} f(x)\omega\omega_s(x)d^*x = \int_K f(x)\omega(x)|x|^{s-1}dx$$

where $s \in \mathbb{C}$ and $\omega_s(x) = |x|^s$ for $x \in K^*$.

PROPOSITION 2.11.

In §2.10 $\zeta(s,\omega,f)$ is absolutely convergent for all $f \in S(K)$ provided that $\mathcal{R}e(s) > 0$. In this range the distribution $f \mapsto \zeta(s,\omega,f)$ defines a non-zero element $\zeta(s,\omega) \in S'(\omega\omega_s)$.

Proof

Consider the zeta integral in its additive manifestation

$$
\begin{aligned}
\zeta(s,\omega,f) \;&=\; \int_K f(x)\omega(x)|x|^{s-1}dx \\[4pt]
&=\; I_K(f(x)\omega(x)\omega_{s-1}(x)) \\[4pt]
&=\; I_{K,r,n}(f(x)\omega(x)\omega_s(x)|x|^{-1}) \\[4pt]
&=\; \textstyle\sum_{x\in\mathcal{P}_K^{-r}/\mathcal{P}_K^n} f(x)\omega(x)q^{-(s-1)v_K(x)}\phi_{K,r,n}(x+\mathcal{P}_K^n) \\[4pt]
&=\; \textstyle\sum_{x\in\mathcal{P}_K^{-r}/\mathcal{P}_K^n} f(x)\omega(x)q^{-(s-1)v_K(x)}q^{(1/2)-n},
\end{aligned}
$$

for $r,n \gg 0$. Here the choice of r and n depend upon f.

The absolute value of $q^{v_K(x)}q^{(1/2)-n}$ as x varies through the terms in the sum is $q^n q^{(1/2)-n} = q^{1/2}$. The absolute value of $f(x)\omega(x)$ is bounded as x varies, since the sum is finite. Therefore, if $\alpha = \mathcal{R}e(s)$ is the real part of s, then the absolute value of $\zeta(s,\omega,f)$ is bounded by a constant times

$$
q^{-\alpha(-r)} + q^{-\alpha(-r+1)} + q^{-\alpha(-r+2)} + \ldots + q^{-\alpha(n-1)} + q^{-\alpha n}
$$

$$
= \tfrac{q^{r\alpha}-q^{-(n+1)\alpha}}{1-q^{-\alpha}},
$$

which shows that $\zeta(s,\omega,f)$ is absolutely convergent when $\mathcal{R}e(s) > 0$, as required.

The eigendistribution condition is equivalent to the relation that

$$
\zeta(s,\omega,f(-\cdot a^{-1})) = \langle \zeta(s,\omega), f(-\cdot a^{-1}) \rangle = \omega(a)\omega_s(a)\zeta(s,\omega,f),
$$

which follows immediately from invariance of the Haar integral under right multiplication by $a^{-1} \in K^*$. $\square$

2.12. The case when ω is unramified

Suppose that ω is unramified (i.e. $\omega(x) = 1$ if $x \in \mathcal{O}_K^*$).

Let $f \in S(K)$ have compact support in K^* and consider the integral

$$
\zeta(s,\omega,f) = \int_{K^*} f(x)\omega\omega_s(x)d^*x
$$

which is absolutely convergent when $\mathcal{R}e(s) > 0$. We would like to modify this into a function which is absolutely convergent over the whole plane. The idea is to kill the support of an arbitrary $f \in S(K)$ by applying a suitable element of the group-ring $\mathbb{Z}[F^*]$ - namely $\tau = 1 - \pi_K^{-1}$.

Since f is constant in a sufficiently small neighbourhood of 0, say $\mathcal{P}_K^r$, we have for any $x \in \mathcal{P}_F^{r+1}$ then

$$
(r(\tau)f)(x) = f(x) - r(\pi_K^{-1})f(x) = f(x) - f(x\pi_K^{-1}) = 0.
$$

Thus there is a distribution $\zeta_0(s,\omega) \in S'(\omega\omega_s)$ defined by

$$\langle \zeta_0(s,\omega), f \rangle = \int_{K^*} (r(\tau)f)(x)\omega\omega_s(x)d^*x.$$

In the half-plane $\mathcal{R}e(s) > 0$ we have

$$\langle \zeta_0(s,\omega), f \rangle$$

$$= \int_{K^*} (f(x) - f(x\pi_K^{-1}))\omega\omega_s(x)d^*x$$

$$= \int_{K^*} f(x)\omega\omega_s(x)d^*x - \int_{K^*} f(x\pi_K^{-1})\omega\omega_s(x)d^*x$$

$$= (1 - \omega(\pi_K)\omega_s(\pi_K))\zeta(s,\omega; f)$$

$$= (1 - \omega(\pi_K)q^{-s})\zeta(s,\omega); f)$$

$$= L(s,\omega)^{-1}\langle \zeta(s,\omega), f \rangle.$$

In terms of distributions this says that for $\mathcal{R}e(s) > 0$

$$\zeta(s,\omega) = L(s,\omega)\zeta_0(s,\omega).$$

PROPOSITION 2.13.
In §2.12 the integral $\int_{K^*} r(\tau)(f)(x)\omega\omega_s(x)d^*x$ is absolutely convergent for all $s \in \mathbb{C}$ and the distribution $\zeta_0(s,\omega)$ is meromorphic analytic for all $s \in \mathbb{C}$.
Also if Φ_X is the characteristic function of X then

$$\langle \zeta_0(s,\omega), \Phi_{\mathcal{O}_K} \rangle = \int_{K^*} \Phi_{\mathcal{O}_K^*}d^*x \neq 0.$$

Proof
Suppose that $h(x) \in S(K)$ is such that for some integer $r > 0$ $h(x) = 0$ when either $x \in \mathcal{P}_K^r$ or $x \in K - \mathcal{P}_K^{-r}$. Then, as in the proof of Proposition 2.11,

$$\zeta(s,\omega,h) = \int_K h(x)\omega(x)|x|^{s-1}dx$$

$$= \sum_{x \in \mathcal{P}_K^{-r}/\mathcal{P}_K^n} h(x)\omega(x)q^{-(s-1)v_K(x)}q^{(1/2)-n}$$

$$= \sum_{x \in \mathcal{P}_K^{-r}/\mathcal{P}_K^r} h(x)\omega(x)q^{-(s-1)v_K(x)}q^{(1/2)-n}$$

for $n >> r > 0$. Here r depends on h. Since n is arbitrarily large we see that the sum is absolutely convergent for all $s \in \mathbb{C}$ and $\zeta_0(s,\omega)$ is meromorphic throughout the complex plane, as required.

Finally

$$\langle \zeta_0(s,\omega), \Phi_{\mathcal{O}_K} \rangle$$

$$= \int_{K^*} \Phi_{\mathcal{O}_K}(x)\omega\omega_s(x)d^*x - \int_{K^*} \Phi_{\mathcal{O}_K}(x\pi_K^{-1})\omega\omega_s(x)d^*x$$

$$= \int_{K^*} \Phi_{\mathcal{O}_K}(x)\omega\omega_s(x)d^*x - \int_{K^*} \Phi_{\mathcal{P}_K}(x)\omega\omega_s(x)d^*x$$

$$= \int_{K^*} \Phi_{\mathcal{O}_K^*} d^*x,$$

as required. $\square$

REMARK 2.14. By Proposition 2.11, §2.12 and Proposition 2.13, when ω is unramified, the right-hand side of the equation $\zeta(s,\omega) = L(s,\omega)\zeta_0(s,\omega)$ gives the meromorphic analytic continuation of $\zeta(s,\omega)$ to the whole s plane. Moreover it interprets the local L-factor as a constant of proportionality between two natural bases of the one-dimensional space $S'(\omega\omega_s)$ away from the poles of $L(s,\omega)$. Thus $\zeta_0(s,\omega)$ is never zero and gives a basis vector for $S'(\omega\omega_s)$ for all s.

2.15. *Ramified local theory*

Suppose that ω is ramified (i.e. non-trivial on $\mathcal{O}^*$). The conductor $c(\omega)$ is the smallest integer c such that ω is trivial on $1+\mathcal{P}_K^c$. Since any $f \in S(K)$ is constant in a neighbourhood of 0 the integral

$$\int_{K^*-\mathcal{P}_K^n} f(x)\omega\omega_s(x)d^*x$$

is independent of n for $n \gg 0$. This is seen as follows.

Recall that the integral over K^* is, for $r, N \gg 0$,

$$\sum_{x \in \mathcal{P}_K^{-r}/\mathcal{P}_K^N} f(x)\phi(x)|x|^{s-1}\phi_{K,r,N}(x + \mathcal{P}_K^N).$$

The integral over $K^* \bigcap \mathcal{P}_K^n$ is

$$\sum_{x \in (K^* \bigcap \mathcal{P}_K^n)/\mathcal{P}_K^N} f(x)\phi(x)|x|^{s-1}\phi_{r,N}(x + \mathcal{P}_K^N).$$

The integral over $K^* \bigcap \mathcal{P}_K^{n+1}$ is

$$\sum_{x \in (K^* \bigcap \mathcal{P}_K^{n+1})/\mathcal{P}_K^N} f(x)\phi(x)|x|^{s-1}\phi_{r,N}(x + \mathcal{P}_K^N).$$

Now $K^* \bigcap \mathcal{P}_K^n$ consists of all the elements of K^* of the form $a\pi_K^m$ with $a \in \mathcal{O}_K^*$ and $m \geq n$. Therefore the difference between these two sums is the sum over $a\pi_K^n$ where a runs through the finite group $\mathcal{O}_K^*/(1 + \mathcal{P}_K^N)$. If n is large enough that $f(x)$ is constant on $K^* \cap \mathcal{P}_K^n$ so the difference between sums is a constant times

$$\sum_{a \in \mathcal{O}_K^*/(1+\mathcal{P}_K^N)} \phi(a)$$

which is zero because ϕ is a non-constant character on the finite group $\mathcal{O}_K^*/(1 + \mathcal{P}_K^N)$.

As in Proposition 2.13, this gives analytic continuation of $\zeta(s, \omega, f)$ to the whole s-plane and hence gives a basis vector

$$\zeta_0(s, \omega) = \zeta(s, \omega)$$

for the one-dimensional space $S'(\omega \omega_s)$ for all s. We set

$$L(s, \omega) = 1$$

in this case. Notice that for

$$f^0(x) = \begin{cases} \omega(x)^{-1} & \text{if } x \in \mathcal{O}^* \\ 0 & \text{otherwise} \end{cases}$$

we have

$$\langle \zeta_0(s, \omega), f^0 \rangle \neq 0.$$

Thus in the ramified case Remark 2.14 also holds.

EXAMPLE 2.16. $G = \langle K^*, u \rangle$

Consider the finite modulo the centre group

$$G = \langle K^*, u \rangle \subset GL_2 K$$

where, as usual,

$$u = \begin{pmatrix} 0 & 1 \\ \pi_K^{-1} & 0 \end{pmatrix}$$

so that $u^2 = \pi_K^{-1} \in K^*$.

In this example we are going to imitate the material of §§2.5–2.15.

Let $S(G)$ equal the set of compactly supported and locally constant functions

$$f : G \longrightarrow \mathbb{C}.$$

Since G is homeomorphic to two copies of K^*, given by the cosets K^* and $K^* u$, we have

$$i : S(K^*) \oplus S(K^*) = C_c^\infty(K^*) \oplus C_c^\infty(K^*) \xrightarrow{\cong} S(G)$$

given by $i(f_1, f_2) = f$ where

$$f\left(\begin{pmatrix} \alpha & 0 \\ 0 & \alpha \end{pmatrix} \right) = f_1(\alpha) \quad f\left(\begin{pmatrix} 0 & \alpha \\ \alpha \pi_K^{-1} & 0 \end{pmatrix} \right) = f_2(\alpha).$$

Hence $f \in S(G)$ if and only if there exists an integer $r \geq 0$, depending on f, such that

(i) $f(x) \neq 0$ implies that $x \in \mathcal{P}_K^{-r} \bigcup \mathcal{P}_K^{-r} u$ and

(ii) if $x, y \in K$ and $x - y \in (M_2 \mathcal{O}_K) \pi_K^r$ then $f(x) = f(y)$, $f(xu) = f(yu)$.

Set $S'(G)$ equal to the space of continuous linear functionals (distributions) $\lambda : S(G) \longrightarrow \mathbb{C}$.

As in §2.6, evaluation $(\lambda, f) \mapsto \lambda(f)$ gives a non-singular pairing

$$\langle -, - \rangle : S'(G) \times S(G) \longrightarrow \mathbb{C}.$$

Now consider the action of G on $S(G)$ given, for $x, g \in G$, by $r(g)(f)(x) = f(xg)$. By duality we have $r'(g) : S'(G) \longrightarrow S'(G)$ satisfying

$$\langle r'(g)(\lambda), f \rangle = \langle \lambda, r(g^{-1})(f) \rangle.$$

Consider

$$r\left(\begin{pmatrix} \alpha^{-1} & 0 \\ 0 & \alpha^{-1} \end{pmatrix} \right) \text{ and } r(u^{-1}) = r\left(\begin{pmatrix} 0 & \pi_K \\ 1 & 0 \end{pmatrix} \right).$$

We have

$$r\left(\begin{pmatrix} \alpha^{-1} & 0 \\ 0 & \alpha^{-1} \end{pmatrix} \right)(f_1, f_2)\left(\begin{pmatrix} \beta & 0 \\ 0 & \beta \end{pmatrix} \right)$$

$$= (f_1, f_2)\left(\begin{pmatrix} \beta\alpha^{-1} & 0 \\ 0 & \beta\alpha^{-1} \end{pmatrix} \right) = f_1(\beta\alpha^{-1}),$$

$$r\left(\begin{pmatrix} \alpha^{-1} & 0 \\ 0 & \alpha^{-1} \end{pmatrix} \right)(f_1, f_2)\left(\begin{pmatrix} 0 & \beta \\ \beta\pi_K^{-1} & 0 \end{pmatrix} \right)$$

$$= (f_1, f_2)\left(\begin{pmatrix} 0 & \beta\alpha^{-1} \\ \beta\alpha^{-1}\pi_K^{-1} & 0 \end{pmatrix} \right) = f_2(\beta\alpha^{-1})$$

$$r\left(\begin{pmatrix} 0 & \pi_K \\ 1 & 0 \end{pmatrix}\right)(f_1, f_2)\left(\begin{pmatrix} \beta & 0 \\ 0 & \beta \end{pmatrix}\right)$$

$$= (f_1, f_2)\left(\begin{pmatrix} 0 & \beta\pi_K \\ \beta & 0 \end{pmatrix}\right) = f_2(\beta\pi_K),$$

$$r\left(\begin{pmatrix} 0 & \pi_K \\ 1 & 0 \end{pmatrix}\right)(f_1, f_2)\left(\begin{pmatrix} 0 & \beta \\ \beta\pi_K^{-1} & 0 \end{pmatrix}\right)$$

$$= (f_1, f_2)\left(\begin{pmatrix} \beta & 0 \\ 0 & \beta \end{pmatrix}\right) = f_1(\beta).$$

Therefore

$$r\left(\begin{pmatrix} \alpha^{-1} & 0 \\ 0 & \alpha^{-1} \end{pmatrix}\right)(f_1, f_2) = (r(\alpha^{-1})(f_1), r(\alpha^{-1})(f_2)$$

and

$$r(u^{-1})(f_1, f_2) = (r(\pi_K)(f_2), f_1).$$

Now suppose we have a continuous character $\phi : G \longrightarrow \mathbb{C}^*$ whose restriction to K^* is the fixed choice of central character $\underline{\phi}$; equivalently $\phi(u)^2 = \underline{\phi}(\pi_K)^{-1}$. Given a space of distributions $S'(G)$ with G-action the space of $\underline{\phi}$-eigendistributions is

$$S'(G)(\phi) = \{\lambda \in S' \mid r'(g)(\lambda) = \phi(g)\lambda \text{ for all } g \in G\}.$$

Therefore we have an isomorphism

$$\mu : C_c^\infty(K^*)'(\underline{\phi}) \xrightarrow{\cong} ((C_c^\infty(K^*)' \oplus C_c^\infty(K^*))'(\phi)$$

given by $\mu(\lambda) = (\lambda, \phi(u)\lambda)$. To check that this is an isomorphism consider, for $\lambda_1, \lambda_2 \in C_c^\infty(K^*)'$ and $f_1, f_2 \in C_c^\infty(K^*)$ the pairing

$$\langle (\lambda_1, \lambda_2), (f_1, f_2) \rangle = \langle \lambda_1, f_1 \rangle \cdot \langle \lambda_2, f_2 \rangle.$$

We have
$$\langle r'(u)(\lambda_1, \lambda_2), (f_1, f_2)\rangle$$

$$= \langle (\lambda_1, \lambda_2), r(u^{-1})(f_1, f_2)\rangle$$

$$= \langle (\lambda_1, \lambda_2), (r(\pi_K)(f_2), f_1)\rangle$$

$$= \langle \lambda_2, f_1\rangle \cdot \langle \lambda_1, r(\pi_K)(f_2)\rangle$$

$$= \langle \lambda_2, f_1\rangle \cdot \langle r'(\pi_K^{-1})(\lambda_1), f_2\rangle$$

$$= \langle (\lambda_2, r'(\pi_K^{-1})(\lambda_1)), (f_1, f_2)\rangle$$

so that
$$(\lambda_2, r'(\pi_K^{-1})(\lambda_1)) = r'(u)(\lambda_1, \lambda_2) = \phi(u)(\lambda_1, \lambda_2)$$
which implies that $\lambda_2 = \phi(u)\lambda_1$ and
$$(r'(\pi_K^{-1})(\lambda_1), r'(\pi_K^{-1})(\lambda_2)) = \phi(u)^2(\lambda_1, \lambda_2) = \underline{\phi}(\pi_K^{-1})(\lambda_1, \lambda_2)$$
so that $\lambda_1, \lambda_2 = \phi(u)\lambda_1 \in C_c^\infty(K^*)'(\underline{\phi})$, as required.
Therefore we have shown the following result:

PROPOSITION 2.17.
In §2.16 $\dim_{\mathbb{C}}(S'(G)(\phi)) = 1$.

EXAMPLE 2.18. *Haar integration on* $G = \langle K^*, u\rangle \subset GL_2K$
Let $G = \langle K^*, u\rangle$ be the group considered in §2.16.
The uniqueness, up to scalars, of the Haar integral means that Haar integration is natural, up to scalars, with respect to subgroups. If we have an inclusion of a closed subgroup $H \subseteq G$ then the composition which takes $f \in C_c^\infty(H)$ to $f_{ext} \in C_c^\infty(G)$ given by extending by zero and then forming $I_G(f_{ext})$ gives a right-invariant integral

$$C_c^\infty(H) \xrightarrow{\text{ext}} C_c^\infty(G) \xrightarrow{I_G} \mathbb{C}$$

which must be a scalar times I_H.
Taking $H = K^*$ we have, for some non-zero scalar λ_1,
$$\lambda_1 I_{K^*}(f) = I_G(f_{ext}).$$
Now $G = K^* \bigcup K^*u$ and $f \in C_c^\infty(G)$ is equal to the sum
$$f = \text{Res}_{K^*}^G(f)_{ext} + \text{Res}_{K^*u}^G(f)_{ext}.$$
Therefore
$$I_G(f)$$

$$= I_G(\text{Res}_{K^*}^G(f)_{ext}) + I_G(\text{Res}_{K^*u}^G(f)_{ext})$$

$$= \lambda_1 I_{K^*}(\text{Res}_{K^*}^G(f)) + I_G(\text{Res}_{K^*u}^G(f)_{ext}).$$

Define a function $f_1 \in C_c^\infty(G)$ by $f_1(x) = f(xu)$ for $x \in K^*$ and $f_1(x) = 0$ otherwise. Hence $f_1 = \mathrm{Res}_{K^*u}^G(f)_{\mathrm{ext}}(- \cdot u)$. For

$$\mathrm{Res}_{K^*u}^G(f)_{\mathrm{ext}}(z) = \begin{cases} f(xu) & \text{if } z = xu, x \in K^*, \\ 0 & \text{if } z \in K^*. \end{cases}$$

Therefore

$$\mathrm{Res}_{K^*u}^G(f)_{\mathrm{ext}}(- \cdot u)(x) = \begin{cases} f(xu) & \text{if } x \in K^*, \\ 0 & \text{if } x \notin K^*. \end{cases}$$

Therefore, by right G-invariance,

$$I_G(f_1) = I_G(\mathrm{Res}_{K^*u}^G(f)_{\mathrm{ext}}).$$

Therefore

$$I_G(f) = \lambda_1(I_{K^*}(\mathrm{Res}_{K^*}^G(f)) + I_{K^*}(\mathrm{Res}_{K^*}^G(f_1))).$$

Henceforth we shall assume that $\lambda_1 = 1$ and that

$$I_G(f) = I_{K^*}(\mathrm{Res}_{K^*}^G(f)) + I_{K^*}(\mathrm{Res}_{K^*}^G(f_1)).$$

This integral is invariant under right multiplication for consider $I_G(f(- \cdot u))$ because we have $\mathrm{Res}_{K^*}^G(f(- \cdot u))(x) = f(xu) = \mathrm{Res}_{K^*}^G(f_1)(x)$ and $\mathrm{Res}_{K^*}^G(f(- \cdot u)_1)(x) = f(xu^2) = \mathrm{Res}_{K^*}^G(f)(- \cdot \pi_K^{-1})(x)$ which implies

$$I_G(f(- \cdot u)) = I_{K^*}(\mathrm{Res}_{K^*}^G(f_1)) + I_{K^*}(\mathrm{Res}_{K^*}^G(f(- \cdot \pi_K^{-1})))$$

$$= I_{K^*}(\mathrm{Res}_{K^*}^G(f_1)) + I_{K^*}(\mathrm{Res}_{K^*}^G(f)),$$

as required.

DEFINITION 2.19. *Zeta integrals on G*

Suppose that $\phi : G \longrightarrow \mathbb{C}^*$ is a continuous character extending the chosen central character $\underline{\phi}$. On G we may extend ω_s to ϕ_s by setting $\phi_s(u) = q^{s/2} = |\det(u)|^{-s/2}$ since $q = \omega_s(\pi_K)^{-1} = \phi_s(u)^2 = |\det(\pi_K I_2)|^{-s/2}$.

For a character ϕ of K^* the local zeta integral is

$$\zeta_G(s, \phi, f)$$

$$= \int_G f(x)\phi(x)\phi_s(x)d^*x$$

$$= I_{K^*}(\mathrm{Res}_{K^*}^G(f\phi\phi_s)) + I_{K^*}(\mathrm{Res}_{K^*}^G((f\phi\phi_s)_1)).$$

The distribution

$$\zeta_G(s, \phi) : f \mapsto \zeta_G(s, \phi, f)$$

is meromorphic on the half-plane $\mathcal{R}e(s) > 0$.

2.20. *A meromorphic distribution in* $S'(G)(\phi)$

Consider

$$I_G((f(x) - \phi(u)\phi_s(u)f(x \cdot u))\phi(x)\phi_s(x)).$$

In the formula of §2.18 for I_G applied to f we obtain $\operatorname{Res}_{K^*}^G(f(x))$ and $\operatorname{Res}_{K^*}^G(f_1(x))$ where f_1 is given by $x \mapsto xu$. In that formula for I_G applied to $f(-\cdot u)$ we obtain $\operatorname{Res}_{K^*}^G(f_1(x))$ and $\operatorname{Res}_{K^*}^G(f(x)\pi_K^{-1})$ (see §2.18). Therefore

$$I_G((f(x) - \phi(u)\phi_s(u)f(x \cdot u))\phi(x)\phi_s(x))$$

$$= I_K(\operatorname{Res}_{K^*}^G(f)(x)\underline{\phi}(x)|x|^{s-1}) + \phi(u)\phi_s(u)I_K(\operatorname{Res}_{K^*}^G(f_1)(x)\underline{\phi}(x)|x|^{s-1})$$

$$-\phi(u)\phi_s(u)I_K(\operatorname{Res}_{K^*}^G(f_1)(x)\underline{\phi}(x)|x|^{s-1})$$

$$-\phi(u)\phi_s(u)I_K(\operatorname{Res}_{K^*}^G(f)(x \cdot \pi_K^{-1})\underline{\phi}(x)|x|^{s-1})$$

$$= I_{K^*}(\operatorname{Res}_{K^*}^G(f)(x)\underline{\phi}(x)|x|^s) - \phi(u)\phi_s(u)I_{K^*}(\operatorname{Res}_{K^*}^G(f)(x \cdot \pi_K^{-1})\underline{\phi}(x)|x|^s)$$

$$= I_{K^*}(\operatorname{Res}_{K^*}^G(f)(x)\underline{\phi}(x)|x|^s)$$

$$-\phi(u)\phi_s(u)\underline{\phi}(\pi_K)|\pi_K|^s)I_{K^*}(\operatorname{Res}_{K^*}^G(f)(x)\underline{\phi}(x)|x|^s))$$

$$= (1 - \phi(u)^{-1}\phi_s(u)^{-1})I_{K^*}(\operatorname{Res}_{K^*}^G(f)(x)\underline{\phi}(x)|x|^s).$$

Therefore the distribution

$$\zeta_{G,0}(s, \phi) : f \mapsto I_G((f(x) - \phi(u)\phi_s(u)f(x \cdot u))\phi(x)\phi_s(x))$$

is $(1 - \phi(u)^{-1}\phi_s(u)^{-1})$ times the distribution

$$f \mapsto I_{K^*}(\operatorname{Res}_{K^*}^G(f)(x)\underline{\phi}(x)|x|^s) = \zeta(s, \underline{\phi}, \operatorname{Res}_{K^*}^G(f)).$$

From §2.12 and §2.15 for $\mathcal{R}e(s) > 0$ we have

$$I_{K^*}(\operatorname{Res}_{K^*}^G(f)(x)\underline{\phi}(x)|x|^s)$$

$$= \zeta(s, \underline{\phi}, \operatorname{Res}_{K^*}^G(f))$$

$$= \begin{cases} L(s, \underline{\phi})\zeta_0(s, \underline{\phi}, \operatorname{Res}_{K^*}^G(f)) & \text{if } \underline{\phi} \text{ is unramified} \\ \zeta_0(s, \underline{\phi}, \operatorname{Res}_{K^*}^G(f)) & \text{if } \underline{\phi} \text{ is ramified.} \end{cases}$$

Therefore for $\mathcal{R}e(s) > 0$ we have relations between distributions

$\zeta_{G,0}(s, \phi)$

$$= \begin{cases} (1 + \phi(u)^{-1}\phi_s(u)^{-1})^{-1}\zeta_0(s, \underline{\phi}, \operatorname{Res}_{K^*}^G(-)) & \text{if } \underline{\phi} \text{ is unramified} \\[2mm] (1 - \phi(u)^{-1}\phi_s(u)^{-1})\zeta_0(s, \underline{\phi}, \operatorname{Res}_{K^*}^G(-)) & \text{if } \underline{\phi} \text{ is ramified} \end{cases}$$

which gives a meromorphic extension of $\zeta_{G,0}(s, \phi)$ to the whole complex plane, as in §2.12 and §2.15, which is a basis for $S'(G)(\phi)$.

DEFINITION 2.21. *Fourier transforms*
Let ψ_K be the non-trivial character of the additive group K which was introduced in §1.1. Identify K with its topological dual character group

$$\hat{K} = \operatorname{Hom}_{\operatorname{conts}}(K, \mathbb{C}^*)$$

by the isomorphism

$$K \xrightarrow{\cong} \hat{K}, \qquad y \mapsto (x \mapsto \psi_K(xy)).$$

The conductor $\nu_K(\psi_K)$ of ψ_K is the largest integer such that ψ_K is trivial on $\mathcal{P}_K^{-\nu_K(\psi_K)}$.

The Fourier transform

$$\hat{f}(x) = \int_K f(y)\psi_K(xy)dy$$

of a function $f \in S(K)$ is well-defined and lies in $S(K)$. This is a fundamental property of the space of Schwartz-Bruhat functions. The map $f \mapsto \hat{f}$ is an isomorphism $(\hat{-}) : S(K) \xrightarrow{\cong} S(K)$. There is a unique choice of the Haar measure, the self-dual measure with respect to ψ_K, such that Fourier inversion satisfies

$$\hat{\hat{f}}(x) = f(-x).$$

For given ψ_K we fix this choice of dx henceforth. The Fourier transform of a distribution λ is defined by

$$\langle \hat{\lambda}, f \rangle = \langle \lambda, \hat{f} \rangle.$$

LEMMA 2.22.
If $\lambda \in S'(\omega)$ is an ω-distribution then $\hat{\lambda} \in S'(\omega^{-1}\omega_1)$ where $\omega_1(x) = |x|$.

By Theorem 2.9, §2.12 and §2.15, $\hat{\zeta}_0(s, \omega)$ is a constant multiple (depending on s, ω and ψ_K) of $\zeta_0(1 - s, \omega^{-1})$. Hence we have shown the following result:

COROLLARY 2.23. *Local functional equation*
(i) There exists a non-zero complex number $\epsilon(s, \omega, \psi_K)$ such that

$$\hat{\zeta}_0(1 - s, \omega^{-1}) = \epsilon(s, \omega, \psi_K) \cdot \zeta_0(s, \omega).$$

(ii) Given $f \in S(K)$ (i) may be written as the relation

$$\frac{\zeta(1-s,\omega^{-1},\hat{f})}{L(1-s,\omega^{-1})} = \epsilon(s,\omega,\psi_K) \cdot \frac{\zeta(s,\omega,f)}{L(s,\omega)}.$$

REMARK 2.24. An alternative formulation of Corollary 2.23(ii) is

$$\zeta(1-s,\omega^{-1},\hat{f}) = \epsilon(s,\omega,\psi_K) \cdot \frac{L(1-s,\omega^{-1})}{L(s,\omega)}\zeta(s,\omega,f)$$

in which the constant of proportionality

$$\gamma(s,\omega,\psi_K) = \epsilon(s,\omega,\psi_K) \cdot \frac{L(1-s,\omega^{-1})}{L(s,\omega)}$$

is called the local gamma factor. It can have zeroes and poles. This is the traditional formulation while that of Corollary 2.23 is emphasised in [**142**].

Various important properties of the ϵ-factor $\epsilon(s,\omega,\psi_K)$ can be deduced from Corollary 2.23. For example, if f^0 is the standard function with respect to ω then, as in §2.15, evaluating both sides of Corollary 2.23(i) on f^0 yields the useful identity

$$\epsilon(s,\omega,\psi_K) = \langle \zeta(1-s,\omega^{-1}), \hat{f}^0 \rangle.$$

By construction the zeta distribution satisfies

$$\zeta(s,\omega\omega_t) = \zeta(s+t,\omega).$$

Since the local L-function is also required to have this property we also have

$$\zeta_0(s,\omega\omega_t) = \zeta_0(s+t,\omega).$$

The standard function f^0 does not change under multiplication by ω_t so that

$$\epsilon(s,\omega\omega_t,\psi_K) = \epsilon(s+t,\omega,\psi_K).$$

For $\beta \in K^*$ let $\psi_{K,\beta}(x) = \psi_K(\beta x)$. If $\hat{f}$ is the Fourier transform of $f \in S(K)$ with respect to ψ_K then $\sqrt{|\beta|}r(\beta)\hat{f}$ is the Fourier transform with respect to $\psi_{K,\beta}$ (using the self-dual measure). Therefore

$$\epsilon(s,\omega,\psi_{K,\beta}) = |\beta|^{s-\frac{1}{2}}\omega(\beta)\epsilon(s,\omega,\psi_K).$$

2.25. Next we give the explicit formulae for the ϵ-factors, observing that by the results of Remark 2.24, it suffices to do this for ψ and ω both in "standard form". Recall that $\nu_K(\psi_K)$ is the conductor of ψ_K, as in Definition 2.21.

PROPOSITION 2.26.
If ϕ is unramified then

$$\epsilon(s,\phi,\psi_K) = \phi(\pi_F^{\nu_K(\psi_K)})q^{(\frac{1}{2}-s)\nu_K(\psi_K)}.$$

If ϕ is ramified with conductor c then

$$\epsilon(s, \phi, \psi_K) = \phi(\pi_K^{\nu_K(\psi_K)+c})q^{(\frac{1}{2}-s)(\nu_K(\psi_K)+c)}\mathcal{G}(\phi, \psi_K)$$

where $\mathcal{G}(\phi, \psi_K)$ is the Gauss sum

$$\mathcal{G}(\phi, \psi_K) = q^{\frac{\nu_K(\psi_K)+c}{2}} \int_{\mathcal{O}_K^*} \phi^{-1}(y)\psi(\pi_K^{-\nu-c}y)dy$$

with dy the self-dual measure on K with respect to ψ_K.

Proof

By Remark 2.24 one has to compute the Fourier transform of the standard function f^0 for ω where the character ψ can be taken as standard.

For example, if ϕ is unramified, f^0 is the characteristic function of $\mathcal{O}_K$ and assuming $\nu_K(\psi_K) = 0$, $\hat{f}^0 = f^0$. This gives $\epsilon(s, \phi, \psi_K) = 1$ in this case. The general unramified case then follows at once from the final formula in Remark 2.24.

If ϕ is ramified f^0 is given as in Example 2.15. Then, following ([**143**] Proposition 13 p. 131)

$$\hat{f}^0(x) = \int_{\mathcal{O}_K^*} \omega^{-1}\psi_K(xy)dy$$

vanishes unless $\mathrm{ord}_K(x) = -\nu_K(\psi_K) - c$ where $c = c(\phi)$ is the conductor of ϕ. In fact

$$\hat{f}^0(x) = \mathcal{G}(\omega, \psi_K)q^{-\frac{\nu_K(\psi_K)+c}{2}} \overline{f^0(\pi_K^{\nu_K(\psi_K)+c}x)}.$$

The ramified result follows from the fact that $\mathcal{G}(\phi, \psi_K)\overline{\mathcal{G}(\phi, \psi_K)} = 1$ and that $\overline{f^0}$ is the standard function for ϕ^{-1}.

More details on Tate's thesis may be found in [**143**] and [**40**]. $\square$

EXAMPLE 2.27. *Haar integration on $M_n K$*
In order to define Haar integrals on $M_n K$ we need functions

$$\phi_{M_n K, r, m} : M_n \mathcal{P}_K^{-r}/M_n \mathcal{P}_K^m \longrightarrow \mathbb{C}.$$

For $m \geq 2, r \geq 0$ define

$$\phi_{M_n K, r, m}(X + M_n \mathcal{P}_K^m) = |\mathcal{O}_K/\mathcal{P}_K|^{-(n^2-1)m}\phi_{K, r, m}(\mathrm{Trace}(X) + \mathcal{P}_K^m).$$

Under the canonical epimorphism, for $m \geq 3$,

$$M_n \mathcal{P}_K^{-r}/M_n \mathcal{P}_K^m \longrightarrow M_n \mathcal{P}_K^{-r}/M_n \mathcal{P}_K^{m-1}$$

each coset has inverse image of order $|\mathcal{O}_K/\mathcal{P}_K|^{n^2}$. Hence there are $|\mathcal{O}_K/\mathcal{P}_K|^{n^2}$ cosets $Y + M_n \mathcal{P}_K^m$ above $X + M_n \mathcal{P}_K^{m-1}$ of which $|\mathcal{O}_K/\mathcal{P}_K|^{(n^2-1)}$ have the same trace modulo $M_n \mathcal{P}_K^{m-1}$.

Therefore for $m \geq 3$

$$\sum_{Y+M_n\mathcal{P}_K^{m+1}\mapsto X+M_n\mathcal{P}_K^m} \phi_{M_nK,r,m+1}(Y+M_n\mathcal{P}_K^{m+1})$$

$$= \sum_{Y+M_n\mathcal{P}_K^{m+1}\mapsto X+M_n\mathcal{P}_K^m} |\mathcal{O}_K/\mathcal{P}_K|^{-(n^2-1)(m+1)} \times$$

$$\phi_{K,r,m+1}(\mathrm{Trace}(Y)+\mathcal{P}_K^{m+1})$$

$$= \sum_{Z+\mathcal{P}_K^{m+1}\mapsto \mathrm{Trace}(X)+\mathcal{P}_K^m} |\mathcal{O}_K/\mathcal{P}_K|^{-(n^2-1)m}\phi_{K,r,m+1}(Z+\mathcal{P}_K^{m+1})$$

$$= |\mathcal{O}_K/\mathcal{P}_K|^{-(n^2-1)m} \sum_{Z+\mathcal{P}_K^{m+1}\mapsto \mathrm{Trace}(X)+\mathcal{P}_K^m} \phi_{K,r,m+1}(Z+\mathcal{P}_K^{m+1})$$

$$= |\mathcal{O}_K/\mathcal{P}_K|^{-(n^2-1)m}\phi_{K,r,m}(\mathrm{Trace}(X)+\mathcal{P}_K^m)$$

$$= \phi_{M_nK,r,m}(X+M_n\mathcal{P}_K^m),$$

as required.

EXAMPLE 2.28. *Local function equation for $\zeta_{G,0}(s,\phi)$ of §2.20*
We could state this example using the Fourier transform on M_2K via the Haar integral of §2.27. However, for convenience, we shall use the Fourier transform on $S'(K)$.

By Lemma 2.8 if $\omega \neq 1$ then the surjection

$$S'(K) \longrightarrow C_c^\infty(K^*)'$$

$$(\hat{-}) : S(K) \xrightarrow{\cong} S(K)$$

induces

$$(\hat{-}) : S'(K) \xrightarrow{\cong} S'(K)$$

which restricts to an isomorphism

$$(\hat{-}) : S'(\omega) \xrightarrow{\cong} S'(\omega^{-1}\omega_1)$$

where $\omega_1(x) = |x| = q^{-v_K(x)}$ for $x \in K^*$.

Hence if $\omega \neq 1, \omega_1$ we have a Fourier isomorphism

$$(\hat{-}) : C_c^\infty(K^*)'(\omega) \xrightarrow{\cong} C_c^\infty(K^*)'(\omega^{-1}\omega_1).$$

As in the notation of §2.20 let ϕ_1 denote the extension of ω_1 to G and for $f \in S(G)$ let $\hat{Res}_{K^*}^G(f)$ denote the Fourier transform of the restriction of f to K^*.

From §2.20 we have an isomorphism

$$S'(G)(\phi) \xrightarrow{\cong} C_c^\infty(K^*)'(\underline{\phi})$$

given by sending $\zeta_{G,0}(s,\phi)$ to $(1+\phi(u)^{-1}\phi_s(u)^{-1})^{-1}\zeta_0(s,\underline{\phi},\mathrm{Res}_{K^*}^G(-))$ if $\underline{\phi}$ is unramified and to $(1-\phi(u)^{-1}\phi_s(u)^{-1})\zeta_0(s,\underline{\phi},\mathrm{Res}_{K^*}^G(-))$ otherwise.

Transporting the Fourier transform via this isomorphism to give

$$(\hat{-}) : S'(G)(\phi) \xrightarrow{\cong} S'(G)(\phi^{-1}\phi_1)$$

the local functional equation transports to give

$$\hat{\zeta}_{G,0}(1 - s, \phi^{-1}) = \frac{(1+\phi(u)^{-1}\phi_s(u)^{-1})}{(1+\phi(u)^{-1}\phi_{1-s}(u)^{-1})}\epsilon(s, \underline{\phi}, \psi_K)\zeta_{G,0}(s, \phi)$$

when ϕ is unramified and

$$\hat{\zeta}_{G,0}(1 - s, \phi^{-1}) = \frac{(1-\phi(u)^{-1}\phi_s(u)^{-1})}{(1-\phi(u)^{-1}\phi_{1-s}(u)^{-1})}\epsilon(s, \underline{\phi}, \psi_K)\zeta_{G,0}(s, \phi)$$

when ϕ is ramified.

3. Monomial resolutions and local function equations

3.1. This short section contains some fundamental questions which arise from the previous sections of this chapter. Here K continues to be a p-adic local field. We continue to study complex characters and complex admissible representations.

Suppose first that H is a subgroup of GL_nK which contains K^* and which is compact open modulo K^*. Let $(H, \phi) \in \mathcal{M}_\phi(H)$. Hence there is an inverse system of quotients H_α of H which contain $\overline{K^*}$, are finite modulo K^* and possess a character $\phi_\alpha : H_\alpha \longrightarrow \mathbb{C}^*$ through which ϕ factorises. An argument analogous to that if §2.16 shows that $\dim_{\mathbb{C}}(S'(H_\alpha)(\phi_\alpha)) = 1$. For two groups H_α and H_β in the inverse system with quotient map

$$\pi_{\alpha,\beta} : H_\alpha \longrightarrow H_\beta$$

we have induced maps

$$\pi^*_{\alpha,\beta} : S(H_\beta) \longrightarrow S(H_\alpha) \text{ and } \pi^*_{\alpha,\beta} : S'(H_\alpha) \longrightarrow S'(H_\beta).$$

The map of eigendistributions will be an isomorphism of one-dimensional complex vector spaces

$$\pi^*_{\alpha,\beta} : S'(H_\alpha)(\phi_\alpha) \longrightarrow S'(H_\beta)(\phi_\beta).$$

The direct limit

$$\underline{S}'(H)(\phi) = \varinjlim_\alpha S'(H_\alpha)(\phi_\alpha)$$

will be a one-dimensional space.

QUESTION 3.2. *Is there a meromorphic distribution in $\underline{S}'(H)(\phi)$?*

Is there always a non-zero distribution in $\zeta_{H,0}(s, \phi) \in \underline{S}'(H)(\phi)$, as in Example 2.20, which is meromorphic over the whole complex plane?

QUESTION 3.3. *Is there a Fourier transform on $\underline{S}'(H)$?*

Assuming an affirmative answer to Question 3.2, is there a Fourier transform, as in Example 2.28, on $\underline{S}'(H)$ which induces an isomorphism of the form

$$(\hat{-}) : \underline{S}'(H)(\phi) \xrightarrow{\cong} \underline{S}'(H)(\phi^{-1}\phi_1)?$$

What is the form of the resulting local functional equation?

QUESTION 3.4. *L-series and ϵ-factors for local admissible representations*

Suppose that we have an affirmative answer to Question 3.2 and Question 3.3. Therefore we may apply the local functional equation to each Line in the monomial resolution for an admissible representation V of $GL_n K$ with central character $\underline{\phi}$. The function equation for any two Lines in $c - \mathrm{Ind}_H^{GL_n K}(\phi)$ should be related and essentially only depend on the induced monomial representation to which they belong. By this token, for each summand $c - \underline{\mathrm{Ind}}_H^{GL_n K}(\phi)$ in the monomial resolution we should obtain a functional equation.

Does the "Euler characteristic" of these functional equations make sense (we might have to capitalise on the finiteness property of Chapter One, §6)? If so, does the result have a connection with the functional equation, L-series and ϵ-factors constructed in the local situation in [66] (see also ([40] §24) for an account in the case of $GL_2 K$).

CHAPTER 7

Hecke operators and monomial resolutions

This chapter recalls how Hecke operators are defined and explains how they fit in with the exact sequences

$$M_*^{((H,\phi))} \longrightarrow V^{(H,\phi)} \longrightarrow 0$$

which originate from a monomial resolution of the representation V. Given two subgroups J and H of G the Hecke operators take the form of k-linear maps

$$[JgH] : V^{(H,\phi)} \longrightarrow V^{(H',\phi')}$$

for $g \in G$. In Example 1.1 I describe how, when G is finite, the operators may be defined via the Double Coset Formula ([126] p. 32) for $\mathrm{Res}_J^G \mathrm{Ind}_H^G(\phi)$.

In Definition 1.2 it is pointed out that the formulae of Example 1.1 apply equally well when J and H are compact open modulo the centre subgroups of a locally profinite group. In particular the latter applies to the adèlic case of an automorphic representation. Then, if J, H are the usual congruence subgroups $\Gamma_0(N), \Gamma_1(N)$, the $[JgH]$'s are the classical Hecke operators ([51] §11.2).

In §1.3 I describe the conditions on H, J and g in order that $[HgJ]$ extends to a chain map from from the exact complex

$$M_*^{((H,\phi))} \longrightarrow V^{(H,\phi)} \longrightarrow 0$$

to

$$M_*^{((J,\phi'))} \longrightarrow V^{(J,\phi')} \longrightarrow 0.$$

The chapter concludes with Example 1.4. This solitary example is included to show that the conditions for $[JgH]$ to extend to the monomial resolution can indeed be satisfied. These extensions should be especially important when V is an automorphic representation and $V^{(H,\phi)}$ is a space of modular forms.

1. Hecke operators for an admissible representation

EXAMPLE 1.1. Let G be a finite group with subgroups H and J. Let k be an algebraically closed field. Let V be a k-representation of G and let

185

$\phi \in \hat{H}, \phi' \in \hat{J}$ be characters. Suppose that $g \in G$ and that JgH is equal to the disjoint union of left cosets of H so that

$$JgH = \bigcup_{i=1}^{n} y_i H.$$

In addition suppose that ϕ' and $(g^{-1})^*(\phi)$ are equal on $J \bigcap gHg^{-1}$ where $(g^{-1})^*(\phi)(ghg^{-1}) = \phi(h)$.

In this situation there exists a linear transformation of k-vector spaces, called a Hecke operator,

$$[JgH] : V^{(H,\phi)} \longrightarrow V^{(J,\phi')}.$$

This transformation may be constructed by means of the double coset formula ([126] Theorem 1.2.40) which in our case takes the form of an isomorphism of $k[J]$-representations

$$\mathrm{Res}_J^G \mathrm{Ind}_H^G(\phi) \xrightarrow{\alpha} \oplus_{z \in J \backslash G/H} \mathrm{Ind}_{J \bigcap zHz^{-1}}^J ((g^{-1})^*(\phi))$$

given by $\alpha(g \otimes_H w) = j \otimes_{J \bigcap zHz^{-1}} hw$ for $g = jzh, j \in J, h \in H$. The inverse of α is given by $\alpha^{-1}(j \otimes_{J \bigcap zHz^{-1}} w) = jz \otimes_H w$.

Since we have isomorphisms

$$\mathrm{Hom}_G(\mathrm{Ind}_J^G(\phi'), V) \cong V^{(J,\phi')} \text{ and } \mathrm{Hom}_G(\mathrm{Ind}_H^G(\phi), V) \cong V^{(H,\phi)}$$

the transformation $[JgH]$ may be induced by pre-composition with a $k[G]$-module homomorphism of the form

$$\mathrm{Ind}_J^G(\phi') \longrightarrow \mathrm{Ind}_H^G(\phi)$$

and by Frobenius reciprocity such a map is equivalent to a $k[J]$-module homomorphism of the form

$$k_{\phi'} \longrightarrow \mathrm{Res}_J^G \mathrm{Ind}_H^G(\phi) \cong \oplus_{z \in J \backslash G/H} \mathrm{Ind}_{J \bigcap zHz^{-1}}^J ((z^{-1})^*(\phi)).$$

Consider the expression

$$\sum_{i=1}^{n} y_i g^{-1} \otimes_{J \bigcap gHg^{-1}} \phi'(y_i g^{-1})^{-1} \in \mathrm{Ind}_{J \bigcap gHg^{-1}}^J ((g^{-1})^*(\phi)).$$

Since $JgHg^{-1} = \bigcup_{i=1}^{n} y_i g^{-1} gHg^{-1}$ the set $\{y_1 g^{-1}, \dots, y_n g^{-1}\}$ is a complete set of coset representatives for $J/J \bigcap gHg^{-1}$. In particular $y_i g^{-1} \in J$. For $j \in J$ there is a permutation $\sigma(j)$ of $1, \dots, n$ such that $jy_i = y_{\sigma(j)(i)} h$ for some $h \in H$. Therefore $jy_i g^{-1} = y_{\sigma(j)(i)} g^{-1} ghg^{-1}$ which implies that $ghg^{-1} \in J \bigcap gHg^{-1}$. Therefore

$$\phi(h) = (g^{-1})^*(\phi)(ghg^{-1}) = \phi'(ghg^{-1}) = \phi'(jy_i g^{-1}) \phi'(y_{\sigma(j)(i)} g^{-1})^{-1}.$$

This implies that in $\operatorname{Ind}_{J \bigcap gHg^{-1}}^{J}((g^{-1})^*(\phi))$

$$\sum_{i=1}^{n} jy_i g^{-1} \otimes_{J \bigcap gHg^{-1}} \phi'(y_i g^{-1})^{-1}$$

$$= \sum_{i=1}^{n} y_{\sigma(j)(i)} g^{-1} ghg^{-1} \otimes_{J \bigcap gHg^{-1}} \phi'(y_i g^{-1})^{-1}$$

$$= \sum_{i=1}^{n} y_{\sigma(j)(i)} g^{-1} \otimes_{J \bigcap gHg^{-1}} \phi'(ghg^{-1})\phi'(y_i g^{-1})^{-1}$$

$$= \sum_{i=1}^{n} y_{\sigma(j)(i)} g^{-1} \otimes_{J \bigcap gHg^{-1}} \phi'(jy_i g^{-1})\phi'(y_{\sigma(j)(i)} g^{-1})^{-1} \phi'(y_i g^{-1})^{-1}$$

$$= \phi'(j) \sum_{i=1}^{n} y_{\sigma(j)(i)} g^{-1} \otimes_{J \bigcap gHg^{-1}} \phi'(y_{\sigma(j)(i)} g^{-1})^{-1}$$

$$= \phi'(j) \sum_{i=1}^{n} y_i g^{-1} \otimes_{J \bigcap gHg^{-1}} \phi'(y_i g^{-1})^{-1}.$$

Therefore $1 \mapsto \sum_{i=1}^{n} y_i g^{-1} \otimes_{J \bigcap gHg^{-1}} \phi'(y_i g^{-1})^{-1}$ gives a $k[J]$-module homomorphism from $k_{\phi'}$ to the summand $\operatorname{Ind}_{J \bigcap gHg^{-1}}^{J}((g^{-1})^*(\phi))$. Applying α^{-1} gives a $k[J]$-module homomorphism

$$k_{\phi'} \longrightarrow \operatorname{Res}_J^G \operatorname{Ind}_H^G(\phi)$$

given by $1 \mapsto \sum_{i=1}^{n} y_i \otimes_H \phi'(y_i g^{-1})^{-1}$. Hence the required $k[G]$-module homomorphism

$$\lambda_{[JgH]} : \operatorname{Ind}_J^G(\phi') \longrightarrow \operatorname{Ind}_H^G(\phi)$$

is given by

$$\lambda_{[JgH]}(g' \otimes_J w) = \sum_{i=1}^{n} g' y_i \otimes_H \phi'(y_i g^{-1})^{-1} w.$$

Since $v \in V^{(H,\phi)}$ corresponds to the $k[G]$-module homomorphism $g' \otimes_H 1 \mapsto g'v \in V$ the composition of this with $\lambda_{[JgH]}$ is given by

$$1 \otimes_J 1 \mapsto \sum_{i=1}^{n} y_i \otimes_H \phi'(y_i g^{-1})^{-1} \mapsto \sum_{i=1}^{n} \phi'(y_i g^{-1})^{-1} y_i v \in V^{(J,\phi')}.$$

In other words, $[JgH](v) = \sum_{i=1}^{n} \phi'(y_i g^{-1})^{-1} y_i v \in V^{(J,\phi')}$.

In particular, if $\phi' = 1$ then $[JgH](v) = \sum_{i=1}^{n} y_i v \in V^{(J,\phi')}$, which is the formula for the classical Hecke operator ([51] §11.2).

DEFINITION 1.2. Let G be a locally profinite group and let $\pi : G \longrightarrow \operatorname{Aut}_k(V)$ be an admissible representation. Let H and J be compact open subgroups with $(H, \phi), (J, \phi') \in \mathcal{M}_G$. The procedure of §1.1 (with $c -$ Ind_H^G replacing Ind_H^G etc. and using Frobenius reciprocity for compactly supported induction) yields a linear transformation of k-vector spaces

$$[JgH] : V^{(H,\phi)} \longrightarrow V^{(J,\phi')}$$

given by the same formula as that of the finite group case in §1.1.

This is the classical Hecke operator of ([**51**] §11.2) when $\phi = 1 = \phi'$.

In the adélic case of an automorphic representation when J, H are the usual congruence subgroups these are the classical Hecke operators on the spaces of modular forms ([**51**] §11.1).

Furthermore, if V has a central character $\underline{\phi}$ and $(H, \phi), (J, \phi') \in \mathcal{M}_{\underline{\phi}}(G)$ the construction and formulae of §1.1 continue to make sense and extend the definition of Hecke operators to that situation.

1.3. *Hecke operators on the monomial resolution*
In the situation of Definition 1.2 suppose that

$$M_* \xrightarrow{\ \epsilon\ } V \longrightarrow 0$$

is a monomial resolution of V in $_{k[G],\underline{\phi}}\mathbf{mon}$. By Frobenius reciporicity in the monomial category ([**19**] Remark 1.5(g)) there are isomorphisms

$$\mathrm{Hom}_{k[G],\underline{\phi}\mathbf{mon}}(c - \underline{\mathrm{Ind}}_H^G(k_\phi), M_i) \cong \mathrm{Hom}_{k[H],\underline{\phi}\mathbf{mon}}(k_\phi),$$

$$\mathrm{Res}_H^G(M_i)) \cong M_i^{((H,\phi))}$$

and

$$\mathrm{Hom}_{k[G],\underline{\phi}\mathbf{mon}}(c - \underline{\mathrm{Ind}}_J^G(k_{\phi'}), M_i) \cong \mathrm{Hom}_{k[J],\underline{\phi}\mathbf{mon}}(k_{\phi'}),$$

$$\mathrm{Res}_J^G(M_i)) \cong M_i^{((J,\phi'))},$$

since H, J are compact open modulo the centre.

In order that the monomial analogue of $\lambda_{[JgH]}$ of §1.1

$$\lambda_{[JgH]} : c - \mathrm{Ind}_J^G(\phi') \longrightarrow c - \mathrm{Ind}_H^G(\phi)$$

should be a monomial morphism it is sufficient that for each y_i we have (see Chapter One §1.8)

$$(J, \phi') \le (y_i H y_i^{-1}, (y_i^{-1})^*(\phi)).$$

In that case $\lambda_{[JgH]}$ is given by the formula

$$\lambda_{[JgH]}(g' \otimes_J w) = \sum_{i=1}^n g'y_i \otimes_H \phi'(y_i g^{-1})^{-1} w.$$

and induces a chain map

$$[JgH] : M_*^{((H,\phi))} \longrightarrow M_*^{((J,\phi'))}.$$

This chain map together with the Hecke operator of §1.1 induces a chain map between the exact complexes of the monomial resolution from

$$\cdots \longrightarrow M_i^{((H,\phi))} \longrightarrow M_{i-1}^{((H,\phi))} \longrightarrow \cdots \longrightarrow M_0^{((H,\phi))} \xrightarrow{\ \epsilon\ } V^{(H,\phi)} \longrightarrow 0$$

to

$$\cdots \longrightarrow M_i^{((J,\phi'))} \longrightarrow M_{i-1}^{((J,\phi'))} \longrightarrow \cdots \longrightarrow M_0^{((J,\phi'))} \overset{\epsilon}{\longrightarrow} V^{(J,\phi')} \longrightarrow 0.$$

The question arises whether or not this ever happens. The following example is related to the Hecke operator $T(p)$ of ([**51**] §3.3 and Remark 11.1.1). The classical operators concern the case when $\phi' = \phi = 1$.

In the context of this section it would be interesting to find out how many of the classical operators described in ([**51**] Theorem 3.3.1) extend to give chain maps of the chain complexes derived from the monomial resolution of V.

EXAMPLE 1.4. The following example satisfies all the conditions of §1.3. Let $J = H = \Gamma_1(N) \subset GL_2\mathbb{Q}_p$ where $N = p^s$ for some $s \geq 1$. Explicitly

$$H = \left\{ \begin{pmatrix} 1 + Na & b \\ Nc & 1 + Nd \end{pmatrix} \mid a, b, c, d \in \mathbb{Z}_p \right\}.$$

Let

$$g = \begin{pmatrix} 1 & 0 \\ 0 & p^{-1} \end{pmatrix} \in GL_2\mathbb{Q}_p$$

so that

$$g \begin{pmatrix} 1 + Na & b \\ Nc & 1 + Nd \end{pmatrix} g^{-1}$$

$$= \begin{pmatrix} 1 & 0 \\ 0 & p^{-1} \end{pmatrix} \begin{pmatrix} 1 + Na & b \\ Nc & 1 + Nd \end{pmatrix} \begin{pmatrix} 1 & 0 \\ 0 & p \end{pmatrix}$$

$$= \begin{pmatrix} 1 + Na & b \\ Ncp^{-1} & (1 + Nd)p^{-1} \end{pmatrix} \begin{pmatrix} 1 & 0 \\ 0 & p \end{pmatrix}$$

$$= \begin{pmatrix} 1 + Na & bp \\ Ncp^{-1} & 1 + Nd \end{pmatrix}.$$

Therefore

$$J \bigcap gHg^{-1} = \left\{ \begin{pmatrix} 1 + Na & bp \\ Nc & 1 + Nd \end{pmatrix} \in H \mid a, b, c, d \in \mathbb{Z}_p \right\}.$$

Therefore we have a group extension

$$J \bigcap gHg^{-1} \longrightarrow H \longrightarrow \mathbb{Z}/p$$

and a coset representatives for $H/J \bigcap gHg^{-1}$ is given by $1, y, y^2, \ldots, y^{p-1}$ where

$$y = \begin{pmatrix} 1 & 1 \\ 0 & 1 \end{pmatrix}.$$

Since

$$\begin{pmatrix} 1 & 1 \\ 0 & 1 \end{pmatrix} \begin{pmatrix} u & v \\ w & x \end{pmatrix} \begin{pmatrix} 1 & -1 \\ 0 & 1 \end{pmatrix} = \begin{pmatrix} u+w & v+x-u-w \\ w & x-w \end{pmatrix}$$

we see that y normalises H.

Therefore if we have $(J, \phi') = (H, \phi)$ and $(H, y^{-1})^*(\phi)) = (H, \phi)$ then automatically we have ϕ' equals ϕ when restricted to $J \bigcap gHg^{-1}$. This happens, for example, for any character of the form

$$\phi_\lambda : \begin{pmatrix} 1+Na & bp \\ & \\ Nc & 1+Nd \end{pmatrix} \mapsto \lambda(Nc)$$

where λ is any character of the form $\lambda : N\mathbb{Z}_p/N^2\mathbb{Z}_p \longrightarrow k^*$.

This gives an example of a Hecke operator chain map $[\Gamma_1(N)g\Gamma_1(N)]$ which is an endomorphism of the complex

$$M_*^{((\Gamma_1(N), \phi_\lambda))} \xrightarrow{\epsilon} V^{(\Gamma_1(N), \phi_\lambda)} \longrightarrow 0$$

coming from the monomial resolution of V.

In the formulae of §1.3 we have $y_i = y^{i-1}g$ and

$$\lambda_{[\Gamma_1(N)g\Gamma_1(N)]}(g' \otimes_{\Gamma_1(N)} w) = \sum_{i=1}^{p} g'y^{i-1}g \otimes_{\Gamma_1(N)} \phi'(y^{i-1})^{-1}w.$$

CHAPTER 8

Could Galois descent be functorial?

1. Morphisms and Shintani descent

1.1. Let us examine some examples of possible "Galois naturality" of Shintani base change [117]. For example, if V is a Galois invariant irreducible of $GL_s\mathbb{F}_{q^n}$ with extension $\tilde{V}$ to $\mathrm{Gal}(\mathbb{F}_{q^n}/\mathbb{F}_q) \propto GL_s\mathbb{F}_{q^n}$, let $H \subseteq GL_s\mathbb{F}_{q^n}$ be a "base change group" for the sub-Galois group $\mathrm{Gal}(\mathbb{F}_{q^n}/\mathbb{F}_{q^d}) \subseteq \mathrm{Gal}(\mathbb{F}_{q^n}/\mathbb{F}_q)$. Examples of such an H include $\mathbb{F}_{q^{ns}}^*$ or $GL_a\mathbb{F}_{q^n} \times GL_{s-a}\mathbb{F}_{q^n}$ [117]. If W is a $\mathrm{Gal}(\mathbb{F}_{q^n}/\mathbb{F}_{q^d})$-invariant irreducible of H with extension $\tilde{W}$ to $\mathrm{Gal}(\mathbb{F}_{q^n}/\mathbb{F}_{q^d}) \propto H$ and an H-map $\lambda : W \longrightarrow V$ is there a functorial $H^{\mathrm{Gal}(\mathbb{F}_{q^n}/\mathbb{F}_{q^d})}$-map of the form $Sh(\lambda) : Sh(W) \longrightarrow Sh(V)$ between the Shintani base change representations of W and V?

In general $\tilde{V}$ can only be constructed as a representation of $\mathrm{Gal}(\mathbb{F}_{q^{2n}}/\mathbb{F}_q) \propto GL_s\mathbb{F}_{q^n}$ (see Chapter Two, §6.1 (footnote) and Theorem 3.11; [117] Theorem 1 p. 406). However the naturality question introduced above also has an analogous formulation in the general case.

A weaker alternative question, potentially almost as useful, would be to ask for a morphism $Sh(\lambda)$ which is functorial up to multiplication by scalars.

EXAMPLE 1.2. In Appendix I §§7–8 we find the formulae to analyse the case for $GL_2\mathbb{F}_4$ and $(H, \phi) = (\mathbb{F}_4^*, \phi)$. Since the fixed subfield of $\mathbb{F}_4$ is $\mathbb{F}_2$ we must have $\phi = 1$. Hence $Sh_{\mathbb{F}_4^*}(1) = 1$ the trivial one-dimensional representation of $\mathbb{F}_2^* = \{1\}$.

In the notation of Appendix I the matrix C generates $\mathbb{F}_4^*$. There are three Galois invariant $GL_2\mathbb{F}_4$-irreducibles — ν_5, ν_4 and 1.

We have $\nu_5^{(\langle C \rangle, 1)} = \mathbb{C}$ the unique C-fixed subspace of the 5-dimensional ν_5 and therefore a non-zero H-map λ. Also $Sh_{GL_2\mathbb{F}_4}(\nu_5) = \chi$, the non-trivial character of order two on $GL_2\mathbb{F}_2 \cong D_6$. Therefore

$$Sh_{GL_2\mathbb{F}_4}(\nu_5)^{(\{1\}, 1)} = 1$$

so, at least up to scalars, we have a unique $\mathbb{F}_2^*$-map which is a candidate for $Sh(\lambda)$. Note that in this example, in the extended representation $\tilde{\nu}_5$ the Frobenius σ also fixes $\nu_5^{(\langle C \rangle, 1)}$.

Now consider $\tilde{\nu}_4$ on which C and Σ, the Frobenius, act by the matrices

$$C = \begin{pmatrix} 1 & 0 & 0 & 0 \\ 0 & -1 & -1 & -2 \\ 0 & 1 & 0 & 0 \\ 0 & 0 & 0 & 1 \end{pmatrix} \text{ and } \Sigma = \begin{pmatrix} 1 & 0 & 0 & 0 \\ 0 & -1 & 0 & 0 \\ 0 & 0 & -1 & 0 \\ 0 & 0 & 0 & -1 \end{pmatrix}.$$

We have $\nu_4^{(\langle C \rangle, 1)} = \mathbb{C}^2$ so that we have a 2-dimensional space of λ's. In this case $Sh(\nu_4) = \nu$, the 2-dimensional irreducible so that

$$Sh_{GL_2\mathbb{F}_4}(\nu_4)^{(\{1\},1)} = 2$$

and we have a 2-dimensional space of potential $Sh(\lambda)$'s.

Perhaps the recipe for assigning $\lambda \mapsto Sh(\lambda)$ should incorporate the fact that the fixed subspace of the Frobenius on the extension $\tilde{\nu}_4$ is 1-dimensional?

Finally, $Sh_{GL_2\mathbb{F}_4}(1) = 1$ and the functorial $Sh(\lambda)$ ought to be the identity map of the complex numbers when λ is non-zero.

EXAMPLE 1.3. *A more elaborate example ([117] p. 412)*

In this section Galois extension is $\mathbb{F}_{q^{2p}}/\mathbb{F}_q$ and the irreducible is $R(\chi_1, \chi_2)$ with $\chi_i : \mathbb{F}_{q^{2p}}^* \longrightarrow \mathbb{C}^*$ and Frobenius action $\Sigma(\chi_1) = \chi_2, \Sigma(\chi_2) = \chi_1$ so that

$$\Sigma^* R(\chi_1, \chi_2) = R(\chi_1, \chi_2).$$

Hence $\text{Gal}(\mathbb{F}_{q^{2p}}/\mathbb{F}_{q^2}) = \langle \Sigma^2 \rangle$ fixes χ_1 and χ_2 so that, by Hilbert's Theorem 90,

$$\chi_1 = \Theta \cdot \text{Norm} : \mathbb{F}_{q^{2p}}^* \longrightarrow \mathbb{F}_{q^2}^* \longrightarrow \mathbb{C}^*$$

and

$$\chi_2 = \Sigma^*(\Theta) \cdot \text{Norm} : \mathbb{F}_{q^{2p}}^* \longrightarrow \mathbb{F}_{q^2}^* \longrightarrow \mathbb{C}^*.$$

Therefore $\Theta \neq \Sigma^*(\Theta)$.

From ([117]; [126], Chapter Two) $Sh(R(\chi_1, \chi_2)) = R(\Theta)$, the Weil representation associated to Θ, which is an irreducible representation of $GL_2\mathbb{F}_q$.

Let B denote the Borel subgroup of $GL_2\mathbb{F}_{q^{2p}}$ consisting of the upper triangular matrices, which surjects in the obvious manner onto the diagonal matrices

$$\text{Diag} \cong \mathbb{F}_{q^{2p}}^* \times \mathbb{F}_{q^{2p}}^*.$$

By definition we have

$$R(\chi_1, \chi_2) = \text{Ind}_B^{GL_2\mathbb{F}_{q^{2p}}} (\text{Inf}_{Diag}^B(\chi_1 \otimes \chi_2)).$$

Therefore a basis for this representation is given by

$$\begin{pmatrix} 1 & 0 \\ u & 1 \end{pmatrix} \otimes_B 1 \text{ with } u \in \mathbb{F}_{q^{2p}}$$

and

$$\begin{pmatrix} 0 & 1 \\ -1 & 0 \end{pmatrix} \otimes_B 1 = \chi_1(-1) \begin{pmatrix} 0 & 1 \\ 1 & 0 \end{pmatrix} \otimes_B 1.$$

From [117] $R(\chi_1, \chi_2)$ is realised by the vector space of functions f on $GL_2\mathbb{F}_{q^{2p}}$ with

$$f(\begin{pmatrix} a & b \\ 0 & d \end{pmatrix} x) = \chi_1(a)\chi_2(d)f(x)$$

and the $GL_2\mathbb{F}_{q^{2p}}$-action is given by $(g\#f)(x) = f(xg)$. Therefore the action first by g_2 and then by g_1 satisfies $(g_1\#(g_2\#f))(x) = (g_2\#f))(xg_1) = f(xg_1g_2)$.

Suppose we map f to $\mathbb{C}[GL_2\mathbb{F}_{q^{2p}}] \otimes_B \mathbb{C}_{\chi_1 \otimes \chi_2}$ by $f \mapsto \sum g^{-1} \otimes_B f(g)$. We have

$$\sum g^{-1} \otimes_B f(g)$$

$$= \sum (\begin{pmatrix} a & b \\ 0 & d \end{pmatrix} g)^{-1} \otimes_B f(\begin{pmatrix} a & b \\ 0 & d \end{pmatrix} g)$$

$$= \sum g^{-1} \begin{pmatrix} a & b \\ 0 & d \end{pmatrix}^{-1} \otimes_B f(\begin{pmatrix} a & b \\ 0 & d \end{pmatrix} g)$$

$$= \sum g^{-1} \otimes_B \chi_1(a)^{-1}\chi_2(d)^{-1}f(\begin{pmatrix} a & b \\ 0 & d \end{pmatrix} g)$$

so that

$$f(\begin{pmatrix} a & b \\ 0 & d \end{pmatrix} x) = \chi_1(a)\chi_2(d)f(x)$$

as required.

Furthermore the left action by $z \in GL_2\mathbb{F}_{q^{2p}}$ is given by

$$z(\sum g^{-1} \otimes_B f(g))$$

$$= \sum zg^{-1} \otimes_B f(g))$$

$$= \sum (gz^{-1})^{-1} \otimes_B f((gz^{-1})z))$$

$$= \sum g^{-1} \otimes_B f(gz))$$

so that $(zf)(g) = f(gz)$.

Shintani has a transformation I on the functions which is the action of the Frobenius in the semi-direct product. It is given by

$$(I \cdot f)(g) = q^{-p} \sum_{u \in \mathbb{F}_{q^{2p}}} f(\begin{pmatrix} 0 & 1 \\ 1 & 0 \end{pmatrix} \begin{pmatrix} 1 & u \\ 0 & 1 \end{pmatrix} \Sigma(g)).$$

Transforming this over to the tensor product followed by left multiplication by $\Sigma(z)$ yields

$$\Sigma(z)(I(\sum g^{-1} \otimes_B f(g)))$$

$$= \Sigma(z)(\sum g^{-1} \otimes_B q^{-p} \sum_{u \in \mathbb{F}_{q^{2p}}} f(\begin{pmatrix} 0 & 1 \\ 1 & 0 \end{pmatrix} \begin{pmatrix} 1 & u \\ 0 & 1 \end{pmatrix} \Sigma(g)))$$

$$= \sum \Sigma(z) g^{-1} \otimes_B q^{-p} \sum_{u \in \mathbb{F}_{q^{2p}}} f(\begin{pmatrix} 0 & 1 \\ 1 & 0 \end{pmatrix} \begin{pmatrix} 1 & u \\ 0 & 1 \end{pmatrix} \Sigma(g))$$

$$= \sum g^{-1} \otimes_B q^{-p} \sum_{u \in \mathbb{F}_{q^{2p}}} f(\begin{pmatrix} 0 & 1 \\ 1 & 0 \end{pmatrix} \begin{pmatrix} 1 & u \\ 0 & 1 \end{pmatrix} \Sigma(gz))$$

whereas

$$I(z(\sum g^{-1} \otimes_B f(g)))$$

$$= I(\sum zg^{-1} \otimes_B f(g))$$

$$= I(\sum g^{-1} \otimes_B f(gz))$$

$$= \sum g^{-1} \otimes_B q^{-p} \sum_{u \in \mathbb{F}_{q^{2p}}} f(\begin{pmatrix} 0 & 1 \\ 1 & 0 \end{pmatrix} \begin{pmatrix} 1 & u \\ 0 & 1 \end{pmatrix} \Sigma(gz)).$$

Hence on $R(\chi_1, \chi_2)$ we have $\Sigma(z) = I \cdot z \cdot I^{-1}$ which, up to a scalar which we shall assume Shintani got right, I acts like the Frobenius in the semi-direct product on $\tilde{R}(\chi_1, \chi_2)$.

Now consider the action of the diagonal torus $H = \mathbb{F}_{q^{2t}}^* \times \mathbb{F}_{q^{2t}}^*$ on $R(\chi_1, \chi_2)$. In the tensor product realisation a basis for this representation is given by

$$(\begin{pmatrix} 1 & 0 \\ u & 1 \end{pmatrix} \otimes_B 1) \text{ with } u \in \mathbb{F}_{q^{2t}} \text{ and } (\begin{pmatrix} 0 & 1 \\ 1 & 0 \end{pmatrix} \otimes_B 1).$$

The matrix relation

$$\begin{pmatrix} \alpha & 0 \\ 0 & \beta \end{pmatrix} \begin{pmatrix} 1 & 0 \\ u & 1 \end{pmatrix} = \begin{pmatrix} \alpha & 0 \\ \beta u & \beta \end{pmatrix} = \begin{pmatrix} 1 & 0 \\ \beta u/\alpha & 1 \end{pmatrix} \begin{pmatrix} \alpha & 0 \\ 0 & \beta \end{pmatrix}$$

shows that the H-action satisfies

$$\begin{pmatrix} \alpha & 0 \\ 0 & \beta \end{pmatrix}((\begin{pmatrix} 1 & 0 \\ u & 1 \end{pmatrix} \otimes_B 1) = \chi_1(\alpha)\chi_2(\beta) \begin{pmatrix} 1 & 0 \\ \beta u/\alpha & 1 \end{pmatrix} \otimes_B 1.$$

We also have

$$\begin{pmatrix} \alpha & 0 \\ 0 & \beta \end{pmatrix}((\begin{pmatrix} 0 & 1 \\ 1 & 0 \end{pmatrix} \otimes_B 1) = \chi_1(\beta)\chi_2(\alpha) \begin{pmatrix} 0 & 1 \\ 1 & 0 \end{pmatrix} \otimes_B 1.$$

Therefore

$$R(\chi_1,\chi_2)^{(T,\chi_1\otimes\chi_2)} = \langle\!\left(\begin{pmatrix} 1 & 0 \\ 0 & 1 \end{pmatrix}\otimes_B 1\right)\rangle \oplus \langle\, \sum_{u\in\mathbb{F}_{q^{2t}}^*} \left(\begin{pmatrix} 1 & 0 \\ u & 1 \end{pmatrix}\otimes_B 1\right)\rangle.$$

Furthermore, since Σ^2 preserves $\chi_1\otimes\chi_2$ we see that $R(\chi_1,\chi_2)^{(H,\chi_1\otimes\chi_2)}$ extends to a two-dimensional representation of $\mathrm{Gal}(\mathbb{F}_{q^{2t}}/\mathbb{F}_{q^2})\propto H$.

In ([**126**] Chapter Two) it is shown that

$$\mathrm{Res}_{\mathbb{F}_{q^2}^*}^{GL_2\mathbb{F}_q} R(\Theta) = \oplus_{\theta=\Theta \text{ on } \mathbb{F}_q^*,\,\theta\neq\Theta,\Sigma(\Theta)}\ \theta,$$

the direct sum of the $q-1$ characters of $\mathbb{F}_{q^2}^*$ which coincide with Θ on the scalar matrices but are not equal to Θ or its Frobenius translate.

Consider the $\mathrm{Gal}(\mathbb{F}_{q^{2t}}/\mathbb{F}_{q^2}$-equivariant map

$$H = \mathbb{F}_{q^{2t}}^* \times \mathbb{F}_{q^{2t}}^* \longrightarrow GL_2\mathbb{F}_{q^{2t}}$$

given by

$$(\alpha,\beta) \mapsto \begin{pmatrix} \alpha^{s_1}\beta^{s_2} & 0 \\ 0 & \alpha^{t_1}\beta^{t_2} \end{pmatrix}.$$

Via this map H acts of $R(\chi_1,\chi_2)^{(H,\chi_1\otimes\chi_2)}$ as multiplication by

$$\chi_1(\alpha)^{s_1}\chi_1(\beta)^{s_2}\chi_2(\alpha)^{t_1}\chi_2(\beta)^{t_2}.$$

This Shintani descends to $\mathbb{F}_{q^2}^* \times \mathbb{F}_{q^2}^*$ acting via $(a,1)$ as multiplication by

$$\Theta(a)^{s_1}\Theta(\Sigma(a))^{t_1} = \Theta(a)^{s_1+t_1}\frac{\Theta(\Sigma(a))^{t_1}}{\Theta(a)^{t_1}}$$

and $(1,b)$ by

$$\Theta(b)^{s_s}\Theta(\Sigma(b))^{t_2} = \Theta(b)^{s_2+t_2}\frac{\Theta(\Sigma(b))^{t_2}}{\Theta(b)^{t_2}}.$$

Therefore, if $s_1 \equiv s_2$ (modulo q^2-1), $t_1 \equiv t_2$ (modulo $q+1$) and $s_1+t_1 \equiv 1$ (modulo q^2-1) then (a,b) acts via multiplication by

$$\Theta(ab)\frac{\Theta(\Sigma(ab))^{t_1}}{\Theta(ab)^{t_1}}.$$

The character of $\hat{\mathbb{F}}_{q^2}^*$ given by $x \mapsto \Theta(x)\frac{\Theta(\Sigma(x))^{t_1}}{\Theta(x)^{t_1}}$ equal to Θ when restricted to $\mathbb{F}_q^*$ but is not equal to Θ or $\Sigma^*(\Theta)$ if t_1 is chosen correctly. Therefore this is one of the characters appearing with multiplicity one in the restriction of $R(\Theta)$ to $\hat{\mathbb{F}}_{q^2}^*$. Therefore, up to a scalar, we have a unique morphism between Shintani descents.

Sometimes, in this example, the only candidate for $Sh(\lambda)$ is zero.

1.4. *The role of the $V^{(H,\phi)}$'s*

Suppose Shintani base change is functorial in the sense of §1.1 and that V is a $\mathrm{Gal}(\mathbb{F}_{q^n}/\mathbb{F}_q)$-invariant irreducible representation of $GL_s\mathbb{F}_{q^n}$ as in [117]. Suppose that $H \subseteq GL_s\mathbb{F}_{q^n}$ is a subgroup preserved by $\mathrm{Gal}(\mathbb{F}_{q^n}/\mathbb{F}_{q^d})$ and that $\phi \in \hat{H}$ is fixed by this Galois action. Then the inclusion

$$\lambda : V^{(H,\phi)} \longrightarrow V$$

is an H-map. If H is a $\mathrm{Gal}(\mathbb{F}_{q^n}/\mathbb{F}_{q^d})$-base change group then functoriality of base change would yield a $H^{\mathrm{Gal}(\mathbb{F}_{q^n}/\mathbb{F}_{q^d})}$-map

$$Sh(\lambda) : Sh(V^{(H,\phi)}) \longrightarrow Sh(V)$$

which is unique (possibly only up to scalars).

If (H, ϕ) is not Galois-fixed we could form the subspace given by

$$W = \sum_{\sigma \in \mathrm{Gal}(\mathbb{F}_{q^n}/\mathbb{F}_{q^d})} V^{(\sigma(H),(\sigma^{-1})^*(\phi))} \subseteq V.$$

Sometimes this inclusion will be a J-map of $\mathrm{Gal}(\mathbb{F}_{q^n}/\mathbb{F}_{q^d})$ irreducible representations of base change groups. In this case functoriality would lead one to expect a $J^{\mathrm{Gal}(\mathbb{F}_{q^n}/\mathbb{F}_{q^d})}$-map

$$Sh(W) \longrightarrow Sh(V),$$

unique up to scalars.

1.5. *The role of monomial resolutions*

The discussion of §1.4 suggests (to an inrepressible optimist) that the functoriality of base change would yield not just a $J^{\mathrm{Gal}(\mathbb{F}_{q^n}/\mathbb{F}_{q^d})}$-map $Sh(W) \longrightarrow Sh(V)$ but also morphisms of the complexes of Chapter One coming from the monomial resolutions and mapping from

$$M(Sh(W))_*^{((J',\chi'))} \longrightarrow Sh(W)^{(J',\chi')}$$

to

$$M(Sh(V))_*^{((J',\chi'))} \longrightarrow Sh(V)^{(J',\chi')}$$

for suitable choices of (J', χ').

1.6. *Is there a canonical choice of morphism?*

Suppose that, in §1.4, there were a $J^{\mathrm{Gal}(\mathbb{F}_{q^n}/\mathbb{F}_{q^d})}$-map

$$Sh(\lambda) : Sh(W) \longrightarrow Sh(V),$$

unique up to scalars. How could it be normalised in order to become unique?

The formulae for examples which I have encountered suggest the Marshall McLuhan-style slogan "the morphism IS the invariant". That is, somehow (I have the local field case in mind here) the formula can be read off the formulae for the epsilon factor or local L-function invariant[1].

1.7. *A simple observation*

Suppose that V is a finite-dimensional complex $\mathrm{Gal}(\mathbb{F}_{q^m}/\mathbb{F}_q)$-invariant irreducible representation of $GL_n\mathbb{F}_{q^m}$ and that

$$\lambda : \mathbb{F}_{q^m}^* \times \mathbb{F}_{q^m}^* \longrightarrow GL_n\mathbb{F}_{q^m}$$

is a $\mathrm{Gal}(\mathbb{F}_{q^m}/\mathbb{F}_q)$-equivariant homomorphism. Let ϕ be a $\mathrm{Gal}(\mathbb{F}_{q^m}/\mathbb{F}_q)$-invariant character and denote by $\mathbb{C}_\phi$ the associated one-dimensional representation.

We have

$$\dim_\mathbb{C}\mathrm{Hom}_{\mathbb{F}_{q^m}^* \times \mathbb{F}_{q^m}^*}(\mathbb{C}_\phi, V)$$

$$= \tfrac{1}{|\mathbb{F}_{q^m}|^2} \textstyle\sum_{(h_1,h_2)\in\mathbb{F}_{q^m}^*\times\mathbb{F}_{q^m}^*} \phi(h_1,h_2)^{-1}\chi_V(h_1,h_2).$$

If $\dim_\mathbb{C}\mathrm{Hom}_{\mathbb{F}_{q^m}^* \times \mathbb{F}_{q^m}^*}(\mathbb{C}_\phi, V) = 1$ and m is prime we may write

$$\mathrm{Res}^{GL_n\mathbb{F}_{q^m}}_{\mathbb{F}_{q^m}^* \times \mathbb{F}_{q^m}^*}(V) = \mathbb{C}_\phi \oplus (\oplus_{i=1}^t \mathbb{C}_{\phi_i}) \oplus (\oplus_{j=1}^s \mathbb{C}_{\psi_j} \oplus \Sigma^*\psi_j \oplus \ldots \oplus \Sigma^{m,*}\psi_j)$$

where ϕ_i for $1 \leq i \leq t$ are $\mathrm{Gal}(\mathbb{F}_{q^m}/\mathbb{F}_q)$-invariant characters different from ϕ. The extension of V to a representation of the semi-direct product $\mathrm{Gal}(\mathbb{F}_{q^m}/\mathbb{F}_q) \propto GL_n\mathbb{F}_{q^m}$ is denote by $\tilde{V}$. The restriction of $\tilde{V}$ to $\mathrm{Gal}(\mathbb{F}_{q^m}/\mathbb{F}_q) \propto (\mathbb{F}_{q^m}^* \times \mathbb{F}_{q^m}^*)$ is given by the sum of $\tilde{\phi}, \tilde{\phi}_1, \ldots, \tilde{\phi}_t$ and the obvious extensions of the $\mathbb{C}_{\psi_j} \oplus \Sigma^*\psi_j \oplus \ldots \oplus \Sigma^{m,*}\psi_j$'s.

Suppose that $h_1, h_2 \in \mathbb{F}_{q^m}^*$ and

$$(w_1, w_2) = (h_1\Sigma(h_1))\ldots\Sigma^{m-1}(h_1), h_2\Sigma(h_2))\ldots\Sigma^{m-1}(h_2)) \in \mathbb{F}_q^* \times \mathbb{F}_q^*$$

and for any w_i there are $|\mathbb{F}_{q^m}^*|/ \, |\mathbb{F}_q^*| \; h_i$'s whose norm is w_i.

---

[1]Perhaps this is getting to be really too optimistic, attributable no doubt to my (like Herbert Marshall McLuhan) spending too much time in Edmonton, Alberta!

The character function on (Σ, h_1, h_2) of the extension to each of the $\mathbb{C}_{\psi_j} \oplus \Sigma^* \psi_j \oplus \ldots \oplus \Sigma^{m,*} \psi_j$'s to the semi-direct product is zero. Therefore

$$\dim_{\mathbb{C}} \mathrm{Hom}_{\mathbb{F}_q^* \times \mathbb{F}_q^*}(Sh(\mathbb{C}_\phi), Sh(V))$$

$$= \tfrac{1}{|\mathbb{F}_q|^2} \sum_{(w_1,w_2) \in \mathbb{F}_q^* \times \mathbb{F}_q^*} Sh(\phi)(w_1, w_2)^{-1} \chi_{Sh(V)}(w_1, w_2)$$

$$= \tfrac{1}{|\mathbb{F}_q|^2} \tfrac{|\,\mathbb{F}_q^*|^2}{|\mathbb{F}_{q^m}^*|^2} \sum_{\substack{(w_1,w_2) \in \mathbb{F}_q^* \times \mathbb{F}_q^* \\ w_i = h_i \Sigma(h_i)) \ldots \Sigma^{m-1}(h_i)}} \phi)(h_1, h_2)^{-1} \chi_{\tilde{V}}(\Sigma, h_1, h_2))$$

$$= \tfrac{1}{|\mathbb{F}_{q^m}|^2} \sum_{(h_1,h_2) \in \mathbb{F}_{q^m}^* \times \mathbb{F}_{q^m}^*} \phi(h_1, h_2)^{-1} (\phi(h_1, h_2) + \sum_{i=1}^t \phi_i(h_1, h_2))$$

$$= 1.$$

This discussion, by induction, establishes the following result, which is a further modest contribution to the topic of functoriality of Shintani base change.

PROPOSITION 1.8.
Let V be a finite-dimensional complex $\mathrm{Gal}(\mathbb{F}_{q^m}/\mathbb{F}_q)$-invariant irreducible representation of $GL_n\mathbb{F}_{q^m}$ and let

$$\lambda : \mathbb{F}_{q^m}^* \times \mathbb{F}_{q^m}^* \longrightarrow GL_n\mathbb{F}_{q^m}$$

be a $\mathrm{Gal}(\mathbb{F}_{q^m}/\mathbb{F}_q)$-equivariant homomorphism. Let ϕ be a $\mathrm{Gal}(\mathbb{F}_{q^m}/\mathbb{F}_q)$-invariant character and denote by $\mathbb{C}_\phi$ the associated one-dimensional representation.
If $\dim_{\mathbb{C}} \mathrm{Hom}_{\mathbb{F}_{q^m}^* \times \mathbb{F}_{q^m}^*}(\mathbb{C}_\phi, V) = 1$ then

$$\dim_{\mathbb{C}} \mathrm{Hom}_{\mathbb{F}_q^* \times \mathbb{F}_q^*}(Sh(\mathbb{C}_\phi), Sh(V)) = 1.$$

EXAMPLE 1.9. *One last base change naturality example*
Let $j : H \longrightarrow GL_t\mathbb{F}_{q^m}$ be a homomorphism where H is a subgroup with a Galois action by $\mathrm{Gal}(\mathbb{F}_{q^m}/\mathbb{F}_q)$ with respect to which H is a "Shintani base change group". For example, H might be the product of several $GL_{a_i}\mathbb{F}_{q^m}$'s. Suppose that j commutes with the $\mathrm{Gal}(\mathbb{F}_{q^m}/\mathbb{F}_q)$-action. Let V_2 be a complex irreducible representation of $\mathrm{Gal}(\mathbb{F}_{q^m}/\mathbb{F}_q)$ which is Galois invariant. Therefore there exists, for each $g \in \mathrm{Gal}(\mathbb{F}_{q^m}/\mathbb{F}_q)$ an automorphism $I_g \in \mathrm{Aut}_{\mathbb{C}}(V_2)$ such that for $v_2 \in V_2, x \in GL_t\mathbb{F}_{q^m}$ we have

$$I_g(xv_2) = g(x)I_g(v_2) \text{ or equivalently } xI_g^{-1}(v_2) = I_g^{-1}(g(x)v_2).$$

Let V_1 be a complex irreducible representation of H which is also $\mathrm{Gal}(\mathbb{F}_{q^m}/\mathbb{F}_q)$-invariant. In addition suppose that

$$\dim_{\mathbb{C}} \mathrm{Hom}_H(V_1, j^*(V_2)) = 1.$$

Therefore there is an H-isomorphisms

$$\Phi : V_2 \xrightarrow{\cong} V_1 \oplus W \text{ and } i : V_1 \longrightarrow V_2$$

such that $\Phi(i(v_1)) = (v_1, 0)$.

There is also $J_g \in \mathrm{Aut}_{\mathbb{C}}(V_1)$ such that, for $v_1 \in V_1$, $J_g(xv_1) = g(x)J_g(v_1)$.

Consider the composition

$$V_1 \xrightarrow{J_g} V_1 \xrightarrow{i} V_2 \xrightarrow{I_g^{-1}} V_2 \xrightarrow{\Phi} V_1 \oplus W.$$

This first maps xv_1 to $J_g(xv_1) = g(x)J_g(v_1)$ and then to $g(x)i(J_g(v_1))$ and thereafter to $I_g^{-1}(g(x)i(J_g(v_1))) = xI_g^{-1}(i(J_g(v_1)))$. If $\Phi(I_g^{-1}(i(J_g(v_1)))) = (\alpha, \beta) \in V_1 \oplus W$ the final image of xv_1 is $(x\alpha, x\beta)$. By Schur's Lemma the first coordinate of $I_g^{-1}(i(J_g()))$ is a scalar multiple of the identity map of V_1 and by the H-multiplicity of V_1 in V_2 the second component of $I_g^{-1}(i(J_g()))$ must be zero. Since the composition is injective the scalar multiple in the first coordinate is non-zero and may be chosen equal to 1.

In other words we may choose the restriction of I_Σ to V_1 to be equal to J_Σ where Σ is the Frobenius Galois automorphism.

Now let $Sh(V_1) \in \mathrm{Irr}(H^{\mathrm{Gal}(\mathbb{F}_{q^m}/\mathbb{F}_q)})$ and $Sh(V_2) \in \mathrm{Irr}(GL_t\mathbb{F}_q)$.

Consider now the dimension

$$\dim_{\mathbb{C}} \mathrm{Hom}_{H^{\mathrm{Gal}(\mathbb{F}_{q^m}/\mathbb{F}_q)}}(Sh(V_1), Sh(V_2))$$

$$\frac{1}{|H^{\mathrm{Gal}(\mathbb{F}_{q^m}/\mathbb{F}_q)}|} \sum_{Y \in H^{\mathrm{Gal}(\mathbb{F}_{q^m}/\mathbb{F}_q)}} \overline{\chi_{Sh(V_1)}(Y)} \chi_{Sh(V_2)}(Y).$$

Let Σ denote the Frobenius automorphism. There is an automorphism $I_\Sigma \in \mathrm{Aut}_{\mathbb{C}}(V_2)$ satisfying, for $x \in GL_t\mathbb{F}_{q^m}, v \in V_2$,

$$I_\Sigma(x \cdot v) = \Sigma(x) \cdot I_\Sigma(v)$$

and $I_\Sigma^m = \pm 1$. Therefore $(\Sigma, x) \cdot v = x \cdot I_\Sigma(v)$ defines an irreducible representation ρ' on V_2 of the semi-direct product $\mathrm{Gal}(\mathbb{F}_q^{2m}/\mathbb{F}_q) \ltimes GL_t\mathbb{F}_{q^m}$[2] When $I_\Sigma^m = 1$ ρ' factors through a representation $\tilde{\rho}$ of $\mathrm{Gal}(\mathbb{F}_q^m/\mathbb{F}_q) \ltimes GL_t\mathbb{F}_{q^m}$. In either case

$$\chi_{\rho'}(\Sigma, x) = \chi_{Sh(V_2)}([x\Sigma(x)\ldots\Sigma^{m-1}(x)]).$$

Here $[x\Sigma(x)\ldots\Sigma^{m-1}(x)]$ denotes the unique conjugacy class in $GL_n\mathbb{F}_q$ given by the intersection of the conjugacy class of $x\Sigma(x)\ldots\Sigma^{m-1}(x)$ in $GL_n\mathbb{F}_{q^m}$ with $GL_n\mathbb{F}_q$.

I shall take the assumption that H is a "Shintani base change group" to mean that the analogous extension ρ'' to $\mathrm{Gal}(\mathbb{F}_q^{2m}/\mathbb{F}_q) \ltimes H$, constructed using J_Σ, satisfies the analogous character formula

$$\chi_{\rho''}(\Sigma, x) = \chi_{Sh(V_1)}([x\Sigma(x)\ldots\Sigma^{m-1}(x)]).$$

---

[2]My convention for the product is given by $(g_1, x_1)(g_2, x_2) = (g_1 g_2, x_1 g_1(x_2))$.

Since J_Σ may be taken to be the restriction of I_Σ to H the conjugacy class in $H^{\mathrm{Gal}(\mathbb{F}_q^{2m}/\mathbb{F}_q)} = H^{\mathrm{Gal}(\mathbb{F}_q^m/\mathbb{F}_q)}$ denoted by $[x\Sigma(x)\ldots\Sigma^{m-1}(x)]$ is the same in the $\chi_{\rho''}(\Sigma,x)$ formula as in the $\chi_{\rho'}(\Sigma,x)$-formula provided that $x \in H$.

Also, since J_Σ may be taken to be the restriction of I_Σ to H, there is a non-zero map of representations from ρ'' to ρ'. Therefore the equation

$$\dim_{\mathbb{C}}\mathrm{Hom}_H(V_1, j^*(V_2)) = 1$$

implies that

$$\dim_{\mathbb{C}}\mathrm{Hom}_{\mathrm{Gal}(\mathbb{F}_q^{2m}/\mathbb{F}_q)\ltimes H}(\rho'', j^*(\rho')) = 1.$$

We have

$$\dim_{\mathbb{C}}\mathrm{Hom}_{\mathrm{Gal}(\mathbb{F}_q^{2m}/\mathbb{F}_q)\ltimes H}(\rho'', j^*(\rho'))$$

$$= \tfrac{1}{|\mathrm{Gal}(\mathbb{F}_q^{2m}/\mathbb{F}_q)\ltimes H|} \sum_{(\Sigma^i,x)\in\mathrm{Gal}(\mathbb{F}_q^{2m}/\mathbb{F}_q)\ltimes H} \overline{\chi_{\rho''}(\Sigma^i,x)}\chi_{\rho'}(\Sigma^i,x)$$

$$= \tfrac{1}{|\mathrm{Gal}(\mathbb{F}_q^{2m}/\mathbb{F}_q)\ltimes H|} \sum_{(1,x)\in\mathrm{Gal}(\mathbb{F}_q^{2m}/\mathbb{F}_q)\ltimes H} \overline{\chi_{\rho''}(1,x)}\chi_{\rho'}(1,x)$$

$$+ \tfrac{1}{|\mathrm{Gal}(\mathbb{F}_q^{2m}/\mathbb{F}_q)\ltimes H|} \sum_{i=1}^{m-1} \sum_{(\Sigma^i,x)\in\mathrm{Gal}(\mathbb{F}_q^{2m}/\mathbb{F}_q)\ltimes H} \overline{\chi_{\rho''}(\Sigma^i,x)}\chi_{\rho'}(\Sigma^i,x)$$

$$= \tfrac{1}{2m}\dim_{\mathbb{C}}\mathrm{Hom}_H(V_1, j^*(V_2))$$

$$+ \tfrac{1}{2m|H|} \sum_{i=1}^{2m-1} \sum_{(\Sigma^i,x)\in\mathrm{Gal}(\mathbb{F}_q^{2m}/\mathbb{F}_q)\ltimes H} \overline{\chi_{\rho''}(\Sigma^i,x)}\chi_{\rho'}(\Sigma^i,x).$$

Next we observe the $I_\Sigma^m = \pm 1 = J_\Sigma^m$ implies that

$$\overline{\chi_{\rho''}(\Sigma^i,x)}\chi_{\rho'}(\Sigma^i,x) = \overline{\chi_{\rho''}(\Sigma^{m+i},x)}\chi_{\rho'}(\Sigma^{m+i},x).$$

Therefore

$$\dim_{\mathbb{C}}\mathrm{Hom}_{\mathrm{Gal}(\mathbb{F}_q^{2m}/\mathbb{F}_q)\ltimes H}(\rho'', j^*(\rho'))$$

$$= \tfrac{1}{m}\dim_{\mathbb{C}}\mathrm{Hom}_H(V_1, j^*(V_2))$$

$$+ \tfrac{1}{m|H|} \sum_{i=1}^{m-1} \sum_{(\Sigma^i,x)\in\mathrm{Gal}(\mathbb{F}_q^{2m}/\mathbb{F}_q)\ltimes H} \overline{\chi_{\rho''}(\Sigma^i,x)}\chi_{\rho'}(\Sigma^i,x).$$

Also, from [117] or the observation of Digne-Michel described in (Appendix I, §10.3), we have the following relation between sizes of conjugacy classes

$$|GL_t\mathbb{F}_{q^m}| \cdot |GL_t\mathbb{F}_q \text{ conjugacy class of } [x\Sigma(x)\ldots\Sigma^{m-1}(x)]|$$

$$= |GL_t\mathbb{F}_q| \cdot |GL_t\mathbb{F}_{q^m} \text{ conjugacy class of } (\Sigma,x)|.$$

This relation is a consequence of Lang's Theorem (see Appendix I §10) and I am assuming that H being a "Shintani base change group" includes the implication that the analogous relation holds for H. That is, for $x \in H$,

$$|H| \cdot |H^{\mathrm{Gal}(\mathbb{F}_{q^m}/\mathbb{F}_q)} \text{ conjugacy class of } [x\Sigma(x)\dots\Sigma^{m-1}(x)]|$$

$$= |H^{\mathrm{Gal}(\mathbb{F}_{q^m}/\mathbb{F}_q)}| \cdot |H \text{ conjugacy class of } (\Sigma, x)|.$$

To count the size of conjugacy classes of (Σ^i, x) in the semi-direct product (rather than just conjugation by elements of H) we observe that

$$(\Sigma, 1)(\Sigma^i, x)\Sigma, 1)^{-1} = (\Sigma^i, \Sigma(x)).$$

Now, for simplification, assume that m is prime so that each Σ^i for $1 \le i \le m-1$ generates the Galois group. In which case each term in the sum of the form

$$\overline{\chi_{Sh(V_1)}[x\Sigma(x)\dots\Sigma^{m-1}(x)]}|\chi_{Sh(V_2)}([x\Sigma(x)\dots\Sigma^{m-1}(x)]|).$$

Therefore the relation between character sums simplifies to

$$1 - \frac{1}{m}$$

$$= \frac{m-1}{m}\dim_{\mathbb{C}}\mathrm{Hom}_{H^{\mathrm{Gal}(\mathbb{F}_q^{2m}/\mathbb{F}_q)}}(Sh(V_1), Sh(V_2))$$

so that

$$\dim_{\mathbb{C}}\mathrm{Hom}_{H^{\mathrm{Gal}(\mathbb{F}_q^m/\mathbb{F}_q)}}(Sh(V_1), Sh(V_2)) = 1$$

as required.

Finally, when m is composite we proceed by induction on m, grouping together the terms

$$\overline{\chi_{\rho''}(\Sigma^i, x)}\chi_{\rho'}(\Sigma^i, x)$$

according to the fixed field of Σ^i, say $\mathbb{F}_{q^d}$, and using the result that

$$\dim_{\mathbb{C}}\mathrm{Hom}_{H^{\mathrm{Gal}(\mathbb{F}_q^m/\mathbb{F}_{q^d})}}(Sh_{\mathbb{F}_q^m/\mathbb{F}_{q^d}}(V_1), Sh_{\mathbb{F}_q^m/\mathbb{F}_{q^d}}(V_2)) = 1$$

for all divisors $d > 1$.

2. Galois base change of automorphic representations

2.1. *Local fields*

Let L/K be a Galois extension of unramified q-adic local fields with residue field extension $\mathbb{F}_{q^{2p}}/\mathbb{F}_q$. Let $\chi_1^+, \chi_2^+ : L^* \longrightarrow \mathbb{C}^*$ be continuous characters such that

$$\mathrm{Res}^{L^*}_{\mathcal{O}_L^*}(\chi_i^+) = \mathrm{Inf}^{\mathcal{O}_L^*}_{\mathbb{F}_{q^{2p}}^*}(\chi_i)$$

in the notation of Example 1.3. In addition suppose that $\Sigma(\chi_1^+) = \chi_2^+$ and $\Sigma(\chi_2^+) = \chi_1^+$ where $\Sigma \in \mathrm{Gal}(L/K) \cong \mathrm{Gal}(\mathbb{F}_{q^{2p}}/\mathbb{F}_q)$ corresponds to the Frobenius of Example 1.3. Then there is a $\mathrm{Gal}(L/K)$-invariant admissible irreducible $R(\chi_1^+, \chi_2^+)$ of GL_2L [63] which is related to $R(\chi_1, \chi_2)$ by c-Ind induction. In fact, the entire base change from $R(\chi_1^+, \chi_2^+)$ to an admissible irreducible of $GL_2 K$ is related to the Shintani base change of Example 1.3 by c-Ind induction.

Hence one is led to conjecture that base change for GL_n of local fields might be functorial in the analogous sense to that of §1.1. The pay-off for such functoriality of local base change could be very interesting. Suppose that Π is a $\mathrm{Gal}(L/K)$-invariant admissible irreducible of $GL_n L$. Let $\mathcal{E}$ be Tammo tom Dieck's space (see Appendix IV) associated to the class of cyclic subgroups of $\mathrm{Gal}(L/K)$. Let σ be a simplex of $\mathcal{E}$ with stabiliser $H_\sigma = Stab_{\mathrm{Gal}(L/K)}(\sigma)$, which is a cyclic group. Assigning to σ the H_σ-base change of Π, functoriality of base change would give a sheaf of admissible representations on $\mathcal{E}$. The Cech complex of this sheaf would be a complex of $\mathrm{Gal}(L/K) \times GL_n K$ admissible representations. Since each fixed-point sub-complex $\mathcal{E}^{H_\sigma}$ is contractible the spectral sequence for computing the Cech sheaf cohomology simplifies and yields an interesting "base change complex" of admissible $\mathrm{Gal}(L/K) \times GL_n K$ representations.

2.2. Automorphic representations

Cyclic Galois base change for automorphic representations of $GL_n \mathbb{A}_F$, F being a number field, was established in [7]. This was accomplished by proving local cyclic base change for GL_n and appealing to the Tensor Product Theorem (Chapter Three, §2). If V is an automorphic representation we have seen (Chapter Three §4 and Chapter Seven §1) that the subspaces $V^{(H,\lambda)}$ can sometimes be spaces of automorphic forms. For example, in the adèlic language, Hecke characters may be interpreted as automorphic forms on GL_1 and modular forms as automorphic forms on GL_2. Therefore Galois base change for automorphic representations of GL_2 is related to a similar base change for modular forms. This was first studied, for GL_2 and quadratic extensions, by Doi and Naganuma in [55] and [56], using Weil's converse to Hecke theory, as did Jacquet in [81]. Saito [107] introduced the use of a twisted trace formula to treat the case of base change for some Hilbert modular forms in cyclic extensions of totally real fields. Saito's method was recast in terms of automorphic forms on adèlic groups by Shintani [118] and Langlands [91].

For further details see ([39] pp. 84–88 and pp. 90–103) and, of course, [7].

Functoriality of base change may fit in with base change for modular forms in the following manner. By the Tensor Product Theorem and the Multiplicity One Theorem ([39], [91]) functoriality of local base change for GL_2 should imply functoriality for base change of adèlic representations. If V in §1.4 were an automorphic representation and

$$\lambda : V^{(H,\phi)} \longrightarrow V$$

is the embedding of a line spanned by a "modular form" then the morphism of base changes, analogous to

$$Sh(\lambda) : Sh(V^{(H,\phi)}) \longrightarrow Sh(V)$$

of §1.4, $Sh(V^{(H,\phi)})$ being a line, would have image spanned by a "base changed modular form".

3. Integrality and the proof of Shintani's theorem

3.1. This section sketches Shintani's proof of Galois base change (Galois descent) for finite-dimensional complex irreducible representations of $GL_n\mathbb{F}_{q^d}$. Shintani's proof is a baby version of the local base change proof of [7], where "baby" means that applications of the highly technical twisted trace formula are replaced by applications of the elementary character (trace) functions for representations of finite groups. That is not to derogate either the complexity or the importance of Shintani's result. Far from it, for [117] served as the insight and the motivation for the fundamental [91] and many subsequent papers.

Since §1 and §2 of this chapter were concerned with speculation about functoriality of base change for GL_n and its subgroups for finite and local fields, I should try to present a sketch proof of the main result of [117] which contains at least one fundamental difference. In what follows the main difference will consist of reducing the proof to an integrality condition related to the Explicit Brauer Induction formula of Appendix I, §5 and therefore, by inference, to the monomial resolutions of Chapter One and Chapter Two.

If G is a finite group recall that $R(G)$ (Appendix I, §5) denotes the complex representation ring of G; i.e. $R(G) = K_0(\mathbb{C}[G])$. An element of $R(G)$ will be called a virtual representation (or simply a character in the terminology of ([69] Theorem 1)). Any finite-dimensional complex representation V of G defines a class in $R(G)$ and two representations V, W become equal in $R(G)$ if and only if they are equivalent. In general the elements of $R(G)$ are formal differences $x = V_1 - V_2$ of finite-dimensional representations V_i of G. The character function of x given by the conjugacy class function on G defined by

$$x(g) \mapsto \text{Trace}(x)(g) = \text{Trace}_{V_1}(g) - \text{Trace}_{V_2}(g) \in \mathbb{C}$$

uniquely characterises $x \in R(G)$ (see [126]).

3.2. Any complex-valued conjugacy class function f on a finite group G defines an element of $R(G) \otimes \mathbb{C}$. Conversely, via its trace function, any element of $R(G)$ gives rise to such a conjugacy class function. In order to describe all the class functions arising from $R(GL_n\mathbb{F}_q)$ we shall need a necessary and sufficient recognition criterion.

Recall that G is an M-group [77] if every finite-dimensional complex irreducible representation of G has the form $\operatorname{Ind}_H^G(\phi)$ for some character $\phi \in \hat{H}$. Following [69] G will be called an elementary group if it is isomorphic to a product $H \times C$ with H a p-group and C a cyclic group whose order is not divisible by the prime p. As in Chapter One, §1, $\mathcal{M}(G)$ denotes the poset of pairs (H, ϕ) with $H \subseteq G$ and $\phi \in \hat{H}$.

PROPOSITION 3.3.
The following conditions on the class function f of §3.2 are equivalent:

(i) $f \in R(G) \subset R(G) \otimes \mathbb{C}$,

(ii) for every M-group $H \subseteq G$

$$\operatorname{Res}_H^G(f) \in R(H) \subset R(H) \otimes \mathbb{C},$$

(iii) for every elementary group $H \subseteq G$

$$\operatorname{Res}_H^G(f) \in R(H) \subset R(H) \otimes \mathbb{C},$$

(iv) for every $(H, \phi) \in \mathcal{M}(G)$

$$\frac{1}{|G|} \sum_{(H,\phi)=(H_0,\phi_0)<\ldots<(H_r,\phi_r)} \sum_{h \in H} \phi(h)^{-1} f(h) \in \mathbb{Z}.$$

Here the sum is taken over all ascending chains on $\mathcal{M}(G)$ starting at (H, ϕ).

Proof

Since the restriction map on representations $R(G) \otimes \mathbb{C} \longrightarrow R(H) \otimes \mathbb{C}$ maps $R(G)$ to $R(H)$ we see that (i) immediately implies (ii) and (iii).

By Brauer's induction theorem [33] every virtual representation $\lambda \in R(G)$ may be written as a $\mathbb{Z}$-linear combination of the form

$$\lambda = \sum_{i=1}^{t} \alpha_i \operatorname{Ind}_{J_i}^{GL_s \mathbb{F}_{q^n}}(\phi_i)$$

where $\phi : J_i \longrightarrow \mathbb{C}^*$ is a character and $J_i \subseteq G$ is an elementary group. In particular take $\lambda = 1$ and multiply the relation by the conjugacy class function f to obtain, via Frobenius reciprocity, a relation between class functions

$$f = \sum_{i=1}^{t} \alpha_i f \cdot \operatorname{Ind}_{J_i}^G(\phi_i) = \alpha_i \operatorname{Ind}_{J_i}^G(\operatorname{Res}_{J_i}^G(f) \cdot \phi_i)$$

which shows that if each $\operatorname{Res}_{J_i}^G(f) \in R(J_i) \subset R(J_i) \otimes \mathbb{C}$ then $f \in R(G) \subset R(G) \otimes \mathbb{C}$, proving (iii).

In ([**126**] Proposition 2.1.17) a topological argument[3] shows that

$$1 = \sum_\alpha \mathrm{Ind}_{H_\alpha}^G(1) \in R(G)$$

in which every H_α is an M-group. The proof of part (iii) applied to this relation proves (ii).

Let a_G and b_G denote the explicit Brauer induction homomorphism of Appendix I §5 and its left inverse. If V is a representation of G then

$$\dim_{\mathbb{C}}(V^{(H,\phi)}) = \dim_{\mathbb{C}}(\mathrm{Hom}_H(\mathbb{C}_\phi, V)$$

$$= \dim_{\mathbb{C}}(\mathrm{Hom}_G(\mathrm{Ind}_H^G(\mathbb{C}_\phi), V)$$

$$= \frac{|H|}{|G|} \sum_{h \in H} \phi(h)^{-1} \chi_V(h)$$

where χ_V is the trace function of V. By additivity the formula is true for any $V \in R(G)$. Therefore, by Appendix I §5, the expression in (iv) is precisely the coefficient of $(H, \phi)^G \otimes 1$ in $(a_G \otimes 1)(f) \in R_+(G) \otimes \mathbb{C}$. Hence if $f \in R(G)$ then this expression has integral coefficients and conversely the integrality of the coefficients implies that $(a_G \otimes 1)(f) \in R_+(G) \subset R_+(G) \otimes \mathbb{C}$ and so $f = b_G \otimes 1(a_G \otimes 1(f)) \in R(G)$, which proves that (iv) is equivalent to (i). □

DEFINITION 3.4. Choose an injective homomorphism $\theta : \overline{\mathbb{F}}_q^* \longrightarrow \mathbb{C}^*$. Let G be a finite group and let

$$\rho : G \longrightarrow GL_n\mathbb{F}_q$$

be a homomorphism. For $X \in G$ let $\{\xi_1(X), \dots, \xi_n(X)\}$ denote the set of eigenvalues of $\rho(X)$ considered as lying in $\overline{\mathbb{F}}_q^*$. Let $S(t_1, \dots, t_n)$ be any symmetric function in n variables.

Define a complex-valued conjugacy class function $\chi_{\rho,S}$ on G by the formula

$$\chi_{\rho,S}(X) = S(\theta(\xi_1(X)), \dots, \theta(\xi_n(X))).$$

THEOREM 3.5. ([**69**] *Theorem 1*)

Let $\chi_{\rho,S} \in R(G) \otimes \mathbb{C}$ correspond to the conjugacy class function of Definition 3.4. Then $\chi_{\rho,S} \in R(G)^4$.

---

[3]The M-group induction result together with a simple case of the cyclotomic number theory (when G is cyclic) used in [**33**] gives a proof of Brauer's Induction Theorem in the form appearing in [**33**]. I tried in [**126**] to be clever and eliminate the cyclotomic number theory with the result of messing up the last step!

[4]Topologists call this result "Brauer Lifting". It appears famously in the classic paper [**105**].

Proof

By Proposition 3.3 it suffices to prove Theorem 3.5 with G replaced by an elementary subgroup of the form $H \times C$ where H is a p-group for some prime p and C is a cyclic group of order prime to p.

Now consider the cyclotomic field $\mathbb{Q}(\xi_{q^m-1})$ where $\xi_a = e^{2\pi\sqrt{-1}/a}$ and m is such that $\mathbb{F}_{q^m}$ contains all the eigenvalues of all the matrices $\rho(X)$ with $X \in H \times C$. Let $\mathcal{P} \lhd \mathbb{Z}[\xi_{q^m-1}]$ be any prime ideal dividing the characteristic of $\mathbb{F}_q$. Then $\mathbb{Z}[\xi_{q^m-1}]/\mathcal{P} \cong \mathbb{F}_{q^m}$ where the isomorphism sends a (q^m-1)-th root of unity x to its residue class $\overline{x}$.

We may choose θ in Definition 3.4 so that $\theta(\overline{x}) = x$.

Now let l be the characteristic of $\mathbb{F}_q$. If l does not divide the order of $H \times C$ then it is well-known that there is a complex representation of the form $\rho_1 : H \times C \longrightarrow GL_s\mathbb{Z}[\xi_{q^m-1}]$ such that reduction modulo $\mathcal{P}$ gives a representation $\overline{\rho}_1$ of $H \times C$ over $\mathbb{F}_{q^m}$ which is equivalent to the $\mathbb{F}_q$-representation ρ. For $g \in H \times C$ the eigenvalues of $\overline{\rho}_1$ are precisely the images under θ of the $\overline{\mathbb{F}}_q$ eigenvalues of g. The i-th elementary symmetric function of the θ-values of the $\overline{\mathbb{F}}_q$ eigenvalues of g is equal to the i-elementary symmetric function of the complex eigenvalues of $\overline{\rho}_1$ which is the trace function of the i-th exterior power representation $\lambda^i(\overline{\rho}_1)$ of $\overline{\rho}_1$.

Now assume that l does divide the order of $H \times C$. We may write $H \times C \cong H_1 \times H_2$ where the order of H_1 is not divisible by l and H_2 is an l-group. If $g \in H \times C$ corresponds to $(h_1, h_2) \in H_1 \times H_2$ then h_1 and h_2 are two commuting elements of $GL_s\mathbb{F}_{q^m}$ and the $\overline{\mathbb{F}}_q$ eigenvalues of h_2 are all equal to 1. Therefore elementary matrix algebra shows that in $GL_s\mathbb{F}_{q^m}$ we may simultaneously conjugate $(h_1, 1)$ and $(1, h_2)$ to upper triangular matrices of the form

$$\begin{pmatrix} \zeta_1 & \cdots & \cdots & \cdots \\ 0 & \zeta_2 & \cdots & \cdots \\ \vdots & \vdots & \vdots & \vdots \\ 0 & 0 & \cdots & \zeta_s \end{pmatrix} \quad \text{and} \quad \begin{pmatrix} 1 & \cdots & \cdots & \cdots \\ 0 & 1 & \cdots & \cdots \\ \vdots & \vdots & \vdots & \vdots \\ 0 & 0 & \cdots & 1 \end{pmatrix}$$

respectively.

Therefore the θ-values of the $\overline{\mathbb{F}}_q$ eigenvalues of $g = (h_1, h_2)$ are exactly those of $(h_1, 1)$. This reduces us, by projection onto H_1 from $H \times C$, to the case in which l does not divide the order of the subgroup, which completes the proof. $\square$

DEFINITION 3.6. Let $X \in GL_n\mathbb{F}_q$ then, for θ as in Definition 3.4, set

$$\chi_{\Sigma_r^s}(X) = \sum_{1 \leq i_1 < \ldots < i_r \leq n} \theta(\lambda_{i_1} \ldots \lambda_{i_r})^s \in \mathbb{C}.$$

Define $\Sigma_r^s \in R(GL_n\mathbb{F}_q)$ to be the virtual representation, given by Theorem 3.5, whose character function is $\chi_{\Sigma_r^s}$.

THEOREM 3.7. ([69] Theorem 5; see also Theorems 12 and 13)

Each irreducible representation of $GL_n\mathbb{F}_q$ is, in $R(GL_n\mathbb{F}_q)$, equal to an integral linear combination of the $\{\Sigma_r^s; 1 \leq r \leq n, s \in \mathbb{Z}\}$ of Definition 3.6 and the irreducible representations constructed in Appendix III, §2 as summands of induction of tensor products from parabolic subgroups (denoted by $D_{n_1,n_2,\ldots,n_r}$ in Appendix III, §2).

3.8. ([117] Lemma 1.4)

Let G be a finite group and let Σ be an automorphism of G of order m. Let $\langle \Sigma \rangle \propto G$ denote the semi-direct product of the cyclic group of order m, generated by Σ, with G. That is, the product $\tilde{G} = \langle \Sigma \rangle \times G$ with the multiplication given by the formula of Appendix I, §3

$$(\Sigma^i, g_1) \cdot (\Sigma^j, g_2) = (\Sigma^{i+j}, \Sigma^i(g_1)g_2).$$

LEMMA 3.9.

Let ρ be a representation of G on a vector space V with character χ_V. Then there exists a representation W of $\tilde{G}$ whose character function with χ_W satisfies, for all $g \in G$,

$$\chi_W(\Sigma, g) = \chi_V(g \cdot \Sigma(g) \cdot \ldots \cdot \Sigma^{m-2}(g) \cdot \Sigma^{m-1}(g)).$$

Proof

Let Σ act on the m-fold tensor product of V with itself by

$$\Sigma(v_1 \otimes v_2 \otimes \ldots \otimes v_m) = v_m \otimes v_1 \otimes v_2 \otimes \ldots \otimes v_{m-1}.$$

Let g act by a formula of the type

$$g(v_1 \otimes v_2 \otimes \ldots \otimes v_m) = \Sigma^{a_1}(g)(v_1) \otimes \Sigma^{a_2}(g)(v_2) \otimes \ldots \otimes \Sigma^{a_m}(g)(v_m).$$

Therefore

$$(\Sigma \cdot g \cdot \Sigma^{-1})(v_1 \otimes v_2 \otimes \ldots \otimes v_m)$$

$$= (\Sigma \cdot g)(v_2 \otimes v_3 \otimes \ldots \otimes v_m \otimes v_1)$$

$$= \Sigma(\Sigma^{a_1}(g)(v_2) \otimes \Sigma^{a_2}(g)(v_3) \otimes \ldots \otimes \Sigma^{a_m}(g)(v_1))$$

$$= \Sigma^{a_m}(g)(v_1) \otimes \Sigma^{a_1}(g)(v_2) \otimes \Sigma^{a_2}(g)(v_3) \otimes \ldots \otimes \Sigma^{a_{m-1}}(g)(v_m))$$

and in order for this to be $\Sigma(g)(v_1 \otimes v_2 \otimes \ldots \otimes v_m)$ we need that

$$a_m = a_1 + 1, a_1 = a_2 + 1, \ldots, a_{m-1} = a_m + 1 \text{ (modulo } m).$$

This works if

$$a_1 = m - 1, a_m = 0, a_{m-1} = 1, a_{m-2} = 2, \ldots, a_2 = m - 2.$$

Therefore, if $\{e_\alpha\}$ is a basis of V, then

$$(\Sigma, g)(e_{\alpha_1} \otimes e_{\alpha_2} \otimes \ldots \otimes e_{\alpha_m})$$

$$= (1, g)(\Sigma, 1)(e_{\alpha_1} \otimes e_{\alpha_2} \otimes \ldots \otimes e_{\alpha_m})$$

$$= \sum_{i_1, i_2, \ldots} \Sigma^{m-1}(g)_{i_1, \alpha_m} e_{i_1} \otimes \Sigma^{m-1}(g)_{i_2, \alpha_1} e_{i_2} \otimes \ldots \otimes g_{i_m, \alpha_{m-1}} e_{i_m}$$

whose trace is given by

$$\sum_{\alpha_1, \alpha_2, \ldots} \Sigma^{m-1}(g)_{\alpha_1, \alpha_m} \Sigma^{m-1}(g)_{\alpha_2, \alpha_1} \cdots g_{\alpha_m, \alpha_{m-1}}$$

$$= \mathrm{Trace}(g \cdot \Sigma(g) \cdot \ldots \cdot \Sigma^{m-2}(g) \cdot \Sigma^{m-1}(g)).$$

$\square$

COROLLARY 3.10.
Let $\Sigma_r^s \in R(GL_n\mathbb{F}_q)$ be the virtual representation of Definition 3.6. Then for any integer $m > 1$ there exists a virtual representation

$$W \in R(\mathrm{Gal}(\mathbb{F}_{q^m}/\mathbb{F}_q) \propto GL_n\mathbb{F}_{q^m})$$

such that

$$\chi_W(\Sigma, g) = \chi_{\Sigma_r^s}(g \cdot \Sigma(g) \cdot \ldots \cdot \Sigma^{m-2}(g) \cdot \Sigma^{m-1}(g))$$

for all $g \in GL_n\mathbb{F}_{q^m}$, where Σ is the Frobenius.

Proof
Lemma 3.9, by additivity, extends to virtual representations. Therefore we may apply Lemma 3.9 to $\Sigma_r^s \in R(GL_n\mathbb{F}_{q^m})$, which is a virtual representation which restricts to $\Sigma_r^s \in R(GL_n\mathbb{F}_q)$. $\square$

THEOREM 3.11. *([117] Theorem 1; see also Lemmas 2.7 and 2.11)*
(i) Let ρ be a finite-dimensional complex irreducible representation of $GL_n\mathbb{F}_q$. Then there exists an irreducible representation $\tilde{\rho}$ of the semi-direct product $\mathrm{Gal}(\mathbb{F}_{q^m}/\mathbb{F}_q) \propto GL_n\mathbb{F}_{q^m}$ which satisfies, for all $g \in GL_n\mathbb{F}_{q^m}$,

$$\chi_{\tilde{\rho}}(\Sigma, g) = \epsilon \chi_\rho([g\Sigma(g) \ldots \Sigma^{m-1}(g)])$$

where $\epsilon = \pm 1$ is independent of g. Here $[g\Sigma(g) \ldots \Sigma^{m-1}(g)]$ denotes the unique conjugacy class in $GL_n\mathbb{F}_q$ given by the intersection of the conjugacy class of $g\Sigma(g) \ldots \Sigma^{m-1}(g)$ in $GL_n\mathbb{F}_{q^m}$ with $GL_n\mathbb{F}_q$.
(ii) The Shintani base change correspondence (see Appendix I, §4)

$$Sh : \mathrm{Irr}(GL_n\mathbb{F}_{q^m})^{\mathrm{Gal}(\mathbb{F}_{q^m}/\mathbb{F}_q)} \xrightarrow{\cong} \mathrm{Irr}(GL_n\mathbb{F}_q)$$

is given by, in the case where ϵ may be chosen to equal 1 in part (i),

$$Sh(\mathrm{Res}_{GL_n\mathbb{F}_{q^m}}^{\mathrm{Gal}(\mathbb{F}_{q^m}/\mathbb{F}_q) \propto GL_n\mathbb{F}_{q^m}}(\tilde{\rho})) = \rho.$$

When $\epsilon = -1$ is the only possibility there is an extension, denoted by ρ', of ρ to $\text{Gal}(\mathbb{F}_{q^{2m}}/\mathbb{F}_q) \propto GL_n\mathbb{F}_{q^m}$ and

$$\chi_{\rho'}(\Sigma, g) = \chi_\rho([g\Sigma(g)\ldots\Sigma^{m-1}(g)])$$

specifies $\chi(\rho)$ in this case.

Sketch Proof of Theorem 3.11

Write $N(g)$ for $[g\Sigma(g)\ldots\Sigma^{m-1}(g)]$, which is a conjugacy class in $GL_n\mathbb{F}_q{}^5$. It is shown in [117] that every conjugacy class occurs as an $N(g)$, which also follows from the observation of Digne-Michel in Appendix I, §10. Observe that $N(gx\Sigma(g)^{-1}) = g^{-1}N(x)g$.

The proof consists of an induction in which one assumes for smaller values $s < n$ that for each $\rho \in \text{Irr}(GL_s\mathbb{F}_q)$ there exists a virtual representation $\hat{\rho} \in R(\text{Gal}(\mathbb{F}_{q^m}/\mathbb{F}_q) \propto GL_s\mathbb{F}_{q^m})$ such that, for all $g \in GL_s\mathbb{F}_{q^m}$,

$$\chi_{\hat{\rho}}(\Sigma, g) = \chi_\rho([g\Sigma(g)\ldots\Sigma^{m-1}(g)]).$$

There, if $(n_1,\ldots,n_r)$ is a partition of n by strictly positive integers, and $\rho_i \in \text{Irr}(GL_{n_i}\mathbb{F}_q)$ there exists a virtual representation

$$\hat{\rho}_i \in R(\text{Gal}(\mathbb{F}_{q^m}/\mathbb{F}_q) \propto GL_{n_i}\mathbb{F}_{q^m})$$

such that, for all $g \in GL_{n_i}\mathbb{F}_{q^m}$,

$$\chi_{\hat{\rho}_i}(\Sigma, g) = \chi_{\rho_i}([g\Sigma(g)\ldots\Sigma^{m-1}(g)]).$$

If $\rho \in \text{Irr}(GL_n\mathbb{F}_q)$ is constructed by parabolic inflation-induction from $\rho_{n_1},\ldots,\rho_{n_r}$, as in Appendix III, §2, $\tilde{\rho}$ is constructible by parabolic inflation-induction from $\tilde{\rho}_{n_1},\ldots,\tilde{\rho}_{n_r}$ ([117] Lemmas 2.8 and 2.9).

By additivity, Theorem 3.7 and Corollary 3.10 for each $\rho \in \text{Irr}(GL_n\mathbb{F}_q)$ there exists a virtual representation $\tilde{\rho}$ satisfying the character condition of part (i). A character calculation then shows that $\pm\tilde{\rho}$ is an irreducible representation and, finally, the counting (as in Appendix I, §10) shows that $\rho \mapsto \tilde{\rho}$ gives a bijection as in part (ii). $\square$

3.12. *Integrality is equivalent to Shintani descent*

I shall close this chapter by recapitulating and elaborating upon the integrality remarks of Appendix I, §9 and finally posing a question about a potentially alternative proof of Theorem 3.11. The sketch proof of Shintani's Theorem starts from the bottom of Galois descent, with $\rho \in \text{Irr}(GL_n\mathbb{F}_q)$. The integrality point of view starts at the top with a Galois invariant irreducible $\pi \in \text{Irr}(GL_n\mathbb{F}_{q^m})^{\text{Gal}(\mathbb{F}_{q^m}/\mathbb{F}_q)}$. As explained in Appendix I, being Galois invariant is equivalent to there existing an extension of π to

$$\tilde{\pi} \in \text{Irr}(\text{Gal}(\mathbb{F}_{q^m}/\mathbb{F}_q) \propto GL_n\mathbb{F}_{q^m})$$

---

[5]$N(g)$ in [117] is constructed from $\Sigma^{m-1}(g)\ldots\Sigma(g)g$ due to the convention used for the multiplication in the semi-direct product (see [117] §1) which differs from mine.

which is unique up to twisting by a one-dimensional character of $\text{Gal}(\mathbb{F}_{q^m}/\mathbb{F}_q)$.

Assigning to the conjugacy class $[g\Sigma(g)\ldots\Sigma^{m-1}(g)]$ in $GL_n\mathbb{F}_q$, as defined in Theorem 3.11(i), the complex number $\chi_{\tilde{\pi}}(\Sigma, g)$ gives a well-defined class function and hence an element $\pm Sh(\pi)$ of $R(GL_n\mathbb{F}_q)\otimes\mathbb{C}$.

As we saw in the sketch-proof of Theorem 3.11, in order to complete the proof it is sufficient to show that $\pm Sh(\pi)\in R(GL_n\mathbb{F}_q)$. As explained in Appendix I, §9, this condition is equivalent to the condition that each coefficient in the Explicit Brauer Induction formula $a_{GL_n\mathbb{F}_q}(\pm Sh(\pi))$ is integral.

If $(H_0, \phi_0)\in\mathcal{M}_{GL_n\mathbb{F}_q}$ the coefficient of the $GL_n\mathbb{F}_q$-conjugacy class of (H_0, ϕ_0) in $a_{GL_n\mathbb{F}_q}(\pm Sh(\pi))$ is given by

$$\sum_{(H_0,\phi_0)<(H_1,\phi_1)<\ldots<(H_r,\phi_r)} (-1)^r \frac{|H_0|}{|GL_s\mathbb{F}_q|} \dim_{\mathbb{C}}(\pm Sh(\pi)^{(H_r,\phi_r)})$$

where the sum is over the set of strictly ascending chains

$$(H_0, \phi_0) < (H_1, \phi_1) < \ldots < (H_r, \phi_r)$$

ending in (H_r, ϕ_r).

Here, as for a virtual representation, $\dim_{\mathbb{C}}(\pm Sh(\pi)^{(H_r,\phi_r)})$ denotes the complex number given by

$$\dim_{\mathbb{C}}(\pm Sh(\pi)^{(H_r,\phi_r)}) = \frac{1}{|H_r|}\sum_{h\in H_r} \phi_r(h)^{-1}\chi_{\pm Sh(\pi)}(h)$$

$$= \frac{1}{|H_r|}\sum_{h=[g\Sigma(g)\ldots\Sigma^{m-1}(g)]\in H_r} \phi_r(h)^{-1}\chi_{\pm Sh(\pi)}(h)$$

$$= \frac{\epsilon}{|H_r|}\sum_{h=[g\Sigma(g)\ldots\Sigma^{m-1}(g)]\in H_r} \phi_r(h)^{-1}\chi_{\tilde{\rho}}(\Sigma, g).$$

Since

$$(-)^{(H_r,\phi_r)} \cong \text{Hom}_{H_r}(\mathbb{C}_{\phi_r}, -) \cong \text{Hom}_{GL_n\mathbb{F}_q}(\text{Ind}_{H_r}^{GL_n\mathbb{F}_q}(\mathbb{C}_{\phi_r}), -)$$

the above dimension formula can be rewritten in terms of the character values of $\text{Ind}_{H_r}^{GL_n\mathbb{F}_q}(\mathbb{C}_{\phi_r})$ using the formula from ([**126**] Theorem 1.2.8) and the sizes of conjugacy classes given in ([**117**] §2.6(ii); see also Appendix I, §10).

QUESTION 3.13. In §3.12 we saw that the integrality of each of certain sums over chains

$$(H_0, \phi_0) < (H_1, \phi_1) < \ldots < (H_r, \phi_r)$$

implies Theorem 3.11. It is straight forward to compare each of those sums with similar sums over chains, at least when m is prime,

$$(\text{Gal}(\mathbb{F}_{q^m}/\mathbb{F}_q)\times H_0, \phi_0) < (\text{Gal}(\mathbb{F}_{q^m}/\mathbb{F}_q)\times H_1, \phi_1)$$

$$< \ldots < (\text{Gal}(\mathbb{F}_{q^m}/\mathbb{F}_q)\times H_r, \phi_r).$$

We know that the Explicit Brauer Induction formula for $\tilde{\pi}$ has integral coefficients which involve sums over chains in $\mathcal{M}(\mathrm{Gal}(\mathbb{F}_{q^m}/\mathbb{F}_q) \propto GL_n\mathbb{F}_{q^m})$ — in particular, sums over chains commencing at $(\mathrm{Gal}(\mathbb{F}_{q^m}/\mathbb{F}_q) \times H_0, \phi_0)$. Could the latter integrality be used to show that of the sum over chains of the form $(\mathrm{Gal}(\mathbb{F}_{q^m}/\mathbb{F}_q) \times H_i, \phi_i)$?

4. Some recreational integer polynomials

4.1. Choose an injective homomorphism $\theta : \overline{\mathbb{F}}_q^* \longrightarrow \mathbb{C}^*$ as in Definition 3.4. Recall from Definition 3.6 that Σ_r^1 (abbreviated here to Σ_r) is equal to the following conjugacy class function on $GL_s\overline{\mathbb{F}}_q$. For $X \in GL_s\overline{\mathbb{F}}_q$ let $\{\lambda_1, \ldots, \lambda_s\}$ denote the set of eigenvalues of X and set

$$\Sigma_r(X) = \sum_{1 \leq i_1 < \ldots < i_r \leq s} \theta(\lambda_{i_1}, \ldots, \lambda_{i_r}) \in \mathbb{C}.$$

These functions may be collected into polynomials if we set

$$\hat{X} = \begin{pmatrix} \theta(\lambda_1) & 0 & \ldots & \ldots & 0 \\ 0 & \theta(\lambda_2) & 0 & \ldots & 0 \\ \vdots & \vdots & \vdots & \vdots & \vdots \\ \vdots & \vdots & \vdots & \vdots & \vdots \\ 0 & 0 & \ldots & 0 & \theta(\lambda_s) \end{pmatrix}$$

so that the characteristic polynomial of $\hat{X}$ satisfies

$$\det(tI_s - \hat{X}) = t^s - \Sigma_1(X)t^{s-1} + \Sigma_2((X)t^{s-2} + \ldots$$

$$+ (-1)^i \Sigma_i((X)t^{s-i} + \ldots + (-1)^s \Sigma_s((X).$$

By Theorem 3.5 the element of $R(GL_s\mathbb{F}_{q^n})[t] \otimes \mathbb{C}$ given by the complex polynomial-valued class function $X \mapsto \det(tI_s - \hat{X})$ lies in $R(GL_s\mathbb{F}_{q^n})[t]$ for all $s, n \geq 1$. Write $P(s, n)$ for this polynomial.

By the discussion of §3.12 this is equivalent to the condition that for each $(H_0, \phi_0) \in \mathcal{M}(GL_s\mathbb{F}_{q^n})$ the sum

$$\sum_{(H_0,\phi_0)<(H_1,\phi_1)<\ldots<(H_r,\phi_r)} (-1)^r \frac{|H_0|}{|GL_s\mathbb{F}_q|} \dim_\mathbb{C}(P(s,n)^{(H_r,\phi_r)}) \in \mathbb{Z}[t].$$

Here the sum is over the set of strictly ascending chains

$$(H_0, \phi_0) < (H_1, \phi_1) < \ldots < (H_r, \phi_r)$$

starting at $(H_0, \phi_0) \in \mathcal{M}(GL_s\mathbb{F}_{q^n})$.

These integer polynomials are given by the coefficients of

$$(H_0, \phi_0)^{GL_s\mathbb{F}_{q^n}} \in R_+(GL_s\mathbb{F}_{q^n})$$

in the (polynomial-valued) Explicit Brauer Induction formula. This formula is natural with respect to restrictions such as the one from $R_+(GL_s\mathbb{F}_{q^{nd}})$ to $R_+(GL_s\mathbb{F}_{q^n})$, as explained in Appendix I, §5.

In addition, restriction via the map $GL_s \longrightarrow GL_{s+t}$ which extends a matrix by adding zeros off the diagonal and 1's along the diagonal sends $P(s+t, n)$ to $P(s, n)(t-1)^t$.

These two types of naturality relate the $P(s, n)$'s to the Brauer lifting of [105].

QUESTION 4.2. Do there exist closed formulae for the integer polynomials of §4.1?

5. Base change functoriality for stable homotopy theorists

5.1. The Shintani correspondence gives a bijection between $\mathrm{Irr}(GL_s\mathbb{F}_q)$ and $\mathrm{Irr}(GL_s\mathbb{F}_{q^n})^{\mathrm{Gal}(\mathbb{F}_{q^n}/\mathbb{F}_q)}$ for complex irreducibles. Since the former is a free $\mathbb{Z}$-basis for the complex representation ring $R(GL_s\mathbb{F}_q)$ the additive extension of the inverse correspondence Sh^{-1} gives a homomorphism

$$Sh^{-1} : R(GL_s\mathbb{F}_q) \longrightarrow R(GL_s\mathbb{F}_{q^n}).$$

Explicitly, if $\nu = \oplus_{i=1}^t \nu_i$ with $\nu_i \in \mathrm{Irr}(GL_s\mathbb{F}_q)$ then

$$Sh^{-1}(\nu) = \sum_{i=1}^t \mathrm{Res}_{GL_s\mathbb{F}_{q^n}}^{\mathrm{Gal}(\mathbb{F}_{q^n}/\mathbb{F}_q)\propto GL_s\mathbb{F}_{q^n}} (\tilde{\rho}_i)$$

where $\tilde{\rho}_i \in \mathrm{Irr}(\mathrm{Gal}(\mathbb{F}_{q^n}/\mathbb{F}_q) \propto GL_s\mathbb{F}_{q^n})$ is an irreducible of the first kind (i.e. [117] Definition 1.1; $\tilde{\rho}_i$ restricts to $\rho_i \in \mathrm{Irr}(GL_s\mathbb{F}_{q^n})$) and $Sh(\rho_i) = \nu_i$ for $1 \le i \le t$.

Let $IR(GL_s\mathbb{F}_q)$ denote the augmentation ideal, generated by virtual representations of dimension zero. Hence $IR(GL_s\mathbb{F}_q)$ has a base consisting of elements of the form $\rho - \dim(\rho)$ for $\rho \in \mathrm{Irr}(GL_s\mathbb{F}_q)$. The addition extension of Sh^{-1} gives a homomorphism

$$Sh^{-1} : IR(GL_s\mathbb{F}_q) \longrightarrow IR(GL_s\mathbb{F}_{q^n})$$

given by $\rho - \dim(\rho) \mapsto Sh^{-1}(\rho) - \dim(Sh^{-1}(\rho))$.

For a finite group G the $IR(G)$-adic completion of $IR(G)$ is isomorphic to $[BG, BU]$, the topological unitary K-theory of BG, which is the set of based homotopy classes of maps between the classifying space BG of G to that of the infinite unitary group $U = \bigcup_n U_n(\mathbb{C})$ [8]. An element $\rho - \dim(\rho)$ is mapped to the homotopy class of the map

$$B\rho : BG \longrightarrow BU_{\dim(\rho)}(\mathbb{C}) \longrightarrow BU.$$

QUESTION 5.2. Is Sh^{-1} continuous in the $IR(G)$-adic topology?

QUESTION 5.3. *A question for homotopy theorists*

If the answer to Question 5.2 is affirmative then one could ask whether or not there is a functorial map induced by Sh^{-1} of the form

$$Sh^{-1} : [BGL_s\mathbb{F}_q, BU] \longrightarrow [BGL_s\mathbb{F}_{q^n}, BU].$$

5.4. More generally than §5.2 and Question 5.5, there is a similar question in stable homotopy theory.

Let X_+ denote a space X with a disjoint base-point adjoined. Let $\{X_+, Y_+\}$ denote the group of stable homotopy classes of maps from X_+ to Y_+ [6]. There is an isomorphism between $\{BG_+, BU_n(\mathbb{C})_+\}$ and a completion of the abelian group on triples $(G \xleftarrow{\alpha} H \xrightarrow{\rho} U_n(\mathbb{C}))$ where ρ is a representation and α is the inclusion of a subgroup. More precisely, if J is a compact Lie group such as $J = U_n(\mathbb{C})$ let $R_+(G, J)$ denote the free abelian group on triples as above. Then $R_+(G, J)$ is a module over the Burnside ring $R_+(G, \{1\})$ and the completion mentioned above is formed with respect to the Burnside ring augmentation ideal topology ([97], see also [102]). Sending a triple $(G \xleftarrow{\alpha} H \xrightarrow{\rho} J)$ to the stable homotopy class of the composition

$$BG_+ \xrightarrow{\text{Transfer}} BH_+ \xrightarrow{B\rho} BJ_+$$

yields the isomorphism

$$R_+(G, J)^\wedge \xrightarrow{\cong} \{BG_+, BJ_+\}.$$

This result was first proved in [125] for $J = S^1$ and was proved in general in [131]. A cosmological[6] generalisation of this result appeared as [97].

QUESTION 5.5. *A question for stable homotopy theorists*

The Euler characteristic of a monomial resolution over $\mathbb{C}$ lies in $R_+(G, S^1)$. Supposing the monomial resolution to be "continuous in the Burnside ring topology", is there a functorial map between stable homotopy groups which coincides with that induced by the Euler characteristic of the monomial resolution of Sh^{-1}?

6. Inverse Shintani bijection and monomial resolutions

6.1. In this section I shall consider the inverse Shintani correspondence and the resulting homomorphism of complex representation rings

$$Sh^{-1} : R(GL_s\mathbb{F}_q) \longrightarrow R(GL_s\mathbb{F}_{q^n})$$

introduced in §5.1.

Continuing the theme of functoriality of base change I shall consider the possibility of starting from the monomial resolution of $\nu \in \text{Irr}(GL_s\mathbb{F}_q)$ and from it constructing a chain complex of representations of the semi-direct product $\text{Gal}(\mathbb{F}_{q^n}/\mathbb{F}_q) \propto GL_s\mathbb{F}_{q^n}$ which, when restricted to $GL_s\mathbb{F}_{q^n}$,

---

[6]In the sense of SETI.

has homology concentrated in dimension zero and equal to $Sh^{-1}(\nu) \in$ $\mathrm{Irr}(GL_s\mathbb{F}_{q^n})$.

Let us begin with an elementary example given by Shintani descent from $GL_2\mathbb{F}_4$ to $GL_2\mathbb{F}_2 \cong D_6$, the dihedral group of order six, which is discussed in detail in Appendix I.

EXAMPLE 6.2. Let A, C be the elements of $GL_2\mathbb{F}_2$ introduced in Appendix I, §1. Therefore

$$D_6 = \langle A, C \mid A^2 = C^3 = 1, ACA = C^2 \rangle.$$

Let ϕ be the character of $\langle C \rangle$ given by $\phi(C) = \xi_3$ as in Appendix I, §6. Set $\nu = \mathrm{Ind}_{\langle C \rangle}^{D_6}(\phi) \in \mathrm{Irr}(D_6)$ as in Appendix I, §4.

A monomial resolution for $\nu = \mathrm{Ind}_{C_3}^{D_6}(\phi)$ over an algebraically closed field of characteristic different from two is

$$M_* : \quad 0 \longrightarrow \underline{\mathrm{Ind}}_{\{1\}}^{D_6}(1) \overset{\partial}{\longrightarrow} \underline{\mathrm{Ind}}_{C_3}^{D_6}(\phi) \oplus \underline{\mathrm{Ind}}_{C_2}^{D_6}(1) \oplus \underline{\mathrm{Ind}}_{C_2}^{D_6}(\mu)) \overset{\epsilon}{\longrightarrow} \nu \longrightarrow 0$$

where μ is the non-trivial character and the differentials are given by

$$\partial(1 \otimes_{\{1\}} 1) = (1 \otimes_{C_3} 1, -(1/2) \otimes_{C_2} 1, -(1/2) \otimes_{C_2} 1),$$

$$\epsilon(1 \otimes_{C_3} 1, 0, 0)) = 1 \otimes_{C_3} 1,$$

$$\epsilon(0, 1 \otimes_{C_2} 1, 0) = 1 \otimes_{C_3} 1 + A \otimes_{C_3} 1,$$

$$\epsilon(0, 0, 1 \otimes_{C_2} 1) = 1 \otimes_{C_3} 1 - A \otimes_{C_3} 1.$$

In Chapter One, §3 one finds the monomial morphism

$$((K, \psi), g, (H, \phi)) : \underline{\mathrm{Ind}}_K^G(k_\psi) \longrightarrow \underline{\mathrm{Ind}}_H^G(k_\phi)$$

given by $g' \otimes_K v \mapsto g'g \otimes_H v$ when $(K, \psi) \leq (gHg^{-1}, (g^{-1})^*(\phi))$. With this notation the morphism ∂ is equal to

$$\partial = ((\{1\}, 1), 1, (\langle C \rangle, \phi)) - 1/2((\{1\}, 1), 1, (\langle A \rangle, 1)) - 1/2((\{1\}, 1), 1, (\langle A \rangle, \mu)).$$

In the notation of Appendix I, the Euler characteristics of $M_*^{((H))}$ are given by

$\chi(M_*^{((D_6))})$	0
$\chi(M_*^{((C_3))})$	$\phi + \phi^2$
$\chi(M_*^{((C_2))})$	$1 + \mu$
$\chi(M_*^{((\{1\}))})$	2

The irreducible representations are $1, \chi, \nu$ whose character values are given by the table

	1	χ	ν
1	1	1	2
A	1	-1	0
C	1	1	-1

The three Galois invariant irreducibles of $GL_2\mathbb{F}_4$ are $1, \nu_4, \nu_5$ explicitly constructed in Appendix I where explicit extensions to the semi-direct product $\mathrm{Gal}(\mathbb{F}_4/\mathbb{F}_2) \propto GL_2\mathbb{F}_4$ ($\tilde{1}, \tilde{\nu}_4, \tilde{\nu}_5$ respectively) are also constructed.

From Appendix I, §9 we have the following table of character values in $\mathrm{Gal}(\mathbb{F}_4/\mathbb{F}_2) \propto GL_2\mathbb{F}_4$

	1	$\tilde{\nu}_5$	$\tilde{\nu}_4$
$(\Sigma, 1)$	1	1	-2
(Σ, B)	1	-1	0
(Σ, C^2)	1	1	1

so that, if λ is the non-trivial character of the form

$$\mathrm{Gal}(\mathbb{F}_4/\mathbb{F}_2) \propto GL_2\mathbb{F}_4 \longrightarrow \mathrm{Gal}(\mathbb{F}_4/\mathbb{F}_2) \xrightarrow{\cong} \{\pm 1\}$$

then Theorem 3.11(i) (with $\epsilon = 1$ in each case) is satisfied by the map $1 \mapsto 1$, $\chi \mapsto \tilde{\nu}_5$ and $\nu \mapsto \lambda \otimes \tilde{\nu}_4$.

As a sequence of D_6-representations $M_1 \longrightarrow M_0$ is isomorphic to

$$1 \oplus \chi \oplus \nu \oplus \nu \longrightarrow \nu \oplus (1 \oplus \nu) \oplus (\chi \oplus \nu).$$

If we form the two $\mathrm{Gal}(\mathbb{F}_4/\mathbb{F}_2) \propto GL_2\mathbb{F}_4$ representations

$$N_1 = 1 \oplus \tilde{\nu}_5 \oplus \tilde{\nu}_4 \oplus \tilde{\nu}_4 \text{ and } N_0 = \lambda \otimes \tilde{\nu}_4 \oplus (1 \oplus \lambda \otimes \tilde{\nu}_4) \oplus (\tilde{\nu}_5 \oplus \lambda \otimes \tilde{\nu}_4)$$

whose summands are irreducibles of the first kind. We may define a homomorphism of representations $N_1 \longrightarrow N_0$ by sending each irreducible summand in N_1 to the copies of itself in N_0 by the same scalar multiplication as occurred between the corresponding summands in $M_1 \longrightarrow M_0$.

In this simple example this procedure gives an injection whose cokernel, restricted to $GL_2\mathbb{F}_4$, is $\nu_4 = Sh^{-1}(\nu)$.

QUESTION 6.3. Suppose that we are in the general Sh^{-1} situation with ρ an irreducible representation of $GL_s\mathbb{F}_q$ and $Sh^{-1}(\rho)$ an irreducible representation of $GL_s\mathbb{F}_{q^n}$. Suppose we form the monomial resolution of ρ and then, in each dimension, form the direct sum of the first kind irreducibles of the semi-direct provided by Theorem 3.11 and finally construct maps of representations, each reducing the degree by 1, in the manner analogous to the simple case of Example 6.2.

As representations of $GL_s\mathbb{F}_{q^n}$ is the resulting sequence of homomorphisms of representations a chain complex and furthermore is its homology concentrated in dimension zero and isomorphic to $Sh^{-1}(\rho)$?

CHAPTER 9

PSH-algebras and the Shintani correspondence

1. PSH-algebras over the integers

1.1. A PSH-algebra is a connected, positive self-adjoint Hopf algebra over $\mathbb{Z}$. The notion was introduced in [**146**]. Let $R = \oplus_{n \geq 0} R_n$ be an augmented graded ring over $\mathbb{Z}$ with multiplication

$$m : R \otimes R \longrightarrow R.$$

Suppose also that R is connected, which means that there is an augmentation ring homomorphism of the form

$$\epsilon : \mathbb{Z} \xrightarrow{\cong} R_0 \subset R.$$

These maps satisfy associativity and unit conditions.
Associativity:

$$m(m \otimes 1) = m(1 \otimes m) : R \otimes R \otimes R \longrightarrow R.$$

Unit:

$$m(1 \otimes \epsilon) = 1 = m(\epsilon \otimes 1); R \otimes \mathbb{Z} \cong R \cong \mathbb{Z} \otimes R \longrightarrow R \otimes R \longrightarrow R.$$

R is a Hopf algebra if, in addition, there exist comultiplication and counit homomorphisms

$$m^* : R \longrightarrow R \otimes R$$

and

$$\epsilon^* : R \longrightarrow \mathbb{Z}$$

such that
Hopf
m^* is a ring homomorphism with respect to the product $(x \otimes y)(x' \otimes y') = xx' \otimes yy'$ on $R \otimes R$ and ϵ^* is a ring homomorphism restricting to an isomorphism on R_0. The homomorphism m is a coalgebra homomorphism with respect to m^*.

The m^* and ϵ^* also satisfy
Coassociativity:

$$(m^* \otimes 1)m^* = (1 \otimes m^*)m^* : R \longrightarrow R \otimes R \otimes R \longrightarrow R \otimes R \otimes R.$$

Counit:

$$m(1 \otimes \epsilon) = 1 = m(\epsilon \otimes 1); R \otimes \mathbb{Z} \cong R \cong \mathbb{Z} \otimes R \longrightarrow R \otimes R \longrightarrow R.$$

R is a cocomutative if

Cocommutative:

$$m^* = T \cdot m^* : R \longrightarrow R \otimes R$$

where $T(x \otimes y) = y \otimes x$ on $R \otimes R$.

Suppose now that each R_n (and hence R by direct-sum of bases) is a free abelian group with a distinguished $\mathbb{Z}$-basis denoted by $\Omega(R_n)$. Hence $\Omega(R)$ is the disjoint union of the $\Omega(R_n)$'s. With respect to the choice of basis the positive elements R^+ of R are defined by

$$R^+ = \{r \in R \mid r = \sum m_\omega \omega, \ m_\omega \geq 0, \omega \in \Omega(R)\}.$$

Motivated by the representation theoretic examples the elements of $\Omega(R)$ are called the irreducible elements of R and if $r = \sum m_\omega \omega \in R^+$ the elements $\omega \in \Omega(R)$ with $m_\omega > 0$ are call the irreducible constituents of r.

Using the tensor products of basis elements as a basis for $R \otimes R$ we can similarly define $(R \otimes R)^+$ and irreducible constituents etc.

Positivity:

R is a positive Hopf algebra if

$$m((R \otimes R)^+) \subset R^+, m^*(R^+) \subset (R \otimes R)^+, \epsilon(\mathbb{Z}^+) \subset R^+, \epsilon^*(R^+) \subset \mathbb{Z}^+.$$

Define inner products $\langle -, - \rangle$ on R, $R \otimes R$ and $\mathbb{Z}$ by requiring the chosen basis ($\Omega(\mathbb{Z}) = \{1\}$) to be an orthonormal basis.

A positive Hopf $\mathbb{Z}$-algebra is self-adjoint if

Self-adjoint:

m and m^* are adjoint to each other and so are ϵ and ϵ^*. That is

$$\langle m(x \otimes y), z \rangle = \langle x \otimes y, m^* z \rangle$$

and similarly for ϵ, ϵ^*.

The subgroup of primitive elements $P \subset R$ is given by

$$P = \{r \in R \mid m^*(r) = r \otimes 1 + 1 \otimes r\}.$$

2. The Decomposition Theorem

Let $\{R_\alpha \mid \alpha \in \mathcal{A}\}$ be a family of PSH algebras. Define the tensor product PSH algebra

$$R = \otimes_{\alpha \in \mathcal{A}} R_\alpha$$

to be the inductive limit of the finite tensor products $\otimes_{\alpha \in S} R_\alpha$ with $S \subset \mathcal{A}$ a finite subset. Define $\Omega(R)$ to be the disjoint union over finite subsets S of $\prod_{\alpha \in S} \Omega(R_\alpha)$.

The following result of the PSH analogue of a structure theorem for Hopf algebras over the rationals due to Milnor-Moore [99].

THEOREM 2.1.

Any PSH algebra R decomposes into the tensor product of PSH algebras with only one irreducible primitive element. Precisely, let $\mathcal{C} = \Omega \cap P$ denote the set of irreducible primitive elements in R. For any $\rho \in \mathcal{C}$ set

$$\Omega(\rho) = \{\omega \in \Omega \mid \langle \omega, \rho^n \rangle \neq 0 \text{ for some } n \geq 0\}$$

and

$$R(\rho) = \oplus_{\omega \in \Omega(\rho)} \mathbb{Z} \cdot \omega.$$

Then $R(\rho)$ is a PSH algebra with set of irreducible elements $\Omega(\rho)$, whose unique irreducible primitive is ρ and

$$R = \otimes_{\rho \in \mathcal{C}} R(\rho).$$

3. The PSH algebra of $\{GL_m\mathbb{F}_q, \ m \geq 0\}$

3.1. Let $R(G)$ denote the complex representation ring of a finite group G. Set $R = \oplus_{m \geq 0} R(GL_m\mathbb{F}_q)$ with the interpretation that $R_0 \cong \mathbb{Z}$, an isomorphism which gives both a choice of unit and counit for R.

Let $U_{k,m-k} \subset GL_m\mathbb{F}_q$ denote the subgroup of matrices of the form

$$X = \begin{pmatrix} I_k & W \\ 0 & I_{m-k} \end{pmatrix}$$

where W is an $k \times (m-k)$ matrix. Let $P_{k,m-k}$ denote the parabolic subgroup of $GL_m\mathbb{F}_q$ given by matrices obtained by replacing the identity matrices I_k and I_{m-k} in the condition for membership of $U_{k,m-k}$ by matrices from $GL_k\mathbb{F}_q$ and $GL_{m-k}\mathbb{F}_q$ respectively. Hence there is a group extension of the form

$$U_{k,m-k} \longrightarrow P_{k,m-k} \longrightarrow GL_k\mathbb{F}_q \times GL_{m-k}\mathbb{F}_q.$$

If V is a complex representation of $GL_m\mathbb{F}_q$ then the fixed points $V^{U_{k,m-k}}$ is a representation of $GL_k\mathbb{F}_q \times GL_{m-k}\mathbb{F}_q$ which gives the $(k, m-k)$ component of

$$m^* : R \longrightarrow R \otimes R.$$

Given a representation W of $GL_k\mathbb{F}_q \times GL_{m-k}\mathbb{F}_q$ so that $W \in R_k \otimes R_{m-k}$ we may form

$$\text{Ind}_{P_{k,m-k}}^{GL_m\mathbb{F}_q} (\text{Inf}_{GL_k\mathbb{F}_q \times GL_{m-k}\mathbb{F}_q}^{P_{k,m-k}} (W))$$

which gives the $(k, m - k)$ component of

$$m : R \otimes R \longrightarrow R.$$

We choose a basis for R_m to be the irreducible representations of $GL_m\mathbb{F}_q$ so that R^+ consists of the classes of representations (rather than virtual ones). Therefore it is clear that $m, m^*, \epsilon, \epsilon^*$ satisfy positivity. The inner product on R is given by the Schur inner product so that for two representations V, W of $GL_m\mathbb{F}_q$ we have

$$\langle V, W \rangle = \dim_{\mathbb{C}}(\mathrm{Hom}_{GL_m\mathbb{F}_q}(V, W))$$

and for $m \neq n$ R_n is orthogonal to R_m. As is well-known, with these choice of inner product, the basis of irreducible representations for R is an orthonormal basis.

The irreducible primitive elements are represented by irreducible complex representations of $GL_m\mathbb{F}_q$ which have no non-zero fixed vector for any of the subgroups $U_{k,m-k}$. These representations are usually called cuspidal.

In the remainder of this section we shall verify that R is a PSH algebra, as is shown in ([146] Chapter III). I believe, in different terminology, this structural result was known to Sandy Green at the time of writing [69] and to his research supervisor Phillip Hall.

THEOREM 3.2. *(Self-adjoint)*
If X, Y, Z are complex representations of $GL_m\mathbb{F}_q, GL_n\mathbb{F}_q, GL_{m+n}\mathbb{F}_q$ respectively then

$$\langle m(X \otimes Y), Z \rangle = \langle X \otimes Y, m^*(Z) \rangle.$$

Also ϵ and ϵ^* are mutually adjoint.

Proof
This follows from Frobenius reciprocity ([126] Theorem 1.2.39) because the Schur inner product is given by

$$\langle m(X \otimes Y), Z \rangle = \dim_{\mathbb{C}}(\mathrm{Hom}_{GL_{m+n}\mathbb{F}_q}(m(X \otimes Y), Z))$$

$$= \dim_{\mathbb{C}}(\mathrm{Hom}_{P_{m,n}} \mathrm{Inf}_{GL_m\mathbb{F}_q \times GL_n\mathbb{F}_q}^{P_{m,n}}(X \otimes Y), Z))$$

$$= \dim_{\mathbb{C}}(\mathrm{Hom}_{P_{m,n}} \mathrm{Inf}_{GL_m\mathbb{F}_q \times GL_n\mathbb{F}_q}^{P_{m,n}}(X \otimes Y), Z^{U_{m,n}})).$$

The adjointness of ϵ and ϵ^* is obvious. $\square$

PROPOSITION 3.3. *(Associativity and coassociativity)*
The coproduct m^* is coassociative and the product m is associative.

Proof
Clearly m^* is coassociative because taking fixed-points $GL_a\mathbb{F}_q \times GL_b\mathbb{F}_q \times GL_c\mathbb{F}_q$ of a $GL_{a+b+c}\mathbb{F}_q$ representation is clearly associative. It follows from Theorem 3.2 that m is associative, since the Schur inner product is non-singular. $\square$

THEOREM 3.4. *(Hopf condition)*
The homomorphism m^* is an algebra homomorphism with respect to m. The homomorphism m is a coalgebra homomorphism with respect to m^*.

Obviously the coalgebra homomorphism assertion follows from the algebra homomorphism assertion by the adjointness property of Theorem 3.2.

The discussion which follows will establish Theorem 3.4. It is rather delicate and involved so I am going to give it in full detail (following ([**146**] p. 167 and p. 173 with minor changes). For notational convenience I shall write $G_n = GL_n\mathbb{F}_q$ for the duration of this discussion.

Recall that we are attempting to show that for each $(\alpha, m - \alpha)$ and $(a, m - a)$ that the $R(G_a) \otimes R(G_{m-a})$-component of $m^* \cdot m$

$$R(G_\alpha) \otimes R(G_{m-\alpha}) \xrightarrow{m} R(G_m) \xrightarrow{m^*} R(G_a) \otimes R(G_{m-a})$$

is equal to the $R(G_a) \otimes R(G_{m-a})$-component

$$R(G_\alpha) \otimes R(G_{m-\alpha}) \xrightarrow{m^* \otimes m^*} R \otimes R \otimes R \otimes R \xrightarrow{1 \otimes T \otimes 1} R \otimes R \otimes R \otimes R \xrightarrow{m \otimes m} R \otimes R.$$

Let Z be a complex representation of G_m then the $(a, m-a)$-component of $m^*(Z)$ is given by

$$Z^{U_{a,m-a}} \in R(G_a \times G_{m-a}) \cong R(G_a) \otimes R(G_{m-a})$$

with the group action given by the induced $P_{a,m-a}/U_{a,m-a}$-action.

If $Z = m(X \otimes Y)$ with X, Y representations of $G_\alpha, G_{m-\alpha}$ respectively then

$$Z = m(X \otimes Y) = \mathrm{Ind}_{P_{\alpha,m-\alpha}}^{G_m}(\mathrm{Inf}_{G_\alpha \times G_{m-\alpha}}^{P_{\alpha,m-\alpha}}(X \otimes Y)).$$

Therefore we must study the restriction

$$\mathrm{Res}_{P_{a,m-a}}^{G_m}(\mathrm{Ind}_{P_{\alpha,m-\alpha}}^{G_m}(\mathrm{Inf}_{G_\alpha \times G_{m-\alpha}}^{P_{\alpha,m-\alpha}}(X \otimes Y)))$$

by means of the Double Coset Formula ([**126**] Theorem 1.2.40; see also Chapter 7, §1). Explicitly the Double Coset Formula in this case gives

$$\sum_{g \in P_{a,m-a} \backslash G_m / P_{\alpha,m-\alpha}} \mathrm{Ind}_{P_{a,m-a} \cap gP_{\alpha,m-\alpha}g^{-1}}^{P_{a,m-a}}((g^{-1})^* \mathrm{Inf}_{G_\alpha \times G_{m-\alpha}}^{P_{\alpha,m-\alpha}}(X \otimes Y))$$

where the $(g^{-1})^*$-action is given by $(ghg^{-1})(w) = hw$.

The Double Coset Formula isomorphism (downwards) is given by

$$z \otimes_{P_{\alpha,m-\alpha}} w \mapsto j \otimes_{P_{a,m-a} \cap gP_{\alpha,m-\alpha}g^{-1}} hw$$

where $z = jgh$ with $j \in P_{a,m-a}, h \in P_{\alpha,m-\alpha}$ with inverse (upwards) given by

$$j \otimes_{P_{a,m-a} \cap gP_{\alpha,m-\alpha}g^{-1}} w \mapsto jg \otimes_{P_{\alpha,m-\alpha}} w.$$

Next let $\Sigma_m \subset GL_m\mathbb{F}_q$ denote the symmetric group on m letters embedded as the subgroup of permutation matrices (i.e. precisely one non-zero entry on each row and column which is equal to 1).

The following result is proved in ([**146**] p. 173; see also [**30**] Chapter IV, §2)

THEOREM 3.5. *(Bruhat Decomposition)*
The inclusion of Σ_m into $GL_m\mathbb{F}_q$ induces a bijection

$$\Sigma_a \times \Sigma_{m-a}\backslash\Sigma_m/\Sigma_\alpha \times \Sigma_{m-\alpha} \xrightarrow{\cong} P_{a,m-a}\backslash G_m/P_{\alpha,m-\alpha}.$$

Now we shall construct a convenient set of double coset representations for the left-hand side of Theorem 3.5.

Consider the double cosets

$$\Sigma_a \times \Sigma_{m-a}\backslash\Sigma_m/\Sigma_\alpha \times \Sigma_{m-\alpha}.$$

On page 171 of [**146**] one finds the assertion that the double cosets in the title of this section are in bijection with the matrices of non-negative integers

$$\begin{pmatrix} k_{1,1} & k_{1,2} \\ k_{2,1} & k_{2,2} \end{pmatrix}$$

which satisfy

$$k_{1,1} + k_{1,2} = \alpha, \ k_{2,1} + k_{2,2} = m - \alpha, \ k_{1,1} + k_{2,1} = a, \ k_{1,2} + k_{2,2} = m - a.$$

Let $w \in \Sigma_m$ be a permutation of $\{1, \ldots, m\}$. Set $I_1 = \{1, 2, \ldots, a\}$, $I_2 = \{a+1, a+2, \ldots, m\}$, $J_1 = \{1, 2, \ldots, \alpha\}$ and $J_2 = \{\alpha+1, \alpha+2, \ldots, m\}$. Therefore if $g \in \Sigma_a \times \Sigma_{m-a}$ and $g' \in \Sigma_\alpha \times \Sigma_{m-\alpha}$ we have, for $t = 1, 2$ and $v = 1, 2$,

$$gwg'(J_t) \bigcap I_v = gw(J_t) \bigcap I_v = g(w(J_t) \bigcap g^{-1}(I_v)) = g(w(J_t) \bigcap I_v).$$

Therefore if we set

$$k_{t,v} = \#(w(J_t) \bigcap I_v)$$

we have a well-defined map of sets from the double cosets to the 2×2 matrices of the form described above because

$$k_{1,v} + k_{2,v} = \#(I_v) = \begin{cases} a & \text{if } v = 1, \\ m - a & \text{if } v = 2 \end{cases}$$

and

$$k_{t,1} + k_{t,2} = \#(J_t) = \begin{cases} \alpha & \text{if } t = 1, \\ m - \alpha & \text{if } t = 2. \end{cases}$$

Next we consider the passage from the matrix of $k_{i,j}$'s to a double coset. Write

$$J_1 = J(k_{1,1}) \bigcup J(k_{1,2}), \ J_2 = J(k_{2,1}) \bigcup J(k_{2,2})$$

where $J(k_{1,1}) = \{1, \ldots, k_{1,1}\}$ and $J(k_{2,1}) = \{\alpha+1, \ldots, \alpha+k_{2,1}\}$. Similarly write

$$I_1 = I(k_{1,1}) \bigcup I(k_{2,1}), \ I_2 = I(k_{1,2}) \bigcup I(k_{2,2})$$

where $I(k_{1,1}) = \{1, \ldots, k_{1,1}\}$ and $I(k_{1,2}) = \{a + 1, \ldots, a + k_{1,2}\}$. Since the orders of $I(k_{i,j})$ and $J(k_{i,j})$ are both equal to $k_{i,j}$ there is a permutation, denoted by $w(k_{*,*})$ which sends $J_1 \bigcup J_2$ to $I_1 \bigcup I_2$ by the identity on $J(k_{1,1}) = I(k_{1,1})$ and $J(k_{2,2}) = I(k_{2,2})$ and interchanges $J(k_{1,2}), J(k_{2,1})$ with $I(k_{1,2}), I(k_{2,1})$ in an order-preserving manner.

Given the permutation $w(k_{*,*})$ we have

$$\#(w(k_{*,*})(J_1) \bigcap I_1) = \#(J(k_{1,1}) \bigcap I(k_{1,1})) = k_{1,1},$$

$$\#(w(k_{*,*})(J_1) \bigcap I_2) = \#(w(k_{*,*})(J(k_{2,1})) \bigcap I(k_{1,2})) = k_{1,2},$$

$$\#(w(k_{*,*})(J_2) \bigcap I_1) = \#(w(k_{*,*})(J(k_{1,2})) \bigcap I(k_{2,1})) = k_{2,1},$$

$$\#(w(k_{*,*})(J_2) \bigcap I_2) = \#(J(k_{2,2}) \bigcap I(k_{2,2})) = k_{2,2}$$

so that the map $k_{*,*} \mapsto \Sigma_a \times \Sigma_{m-a} w(k_{*,*}) \Sigma_\alpha \times \Sigma_{m-\alpha}$ is a split injection. In addition it is straightforward to verify that any permutation whose $k_{*,*}$-matrix equals that of $w(k_{*,*})$ belongs to the same double coset as $w(k_{*,*})$. Hence the map is a bijection.

For example when $a = 3, \alpha = 4, k_{11} = 1 = k_{22}, k_{21} = 2, k_{12} = 3$

$$w(k_{*,*})^{-1} = \begin{pmatrix} 1 & 0 & 0 & 0 & 0 & 0 & 0 \\ 0 & 0 & 0 & 1 & 0 & 0 & 0 \\ 0 & 0 & 0 & 0 & 1 & 0 & 0 \\ 0 & 0 & 0 & 0 & 0 & 1 & 0 \\ 0 & 1 & 0 & 0 & 0 & 0 & 0 \\ 0 & 0 & 1 & 0 & 0 & 0 & 0 \\ 0 & 0 & 0 & 0 & 0 & 0 & 1 \end{pmatrix}$$

$$w(k_{*,*}) = \begin{pmatrix} 1 & 0 & 0 & 0 & 0 & 0 & 0 \\ 0 & 0 & 0 & 0 & 1 & 0 & 0 \\ 0 & 0 & 0 & 0 & 0 & 1 & 0 \\ 0 & 1 & 0 & 0 & 0 & 0 & 0 \\ 0 & 0 & 1 & 0 & 0 & 0 & 0 \\ 0 & 0 & 0 & 1 & 0 & 0 & 0 \\ 0 & 0 & 0 & 0 & 0 & 0 & 1 \end{pmatrix}.$$

This permutation arises in another way as a permutation of the basis elements of tensor products of four vector spaces. Let

$$V_1 = \mathbb{F}_q^{k_{11}} \oplus \mathbb{F}_q^{k_{12}} \oplus \mathbb{F}_q^{k_{21}} \oplus \mathbb{F}_q^{k_{22}} \text{ and } V_2 = \mathbb{F}_q^{k_{11}} \oplus \mathbb{F}_q^{k_{21}} \oplus \mathbb{F}_q^{k_{12}} \oplus \mathbb{F}_q^{k_{22}}.$$

We have the linear map

$$1 \oplus T(k_{*,*}) \oplus 1 : V_1 \longrightarrow V_2$$

which interchanges the order of the two central direct sum factors. The basis for V_1 is made in the usual manner from ordered bases $\{e_1, \ldots e_{k_{11}}\}$, $\{e_{k_{11}+1}, \ldots e_{k_{11}+k_{12}}\}$, $\{e_{k_{11}+k_{12}+1}, \ldots e_{k_{11}+k_{12}+k_{21}}\}$ and

$\{e_{k_{11}+k_{12}+k_{21}+1}, \cdots e_m\}$ of $\mathbb{F}_q^{k_{11}}$, $\mathbb{F}_q^{k_{12}}$, $\mathbb{F}_q^{k_{21}}$ and $\mathbb{F}_q^{k_{22}}$ respectively. Similarly the basis for V_2 is made in the usual manner from ordered bases $\{v_1, \cdots v_{k_{11}}\}$, $\{v_{k_{11}+1}, \cdots v_{k_{11}+k_{21}}\}$, $\{v_{k_{11}+k_{21}+1}, \cdots v_{k_{11}+k_{21}+k_{12}}\}$ and $\{v_{k_{11}+k_{21}+k_{12}+1}, \cdots v_m\}$ of $\mathbb{F}_q^{k_{11}}$, $\mathbb{F}_q^{k_{21}}$, $\mathbb{F}_q^{k_{12}}$ and $\mathbb{F}_q^{k_{22}}$ respectively.

The linear map $1 \oplus T(k_{*,*}) \oplus 1$ sends the ordered set $\{e_1, \ldots, e_m\}$ to the order set $\{v_1, \ldots, v_m\}$ by $e_j \mapsto v_{w(k_{*,*})(j)}$.

Clearly

$$w(k_{*,*}) G_{k_{11}} \times G_{k_{12}} \times G_{k_{21}} \times G_{k_{22}} w(k_{*,*})^{-1} = G_{k_{11}} \times G_{k_{21}} \times G_{k_{12}} \times G_{k_{22}}$$

from which is it easy to see that

$$w(k_{*,*}) G_\alpha \times G_{m-\alpha} w(k_{*,*})^{-1} \bigcap G_a \times G_{m-a} = G_{k_{11}} \times G_{k_{21}} \times G_{k_{12}} \times G_{k_{22}}.$$

So far we have shown that the $(a, m-a)$-component of $m^*(m(X \otimes Y))$ is the sum of terms, one for each $w(k_{*,*})$, given by the induced $G_a \times G_{m-a}$-action on

$$\mathrm{Ind}_{P_{a,m-a} \cap w(k_*,* P_{\alpha,m-\alpha} w(k_{*,*})^{-1}}^{P_{a,m-a}} ((w(k_{*,*})^{-1})^* \mathrm{Inf}_{G_\alpha \times G_{m-\alpha}}^{P_{\alpha,m-\alpha}} (X \otimes Y)).$$

On the other hand, for each $w(k_{*,*})$ there is a $(a, m-a)$-component of the other composition we are studying given by

$$R(G_\alpha \times G_{m-\alpha}) \xrightarrow{(-)^{U_{k_{11},k_{12}} \times U_{k_{21},k_{22}}}} R(G_{k_{11}} \times G_{k_{12}} \times G_{k_{21}} \times G_{k_{22}})$$

$$\xrightarrow{1 \otimes T(k_{*,*}) \otimes 1} R(G_{k_{11}} \times G_{k_{21}} \times G_{k_{12}} \times G_{k_{22}})$$

$$\xrightarrow{\mathrm{IndInf} \times \mathrm{IndInf}} R(G_a \times G_{m-a}).$$

Composing this second route with the split surjection

$$R(G_a \times G_{m-a}) \xrightarrow{\mathrm{Inf}} R(P_{a,m-a})$$

is equal to the composition

$$R(G_\alpha \times G_{m-\alpha}) \xrightarrow{(-)^{U_{k_{11},k_{12}} \times U_{k_{21},k_{22}}}} R(G_{k_{11}} \times G_{k_{12}} \times G_{k_{21}} \times G_{k_{22}})$$

$$\xrightarrow{1 \otimes T(k_{*,*}) \otimes 1} R(G_{k_{11}} \times G_{k_{21}} \times G_{k_{12}} \times G_{k_{22}}) \xrightarrow{\mathrm{Inf}}$$

$$R(P_{k_{11},k_{21},k_{12},k_{22}}) \xrightarrow{\mathrm{Ind}} R(P_{a,m-a})$$

because the kernels of the quotient maps $P_{k_{11},k_{21},k_{12},k_{22}} \longrightarrow P_{k_{11},k_{21}}$ and $P_{a,m-a} \longrightarrow G_a \times G_{m-a}$ are both equal to $U_{a,m-a}$.

This composition takes the $U_{k_{11},k_{12}} \times U_{k_{21},k_{22}}$-fixed points of $X \otimes Y$ with the $G_{k_{11}} \times G_{k_{12}} \times G_{k_{21}} \times G_{k_{22}}$-action and then conjugates it by $w(k_{*,*})$.

Alternatively it takes the $w(k_{*,*})U_{k_{11},k_{12}} \times U_{k_{21},k_{22}}w(k_{*,*})^{-1}$-fixed points of $(w(k_{*,*})^{-1})^*(X \otimes Y)$ with the $G_{k_{11}} \times G_{k_{21}} \times G_{k_{12}} \times G_{k_{22}}$-action. Now

$$w(k_{*,*})U_{k_{11},k_{12}} \times U_{k_{21},k_{22}}w(k_{*,*})^{-1} \subset U_{a,m-a}.$$

For example, in the small example given in the last section of this chapter, $U_{k_{11},k_{12}} \times U_{k_{21},k_{22}}$ consists of matrices of the form

$$D = \begin{pmatrix} 1 & a_{12} & a_{13} & a_{14} & 0 & 0 & 0 \\ 0 & 1 & 0 & 0 & 0 & 0 & 0 \\ 0 & 0 & 1 & 0 & 0 & 0 & 0 \\ 0 & 0 & 0 & 1 & 0 & 0 & 0 \\ 0 & 0 & 0 & 0 & 1 & 0 & a_{57} \\ 0 & 0 & 0 & 0 & 0 & 1 & a_{67} \\ 0 & 0 & 0 & 0 & 0 & 0 & 1 \end{pmatrix}$$

so that $w(k_{*,*})U_{k_{11},k_{12}} \times U_{k_{21},k_{22}}w(k_{*,*})^{-1}$ consists of matrices

$$w(k_{*,*})Dw(k_{*,*})^{-1} = \begin{pmatrix} 1 & 0 & 0 & a_{12} & a_{13} & a_{14} & 0 \\ 0 & 1 & 0 & 0 & 0 & 0 & a_{57} \\ 0 & 0 & 1 & 0 & 0 & 0 & a_{67} \\ 0 & 0 & 0 & 1 & 0 & 0 & 0 \\ 0 & 0 & 0 & 0 & 1 & 0 & 0 \\ 0 & 0 & 0 & 0 & 0 & 1 & 0 \\ 0 & 0 & 0 & 0 & 0 & 0 & 1 \end{pmatrix}.$$

Since $w(k_{*,*})Dw(k_{*,*})^{-1}$'s act trivially we may inflate the representation to $P_{k_{11},k_{21},k_{12},k_{22}}$ (i.e. extending the action trivially on $U_{k_{11},k_{21},k_{12},k_{22}}$) and then induce up to a representation of $P_{a,m-a}$.

Now let us describe the isomorphism between the result of sending $X \otimes Y$ via the second route and the $U_{a,m-a}$-fixed subspace of

$$\text{Ind}_{P_{a,m-a} \cap w(k_{*,*}P_{a,m-a}w(k_{*,*})^{-1}}^{P_{a,m-a}}((w(k_{*,*})^{-1})^* \text{Inf}_{G_\alpha \times G_{m-\alpha}}^{P_{\alpha,m-\alpha}}(X \otimes Y)).$$

There are inclusions

$$w(k_{*,*})P_{\alpha,m-\alpha}w(k_{*,*})^{-1} \bigcap P_{a,m-a} \subset P_{k_{11},k_{21},k_{12},k_{22}} \subset P_{a,m-a}.$$

For example, $w(k_{*,*})P_{\alpha,m-\alpha}w(k_{*,*})^{-1} \bigcap P_{3,4}$, in the small example of the Appendix, consists of the matrices of the form

$$E' = \begin{pmatrix} a_{11} & a_{15} & a_{16} & a_{12} & a_{13} & a_{14} & a_{17} \\ 0 & a_{55} & a_{56} & 0 & 0 & 0 & a_{57} \\ 0 & a_{65} & a_{66} & 0 & 0 & 0 & a_{67} \\ 0 & 0 & 0 & a_{22} & a_{23} & a_{24} & a_{27} \\ 0 & 0 & 0 & a_{32} & a_{33} & a_{34} & a_{37} \\ 0 & 0 & 0 & a_{42} & a_{43} & a_{44} & a_{47} \\ 0 & 0 & 0 & 0 & 0 & 0 & a_{77} \end{pmatrix}$$

and, as we noted above, $w(k_{*,*})Dw(k_{*,*})^{-1}$ consists of the matrices

$$\{(b_{ij}) \in w(k_{*,*})P_{\alpha,m-\alpha}w(k_{*,*})^{-1} \bigcap U_{3,4} \mid b_{17} = 0\}.$$

There is a bijection of cosets

$$P_{k_{11},k_{21},k_{12},k_{22}}/w(k_{*,*})P_{\alpha,m-\alpha})w(k_{*,*})^{-1} \bigcap P_{a,m-a}$$

$$\cong U_{a,m-a}/w(k_{*,*})P_{\alpha,m-\alpha})w(k_{*,*})^{-1} \bigcap U_{3,4}.$$

Therefore we may take the coset representations X_α to lie in the abelian group $U_{a,m-a}$. In the small example the X_α's may be taken to be of the form

$$X_\alpha = \begin{pmatrix} 1 & 0 & 0 & 0 & 0 & 0 & a \\ 0 & 1 & 0 & b & c & d & 0 \\ 0 & 0 & 1 & e & f & g & 0 \\ 0 & 0 & 0 & 1 & 0 & 0 & 0 \\ 0 & 0 & 0 & 0 & 1 & 0 & 0 \\ 0 & 0 & 0 & 0 & 0 & 1 & 0 \\ 0 & 0 & 0 & 0 & 0 & 0 & 1 \end{pmatrix}.$$

The isomorphism from the image of $X \otimes Y$ via the second route to the $U_{a,m-a}$-fixed subspace of

$$\mathrm{Ind}_{P_{a,m-a}\cap w(k_{*,*}P_{\alpha,m-\alpha}w(k_{*,*})^{-1}}^{P_{a,m-a}}((w(k_{*,*})^{-1})^*\mathrm{Inf}_{G_\alpha \times G_{m-\alpha}}^{P_{\alpha,m-\alpha}}(X \otimes Y))$$

is given by

$$g \otimes_{P_{k_{11},k_{21},k_{12},k_{22}}} v \mapsto \sum_{X_\alpha} gX_\alpha \otimes_{w(k_{*,*})P_{\alpha,m-\alpha})w(k_{*,*})^{-1} \bigcap P_{a,m-a}} v.$$

This concludes the proof of Theorem 3.4. The remainder of the Hopf condition is given by the following result, which is proved in a similar manner to Theorem 3.4 (see [146] p. 175).

THEOREM 3.6.

In the notation of §1, ϵ^* is a ring homomorphism restricting to an isomorphism on R_0.

4. Semi-direct products $\mathrm{Gal}(\mathbb{F}_{q^n}/\mathbb{F}_q) \propto GL_t\mathbb{F}_{q^n}$

4.1. Let V be an irreducible representation of $GL_t\mathbb{F}_{q^n}$ and let $\Sigma \in \mathrm{Gal}(\mathbb{F}_{q^n}/\mathbb{F}_q)$ denote the Frobenius substitution. Hence the representation $\Sigma^i(V)$ given by transporting the $GL_t\mathbb{F}_{q^n}$-action by the i-th power of Σ is another irreducible representation. Suppose that $n = sd$ and that

$$V, \Sigma(V), \Sigma^2(V), \dots, \Sigma^{s-1}(V)$$

are inequivalent $GL_t\mathbb{F}_{q^n}$-irreducibles but that V and $\Sigma^s(V)$ are equivalent $GL_t\mathbb{F}_{q^n}$-irreducibles.

Therefore V (cf. Chapter Two, §6; see also Chapter 8 and Appendix I, §3) extends to an irreducible representation $\tilde{V}$ of the semi-direct product $\text{Gal}(\mathbb{F}_{q^{bn}}/\mathbb{F}_{q^s}) \propto GL_t\mathbb{F}_{q^n}$ for some $b \geq 1$, where $\text{Gal}(\mathbb{F}_{q^{bn}}/\mathbb{F}_{q^s})$ acts via first projecting onto $\text{Gal}(\mathbb{F}_{q^n}/\mathbb{F}_{q^s})$.

In discussions of the semi-direct product I shall attempt to follow the notational conventions of ([125] p. 36) and Appendix I, §3.1 as opposed to those of [117]. Explicitly, if C acts on G via $\lambda : C \longrightarrow \text{Aut}(G)$ then the semi-direct product $C \propto G$ is the group whose underlying set is $C \times G$ with multiplication given by

$$(c_1, g_1) \cdot (c_2, g_2) = (c_1 c_2, g_1 \lambda(c_1)(g_2)), \qquad c_i \in C, g_i \in G.$$

We may form the induced representation

$$\hat{V} = \text{Ind}_{\text{Gal}(\mathbb{F}_{q^{bn}}/\mathbb{F}_{q^s}) \propto GL_t\mathbb{F}_{q^n}}^{\text{Gal}(\mathbb{F}_{q^{bn}}/\mathbb{F}_q) \propto GL_t\mathbb{F}_{q^n}} (\tilde{V})$$

which restricts to give

$$\oplus_{i=0}^{s-1} \Sigma^i(V) \in R(GL_t\mathbb{F}_{q^n}).$$

Also

$$\text{Hom}_{\text{Gal}(\mathbb{F}_{q^{bn}}/\mathbb{F}_q) \propto GL_t\mathbb{F}_{q^n}}(\hat{V}, \hat{V})$$

$$\cong \text{Hom}_{\text{Gal}(\mathbb{F}_{q^{bn}}/\mathbb{F}_{q^s}) \propto GL_t\mathbb{F}_{q^n}}(\tilde{V}, \text{Ind}_{\text{Gal}(\mathbb{F}_{q^{bn}}/\mathbb{F}_{q^s}) \propto GL_t\mathbb{F}_{q^n}}^{\text{Gal}(\mathbb{F}_{q^{bn}}/\mathbb{F}_q) \propto GL_t\mathbb{F}_{q^n}}(\tilde{V}))$$

$$\cong \oplus_{i=0}^{s-1} \text{Hom}_{\text{Gal}(\mathbb{F}_{q^{bn}}/\mathbb{F}_{q^s}) \propto GL_t\mathbb{F}_{q^n}}(\tilde{V}, \Sigma^i(\tilde{V}))$$

$$= \text{Hom}_{\text{Gal}(\mathbb{F}_{q^{bn}}/\mathbb{F}_{q^s}) \propto GL_t\mathbb{F}_{q^n}}(\tilde{V}, \tilde{V})$$

as is seen by restricting representations to $GL_t\mathbb{F}_{q^n}$. Since $\tilde{V}$ is irreducible its endomorphism ring is 1-dimensional and so therefore is that of $\hat{V}$.

In the terminology of [117] when $s > 1$ $\hat{V}$ is called an irreducible representation of the second kind and when $s = 1$ it is called an irreducible representation of the first kind.

Let $\theta : \text{Gal}(\mathbb{F}_{q^{bn}}/\mathbb{F}_q) \longrightarrow \mathbb{C}^*$ be a character which is trivial on $\text{Gal}(\mathbb{F}_{q^{bn}}/\mathbb{F}_{q^s})$. Then

$$\theta \cdot \hat{V} = \text{Ind}_{\text{Gal}(\mathbb{F}_{q^{bn}}/\mathbb{F}_{q^s}) \propto GL_t\mathbb{F}_{q^n}}^{\text{Gal}(\mathbb{F}_{q^{bn}}/\mathbb{F}_q) \propto GL_t\mathbb{F}_{q^n}} (\theta \cdot \tilde{V}) = \hat{V}.$$

All the irreducibles of $\text{Gal}(\mathbb{F}_{q^{bn}}/\mathbb{F}_q) \propto GL_t\mathbb{F}_{q^n}$ are of the form $\hat{V}$ for some s dividing n. For if W is an irreducible of $\text{Gal}(\mathbb{F}_{q^{bn}}/\mathbb{F}_q) \propto GL_t\mathbb{F}_{q^n}$ then its restriction to $GL_t\mathbb{F}_{q^n}$ must have the form

$$mV_1 \oplus m\Sigma(V_1) \oplus \ldots \oplus m\Sigma^{s-1}(V_1)$$

with V_1 irreducible and $\Sigma^s(V_1) = V_1$. Therefore V_1 extends to an irreducible $\tilde{V}_1$ and there is a non-zero map of representations $\hat{V}_1 \longrightarrow W$ which must be an isomorphism (and so $m = 1$).

Twisting $\hat{V}$ by a character θ which is non-trivial on $\mathrm{Gal}(\mathbb{F}_{q^{bn}}/\mathbb{F}_{q^s})$ gives a distinct irreducible. There are bn/s cosets of such θ's so we have bn/s distinct irreducibles

$$\theta_1\hat{V}, \theta_2\hat{V}, \ldots, \theta_{bn/s}\hat{V}$$

each restricting to

$$\oplus_{i=0}^{s-1} \Sigma^i(V) \in R(GL_t\mathbb{F}_{q^n}).$$

By Shintani base change for finite general linear groups ([**117**]; see also Chapter 8 and Appendix I, §3) there is a bijection between $GL_t\mathbb{F}_{q^n}$-irreducibles V such that $\Sigma^s(V) = V$ and the irreducibles of $GL_t\mathbb{F}_{q^s}$. The V's in the construction of $\hat{V}$ are those which are fixed by Σ^s but by no Σ^u with u a proper divisor of s.

En route to the base change result one finds ([**117**] Theorem 1) that $b = 1$ or $b = 2$ suffices for the extension to the semi-direct product which was discussed in this section. This is explained in §6 just after the statement of Theorem 6.2.

5. $\tilde{R}$ and R''

5.1. Let $K = R(\mathrm{Gal}(\mathbb{F}_{q^n}/\mathbb{F}_q))$ which is the ring of integral linear combinations of characters $\chi : \mathrm{Gal}(\mathbb{F}_{q^n}/\mathbb{F}_q) \longrightarrow \mathbb{C}^*$. Suppose that G is a subgroup of $GL_t\mathbb{F}_{q^n}$ which is preserved by the $\mathrm{Gal}(\mathbb{F}_{q^n}/\mathbb{F}_q)$-action. Let $\mathcal{S}(G)$ denote the subset of the irreducibles $\mathrm{Irr}(\mathrm{Gal}(\mathbb{F}_{q^n}/\mathbb{F}_q) \propto G)$ of the first kind (i.e. representations which are irreducible when restricted to G). Tensoring with a Galois character χ permutes the set $\mathcal{S}(G)$ making $\mathbb{Z}[\mathcal{S}(G)]$ into a free K-module.

Define $\tilde{R} = \oplus_{t \geq 0} \tilde{R}_t$ where $\tilde{R}_t = R(\mathrm{Gal}(\mathbb{F}_{q^n}/\mathbb{F}_q) \propto GL_t\mathbb{F}_{q^n})$ for a fixed choice of n. Define $R'' = \oplus_{t \geq 0} \mathcal{S}(GL_t\mathbb{F}_{q^n}) \subset \tilde{R}$.

For each $\mathrm{Gal}(\mathbb{F}_{q^n}/\mathbb{F}_q)$-invariant irreducible $V \in \mathrm{Irr}(GL_t\mathbb{F}_{q^n})^{\mathrm{Gal}(\mathbb{F}_{q^n}/\mathbb{F}_q)}$ choose an irreducible $\tilde{V} \in \tilde{R}_t$ which restricts to V. The set of irreducibles which restrict to V are given by $\{\chi \otimes \tilde{V}\}$ as χ varies through Galois characters. Therefore there is an isomorphism of K-modules, depending on the choice of $\hat{V}$'s, of the form

$$\lambda_{GL_t\mathbb{F}_{q^n}} : K[\mathrm{Irr}(GL_t\mathbb{F}_{q^n})^{\mathrm{Gal}(\mathbb{F}_{q^n}/\mathbb{F}_q)}] \overset{\cong}{\longrightarrow} R''_t = \mathcal{S}(GL_t\mathbb{F}_{q^n})$$

given by sending $\chi \otimes V$ to $\chi \otimes \tilde{V}$. Hence $\tilde{R}$ is a K-module of which $R'' = \oplus_{t \geq 0} \mathcal{S}(GL_t\mathbb{F}_{q^n})$ is a free K-submodule.

The first objective of this section is to make $\tilde{R}$ into a connected, graded K-algebra[1] of which R'' is a connected, graded K-subalgebra. Clearly we have an isomorphism $\epsilon : K \longrightarrow \tilde{R}_0 = R_0''$ which shows that these are connected K-algebras.

Multiplication:

Let $P_{a,b}$ be the usual parabolic subgroup of $GL_{a+b}\mathbb{F}_{q^n}$. Then inflation induces a K-module homomorphism of representation rings

$$\text{Inf}_{a,b} : R(\text{Gal}(\mathbb{F}_{q^n}/\mathbb{F}_q) \propto (GL_a\mathbb{F}_{q^n} \times GL_b\mathbb{F}_{q^n})) \longrightarrow R(\text{Gal}(\mathbb{F}_{q^n}/\mathbb{F}_q) \propto P_{a,b}).$$

We also have induction maps

$$\text{Ind}_{a,b} : R(\text{Gal}(\mathbb{F}_{q^n}/\mathbb{F}_q) \propto P_{a,b}) \longrightarrow R(\text{Gal}(\mathbb{F}_{q^n}/\mathbb{F}_q) \propto GL_{a+b}\mathbb{F}_{q^n}).$$

Let V and W be representations of $\text{Gal}(\mathbb{F}_{q^n}/\mathbb{F}_q) \propto GL_a\mathbb{F}_{q^n})$ and $\text{Gal}\mathbb{F}_{q^n}/\mathbb{F}_q) \propto GL_b\mathbb{F}_{q^n})$ respectively. Define a representation $V \otimes' W$ of $\mathbb{F}_{q^n}/\mathbb{F}_q) \propto (GL_a\mathbb{F}_{q^n} \times GL_b\mathbb{F}_{q^n})$ on the underlying vector space of $V \otimes W$ by

$$(g, X, Y)(v \otimes w) = (g, X)(v) \otimes (g, Y)(w).$$

The multiplication on $\tilde{R}$ is defined, following the $GL\mathbb{F}_{q^n}$-case, by

$$m(V \otimes W) = \text{Ind}_{a,b}(\text{Inf}_{a,b}(V \otimes' W)) \in \tilde{R}_{a+b}.$$

If χ belongs to the character group of $\text{Gal}(\mathbb{F}_{q^n}/\mathbb{F}_q)$ then

$$m(\chi V \otimes W) = m(V \otimes \chi W) = \chi m(V \otimes W) \in \tilde{R}_{a+b}.$$

If V and W are representations in R_a'' and R_b'' respectively we shall show that $m(V \otimes W) \in R_{a+b}''$.

By additivity it suffices to assume that V, W are irreducibles (of the first kind). Then, by the construction of all the irreducibles of the finite general linear groups which first appears in [69] and is reiterated in ([93] §1) and Appendix III, §2, $m(V \otimes W)$ is irreducible when restricted to $GL_{a+b}\mathbb{F}_{q^n}$ unless $W = \chi \otimes V$. In that case $m(V \otimes (\chi \otimes W)) = \chi \otimes m(V \otimes V)$. Restricted to $GL_{a+b}\mathbb{F}_{q^n}$ the latter is known to be the sum of two irreducibles picked out by the idempotents of the symmetric group on two letters[2]. However these idempotents also decompose $m(V \otimes V)$ into two irreducibles of the first kind, in the same way.

Note that m factorises through

$$m : \tilde{R} \otimes_K \tilde{R} \longrightarrow \tilde{R}$$

which is a K-module homomorphism. Also m is associative.

This discussion established the following result.

---

[1]The structure map giving the multiplication in this algebra first appeared in ([117] Definition 2.4).

[2]These idempotents will show up again in §6.

THEOREM 5.2.

With the notation introduced above $\tilde{R}$ is a graded K-algebra of which R'' is a graded K-subalgebra.

Let V be an irreducible of $\mathrm{Gal}(\mathbb{F}_{q^n}/\mathbb{F}_q) \propto GL_{a+b}\mathbb{F}_{q^n}$. We are not going to define a comultiplication on $\tilde{R}$. However, we close this section with the observation that sending V to its $U_{a,b}$-fixed points yields a homomorphism

$$m^* : \tilde{R}_{a+b} \longrightarrow R(\mathrm{Gal}(\mathbb{F}_{q^n}/\mathbb{F}_q) \propto (GL_a\mathbb{F}_{q^n} \times GL_b\mathbb{F}_{q^n}))$$

which covers (via the restriction to general linear groups) the comultiplication defined in §3 on the PSH algebra $\oplus_{t \geq 0} R(GL_t\mathbb{F}_{q^n})$.

6. Shintani base change

6.1. Let us recall the main result of [117] which, for our notation for the semi-direct product, is stated in the following form:

THEOREM 6.2. ([117] Theorem 1; see also Lemmas 2.7 and 2.11)

(i) Let ρ be a finite-dimensional complex irreducible representation of $GL_n\mathbb{F}_q$. Then there exists an irreducible representation $\tilde{\rho}$ of the semi-direct product $\mathrm{Gal}(\mathbb{F}_{q^m}/\mathbb{F}_q) \propto GL_n\mathbb{F}_{q^m}$ which satisfies, for all $g \in GL_n\mathbb{F}_{q^m}$,

$$\chi_{\tilde{\rho}}(\Sigma, g) = \epsilon \chi_\rho([g\Sigma(g) \ldots \Sigma^{m-1}(g)])$$

where $\epsilon = \pm 1$ is independent of g. Here $[g\Sigma(g) \ldots \Sigma^{m-1}(g)]$ denotes the unique conjugacy class in $GL_n\mathbb{F}_q$ given by the intersection of the conjugacy class of $g\Sigma(g) \ldots \Sigma^{m-1}(g)$ in $GL_n\mathbb{F}_{q^m}$ with $GL_n\mathbb{F}_q$.

(ii) The Shintani base change correspondence (see Appendix I, §4)

$$Sh : \mathrm{Irr}(GL_n\mathbb{F}_{q^m})^{\mathrm{Gal}(\mathbb{F}_{q^m}/\mathbb{F}_q)} \overset{\cong}{\longrightarrow} \mathrm{Irr}(GL_n\mathbb{F}_q)$$

is given by, in the case where ϵ may be chosen to equal 1 in part (i),

$$Sh(\mathrm{Res}_{GL_n\mathbb{F}_{q^m}}^{\mathrm{Gal}(\mathbb{F}_{q^m}/\mathbb{F}_q) \propto GL_n\mathbb{F}_{q^m}}(\tilde{\rho})) = \rho.$$

When $\epsilon = -1$ is the only possibility there is an extension, denoted by ρ', of ρ to $\mathrm{Gal}(\mathbb{F}_{q^{2m}}/\mathbb{F}_q) \propto GL_n\mathbb{F}_{q^m}$ and

$$\chi_{\rho'}(\Sigma, g) = \chi_\rho([g\Sigma(g) \ldots \Sigma^{m-1}(g)])$$

specifies $\chi(\rho)$ in this case.

In this Theorem χ_ρ denotes the character function of ρ. In part (ii) of the theorem it should be noted that $\tilde{\rho}$ is an irreducible of the first kind because the $\chi_{\tilde{\rho}}(\Sigma, g)$'s are not identically zero ([117] Lemma 1.1(i)) and therefore $\mathrm{Res}_{GL_n\mathbb{F}_{q^m}}^{\mathrm{Gal}(\mathbb{F}_{q^m}/\mathbb{F}_q) \propto GL_n\mathbb{F}_{q^m}}(\tilde{\rho})$ is an irreducible representation.

Given $\tilde{\rho}$ as in part (i) of the theorem write $\tilde{\rho}(z, 1) = X_z$ for $z \in \mathrm{Gal}(\mathbb{F}_{q^m}/\mathbb{F}_q)$ and $\tilde{\rho}(1, g) = \hat{\rho}(g)$ for $g \in GL_n\mathbb{F}_{q^m}$. Since $(1, g)(z, 1) = (z, g)$ we have

$$X_z \hat{\rho}(g) = \hat{\rho}(z(g))X_z$$

so that $\chi_{\tilde{\rho}}(\Sigma, g) = \text{Trace}(\tilde{\rho}(1, g)X_\Sigma)$ (see [117] Theorem 1)[3].

For $\tilde{\rho}$ and $\hat{\rho}$ as in Theorem 6.2 the matrix X_Σ will satisfy $X_\Sigma^m = 1$. However, as mentioned in the statement of ([117] Theorem 1), for a general Galois invariant $\hat{\rho}$ there exists a choice satisfying $X_\Sigma^m = \pm 1$. When $X_\Sigma^m = 1$ the extension $\tilde{\rho}$ of $\hat{\rho}$ may be constructed as in Theorem 6.2 but when $X_\Sigma^m = -1$ the extension of $\hat{\rho}$ must be a representation of $\text{Gal}(\mathbb{F}_{q^{2m}}/\mathbb{F}_q) \propto GL_n\mathbb{F}_{q^m}$.

Given a choice of $\hat{\rho}$ the irreducible extension $\tilde{\rho}$ to the semi-direct product, which we may take to be $\text{Gal}(\mathbb{F}_{q^{2m}}/\mathbb{F}_q) \propto GL_n\mathbb{F}_{q^m}$ in general, is unique up to twists by Galois characters.

Next we shall examine the multiplicative property of the Shintani correspondence.

Suppose that $\rho_1 \in \text{Irr}(GL_a\mathbb{F}_q)$ and $\rho_2 \in \text{Irr}(GL_b\mathbb{F}_q)$. By Theorem 6.2 there exist $\tilde{\rho}_1 \in \text{Irr}(\text{Gal}(\mathbb{F}_{q^m}/\mathbb{F}_q) \propto GL_a\mathbb{F}_q)$ and $\tilde{\rho}_2 \in \text{Irr}(\text{Gal}(\mathbb{F}_{q^m}/\mathbb{F}_q) \propto GL_b\mathbb{F}_q)$ such that for $i = 1, 2$

$$\chi_{\tilde{\rho}_i}(\Sigma, g) = \epsilon_i \chi_{\rho_i}([g\Sigma(g)\dots\Sigma^{m-1}(g)])$$

where $\epsilon_i = \pm 1$ is independent of g.

Therefore, by ([117] Definition 2.4 and Lemma 2.9),

$$\chi_{m(\tilde{\rho}_1,\tilde{\rho}_2)}(\Sigma, g) = \epsilon_1\epsilon_2\chi_{m(\rho_1,\rho_2)}([g\Sigma(g)\dots\Sigma^{m-1}(g)])$$

where on the left-hand side m denotes the multiplication in $\tilde{R}$ of §5 and on the right-hand side the multiplication in R of §3.

If $\rho_1 \neq \rho_2$ then by the Shintani correspondence the restrictions of $\tilde{\rho}_1$ and $\tilde{\rho}_2$ to the general linear groups are distinct so that $\tilde{\rho}_1$ and $\tilde{\rho}_2$ are distinct irreducible representations. Therefore $m(\tilde{\rho}_1, \tilde{\rho}_2)$ and $m(\rho_1, \rho_2)$ are both irreducible and

$$Sh(\text{Res}_{GL_{a+b}\mathbb{F}_{q^m}}^{\text{Gal}(\mathbb{F}_{q^m}/\mathbb{F}_q)\propto GL_{a+b}\mathbb{F}_{q^m}}(m(\tilde{\rho}_1, \tilde{\rho}_2))) = m(\rho_1, \rho_2).$$

However

$$\text{Res}_{GL_{a+b}\mathbb{F}_{q^m}}^{\text{Gal}(\mathbb{F}_{q^m}/\mathbb{F}_q)\propto GL_{a+b}\mathbb{F}_{q^m}}(m(\tilde{\rho}_1, \tilde{\rho}_2))$$

$$= m(\text{Res}_{GL_a\mathbb{F}_{q^m}}^{\text{Gal}(\mathbb{F}_{q^m}/\mathbb{F}_q)\propto GL_a\mathbb{F}_{q^m}}(\tilde{\rho}_1), \text{Res}_{GL_b\mathbb{F}_{q^m}}^{\text{Gal}(\mathbb{F}_{q^m}/\mathbb{F}_q)\propto GL_b\mathbb{F}_{q^m}}(\tilde{\rho}_2)).$$

Therefore, if $\hat{\rho}_1 = \text{Res}_{GL_a\mathbb{F}_{q^m}}^{\text{Gal}(\mathbb{F}_{q^m}/\mathbb{F}_q)\propto GL_a\mathbb{F}_{q^m}}(\tilde{\rho}_1)$ and

$$\hat{\rho}_2 = \text{Res}_{GL_b\mathbb{F}_{q^m}}^{\text{Gal}(\mathbb{F}_{q^m}/\mathbb{F}_q)\propto GL_b\mathbb{F}_{q^m}}(\tilde{\rho}_2),$$

then

$$Sh(m(\hat{\rho}_1, \hat{\rho}_2)) = m(Sh(\hat{\rho}_1), Sh(\hat{\rho}_2)).$$

---

[3]The formula of [117] differs from mine because we have used different formulae for the multiplication in a semi-direct product.

If $a = b$ and $\rho_1 = \rho_2$ and (see §5 on "multiplication") $m(\rho_1, \rho_1)$ is not irreducible but there is an idempotent e of the symmetric group on two letters such that

$$m(\rho_1, \rho_1) = em(\rho_1, \rho_1) + (1 - e)m(\rho_1, \rho_1)$$

and the two summands on the right are irreducible. Similarly

$$m(\tilde{\rho}_1, \tilde{\rho}_1) = em(\tilde{\rho}_1, \tilde{\rho}_1) + (1 - e)m(\tilde{\rho}_1, \tilde{\rho}_1)$$

where the two summands on the right are irreducible. In addition

$$\chi_{em(\tilde{\rho}_1, \tilde{\rho}_1)}(\Sigma, g) = \epsilon_1 \epsilon_1 \chi_{em(\rho_1, \rho_1)}([g\Sigma(g) \ldots \Sigma^{m-1}(g)])$$

and

$$\chi_{(1-e)m(\tilde{\rho}_1, \tilde{\rho}_1)}(\Sigma, g) = \epsilon_1 \epsilon_1 \chi_{(1-e)m(\rho_1, \rho_1)}([g\Sigma(g) \ldots \Sigma^{m-1}(g)]).$$

Therefore

$$Sh(em(\hat{\rho}_1, \hat{\rho}_1)) = em(Sh(\hat{\rho}_1), Sh(\hat{\rho}_1))$$

and

$$Sh((1 - e)m(\hat{\rho}_1, \hat{\rho}_1)) = (1 - e)m(Sh(\hat{\rho}_1), Sh(\hat{\rho}_1))$$

and adding these relations yields

$$Sh(m(\hat{\rho}_1, \hat{\rho}_1)) = m(Sh(\hat{\rho}_1), Sh(\hat{\rho}_1)).$$

Set $R' = \oplus_{t \geq 0} R'_t$ where $R'_t = \mathbb{Z}[\mathrm{Irr}(GL_t \mathbb{F}_{q^m})^{\mathrm{Gal}(\mathbb{F}_{q^m}/\mathbb{F}_q)}]$, the subgroup of $R(GL_t \mathbb{F}_{q^m})$ spanned by $\mathbb{Z}$-linear combination of irreducible representations which are invariant under the Galois action. In Theorem 5.2 we saw that R'' is a subalgebra of $\tilde{R}$ and a similar argument shows that R' is a subalgebra of $\oplus_{t \geq 0} R(GL_t \mathbb{F}_{q^m})$.

THEOREM 6.3.
With the notation introduced above R' is a graded subalgebra of the algebra $\oplus_{t \geq 0} R(GL_t \mathbb{F}_{q^m})$. Furthermore the restriction map

$$\tilde{R} = \oplus_{t \geq 0} R(\mathrm{Gal}(\mathbb{F}_{q^m}/\mathbb{F}_q) \propto GL_t \mathbb{F}_{q^m}) \longrightarrow \oplus_{t \geq 0} R(GL_t \mathbb{F}_{q^m})$$

restricts to a surjective algebra homomorphism of the form $R'' \longrightarrow R'$.

The Shintani correspondence of Theorem 6.2 is a bijection of set of irreducible representations. Extending it by additivity yields an isomorphism of abelian groups

$$Sh : R' \xrightarrow{\cong} R.$$

The preceding discussion concerning the multiplicativity of the Shintani correspondence establishes the following result.

THEOREM 6.4.

The $\mathbb{Z}$-linear extension of the Shintani correspondence of Theorem 6.2 yields an algebra isomorphism

$$Sh : R' \xrightarrow{\cong} R$$

between the Hopf algebras R' introduced above and R of §3.

6.5. The algebra isomorphism Sh^{-1} of Theorem 6.4 yields an injective algebra homomorphism

$$R \xrightarrow{\cong} R' \subset \oplus_{t \geq 0} R(GL_t \mathbb{F}_{q^m})$$

between two Hopf algebras is *not* a Hopf algebra homomorphism. This is illustrated by the following $GL_2 \mathbb{F}_{q^{2p}}$ example of ([117] p. 412; see also Chapter Eight, §1.3).

Suppose that $m = 2p$. Consider the Galois extension $\mathbb{F}_{q^{2p}}/\mathbb{F}_q$ and the irreducible representation of $GL_2 \mathbb{F}_{q^{2p}}$ given by $m(\chi_1, \chi_2)$ (sometimes denoted by is $R(\chi_1, \chi_2)$) with $\chi_i : \mathbb{F}_{q^{2p}}^* \longrightarrow \mathbb{C}^*$ and Frobenius action $\Sigma(\chi_1) = \chi_2, \Sigma(\chi_2) = \chi_1$ so that

$$\Sigma^* R(\chi_1, \chi_2) = R(\chi_1, \chi_2).$$

This is decomposable in the Hopf algebra $\oplus_{t \geq 0} R(GL_t \mathbb{F}_{q^{2p}})$ and therefore is not primitive and therefore it is not primitive in R.

Hence $\text{Gal}(\mathbb{F}_{q^{2p}}/\mathbb{F}_{q^2}) = \langle \Sigma^2 \rangle$ fixes χ_1 and χ_2 so that, by Hilbert's Theorem 90,

$$\chi_1 = \Theta \cdot \text{Norm} : \mathbb{F}_{q^{2p}}^* \longrightarrow \mathbb{F}_{q^2}^* \longrightarrow \mathbb{C}^*$$

and

$$\chi_2 = \Sigma^*(\Theta) \cdot \text{Norm} : \mathbb{F}_{q^{2p}}^* \longrightarrow \mathbb{F}_{q^2}^* \longrightarrow \mathbb{C}^*.$$

Therefore $\Theta \neq \Sigma^*(\Theta)$.

From ([117]; [126], Chapter Two) $Sh(R(\chi_1, \chi_2)) = R(\Theta)$, the Weil representation associated to Θ, which is an irreducible representation of $GL_2 \mathbb{F}_q$. However the Weil representation is an example of a irreducible cuspidal representation of $GL_2 \mathbb{F}_q$ and, as explained in [146], these are the same as the positive primitive irreducibles in the PSH-algebra for $GL\mathbb{F}_q$.

The character Θ is an example of a regular character of the multiplicative group of a finite field. In fact, as a consequence of the Shintani correspondence together with ([133] Theorem 8-6), the character functions of all the cuspidal representations of the $GL_t \mathbb{F}_q$'s are calculated in ([117] Theorem 2) and, in addition, these cuspidals are shown to be in one-one correspondence with regular characters[4].

---

[4]There are other ways to prove ([117] Theorem 2). For example, in ([69] Theorem 13 p. 439) the cuspidals are classified and denoted by g^λ's. Also the result can be derived from ([41] Theorem 9.3.2) which asserts that the irreducible Deligne-Lusztig characters $\pm R_T^G(\theta)$, for regular θ will be cuspidal if and only if the torus T does not lie in any

7. Counting cuspidals irreducibles of $GL_n\mathbb{F}_q$

7.1. This comes from ([117] §3) which culminates in the proof of ([117] Theorem 2).

Let $B_t \subset GL_t\mathbb{F}_{q^n}$ denote the Borel subgroup of upper triangular matrices so that $B_t = D_tU_t$, the semi-direct product of the diagonal and the unitriangular matrices, D_t and U_t, respectively.

Suppose that $\mathbb{F}_q \subset \mathbb{F}_{q^n} \subset \mathbb{F}_{q^m}$ where $m = nd$. A character $\chi : \mathbb{F}_{q^n}^* \longrightarrow \mathbb{C}^*$ is regular if $g \in \mathrm{Gal}(\mathbb{F}_{q^n}/\mathbb{F}_q)$ and $g(\chi) = \chi$ implies that $g = 1$ (i.e. $\chi \neq \Sigma^l(\chi)$ for $l = 1, \dots, n-1$).

Define $\tilde{\chi}$ to be the composition $\tilde{\chi} = \chi \cdot \mathrm{Norm}_{\mathbb{F}_{q^m}/\mathbb{F}_{q^n}} : \mathbb{F}_{q^m}^* \longrightarrow \mathbb{C}^*$.

Define a character $\phi_\chi : B_n\mathbb{F}_{q^m} \longrightarrow \mathbb{C}^*$ by the formula

$$\phi_\chi(X_{i,j}) = \prod_{i=1}^{n} \Sigma^{i-1}(\chi) \cdot \mathrm{Norm}_{\mathbb{F}_{q^m}/\mathbb{F}_{q^n}}(X_{i,i}).$$

Therefore, by the regularity of χ, the character ϕ_χ is regular in the sense that $\phi_\chi \neq w^*(\phi_\chi)$, the conjugate of ϕ_χ by a permutation matrix w.

Define a function ψ_χ on $GL_n\mathbb{F}_{q^m}$ by the formula

$$\psi_\chi(g) = \begin{cases} \phi_\chi(X_{i,j}) & \text{if } g = (U_{i,j})w(X_{i,j}) \\ 0 & \text{otherwise} \end{cases}$$

where $(U_{i,j}) \in U_n, (X_{i,j}) \in B$ and w is the permutation matrix given by

$$w^{-1} = \begin{pmatrix} 0 & 1 & 0 & \cdots & \cdots & 0 \\ 0 & 0 & 1 & & \cdots & 0 \\ \vdots & \vdots & \vdots & \vdots & \vdots & \vdots \\ 0 & 0 & 0 & \cdots & \cdots & 1 \\ 1 & 0 & 0 & \cdots & \cdots & 0 \end{pmatrix}.$$

THEOREM 7.2. ([117] Theorem 2)[5]
(i) If χ is a regular character of $\mathbb{F}_{q^n}^*$ there exists an m-th root of unity ξ_m and an irreducible cuspidal representation ρ_χ of $GL_n\mathbb{F}_q$ such that

$$\xi_m\rho_\chi(N_{\mathbb{F}_{q^n}/\mathbb{F}_q}(g)) = \frac{q^{-m(n-1)/2}}{|B_n\mathbb{F}_{q^m}|} \sum_{X \in GL_n\mathbb{F}_{q^m}} \psi_\chi(Xg\Sigma(X)^{-1}).$$

(ii) For two regular characters χ_1, χ_2 $\rho_{\chi_1} = \rho_{\chi_2}$ if and only if $\chi_1 = \Sigma^l(\chi_2)$ for some l. Moreover any cuspidal of $GL_n\mathbb{F}_q$ is equal to ρ_χ for some regular character χ of $\mathbb{F}_{q^n}^*$.

---

proper Frobenius-stable Levi subgroup of G. I am grateful to Alexander Stasinski for explaining the latter argument to me.

[5]This result is stated in the conventions for semi-direct products, GL-norms etc. of Appendix I rather than those of [117].

7.3. Sketch proof of Theorem 7.2

By Mackey's irreducibility criterion (proved by Frobenius reciprocity and the Double Coset Formula for the restriction of an induced representation) $\text{Ind}_{B_n\mathbb{F}_{q^m}}^{GL_n\mathbb{F}_{q^m}}(\phi_\chi)$ is irreducible. This uses the fact that the permutation matrices are the double coset representatives of $B_n\mathbb{F}_{q^m}\backslash GL_n\mathbb{F}_{q^m}/B_n\mathbb{F}_{q^m}$.

In the tensor product notation for this induced representation as a left $GL_n\mathbb{F}_{q^m}$ we have $g\otimes_{B_n\mathbb{F}_{q^m}} 1 = gb\otimes_{B_n\mathbb{F}_{q^m}} \phi_\chi(b^{-1})$ so that we may think of $g\otimes_{B_n\mathbb{F}_{q^m}} 1$ as the complex-valued function f_g which is defined by $f(x) = 0$ unless $x\in gB_n\mathbb{F}_{q^m}$ and if $x = gb$ with $b\in B_n\mathbb{F}_{q^m}$ then $f_g(x) = \phi_\chi(x^{-1}g) = \phi_\chi(b^{-1})$.

This makes sense because

$$x\otimes_{B_n\mathbb{F}_{q^m}} f_g(x) = gb\otimes_{B_n\mathbb{F}_{q^m}} \phi_\chi(b^{-1}) = g\otimes_{B_n\mathbb{F}_{q^m}} 1.$$

Note that if $b'\in B_n\mathbb{F}_{q^m}$ then $f_g(xb') = f_g(gbb') = \phi_\chi((b')^{-1}b^{-1}g^{-1}g) = \phi_\chi(b')^{-1}f_g(x)$.

In the tensor-product notation for the induced representation the function f_g, transforming as above, corresponds to

$$g\otimes_{B_n\mathbb{F}_{q^m}} 1 = \sum_{h\in GL_n\mathbb{F}_{q^m}/B_n\mathbb{F}_{q^m}} h\otimes_{B_n\mathbb{F}_{q^m}} f_g(h) \in \mathbb{C}[GL_n\mathbb{F}_{q^m}]\otimes_{B_n\mathbb{F}_{q^m}}\mathbb{C}_{\phi_\chi}.$$

To switch from Shintani's conventions to mine we need to define a function $f_{g^{-1}}^{sh}$ by $f_{g^{-1}}^{sh}(x) = f_g(x^{-1})$. Then, if $b, b'\in B_n\mathbb{F}_{q^m}$, $f_{g^{-1}}^{sh}(bx) = f_g(x^{-1}b^{-1})$ is zero unless $x^{-1}b^{-1} = gb'$ and in the latter case

$$f_{g^{-1}}^{sh}(bx) = f_g(x^{-1}b^{-1}) = \phi_\chi(bxg) = \phi_\chi(b)f_g(x^{-1}) = \phi_\chi(b)f_{g^{-1}}^{sh}(x).$$

Following Shintani if w is a permutation matrix write $U_w^- = U\cap w^{-1}U^-w$ where $U = U_n\mathbb{F}_{q^m}$ and U^- is the transpose of U (i.e. the lower unitriangular matrices). For the permutation matrix introduced above one finds that

$$U_{w^{-1}}^- = \left\{ u = \begin{pmatrix} 1 & u_{1,2} & u_{1,3} & \cdots & \cdots & u_{1,n} \\ 0 & 1 & 0 & \cdots & \cdots & 0 \\ 0 & 0 & 1 & \cdots & \cdots & 0 \\ \vdots & \vdots & \vdots & \vdots & \vdots & \vdots \\ 0 & 0 & 0 & \cdots & \cdots & 1 \end{pmatrix} \right\}.$$

Now consider the product of a matrix in $U_{w^{-1}}^-$ and an unitriangular matrix

$$\begin{pmatrix} 1 & \alpha \\ 0 & I_{n-1} \end{pmatrix}\begin{pmatrix} 1 & \beta \\ 0 & B \end{pmatrix} = \begin{pmatrix} 1 & \beta + \alpha B \\ 0 & B \end{pmatrix} = \begin{pmatrix} 1 & 0 \\ 0 & B \end{pmatrix}\begin{pmatrix} 1 & \beta + \alpha B \\ 0 & I_{n-1} \end{pmatrix}$$

where $\beta, \underline{\alpha}$ are row vectors and B is upper unitriangular $(n-1) \times (n-1)$ matrix. Also we have the matrix relation

$$w^{-1} \begin{pmatrix} 1 & 0 \\ 0 & B \end{pmatrix} w = \begin{pmatrix} B & 0 \\ 0 & 1 \end{pmatrix}.$$

Suppose that b is the upper triangular matrix

$$b = D \begin{pmatrix} 1 & \beta \\ 0 & B \end{pmatrix}$$

where D is the diagonal matrix $D = \mathrm{diag}(d_1, d_2, \ldots, d_n)$. Then there exist matrices $u, u', u'' \in U_{w^{-1}}^-$ where

$$u = \begin{pmatrix} 1 & \alpha \\ 0 & I_{n-1} \end{pmatrix}$$

such that

$$w^{-1} u \Sigma(b)$$

$$= w^{-1} \Sigma(D) w w^{-1} u' \begin{pmatrix} 1 & \Sigma(\beta) \\ 0 & \Sigma(B) \end{pmatrix}$$

$$= w^{-1} \Sigma(D) w \begin{pmatrix} \Sigma(B) & 0 \\ 0 & 1 \end{pmatrix} w^{-1} u''.$$

Notice that

$$w^{-1} \Sigma(D) w \begin{pmatrix} \Sigma(B) & 0 \\ 0 & 1 \end{pmatrix} \in B_n \mathbb{F}_{q^m}$$

and that

$$\phi_\chi \left(w^{-1} \Sigma(D) w \begin{pmatrix} \Sigma(B) & 0 \\ 0 & 1 \end{pmatrix} \right) = \phi_\chi(b).$$

In addition, as u runs through $U_{w^{-1}}^-$ so does u''.

Then Shintani defines I_Σ by the formula

$$(I_\Sigma f_g^{sh})(x) = q^{-m(n-1)/2} \sum_{u \in U_{w^{-1}}^-} f_g^{sh}(w^{-1} u \Sigma(x))$$

and the above discussion explains why $(I_\Sigma f_g^{sh})(bx) = \phi_\chi(b)(I_\Sigma f_g^{sh})(x)$.

Therefore, in my conventions, the right-hand side of the above equation is

$$(I_\Sigma f_{g^{-1}})(x^{-1}) = q^{-m(n-1)/2} \sum_{u \in U_{w^{-1}}^-} f_{g^{-1}}(\Sigma(x^{-1})uw).$$

In the tensor product notation this is equivalent to

$$I_\Sigma(g \otimes_{B_n \mathbb{F}_{q^m}} 1) = \sum_{h \in GL_n \mathbb{F}_{q^m}/B_n \mathbb{F}_{q^m}} \sum_{u \in U_{w^{-1}}^-} h \otimes_{B_n \mathbb{F}_{q^m}} f_g(\Sigma(h)uw).$$

Therefore

$$g' I_\Sigma(g \otimes_{B_n \mathbb{F}_{q^m}} 1)$$

$$= \sum_{h \in GL_n \mathbb{F}_{q^m}/B_n \mathbb{F}_{q^m}} \sum_{u \in U^-_{w-1}} g' h \otimes_{B_n \mathbb{F}_{q^m}} f_g(\Sigma(h)uw)$$

$$= \sum_{h' \in GL_n \mathbb{F}_{q^m}/B_n \mathbb{F}_{q^m}} \sum_{u \in U^-_{w-1}} h' \otimes_{B_n \mathbb{F}_{q^m}} f_g(\Sigma((g')^{-1})\Sigma(h')uw)$$

$$= \sum_{h' \in GL_n \mathbb{F}_{q^m}/B_n \mathbb{F}_{q^m}} \sum_{u \in U^-_{w-1}} h' \otimes_{B_n \mathbb{F}_{q^m}} f_{\Sigma(g')g}(\Sigma(h')uw)$$

$$= I_\Sigma(\Sigma(g')g \otimes_{B_n \mathbb{F}_{q^m}} 1).$$

Set $\rho = \text{Ind}_{B_n \mathbb{F}_{q^m}}^{GL_n \mathbb{F}_{q^m}}(\phi_\chi)$. By ([**117**] Lemma 3.2) $I_\Sigma^m = 1$ and

$$\rho(\Sigma(g')) \cdot I_\Sigma^{-1} = I_\Sigma^{-1} \cdot \rho(g').$$

The multiplication in my convention for semi-direct products is given by $(c,g)(c',g') = (cc', gc(g'))$. With this convention

$$(\Sigma, 1)(1, g) = (\Sigma, \Sigma(g)) = (1, \Sigma(g))(\Sigma, 1)$$

so the irreducible representation ρ extends to an irreducible $\tilde{\rho}$ on $\text{Gal}(\mathbb{F}_{q^m}/\mathbb{F}_q) \propto GL_n \mathbb{F}_{q^m}$ in which (Σ, g) acts via $I_\Sigma^{-1} \cdot \rho(g) = I_{\Sigma^{-1}} \cdot \rho(g)$. Therefore, by ([**117**] p. 409),

$$\text{Trace}(\tilde{\rho}(\Sigma, g) = \frac{q^{-m(n-1)/2}}{|B_n \mathbb{F}_{q^m}|} \sum_{X \in GL_n \mathbb{F}_{q^m}} \psi_\chi(Xg\Sigma(X)^{-1}).$$

To show that $Sh(\rho) = \rho_\chi$ is a cuspidal irreducible of $GL_n \mathbb{F}_q$ it suffices to show that for any pair of irreducibles $\rho_1 \in \text{Irr}(GL_a \mathbb{F}_q)$ and $\rho_2 \in \text{Irr}(GL_{n-a}\mathbb{F}_q)$ that ρ_χ is not an irreducible constituent of $m(\rho_1 \otimes \rho_2)$. By Theorem 6.4, applying the Shintani correspondence up to $\mathbb{F}_{q^m}$, $Sh^{-1}(\rho_1)$ must be equivalent to the PSH algebra product of $\Sigma^{i_j}(\chi)$'s as i_j ranges over some a proper subset of $1, 2, \ldots, n$. However this is impossible because any such product is not Galois invariant.

Similarly, applying the Shintani correspondence up to $\mathbb{F}_{q^m}$, $Sh^{-1}(\rho_{\chi_1}) = Sh^{-1}(\rho_{\chi_2})$ implies that the Galois orbits of χ_1 and χ_2 coincide.

Finally, the discussion shows that the number of distinct regular characters of $\mathbb{F}_{q^n}^*$ is less than or equal to the number of inequivalent irreducible cuspidal representations of $GL_n \mathbb{F}_q$. The fact that these numbers are in fact equal follows from a counting argument given in ([**133**] Theorem 8.6). $\square$

8. An example of $w(k_{*,*})P_{\alpha,m-\alpha}w(k_{*,*})^{-1} \bigcap P_{a,m-a}$

8.1. In the notation of the discussion of Double Cosets in §3 let

$$m = 7, a = 3, \alpha = 4, k_{11} = 1 = k_{22}, k_{21} = 2, k_{12} = 3$$

and consider the double coset representative

$$
w(k_{*,*}) = \begin{pmatrix}
1 & 0 & 0 & 0 & 0 & 0 & 0 \\
0 & 0 & 0 & 0 & 1 & 0 & 0 \\
0 & 0 & 0 & 0 & 0 & 1 & 0 \\
0 & 1 & 0 & 0 & 0 & 0 & 0 \\
0 & 0 & 1 & 0 & 0 & 0 & 0 \\
0 & 0 & 0 & 1 & 0 & 0 & 0 \\
0 & 0 & 0 & 0 & 0 & 0 & 1
\end{pmatrix}.
$$

If $(a_{ij}) \in G_7$ then the conjugate by $w = w(k_{*,*})$ takes the form

$$
w(a_{ij}) = \begin{pmatrix}
a_{11} & a_{15} & a_{16} & a_{12} & a_{13} & a_{14} & a_{17} \\
a_{51} & a_{55} & a_{56} & a_{52} & a_{53} & a_{54} & a_{57} \\
a_{61} & a_{65} & a_{66} & a_{62} & a_{63} & a_{64} & a_{67} \\
a_{21} & a_{25} & a_{26} & a_{22} & a_{23} & a_{24} & a_{27} \\
a_{31} & a_{35} & a_{36} & a_{32} & a_{33} & a_{34} & a_{37} \\
a_{41} & a_{45} & a_{46} & a_{42} & a_{43} & a_{44} & a_{47} \\
a_{71} & a_{75} & a_{76} & a_{72} & a_{73} & a_{74} & a_{77}
\end{pmatrix}.
$$

In order that $w(a_{ij}) = w(a_{ij})w^{-1}$ lies in $wG_4 \times G_3 w^{-1}$ it must have the form

$$
w(a_{ij}) = \begin{pmatrix}
a_{11} & 0 & 0 & a_{12} & a_{13} & a_{14} & 0 \\
0 & a_{55} & a_{56} & 0 & 0 & 0 & a_{57} \\
0 & a_{65} & a_{66} & 0 & 0 & 0 & a_{67} \\
a_{21} & 0 & 0 & a_{22} & a_{23} & a_{24} & 0 \\
a_{31} & 0 & 0 & a_{32} & a_{33} & a_{34} & 0 \\
a_{41} & 0 & 0 & a_{42} & a_{43} & a_{44} & 0 \\
0 & a_{75} & a_{76} & 0 & 0 & 0 & a_{77}
\end{pmatrix}
$$

and so to lie in the intersection $w(G_4 \times G_3) \bigcap G_3 \times G_4$ it must have the form

$$
w(a_{ij}) = \begin{pmatrix}
a_{11} & 0 & 0 & 0 & 0 & 0 & 0 \\
0 & a_{55} & a_{56} & 0 & 0 & 0 & 0 \\
0 & a_{65} & a_{66} & 0 & 0 & 0 & 0 \\
0 & 0 & 0 & a_{22} & a_{23} & a_{24} & 0 \\
0 & 0 & 0 & a_{32} & a_{33} & a_{34} & 0 \\
0 & 0 & 0 & a_{42} & a_{43} & a_{44} & 0 \\
0 & 0 & 0 & 0 & 0 & 0 & a_{77}
\end{pmatrix} = A''.
$$

Therefore in this example

$$
wG_4 \times G_3 w^{-1} \bigcap G_3 \times G_4 = G_{k_{11}} \times G_{k_{21}} \times G_{k_{12}} \times G_{k_{22}}.
$$

In order that $w(a_{ij}) = w(a_{ij})w^{-1}$ lies in $wP_{4,3}w^{-1}$ it must have the form

$$w(a_{ij}) = \begin{pmatrix} a_{11} & a_{15} & a_{16} & a_{12} & a_{13} & a_{14} & a_{17} \\ 0 & a_{55} & a_{56} & 0 & 0 & 0 & a_{57} \\ 0 & a_{65} & a_{66} & 0 & 0 & 0 & a_{67} \\ a_{21} & a_{25} & a_{26} & a_{22} & a_{23} & a_{24} & a_{27} \\ a_{31} & a_{35} & a_{36} & a_{32} & a_{33} & a_{34} & a_{37} \\ a_{41} & a_{45} & a_{46} & a_{42} & a_{43} & a_{44} & a_{47} \\ 0 & a_{75} & a_{76} & 0 & 0 & 0 & a_{77} \end{pmatrix}$$

and so to lie in the intersection $wP_{4,3}w^{-1}\bigcap P_{3,4}$ it must have the form

$$w(a_{ij}) = \begin{pmatrix} a_{11} & a_{15} & a_{16} & a_{12} & a_{13} & a_{14} & a_{17} \\ 0 & a_{55} & a_{56} & 0 & 0 & 0 & a_{57} \\ 0 & a_{65} & a_{66} & 0 & 0 & 0 & a_{67} \\ 0 & 0 & 0 & a_{22} & a_{23} & a_{24} & a_{27} \\ 0 & 0 & 0 & a_{32} & a_{33} & a_{34} & a_{37} \\ 0 & 0 & 0 & a_{42} & a_{43} & a_{44} & a_{47} \\ 0 & 0 & 0 & 0 & 0 & 0 & a_{77} \end{pmatrix} = C.$$

A matrix in $wP_{4,3}w^{-1}\bigcap G_3 \times G_4$ has the form

$$w(a_{ij}) = \begin{pmatrix} a_{11} & a_{15} & a_{16} & 0 & 0 & 0 & 0 \\ 0 & a_{55} & a_{56} & 0 & 0 & 0 & 0 \\ 0 & a_{65} & a_{66} & 0 & 0 & 0 & 0 \\ 0 & 0 & 0 & a_{22} & a_{23} & a_{24} & a_{27} \\ 0 & 0 & 0 & a_{32} & a_{33} & a_{34} & a_{37} \\ 0 & 0 & 0 & a_{42} & a_{43} & a_{44} & a_{47} \\ 0 & 0 & 0 & 0 & 0 & 0 & a_{77} \end{pmatrix} = A$$

and a matrix in $wP_{4,3}w^{-1}\bigcap U_{3,4}$ has the form

$$w(a_{ij}) = \begin{pmatrix} 1 & 0 & 0 & b_{12} & b_{13} & b_{14} & b_{17} \\ 0 & 1 & 0 & 0 & 0 & 0 & b_{57} \\ 0 & 0 & 1 & 0 & 0 & 0 & b_{67} \\ 0 & 0 & 0 & 1 & 0 & 0 & 0 \\ 0 & 0 & 0 & 0 & 1 & 0 & 0 \\ 0 & 0 & 0 & 0 & 0 & 1 & 0 \\ 0 & 0 & 0 & 0 & 0 & 0 & 1 \end{pmatrix} = B.$$

Choosing

$$b_{12} = \frac{a_{12}}{a_{11}}, b_{13} = \frac{a_{13}}{a_{11}}, b_{14} = \frac{a_{14}}{a_{11}}$$

$$X = \begin{pmatrix} a_{11} & a_{15} & a_{16} \\ 0 & a_{55} & a_{56} \\ 0 & a_{65} & a_{66} \end{pmatrix}, \quad X\begin{pmatrix} b_{17} \\ b_{57} \\ b_{67} \end{pmatrix} = \begin{pmatrix} a_{17} \\ a_{57} \\ a_{67} \end{pmatrix}$$

shows that $AB = C$ and therefore

$$P_{3,4} \cap wP_{4,3}w^{-1} = ((G_3 \times G_4) \cap wP_{4,3}w^{-1} \cdot (U_{3,4} \cap wP_{4,3}w^{-1}).$$

In order that $w(a_{ij}) = w(a_{ij})w^{-1}$ lies in $wU_{4,3}w^{-1}$ it must have the form

$$w(a_{ij}) = \begin{pmatrix} 1 & a_{15} & a_{16} & 0 & 0 & 0 & a_{17} \\ 0 & 1 & 0 & 0 & 0 & 0 & 0 \\ 0 & 0 & 1 & 0 & 0 & 0 & 0 \\ 0 & a_{25} & a_{26} & 1 & 0 & 0 & a_{27} \\ 0 & a_{35} & a_{36} & 0 & 1 & 0 & a_{37} \\ 0 & a_{45} & a_{46} & 0 & 0 & 1 & a_{47} \\ 0 & 0 & 0 & 0 & 0 & 0 & 1 \end{pmatrix}$$

and to lie in $wU_{4,3}w^{-1} \cap G_3 \times G_4$ it must have the form

$$w(a_{ij}) = \begin{pmatrix} 1 & a_{15} & a_{16} & 0 & 0 & 0 & 0 \\ 0 & 1 & 0 & 0 & 0 & 0 & 0 \\ 0 & 0 & 1 & 0 & 0 & 0 & 0 \\ 0 & 0 & 0 & 1 & 0 & 0 & a_{27} \\ 0 & 0 & 0 & 0 & 1 & 0 & a_{37} \\ 0 & 0 & 0 & 0 & 0 & 1 & a_{47} \\ 0 & 0 & 0 & 0 & 0 & 0 & 1 \end{pmatrix} = A'.$$

To lie in $wU_{4,3}w^{-1} \cap U_{3,4}$ a matrix must have the form

$$w(b_{ij}) = \begin{pmatrix} 1 & 0 & 0 & 0 & 0 & 0 & b_{17} \\ 0 & 1 & 0 & 0 & 0 & 0 & 0 \\ 0 & 0 & 1 & 0 & 0 & 0 & 0 \\ 0 & 0 & 0 & 1 & 0 & 0 & 0 \\ 0 & 0 & 0 & 0 & 1 & 0 & 0 \\ 0 & 0 & 0 & 0 & 0 & 1 & 0 \\ 0 & 0 & 0 & 0 & 0 & 0 & 1 \end{pmatrix} = B'$$

and to lie in $wU_{4,3}w^{-1} \cap P_{3,4}$ it must have the form

$$w(a_{ij}) = \begin{pmatrix} 1 & a_{15} & a_{16} & 0 & 0 & 0 & a_{17} \\ 0 & 1 & 0 & 0 & 0 & 0 & 0 \\ 0 & 0 & 1 & 0 & 0 & 0 & 0 \\ 0 & 0 & 0 & 1 & 0 & 0 & a_{27} \\ 0 & 0 & 0 & 0 & 1 & 0 & a_{37} \\ 0 & 0 & 0 & 0 & 0 & 1 & a_{47} \\ 0 & 0 & 0 & 0 & 0 & 0 & 1 \end{pmatrix} = C'.$$

Therefore, choosing A', B', C' in a similar manner to the case of A, B, C shows that

$$P_{3,4} \cap wU_{4,3}w^{-1} = ((G_3 \times G_4) \cap wU_{4,3}w^{-1}) \cdot (U_{3,4} \cap wU_{4,3}w^{-1}).$$

From the matrix immediately preceding A'' in order that a matrix lies in $wG_4 \times G_3w^{-1} \bigcap U_{3,4}$ it must have the form

$$w(b_{ij}) = \begin{pmatrix} 1 & 0 & 0 & b_{12} & b_{13} & b_{14} & 0 \\ 0 & 1 & 0 & 0 & 0 & 0 & b_{57} \\ 0 & 0 & 1 & 0 & 0 & 0 & b_{67} \\ 0 & 0 & 0 & 1 & 0 & 0 & 0 \\ 0 & 0 & 0 & 0 & 1 & 0 & 0 \\ 0 & 0 & 0 & 0 & 0 & 1 & 0 \\ 0 & 0 & 0 & 0 & 0 & 0 & 1 \end{pmatrix} = B''$$

and to lie in to lie in $wG_4 \times G_3w^{-1} \bigcap P_{3,4}$ it must have the form the form

$$w(a_{ij}) = \begin{pmatrix} a_{11} & 0 & 0 & a_{12} & a_{13} & a_{14} & 0 \\ 0 & a_{55} & a_{56} & 0 & 0 & 0 & a_{57} \\ 0 & a_{65} & a_{66} & 0 & 0 & 0 & a_{67} \\ 0 & 0 & 0 & a_{22} & a_{23} & a_{24} & 0 \\ 0 & 0 & 0 & a_{32} & a_{33} & a_{34} & 0 \\ 0 & 0 & 0 & a_{42} & a_{43} & a_{44} & 0 \\ 0 & 0 & 0 & 0 & 0 & 0 & a_{77} \end{pmatrix} = C''.$$

Therefore, choosing A'', B'', C'' in a similar manner to the case of A, B, C shows that

$$P_{3,4} \bigcap wG_4 \times G_3w^{-1}$$

$$= ((G_3 \times G_4) \bigcap wG_4 \times G_3w^{-1}) \cdot (U_{3,4} \bigcap wG_4 \times G_3w^{-1}).$$

APPENDIX I

Galois descent of representations

This Appendix contains more detail than any reader might conceivable want concerning Shintani descent from Galois invariant complex irreducible representations of $GL_2\mathbb{F}_4$ to $GL_2\mathbb{F}_2$. In §1 and §2 explicit matrix formulae are given for the 4-dimensional and 5-dimensional Galois invariant irreducibles. In §3 matrix formulae are given for extensions of these representations to the semi-direct product of $GL_2\mathbb{F}_4$ with the Galois group $\mathrm{Gal}(\mathbb{F}_4/\mathbb{F}_2)$. §4 describes the characterisation of the Shintani correspondence for finite general linear groups [117]. §5 recalls the explicit Brauer induction formula which gives the Euler characteristic of the monomial resolution of the representations under consideration without my having to write down the entire bar-monomial resolution. §6 gives the data needed for the explicit Brauer induction formula of the semi-direct product representations. §7 and §8 give a "descent algorithm" which one applies to the monomial resolution of the semi-direct product extensions in order to obtain a monomial complex all of whose Line-stabilisers lie in the product of the Galois group with the subgroup of Galois-fixed points. In this simple example it is shown how the Euler characteristic of the monomial complex resulting from the "descent algorithm" is related to the outcome of Shintani descent. In §9 are explained the necessary and sufficient conditions on the integers $\dim_{\mathbb{C}}(V^{(H,\phi)})$ which ensure that a representation exists which is the Shintani correspondent of a Galois invariant irreducible V.

Note that knowing all the subspaces $V^{(H,\phi)}$ is sufficient to write down the bar-monomial resolution for the Shintani correspondent of V.

1. Subgroups and elements of A_5 via $PGL_2\mathbb{F}_4$

1.1. Let A_5 denote the alternating group consisting of even permutations of the set with five elements and let $GL_2\mathbb{F}_4$ denote the group of 2×2 invertible matrices with entries in the field with four elements.

243

In $GL_2\mathbb{F}_4$, if ξ is a cube root of unity in $\mathbb{F}_4$, then we have matrices

$$X_\xi = \begin{pmatrix} 0 & 1 \\ 1 & \xi \end{pmatrix}, X_\xi^2 = \begin{pmatrix} 0 & 1 \\ 1 & \xi \end{pmatrix}\begin{pmatrix} 0 & 1 \\ 1 & \xi \end{pmatrix} = \begin{pmatrix} 1 & \xi \\ \xi & \xi \end{pmatrix}$$

$$X_\xi^3 = \begin{pmatrix} 0 & 1 \\ 1 & \xi \end{pmatrix}\begin{pmatrix} 1 & \xi \\ \xi & \xi \end{pmatrix} = \begin{pmatrix} \xi & \xi \\ \xi & 1 \end{pmatrix}$$

$$X_\xi^4 = \begin{pmatrix} 0 & 1 \\ 1 & \xi \end{pmatrix}\begin{pmatrix} \xi & \xi \\ \xi & 1 \end{pmatrix} = \begin{pmatrix} \xi & 1 \\ 1 & 0 \end{pmatrix}$$

$$X_\xi^5 = \begin{pmatrix} 0 & 1 \\ 1 & \xi \end{pmatrix}\begin{pmatrix} \xi & 1 \\ 1 & 0 \end{pmatrix} = \begin{pmatrix} 1 & 0 \\ 0 & 1 \end{pmatrix}.$$

This matrix gives a cyclic permutation of the projective line over $\mathbb{F}_4$. In fact, as we shall see from the elements described below, the projective general linear group $PGL_2\mathbb{F}_4 = GL_2\mathbb{F}_4/\mathbb{F}_4^*$ is isomorphic to A_5. Acting via right multiplication on row vectors, $PGL_2\mathbb{F}_4$ permutes via the points of the projective line

$$\mathbb{P}^1(\mathbb{F}_4) = \{(0,1),(1,0),(1,\xi),(\xi,1),(1,1)\}.$$

For example, X_ξ yields the 5-cycle $((0,1),(1,\xi),(1,1),(\xi,1),(1,0))$.

The 2-Sylow subgroup of A_5 is the Klein 4-group V_4 generated by the images in $PGL_2\mathbb{F}_4$ of the matrices A and B given by

$$A = \begin{pmatrix} 0 & 1 \\ 1 & 0 \end{pmatrix}, \qquad B = \begin{pmatrix} 1 & \xi \\ \xi & 1 \end{pmatrix}$$

since $A^2 = I, B^2 = \xi \cdot I$ and

$$AB = \begin{pmatrix} 0 & 1 \\ 1 & 0 \end{pmatrix}\begin{pmatrix} 1 & \xi \\ \xi & 1 \end{pmatrix} = \begin{pmatrix} \xi & 1 \\ 1 & \xi \end{pmatrix} = \begin{pmatrix} 1 & \xi \\ \xi & 1 \end{pmatrix}\begin{pmatrix} 0 & 1 \\ 1 & 0 \end{pmatrix} = BA.$$

As even permutations of the projective line both A and B fix $(1,1)$ since

$$(1,1)A = (1,1)\begin{pmatrix} 0 & 1 \\ 1 & 0 \end{pmatrix} = (1,1),$$

$$(1,1)B = (1,1)\begin{pmatrix} 1 & \xi \\ \xi & 1 \end{pmatrix} = (\xi^2,\xi^2) = (1,1).$$

The 3-Sylow subgroup consists of the images of I, C, C^2 where

$$C = \begin{pmatrix} 0 & 1 \\ 1 & 1 \end{pmatrix}, C^2 = \begin{pmatrix} 1 & 1 \\ 1 & 0 \end{pmatrix}, C^3 = I.$$

Since the order of $PGL_2\mathbb{F}_4$ equals $\frac{(4^2-1)(4^2-4)}{3} = 60$ which is the order of A_5 and $PGL_2\mathbb{F}_4$ sits inside Σ_5 the above calculations with matrices show that $PGL_2\mathbb{F}_4 = A_5 \subset \Sigma_5$.

The subgroup A_4 has index five in A_5 and can be realised as the images of the matrices which fix $(1, 1)$ in the projective line. Also setting

$$Y = \begin{pmatrix} 1 & \xi \\ \xi^2 & 0 \end{pmatrix}$$

which satisfies

$$(1,1)Y = (1,1) \begin{pmatrix} 1 & \xi \\ \xi^2 & 0 \end{pmatrix} = (\xi, \xi) = (1,1)$$

and furthermore $Y^3 = I$. Finally Y normalises $V_4 = \langle A, B \rangle$ since we have

$$YAY^2 = \begin{pmatrix} 1 & \xi \\ \xi^2 & 0 \end{pmatrix} \begin{pmatrix} 0 & 1 \\ 1 & 0 \end{pmatrix} \begin{pmatrix} 0 & \xi \\ \xi^2 & 1 \end{pmatrix}$$

$$= \begin{pmatrix} \xi & 1 \\ 0 & \xi^2 \end{pmatrix} \begin{pmatrix} 0 & \xi \\ \xi^2 & 1 \end{pmatrix}$$

$$= \xi AB,$$

$$YBY^2 = \begin{pmatrix} 1 & \xi \\ \xi^2 & 0 \end{pmatrix} \begin{pmatrix} 1 & \xi \\ \xi & 1 \end{pmatrix} \begin{pmatrix} 0 & \xi \\ \xi^2 & 1 \end{pmatrix}$$

$$= \begin{pmatrix} \xi & 0 \\ \xi^2 & 1 \end{pmatrix} \begin{pmatrix} 0 & \xi \\ \xi^2 & 1 \end{pmatrix}$$

$$= \xi^2 A.$$

Hence

$$A_4 = \langle Y, A, B \rangle \subseteq A_5.$$

By Sylow's theorem $\langle Y \rangle$ and $\langle C \rangle$ are conjugate in A_5. Explicitly we have

$$\begin{pmatrix} 1 & 0 \\ \xi^2 & \xi^2 \end{pmatrix} C = \begin{pmatrix} 0 & 1 \\ \xi^2 & 0 \end{pmatrix} = Y \begin{pmatrix} 1 & 0 \\ \xi^2 & \xi^2 \end{pmatrix}.$$

One easily verifies that

$$Y = X_\xi^2 BCBX_\xi^3 \text{ and } X_\xi = C^2 BC^2.$$

Since A_5 has order 60 its proper subgroups must have orders in the set $\{2, 3, 4, 5, 6, 10, 12, 20\}$. In fact there is no subgroup of order 20. For suppose that a subgroup contains $V_4 = \langle A, B \rangle$ and a 5-cycle. Then conjugating the elements of V_4 by the 5-cycle and multiplying the results by A, B, AB one finds that the subgroup must also contain a 3-cycle and hence equals A_5. Similarly, there is no subgroup of order 30.

The following table shows all the conjugacy class representatives of subgroups of A_5.

1.2. *Conjugacy classes of subgroups H of A_5*

H	Order	Generators	Number in conjugacy class
A_5	60	A, B, Y, X_ξ	1
A_4	12	A, B, Y	5
D_{10}	10	X_ξ, A	6
D_6	6	A, C	10
C_5	5	X_ξ	6
V_4	4	A, B	5
C_3	3	C	10
C_2	2	A	15
$\{1\}$	1	I	1

A simple argument using Sylow's theorems shows that each subgroup of A_5 is determined up to conjugacy by its order.

The classification of irreducible, finite-dimensional complex representations of $GL_2\mathbb{F}_4$ given in ([126] §3.2 p. 89) shows that there are five irreducible representations of the quotient group $PGL_2\mathbb{F}_4 \cong A_5$. Therefore there are five conjugacy classes of elements of A_5 of orders $1, 2, 3, 5$ and 5. To see that there are two distinct conjugacy classes of order 5 observe that only one conjugacy class implies either that C_5 is normal in A_5 or there is a subgroup of order 20 or 30.

In the character table of A_5 given below the conjugacy classes of elements are labelled $1, 2, 3, 5^1$ and 5^2 and are represented by elements having orders $1, 2, 3, 5$ and 5, respectively. The α_i are real numbers given by $\alpha_1 = (1 + \sqrt{5})/2$ and $\alpha_2 = (1 - \sqrt{5})/2$.

1.3. Character table for A_5

	1	2	3	5^1	5^2
1	1	1	1	1	1
$\nu_{3,1}$	3	-1	0	α_1	α_2
$\nu_{3,2}$	3	-1	0	α_2	α_1
ν_4	4	0	1	-1	-1
ν_5	5	1	-1	0	0

2. Complex irreducible representations of A_5

2.1. The action of $\mathrm{Gal}(\mathbb{F}_4/\mathbb{F}_2)$

Let σ denote the Frobenius automorphism of $\mathbb{F}_4$ given by $\sigma(z) = z^2$. Applying σ to the matrix entries given an involution on $GL_2\mathbb{F}_4$ and its quotient $PGL_2\mathbb{F}_4 \cong A_5$. Therefore, if ρ is a finite-dimensional complex irreducible representation of A_5 then so is $\sigma^*(\rho)$, the composition of ρ with the σ.

It is straightforward to verify that σ applied to the conjugacy class 5^1 gives 5^2 so that $\sigma^*(\nu_{3,1}) = \nu_{3,2}$. On the other hand, since σ^* preserves dimension, we must have

$$\sigma^*(1) = 1, \ \sigma^*(\nu_4) = \nu_4 \text{ and } \sigma^*(\nu_5) = \nu_5.$$

2.2. Explicit models for ν_4 and ν_5

The Borel subgroup of upper triangular matrices in $GL_2\mathbb{F}_4$ has order 36 so its image in $PGL_2\mathbb{F}_4$ has order 12 so is conjugate to A_4. Denote by B the image of the Borel subgroup in $PGL_2\mathbb{F}_4$ and also, when there is no confusion, the Borel subgroup of $GL_2\mathbb{F}_4$. This enables us to describe the irreducible representation ν_4 and ν_5 explicitly, following the description of irreducibles given in ([**126**] §3.2 p. 89).

There is a short exact sequence representations of $GL_2\mathbb{F}_4$ of the form

$$0 \longrightarrow \nu_4 = S(1) \longrightarrow \mathrm{Ind}_B^{GL_2\mathbb{F}_4}(1) \longrightarrow L(1) \longrightarrow 0$$

where $L(1) = 1$, the one-dimensional trivial representation, and $S(1)$ is irreducible. Each representation in the short exact sequence factorises through $A_5 = PGL_2\mathbb{F}_4$ and $S(1)$ factorises through the irreducible representation ν_4.

If λ is a non-trivial character of $\mathbb{F}_4^*$ we know from ([**126**] §3.2 p. 89) that $\mathrm{Ind}_B^{GL_2\mathbb{F}_4}(\mathrm{Inf}_T^B(\lambda \otimes \lambda^2))$ is irreducible. Also $\lambda \otimes \lambda^2$ is trivial on the scalar matrices since $\lambda^3 = 1$ so that $\lambda \otimes \lambda^2$ factorises to give a non-trivial character which is conjugate to $\phi : B \cong A_4 \to A_4/V_4 \to \mathbb{C}^*$. This irreducible factorises through

$$\nu_5 = \mathrm{Ind}_B^{PGL_2\mathbb{F}_4}(\phi).$$

2.3. Bases for ν_4 and ν_5

If $\mu : B \to \mathbb{C}^*$ be a character. The standard basis for $\mathrm{Ind}_B^{PGL_2\mathbb{F}_4}(\mu)$ is

$$I \otimes_B 1, \ X_\xi \otimes_B 1, \ X_\xi^2 \otimes_B 1, \ X_\xi^3 \otimes_B 1, \ X_\xi^4 \otimes_B 1.$$

Define a basis V_1, W_1, W_2, W_3 for ν_4 by

$$V_1 = (I \otimes_B 1 - X_\xi^2 \otimes_B 1) - (I \otimes_B 1 - X_\xi^4 \otimes_B 1),$$

$$W_1 = I \otimes_B 1 - X_\xi \otimes_B 1,$$

$$W_2 = I \otimes_B 1 - X_\xi^3 \otimes_B 1,$$

$$W_3 = (I \otimes_B 1 - X_\xi^2 \otimes_B 1) + (I \otimes_B 1 - X_\xi^4 \otimes_B 1).$$

If $\xi_3 = e^{2\pi\sqrt{-1}/3}$ define a basis v, w_1, w_2, w_3, w_4 for ν_5 by

$$v = I \otimes_B 1 + X_\xi \otimes_B 1 + \xi_3 X_\xi^3 \otimes_B 1$$

$$w_1 = -\xi_3 I \otimes_B 1 - \xi_3 X_\xi \otimes_B 1 + 2\xi_3^2 X_\xi^3 \otimes_B 1 - X_\xi^4 \otimes_B 1 - X_\xi^2 \otimes_B 1$$

$$w_2 = (1 - \xi_3^2)I \otimes_B 1 + (\xi_3^2 - 1)X_\xi \otimes_B 1 - X_\xi^4 \otimes_B 1 + X_\xi^2 \otimes_B 1$$

$$w_3 = -\xi_3 I \otimes_B 1 - \xi_3 X_\xi \otimes_B 1 + 2\xi_3^2 X_\xi^3 \otimes_B 1 + 3X_\xi^4 \otimes_B 1 + 3X_\xi^2 \otimes_B 1$$

$$w_4 = (1 - \xi_3^2)I \otimes_B 1 + (\xi_3^2 - 1)X_\xi \otimes_B 1 + 3X_\xi^4 \otimes_B 1 - 3X_\xi^2 \otimes_B 1.$$

2.4. The A_5-action on ν_4

In terms of the ordered basis $\{V_1, W_1, W_2, W_3\}$ the matrices for the action of A, B, C, X_ξ on ν_4 are given by

$$A = \begin{pmatrix} -1 & 0 & 0 & 0 \\ 0 & -1 & -1 & -2 \\ 0 & 0 & 1 & 0 \\ 0 & 0 & 0 & 1 \end{pmatrix}$$

$$B = \begin{pmatrix} 0 & -1 & -\frac{1}{2} & -1 \\ -1 & 0 & 0 & 1 \\ 0 & 0 & 1 & 0 \\ 0 & 0 & -\frac{1}{2} & -1 \end{pmatrix}$$

$$C = \begin{pmatrix} 1 & 0 & 0 & 0 \\ 0 & -1 & -1 & -2 \\ 0 & 1 & 0 & 0 \\ 0 & 0 & 0 & 1 \end{pmatrix}$$

$$X_\xi = \begin{pmatrix} 0 & \frac{1}{2} & -\frac{1}{2} & 0 \\ 0 & -1 & -1 & -2 \\ 1 & 0 & 0 & 1 \\ 0 & \frac{1}{2} & \frac{1}{2} & 0 \end{pmatrix}.$$

2.5. The A_5-action on ν_5

In terms of the ordered basis $\{v, w_1, w_2, w_3, w_4\}$ the matrices for the action of A, B, C, X_ξ on ν_5 are given by

$$A = \begin{pmatrix} 1 & 0 & 0 & 0 & 0 \\ 0 & 1 & 0 & 0 & 0 \\ 0 & 0 & -1 & 0 & 0 \\ 0 & 0 & 0 & 1 & 0 \\ 0 & 0 & 0 & 0 & -1 \end{pmatrix}$$

$$B = \begin{pmatrix} \frac{1}{3} & 0 & 0 & \frac{8}{3}\xi_3 & 0 \\ 0 & 1 & 0 & 0 & 0 \\ 0 & 0 & 0 & 0 & \frac{\xi_3 - \xi_3^2}{3} \\ \frac{\xi_3^2}{3} & 0 & 0 & -\frac{1}{3} & 0 \\ 0 & 0 & \xi_3^2 - \xi_3 & 0 & 0 \end{pmatrix}$$

$$C = \begin{pmatrix} 1 & 0 & 0 & 0 & 0 \\ 0 & -\frac{1}{2} & \frac{1}{2} + \xi_3 & 0 & 0 \\ 0 & \frac{\xi_3}{2} - \frac{\xi_3^2}{2} & -\frac{1}{2} & 0 & 0 \\ 0 & 0 & 0 & -\frac{1}{2} & \frac{\xi_3}{2} - \frac{\xi_3^2}{2} \\ 0 & 0 & 0 & \frac{1}{2} + \xi_3 & -\frac{1}{2} \end{pmatrix}$$

$$X_\xi = \begin{pmatrix} \frac{1}{3} & 0 & 0 & -\frac{4\xi_3}{3} & \frac{8}{3} + \frac{4\xi_3}{3} \\ 0 & \frac{1}{4} & \frac{1}{4} + \frac{\xi_3}{2} & -\frac{1}{4} - \frac{\xi_3}{2} & -\frac{1}{4} \\ 0 & \frac{1}{4} + \frac{\xi_3}{2} & -\frac{3}{4} & -\frac{1}{4} & \frac{\xi_3}{12} - \frac{\xi_3^2}{12} \\ -\frac{\xi_3^2}{6} & \frac{3\xi_3}{4} - \frac{3\xi_3^2}{4} & \frac{3}{4} & -\frac{1}{12} & \frac{\xi_3^2}{12} - \frac{\xi_3}{12} \\ \frac{\xi_3}{6} - \frac{1}{6} & \frac{3}{4} & \frac{\xi_3^2}{4} - \frac{\xi_3}{4} & -\frac{1}{12} - \frac{\xi_3}{6} & \frac{1}{4} \end{pmatrix}.$$

3. Semi-direct products

3.1. Following the notational conventions of ([**125**] p. 36), if C acts on G via $\lambda : C \longrightarrow \mathrm{Aut}(G)$ then the semi-direct product $C \propto G$ is the group whose underlying set is $C \times G$ with multiplication given by

$$(c_1, g_1) \cdot (c_2, g_2) = (c_1 c_2, g_1 \lambda(c_1)(g_2)), \qquad c_i \in C, g_i \in G.$$

Let $\mathcal{H}$ denote a complex vector space and let $\rho : G \longrightarrow \mathrm{Aut}_{\mathbb{C}}(\mathcal{H})$ denote a representation of G on $\mathcal{H}$. For $c \in C$ denote by $c^*(\rho) : G \longrightarrow \mathrm{Aut}_{\mathbb{C}}(\mathcal{H})$ the representation of G given by the formula $c^*(\rho)(g)(h) = \lambda(c)(g)(h)$ for $c \in C, g \in G, h \in \mathcal{H}$.

Suppose that ρ is a representation for which Schur's Lemma holds; that is, $\mathrm{End}_{\mathbb{C}[G]}(\mathcal{H}) = \mathbb{C}$, the ring of scalar endomorphisms. Assume in addition that $c^*(\rho)$ and ρ are equivalent representations for each $c \in C$. Therefore for each $c \in C$ there exists

$$U_c \in \mathrm{Aut}_{\mathbb{C}}(\mathcal{H})$$

such that, for all $c \in C, g \in G$,

$$c^*(\rho)(g) = \rho(\lambda(c)(g)) = U_c \cdot \rho(g) \cdot U_c^{-1} \in \mathrm{Aut}_{\mathbb{C}}(\mathcal{H}).$$

If $V_c \in \mathrm{Aut}_{\mathbb{C}}(\mathcal{H})$ satisfies $U_c \cdot \rho(g) \cdot U_c^{-1} = V_c \cdot \rho(g) \cdot V_c^{-1}$ then, by the Schur Lemma condition, $U_c = V_c \in \mathrm{Aut}_{\mathbb{C}}(\mathcal{H})/\mathbb{C}^* = \mathrm{ProjAut}_{\mathbb{C}}(\mathcal{H})$, the group of projective automorphisms of $\mathcal{H}$.

PROPOSITION 3.2.
Let G and C be as in §3.1. Let ρ be a representation of G for which Schur's Lemma holds. Assume in addition that $c^*(\rho)$ and ρ are equivalent representations for each $c \in C$. Then, in the notation of §3.1, there is a homomorphism of the form

$$\tilde{\rho} : C \ltimes G \longrightarrow \mathrm{ProjAut}_{\mathbb{C}}(\mathcal{H})$$

given by the formula $\tilde{\rho}(c, g) = \rho(g)U_c$.

Proof

Since U_g is unique in the group of projective automorphisms we have $U_g U_{g_1} = U_{gg_1}$ in this group and therefore

$$\tilde{\rho}(cc_1, g\lambda(c)(g_1)) = \rho(g\lambda(c)(g_1))U_{cc_1}$$

$$= \rho(g)\rho(\lambda(c)(g_1))U_c U_{c_1}$$

$$= \rho(g)U_c\rho(g_1)U_c^{-1}U_c U_{c_1}$$

$$= \rho(c)U_c\rho(g_1)U_{c_1}$$

$$= \tilde{\rho}(c, g)\tilde{\rho}(c_1, g_1).$$

$\square$

EXAMPLE 3.3. Let $G = PGL_2\mathbb{F}_4$ and let $C = \mathrm{Gal}(\mathbb{F}_4/\mathbb{F}_2)$ generated by the involution given by σ, the Frobenius automorphism. From §2.1 we know that

$$\sigma^*(1) = 1, \ \sigma^*(\nu_4) = \nu_4 \text{ and } \sigma^*(\nu_5) = \nu_5.$$

(i)   When $\rho = 1$, the trivial one-dimensional representation, then $U_\sigma = 1$ and the homomorphism of Proposition 3.2 is trivial. The trivial projective representation factors through each of the two one-dimensional representations $C \propto GL_2\mathbb{F}_4 \longrightarrow \mathbb{C}^*$ of the form

$$C \propto GL_2\mathbb{F}_4 \longrightarrow C \longrightarrow \mathbb{C}^*.$$

(ii) When $\rho = \nu_4$ we have $\mathcal{H} = \langle V_1, W_1, W_2, W_3 \rangle$. Let U_σ be the linear involution on $\mathcal{H}$ given by

$$U_\sigma(V_1) = V_1, U_\sigma(W_1) = -W_1, U_\sigma(W_2) = -W_2, U_\sigma(W_3) = -W_3.$$

With this choice of U_σ and $U_1 = 1$ the projective homomorphism $\tilde{\rho}$ of Proposition 3.2 lifts to a representation

$$\tilde{\nu}_4 : \mathrm{Gal}(\mathbb{F}_4/\mathbb{F}_2) \propto GL_2\mathbb{F}_4 \longrightarrow \mathrm{Aut}_{\mathbb{C}}(\mathcal{H})$$

given by $(c, g) \mapsto \nu_4(g)U_c$. This is easily verified using the relations $\sigma(A) = A, \sigma(C) = C, \sigma(X_\xi) = X_\xi AC, \sigma(B) = BA$ in $PGL_2\mathbb{F}_4$.

In terms of matrices σ acts on $\tilde{\nu}_4$ as

$$\sigma = \begin{pmatrix} 1 & 0 & 0 & 0 \\ 0 & -1 & 0 & 0 \\ 0 & 0 & -1 & 0 \\ 0 & 0 & 0 & -1 \end{pmatrix}.$$

The other lift of $\tilde{\rho}$ to a linear representation is the tensor product of $\tilde{\nu}_4$ with the non-trivial one-dimensional representation of the form

$$\mathrm{Gal}(\mathbb{F}_4/\mathbb{F}_2) \propto GL_2\mathbb{F}_4 \longrightarrow \mathrm{Gal}(\mathbb{F}_4/\mathbb{F}_2) \longrightarrow \mathbb{C}^*.$$

(iii) When $\rho = \nu_5$ we have $\mathcal{H} = \langle v, w_1, w_2, w_3, w_4 \rangle$. Let U_σ be the linear involution on $\mathcal{H}$ given by

$$U_\sigma(v) = v, U_\sigma(w_1) = -w_1, U_\sigma(w_2) = -w_2, U_\sigma(w_3) = w_3, U_\sigma(w_4) = w_4.$$

With this choice of U_σ and $U_1 = 1$ the projective homomorphism $\tilde{\rho}$ of Proposition 3.2 lifts to a representation[1]

$$\tilde{\nu}_5 : \mathrm{Gal}(\mathbb{F}_4/\mathbb{F}_2) \propto GL_2\mathbb{F}_4 \longrightarrow \mathrm{Aut}_{\mathbb{C}}(\mathcal{H})$$

given by $(c, g) \mapsto \nu_5(g)U_c$.

In terms of matrices σ acts on $\tilde{\nu}_5$ as

$$\sigma = \begin{pmatrix} 1 & 0 & 0 & 0 & 0 \\ 0 & -1 & 0 & 0 & 0 \\ 0 & 0 & -1 & 0 & 0 \\ 0 & 0 & 0 & 1 & 0 \\ 0 & 0 & 0 & 0 & 1 \end{pmatrix}.$$

The other linear lift of $\tilde{\rho}$ is constructed by tensoring with a non-trivial quadratic character, as in (ii).

---

[1] In this case we are fortunate to be able to construct $\tilde{\rho} = \tilde{\nu}_4, \tilde{\nu}_5$ with $E = K = \mathbb{F}_4$ in the notation of Chapter Two, §6.1 and its attendant footnote.

3.4. *Conjugacy classes of subgroups H of* $\mathrm{Gal}(\mathbb{F}_4/\mathbb{F}_2) \propto PGL_2\mathbb{F}_4$

The following table shows the conjugacy classes of subgroups H which are not conjugate to a subgroup of $PGL_2\mathbb{F}_4 \cong A_5$.

H	Order	Generators	Number in conjugacy class
$\langle(\sigma,1),A_5\rangle$	120	$(\sigma,1),A,B,Y,X_\xi$	1
$\langle(\sigma,1),A_4\rangle$	24	$(\sigma,1),A,B,Y$	5
$C_4 \propto C_5$	20	$(\sigma,B),X_\xi$	6
$C_2 \times D_6$	12	$(\sigma,1),A,C$	10
$\langle(\sigma,1),V_4\rangle$	8	$(\sigma,1),A,B$	15
D_6	6	$(\sigma,A),C$	10
$C_2 \times C_3$	6	$(\sigma,1),C$	10
C_4	4	(σ,B)	15
$C_2 \times C_2$	4	$(\sigma,1),A$	15
C_2	2	$(\sigma,1)$	10

To determine the subgroups $J \subseteq C_2 \propto PGL_2\mathbb{F}_4$ up to conjugacy whose projection $J \subseteq C_2 \propto PGL_2\mathbb{F}_4 \longrightarrow C_2$ is non-trivial we consider the kernel of the projection. This a subgroup J' of index two in J which we may assume is one appearing in the table of §1.2. A laborious analysis of the possibilities yields the results of the above table.

In addition, let us record the action of the Frobenius on $Y \in PGL_2\mathbb{F}_4$

$$\sigma(Y) = AY^2A.$$

4. The Shintani correspondence for $GL_n\mathbb{F}_{q^d}$

4.1. Let C denote the cyclic group of order d given by the Galois group of $\mathbb{F}_{q^d}/\mathbb{F}_q$ generated by the Frobenius automorphism, σ. Let $\mathrm{Irr}(GL_n\mathbb{F}_{q^d})^C$ denote the set of finite-dimensional, irreducible complex representations ρ of $GL_n\mathbb{F}_{q^d}$ such that $\sigma^*(\rho)$ is equivalent to ρ. Let $\mathrm{Irr}(GL_n\mathbb{F}_q)$ denote the set of finite-dimensional, irreducible complex representations of $GL_n\mathbb{F}_q$.

The Shintani correspondence [**117**] is a bijection of the form

$$\mathrm{Sh} : \mathrm{Irr}(GL_n\mathbb{F}_{q^d})^C \xrightarrow{\cong} \mathrm{Irr}(GL_n\mathbb{F}_q).$$

This correspondence is characterised in the following manner. Let $\chi_{\mathrm{Sh}(\rho)}$ denote the trace function of the irreducible representation $\mathrm{Sh}(\rho)$. Then there exists an irreducible linear representation $\tilde{\rho}$ of the semi-direct product $C \propto GL_n\mathbb{F}_{q^d}$ which is a lift of the projective homomorphism of Proposition 3.2. Let $\chi_{\tilde{\rho}}$ denote the trace function of $\tilde{\rho}$. This linear lift may be chosen in such a way that, for all $g \in GL_n\mathbb{F}_{q^d}$,

$$\epsilon\chi_{\mathrm{Sh}(\rho)}((\sigma,g)^d) = \chi_{\tilde{\rho}}(\sigma,g)$$

where $\epsilon = \pm 1$ is a sign which is independent of g. The left side of this relation is interpreted in the following manner. The $C \propto GL_n\mathbb{F}_{q^d}$-conjugacy

class of the element
$$(\sigma, g)^d = (1, g\sigma(g) \ldots \sigma^{d-1}(g))$$
intersects $GL_n\mathbb{F}_q$ in a unique $GL_n\mathbb{F}_q$-conjugacy class. The left side of the characterising relation denotes $\chi_{\mathrm{Sh}(\rho)}$ applied to any element of this $GL_n\mathbb{F}_q$-conjugacy class.

EXAMPLE 4.2. $GL_2\mathbb{F}_4$

The group $GL_2\mathbb{F}_2$ is the dihedral group of order six whose irreducible representation consist of two one-dimensions, 1 and χ and a two-dimensional irreducible ν. Therefore the Shintani correspondence takes the form (see §9.6)
$$\mathrm{Irr}(GL_2\mathbb{F}_4)^C = \{1, \nu_4, \nu_5\} \leftrightarrow \mathrm{Irr}(GL_2\mathbb{F}_2) = \{1, \nu, \chi\}.$$
Setting $g = 1$ in the characterising relation we find from Example 3.3(iii) that
$$\dim(\mathrm{Sh}(\nu_5)) = \chi_{\mathrm{Sh}(\nu_5)}((1,1)) = \chi_{\tilde{\nu}_5}(\sigma, 1) = 1.$$
Since $\mathrm{Sh}(1) = 1$ we must have $\mathrm{Sh}(\nu_5) = \chi$ and $\mathrm{Sh}(\nu_4) = \nu$. This agrees with Example 3.3(ii) since $\chi_{\tilde{\nu}_4}(\sigma, 1) = \pm 2$ for the two choices of U_σ.

5. Explicit Brauer Induction a_G

The homomorphism a_G is an explicit formula for Brauer's Induction Theorem, discovered by Robert Boltje [17]. The first such explicit formula (a derivation rather than a homomorphism) appeared in [122] and a topological formula for a_G, analogous to that of [122], was given by Peter Symonds [134]. The material of this section and its notation is taken from [126].

DEFINITION 5.1. Let G be a finite group and let $R_+(G)$ denote the free abelian group on G-conjugacy classes of characters, $\phi : H \longrightarrow \mathbb{C}^*$, where $H \subseteq G$. We shall denote this character by (H, ϕ) and its G-conjugacy class by $(H, \phi)^G \in R_+(G)$.

If $J \subseteq G$ we define a restriction homomorphism $\mathrm{Res}_J^G : R_+(G) \longrightarrow R_+(J)$ by the double coset formula ([126] p. 32)
$$\mathrm{Res}_J^G((H, \phi)^G) = \sum_{z \in J \backslash G / H} (J \cap zHz^{-1}, (z^{-1})^*(\phi))^J$$
where $(z^{-1})^*(\phi)(u) = \phi(z^{-1}uz) \in \mathbb{C}^*$. If $\pi : J \longrightarrow G$ is a surjection we define an inflation homomorphism $\pi^* : R_+(G) \longrightarrow R_+(J)$ by $\pi^*((H, \phi)^G) = (\pi^{-1}(H), \phi\pi)^J$. These maps make $R_+(-)$ into a contravariant functor from finite groups to abelian groups.

Define a homomorphism $b_G : R_+(G) \longrightarrow R(G)$ by $b_G((H, \phi)^G) = \mathrm{Ind}_H^G(\phi)$, the representation of G obtained from ϕ by induction. Then b_G is surjective because $b_G \cdot a_G = 1$.

5.2. *Axioms for* a_G The homomorphism a_G is uniquely characterised by functoriality and a normalisation property on one-dimensional characters of G.

(i) For $H \leq G$ the following diagram commutes.

$$
\begin{array}{ccc}
R(G) & \xrightarrow{\ a_G\ } & R_+(G) \\[2pt]
\Big\downarrow {\scriptstyle Res_H^G} & & \Big\downarrow {\scriptstyle Res_H^G} \\[18pt]
R(H) & \xrightarrow{\ a_H\ } & R_+(H)
\end{array}
$$

(ii) Let $\rho : G \longrightarrow GL_n(\mathbf{C})$ be a representation and suppose that

$$
a_G(\rho) = \sum \alpha_{(H,\phi)^G}(H,\phi)^G \in R_+(G)
$$

then $\alpha_{(G,\phi)^G} = \langle \rho, \phi \rangle$ for each $(H,\phi)^G$ such that $H = G$. In particular, if ρ is one-dimensional then $a_G(\rho) = (G,\rho)^G$.

5.3. *The formula for* $a_G(\rho)$
The formula for $a_G(\rho)$ is given by ([**126**] Theorem 2.3.15 p. 48)

$$
a_G(\rho)
$$

$$
= \tfrac{1}{|G|} \sum\nolimits_{(H,\phi) \leq (H',\phi') \text{ in } \mathcal{M}_G} |H| \mu^{\mathcal{M}_G}_{(H,\phi),(H',\phi')} \cdot \langle \phi', Res_{H'}^G(\rho) \rangle_{H'} \cdot (H,\phi)^G.
$$

Here $\langle \phi', Res_{H'}^G(\rho) \rangle_{H'}$ is the Schur inner product of ϕ' and the restriction of ρ as representations of H' and $\mathcal{M}_G$ denotes the poset of pairs (*not* G-conjugacy classes of pairs) (H,ϕ). The Möbius function of the ordered pair $((H,\phi),(H',\phi'))$ in $\mathcal{M}_G$ is the integer defined by the alternating sum of the number chains in $\mathcal{M}_G$ from (H,ϕ) to (H',ϕ')

$$
\mu^{\mathcal{M}_G}_{(H,\phi),(H',\phi')}
$$

$$
= \sum\nolimits_i (-1)^i \#\{\text{chains of length i with } (H_0,\phi_0)
$$

$$
= (H,\phi), (H_i,\phi_i) = (H',\phi')\}.
$$

A chain of length i is a totally ordered subset of $\mathcal{M}_G$ of the form

$$
(H_0,\phi_0) \overset{<}{\neq} (H_1,\phi_1) \overset{<}{\neq} \ldots \overset{<}{\neq} (H_i,\phi_i).
$$

6. Explicit Brauer Induction data for $C_2 \propto PGL_2\mathbb{F}_4$

6.1. $\mathcal{M}_{C_2 \propto PGL_2\mathbb{F}_4}$

In order to compute the coefficients in the formula of §5.3 for

$$a_{PGL_2\mathbb{F}_4}(\nu_4) \quad a_{PGL_2\mathbb{F}_4}(\nu_5) \qquad\qquad a_{\mathrm{Gal}(\mathbb{F}_4/\mathbb{F}_2)\propto PGL_2\mathbb{F}_4}(\tilde{\nu}_4)$$

and $\qquad a_{\mathrm{Gal}(\mathbb{F}_4/\mathbb{F}_2)\propto PGL_2\mathbb{F}_4}(\tilde{\nu}_5)$

we need to tabulate all the $\mathrm{Gal}(\mathbb{F}_4/\mathbb{F}_2) \propto PGL_2\mathbb{F}_4$-conjugacy classes of pairs $(H, \lambda) \in \mathcal{M}_{\mathrm{Gal}(\mathbb{F}_4/\mathbb{F}_2)\propto PGL_2\mathbb{F}_4}$. This are given in the following table where $\lambda \sim \lambda'$ indicates conjugacy, $\xi_n = e^{2\pi\sqrt{-1}/n}$ and $\hat{H} = \mathrm{Hom}(H, \mathbb{C}^*)$, the group of characters of H.

H	$\hat{H}$	formulae
A_5	1	$-$
A_4	$1, \phi \sim \phi^2$	$\phi(Y) = \xi_3,$ $\phi(A^i B^j) = 1$
D_{10}	$1, \phi$	$\phi(A) = -1,$ $\phi(X_\xi) = 1$
D_6	$1, \phi$	$\phi(A) = -1,$ $\phi(C) = 1$
C_5	$1, \phi \sim \phi^4 \sim \phi^2 \sim \phi^3$	$\phi(X_\xi) = \xi_5$
V_4	$1, \mu_1 \sim \mu_2 \sim \mu_3$	$\mu_1(A) = -1,$ $\mu_1(B) = 1$
C_3	$1, \phi \sim \phi^2$	$\phi(C) = \xi_3$
C_2	$1, \phi$	$\phi(A) = -1$
$\{1\}$	1	$-$
$\langle(\sigma,1)\rangle \cong C_2$	$1, \tau$	$\tau((\sigma,1)) = -1$
$\langle(\sigma,1), A\rangle \cong C_2 \times C_2$	$1, \tau, \phi \sim \tau\phi$	$\tau((\sigma,1)) = -1$ $-1 = \phi(A)$

The table is continued on the next page.

H	$\hat{H}$	formulae
C_4	$1, \phi \sim \phi^3, \phi^2$	$\phi((\sigma, B)) = \xi_4$
$\langle(\sigma,1),C\rangle \cong C_2 \times C_3$	$1, \tau, \phi \sim \phi^2, \tau\phi \sim \tau\phi^2$	$\tau((\sigma,1)) = -1,$ $\phi(C) = \xi_3$
$\langle(\sigma,A),C\rangle \cong D_6$	$1, \tau$	$\tau((\sigma,A)) = -1,$ $\tau(C) = 1$
$\langle(\sigma,1),V_4\rangle$	$1, \tau, \mu, \tau\mu$	$\tau((\sigma,1)) = -1$ $-1 = \mu(A^i B)$
$\langle(\sigma,1),A,C\rangle \cong C_2 \times D_6$	$1, \tau, \phi, \tau\phi$	$\tau((\sigma,1)A^i) = -1$ $\phi((\sigma,1)^i A) = -1$
$\langle(\sigma,B),X_\xi\rangle \cong C_4 \propto C_5$	$1, \tau, \tau^2, \tau^3$	$\tau((\sigma,B)) = \sqrt{-1},$ $\tau(X_\xi) = 1$
$\langle(\sigma,1),A_4\rangle$	$1, \tau, \phi \sim \phi^2, \tau\phi \sim \tau\phi^2$	$\tau((\sigma,1)) = -1,$ $\phi(Y) = \xi_3,$ $\phi(A^i B^j) = 1$
$\langle(\sigma,1),A_5\rangle$	$1, \tau$	$\tau((\sigma,1)) = -1$

Note: (A_4,ϕ) and (A_4,ϕ^2) are conjugate in $C_2 \propto PGL_2\mathbb{F}_4$ (as is seen from the relation at the end of §3.4) but *not* in $PGL_2\mathbb{F}_4$. Similarly (C_5,ϕ^i) for $i = 1,2,3,4$ are all conjugate in $C_2 \propto PGL_2\mathbb{F}_4$ (via (σ,B)) but only $(C_5,\phi^i) \sim (C_5,\phi^{4i})$ in $PGL_2\mathbb{F}_4$.

6.2. $\mathrm{Res}_H^{C_2 \propto PGL_2\mathbb{F}_4}(\tilde{\nu}_4)_{ab}$ and $\mathrm{Res}_H^{C_2 \propto PGL_2\mathbb{F}_4}(\tilde{\nu}_5)_{ab}$

In order to compute the Schur inner products which appear in the formula of §5.3 for $\rho = \tilde{\nu}_4$ and $\rho = \tilde{\nu}_5$ we need to know the multiplicities of each of characters of $\hat{H}$ which appear in the restriction of $\tilde{\nu}_4$ and $\tilde{\nu}_5$ to H. We denote the sum of these one-dimensional representations (with multiplicities) by $\mathrm{Res}_H^{C_2 \propto PGL_2\mathbb{F}_4}(\tilde{\nu}_4)_{ab}$ and $\mathrm{Res}_H^{C_2 \propto PGL_2\mathbb{F}_4}(\tilde{\nu}_5)_{ab}$. These "abelian parts" of the restrictions to H are given in the table below.

H	$\tilde{\nu}_4$	$\tilde{\nu}_5$
A_5	0	0
A_4	1	$\phi + \phi^2$
D_{10}	0	1
D_6	$1 + \phi$	1
C_5	$\phi + \phi^2 + \phi^3 + \phi^4$	$1 + \phi + \phi^2 + \phi^3 + \phi^4$
V_4	$1 + \mu_1 + \mu_2 + \mu_3$	$2 \cdot 1 + \mu_1 + \mu_2 + \mu_3$
C_3	$2 \cdot 1 + \phi + \phi^2$	$1 + 2\phi + 2\phi^2$
C_2	$2 \cdot 1 + 2 \cdot \phi$	$3 \cdot 1 + 2\phi$
$\{1\}$	$4 \cdot 1$	$5 \cdot 1$
$\langle(\sigma, 1)\rangle \cong C_2$	$1 \cdot 1 + 3 \cdot \tau$	$3 \cdot 1 + 2\tau$
$\langle(\sigma, 1), A\rangle \cong C_2 \times C_2$	$2 \cdot \tau + \phi + \tau\phi$	$2 \cdot 1 + \tau + \phi + \tau\phi$
C_4	$\phi + \phi^3 + \phi^2 + 1$	$1 + 2\phi^2 + \phi + \phi^3$
$\langle(\sigma, 1), C\rangle \cong C_2 \times C_3$	$1 + \tau + \tau\phi + \tau\phi^2$	$1 + \phi + \tau\phi + \phi^2 + \tau\phi^2$
$\langle(\sigma, A), C\rangle \cong D_6$	$2 \cdot \tau$	1
$\langle(\sigma, 1), V_4\rangle$	$\tau + \tau\mu$	$1 + \mu + \tau$
$\langle(\sigma, 1), A, C\rangle \cong C_2 \times D_6$	$\tau + \phi$	1
$\langle(\sigma, B), X_\xi\rangle \cong C_4 \propto C_5$	0	τ^2
$\langle(\sigma, 1), A_4\rangle$	τ	0
$\langle(\sigma, 1), A_5\rangle$	0	0

6.3. $a_{PGL_2\mathbb{F}_4}(\nu_4)$ and $a_{PGL_2\mathbb{F}_4}(\nu_5)$

Recalling that $A_5 \cong PGL_2\mathbb{F}_4$, a calculation of the formulae for $a_{PGL_2\mathbb{F}_4}(\nu_{3,1})$ may be found in ([**126**] p. 50). Similar calculations using the data from the tables of §6.1 and §6.2 yield the following formulae:

$$a_{PGL_2\mathbb{F}_4}(\nu_4) = (A_4, 1)^{PGL_2\mathbb{F}_4} + (D_6, 1)^{PGL_2\mathbb{F}_4} + (D_6, \phi)^{PGL_2\mathbb{F}_4}$$
$$+ (C_5, \phi)^{PGL_2\mathbb{F}_4}$$

$$+ (C_5, \phi^2)^{PGL_2\mathbb{F}_4} - (C_3, 1)^{PGL_2\mathbb{F}_4} + (C_3, \phi)^{PGL_2\mathbb{F}_4}$$

$$+ (V_4, \mu_1)^{PGL_2\mathbb{F}_4} - (C_2, 1)^{PGL_2\mathbb{F}_4} - (C_2, \phi)^{PGL_2\mathbb{F}_4}$$

and

$$a_{PGL_2\mathbb{F}_4}(\nu_5) = (A_4, \phi)^{PGL_2\mathbb{F}_4} + (A_4, \phi^2)^{PGL_2\mathbb{F}_4} + (D_{10}, 1)^{PGL_2\mathbb{F}_4}$$
$$+ (D_6, 1)^{PGL_2\mathbb{F}_4}$$

$$+ (C_5, \phi)^{PGL_2\mathbb{F}_4} + (C_5, \phi^2)^{PGL_2\mathbb{F}_4}$$

$$+ (V_4, \mu_1)^{PGL_2\mathbb{F}_4} - 2(C_2, 1)^{PGL_2\mathbb{F}_4}.$$

6.4. $a_{C_2 \propto PGL_2\mathbb{F}_4}(\tilde{\nu}_4)$ and $a_{C_2 \propto PGL_2\mathbb{F}_4}(\tilde{\nu}_5)$

Setting $G = C_2 \propto PGL_2\mathbb{F}_4$, the formula of §5.3 together with the tabulated data of §6.1 and §6.2 yield the following form for $a_G(\tilde{\nu}_5)$, which follows from the detailed calculations of the Appendix of [**130**]:

$$a_G(\tilde{\nu}_5)$$

$$= (A_4, \phi)^G + (\langle (\sigma, B), X_\xi \rangle, \tau^2)^G + (\langle (\sigma, 1), A, C \rangle, 1)^G$$

$$+ (\langle (\sigma, 1), V_4 \rangle, 1)^G + (\langle (\sigma, 1), V_4 \rangle, \tau)^G + (\langle (\sigma, 1), V_4 \rangle, \mu)^G$$

$$+ (\langle (\sigma, 1), C \rangle, \phi)^G + (\langle (\sigma, 1), C \rangle, \tau\phi)^G + (C_5, \phi)^G$$

$$- (V_4, 1)^G - (C_3, \phi)^G - (C_2, \phi)^G$$

$$+ (\{1\}, 1)^G - (\langle (\sigma, 1) \rangle, 1)^G - (\langle (\sigma, 1) \rangle, \tau)^G - (\langle (\sigma, 1), A \rangle, 1)^G$$

$$+ (\langle (\sigma, 1), A \rangle, \phi)^G + (C_4, \phi)^G - (C_4, \phi^2)^G.$$

Similarly for $\tilde{\nu}_4$ we obtain

$$a_G(\tilde{\nu}_4)$$

$$= (\langle (\sigma, 1), A_4 \rangle, \tau)^G + (C_2 \times D_6, \tau)^G + (C_2 \times D_6, \phi)^G$$

$$+ (C_5, \phi)^G + (\langle (\sigma, 1), V_4 \rangle, \tau\mu)^G + (\langle (\sigma, 1), C \rangle, \tau\phi)^G$$

$$- (C_2, \phi)^G - (\langle (\sigma, 1), A \rangle, \tau)^G$$

$$+ (C_4, \phi)^G - (\langle (\sigma, 1), C \rangle, \tau)^G.$$

7. The weak descent algorithm

7.1. *Descent of representations*

In this section we are interested in the following construction. Suppose given a finite-dimensional irreducible, complex representation ρ of a group G which is invariant under under the action of a subgroup $C \subseteq \mathrm{Aut}(G)$. This gives rise to a projective representation

$$\tilde{\rho} : C \propto G \longrightarrow \mathrm{ProjAut}_{\mathbb{C}}(\mathcal{H})$$

where $\mathcal{H}$ is the underlying vector space of ρ.

Suppose that there is a linear lift of $\tilde{\rho}$ (which we shall also denote by $\tilde{\rho}$)

$$\tilde{\rho} : C \propto G \longrightarrow \mathrm{GL}(\mathcal{H}).$$

There are several choices of this linear lift but any two will differ by a twist by a one-dimensional representation of the form $\tau : C \propto G \longrightarrow C \longrightarrow \mathbb{C}^*$.

The same is true for $a_{C \propto G}(\tilde{\rho}) \in R_+(C \propto G)$ since $a_{C \propto G}$ commutes with twisting by one-dimensional representations.

Let G^C denote the subgroup of C-fixed points of G. Consider the endomorphism $\mathcal{E}$ of $R_+(C \propto G)$ which is given on the free generators by

$$\mathcal{E}(H, \phi)^{C \propto G} = \begin{cases} (H, \phi)^{C \propto G} & \text{if H is conjugate to } H' \subseteq C \times G^C \\ 0 & \text{otherwise.} \end{cases}$$

Therefore we may write

$$\mathcal{E}(a_{C \propto G}(\tilde{\rho})) = \sum_{H \subseteq C \times G^C} \alpha_{(H, \phi)^{C \propto G}} \cdot (H, \phi)^{C \propto G},$$

which is well-defined up to twists by a one-dimensional representations τ. Therefore

$$\mathcal{E}(a_{C \propto G}(\tilde{\rho})) \in \text{Im}(\text{Ind}_{C \times G^C}^{C \propto G} : R_+(C \times G^C) \longrightarrow R_+(C \propto G)),$$

well-defined up to one-dimensional twists.

Similarly in terms of representations

$$b_{C \propto G}(\mathcal{E}(a_{C \propto G}(\tilde{\rho}))) \in \text{Im}(\text{Ind}_{C \times G^C}^{C \propto G} : R(C \times G^C) \xrightarrow{} R(C \propto G)),$$

well-defined up to one-dimensional twists.

We have a commutative diagram

$$
\begin{array}{ccc}
R(C \times G^C) & \xrightarrow{\ \ \text{Ind}_{C \times G^C}^{C \propto G}\ \ } & R(C \propto G) \\[2mm]
\Big\downarrow{\scriptstyle \text{Res}_{G^C}^{C \times G^C}} & & \Big\downarrow{\scriptstyle \text{Res}_G^{C \propto G}} \\[2mm]
R(G^C) & \xrightarrow{\ \ \text{Ind}_{G^C}^G\ \ } & R(G)
\end{array}
$$

so that

$$\text{Res}_G^{C \propto G}(b_{C \propto G}(\mathcal{E}(a_{C \propto G}(\tilde{\rho})))) \in \text{Im}(\text{Ind}_{G^C}^G : R(G^C) \longrightarrow R(G))$$

is a well-defined element depending only on ρ.

For example, if C is cyclic then the linear lift always exists.

Finally, if $\text{Ind}_{G^C}^G$ is injective, we have a construction of the form

$$\text{Irr}(G)^C \longrightarrow R(G^C).$$

The following result, proved in Appendix I §11, will be required in Example 7.3.

LEMMA 7.2. (i) The induction homomorphism

$$\mathrm{Ind}_{D_6}^{PGL_2\mathbb{F}_4} : R(D_6) \longrightarrow R(PGL_2\mathbb{F}_4)$$

is injective.

(ii) The kernel of the induction homomorphism

$$\mathrm{Ind}_{C_2 \times D_6}^{C_2 \propto PGL_2\mathbb{F}_4} : R(C_2 \times D_6) \longrightarrow R(C_2 \propto PGL_2\mathbb{F}_4)$$

is equal to $\langle (1-\tau) \otimes (1+2\chi+3\nu) \rangle$ where ν is the 2-dimensional irreducible representation of D_6 and τ, χ are the non-trivial quadratic characters of C_2, D_6, respectively.

EXAMPLE 7.3. $G = GL_2\mathbb{F}_4$ and $C = \mathrm{Gal}(\mathbb{F}_4/\mathbb{F}_2)$

In this example $G^C = GL_2\mathbb{F}_2 = PGL_2\mathbb{F}_2 \cong D_6$. By Lemma 7.2

$$\mathrm{Ind}_{PGL_2\mathbb{F}_2}^{PGL_2\mathbb{F}_4} : R(PGL_2\mathbb{F}_2) \longrightarrow R(PGL_2\mathbb{F}_4)$$

is injective and so is the analogous homomorphism for GL_2.

From the formulae of §6.4 one finds that

$$b_{C \propto G}(\mathcal{E}(a_{C_2 \propto G}(\tilde{\nu}_5))) = \mathrm{Ind}_{C_2 \times C^C}^{C_2 \propto G}(-\tau \otimes \chi - (1+\tau) \otimes \nu)$$

and

$$b_{C \propto G}(\mathcal{E}(a_{C_2 \propto G}(\tilde{\nu}_4))) = \mathrm{Ind}_{C_2 \times C^C}^{C_2 \propto G}(-(1+\tau) \otimes \nu - \tau \otimes (1+2\chi)).$$

These elements are only determined by ν_4 and ν_5 up to tensoring with τ so to obtain elements determined which are uniquely by ν_4 and ν_5 we should form

$$(1+\tau) \cdot b_{C \propto G}(\mathcal{E}(a_{C_2 \propto G}(\tilde{\nu}_5))) = \mathrm{Ind}_{C_2 \times C^C}^{C_2 \propto G}(-(1+\tau) \otimes (\chi + 2 \otimes \nu))$$

and

$$(1+\tau) \cdot b_{C \propto G}(\mathcal{E}(a_{C_2 \propto G}(\tilde{\nu}_4))) = \mathrm{Ind}_{C_2 \times C^C}^{C_2 \propto G}(-(1+\tau) \otimes (2\nu + 1 + 2\chi)).$$

Note that, by Lemma 7.2(ii), the homomorphism

$$(1+\tau) \cdot \mathrm{Ind}_{C_2 \times D_6}^{C_2 \propto PGL_2\mathbb{F}_4} : R(C_2 \times D_6) \longrightarrow R(C_2 \propto PGL_2\mathbb{F}_4)$$

is injective.

REMARK 7.4. If we were to apply a similar algorithm — deleting terms which are not subconjugate to $PGL_2\mathbb{F}_2$ — to $a_{PGL_2\mathbb{F}_4}(\nu_4)$ and $a_{PGL_2\mathbb{F}_4}(\nu_5)$ we would obtain the following results. The irreducible ν_5 would yield $-1-2\nu$ and for ν_4 we would obtain $-1-\chi-\nu$. Applying $(1+\tau) \cdot \mathrm{Ind}_{C_2 \times D_6}^{C_2 \propto PGL_2\mathbb{F}_4}$ to these elements yields

$$(1+\tau) \otimes (-1-2\nu) \text{ and } (1+\tau) \otimes (-1-\chi-\nu),$$

respectively.

Subtracting these elements from the elements of §7.3 yields

$$\nu_5 \mapsto -(1+\tau) \otimes (\chi - 1) \text{ and } \nu_4 \mapsto -(1+\tau) \otimes (\nu + \chi).$$

It is interesting to compare these values with the Shintani correspondence of §4.2.

8. The strong descent algorithm

8.1. *Monomial complexes*

This section is merely a sketch and is suitable only for those familiar with the monomial complexes and monomal resolutions of [19] (see also [128]). In the derived category of monomial complexes there exists a unique monomial resolution for $\tilde{\rho}$ which possesses an Euler characteristic equal to $a_{C \rtimes G}(\tilde{\rho}) \in R_+(C \propto G)$. The Lines of the form $\underline{\text{Ind}}_H^{C \rtimes G}(\phi)$ with H not subconjugate to $C \times G^C$ form a sub-monomial complex. The quotient monomial complex has an Euler characteristic in

$$\text{Im}(\text{Ind}_{C \times G^C}^{C \rtimes G} : R_+(C \times G^C) \longrightarrow R_+(C \propto G))$$

which is the one featured in §7.1 and Example 7.2.

Since I have not computed the monomial resolution in this paper I cannot give here the monomial complex computation which is analogous to §7.

REMARK 8.2. The induction map

$$\text{Ind}_H^G : R_+(H) \longrightarrow R_+(G)$$

is not injective in general for finite groups since $(J, \phi) - (xJx^{-1}, x^*(\phi))$ maps to zero but could be non-zero if $x \in G - H$ and $xJx^{-1} \subseteq H$.

8.3. *Historical note*

Most of this Appendix (§§1–10) was written several years ago. I still have not written down the monomial resolutions M_* for the complex representations $\tilde{\nu}_4$ and $\tilde{\nu}_5$ of $\text{Gal}(\mathbb{F}_4/\mathbb{F}_2) \propto PGL_2\mathbb{F}_4$. However, from the calculation of $a_G(\tilde{\nu}_i)$ in §7, one may calculate the invariants given by the Euler characteristic of $M_*^{((H,\lambda))}$ for each (H, λ).

For use in that calculation, which appears in Chapter Two §§7–8, I have added an extra section (§11) to this Appendix. It contains several tables of $(-)^{((H,\lambda))}$ data and their derivation.

9. The role of the integers $\dim_{\mathbb{C}}(V^{(H,\phi)})$ in Shintani descent

9.1. In the context of finite groups and complex representations we have

$$\dim_{\mathbb{C}}(V^{(H,\phi)}) = \dim_{\mathbb{C}}(\text{Hom}_G(\text{Ind}_H^G(\phi), V)) = \frac{1}{|H|} \sum_{h \in H} \phi(h)^{-1} \chi_V(h).$$

The left-hand factor in the tensor product which features in the bar-monomial resolution is the direct sum of vector spaces of the form $V^{(H,\phi)}$. In the Shintani base change situation the formula for the base-change character values gives the $\dim_{\mathbb{C}}(V^{(H,\phi)})$'s for the base-change correspondent representation.

If we knew that these values come from an actual representation (or even a virtual one) we could form the bar-monomial resolution for it. Therefore the existence of the Shintani correspondents, as described in this Appendix §4.1, is equivalent to an affirmative answer to the following question:

QUESTION 9.2. Suppose that ρ is a complex irreducible representation lying in the Galois invariants

$$\rho \in \mathrm{Irr}(GL_n\mathbb{F}_{q^d})^{\mathrm{Gal}(\mathbb{F}_{q^d}/\mathbb{F}_q)}.$$

Does there exist a representation V of $GL_n\mathbb{F}_q$ such that each integer $\dim_{\mathbb{C}}(V^{(H,\phi)})$ is as predicted by the character formula of [117]?

If such a V exists it will automatically be irreducible.

9.3. The necessary and sufficient condition for a set of candidate integers $\dim_{\mathbb{C}}(V^{(H,\phi)})$, as (H,ϕ) varies through $\mathcal{M}_{GL_n\mathbb{F}_q}$ to come from a virtual representation (i.e. a difference of two representations in the complex representation ring $R(GL_n\mathbb{F}_q) = K_0(\mathbb{C}[GL_n\mathbb{F}_q])$) is equivalent to the condition that of the rational numbers

$$\sum_{(H_0,\phi_0)<(H_1,\phi_1)<...<(H_r,\phi_r)} (-1)^r \frac{|H_0|}{|GL_n\mathbb{F}_q|} \dim_{\mathbb{C}}(V^{(H_r,\phi_r)})$$

is an integer. Here $(H_0,\phi_0) < (H_1,\phi_1) < \ldots < (H_r,\phi_r)$ runs through all strictly ascending chains in $\mathcal{M}_{GL_n\mathbb{F}_q}$. This is because these are the coefficients in the Explicit Brauer Induction formula of ([126] §2.3; see also [122], [17], [20], [102]) for V. The Explicit Brauer Induction formula gives an expression for V as an integer linear combination of induced representations of the form $\mathrm{Ind}_{H_r}^{GL_n\mathbb{F}_q}(k_{\phi_r})$. It is clearly the formula for a virtual representation.

The i-th exterior power of a representation λ^i satisfies the relation in $R(G)$

$$\lambda^i(V_1 \oplus V_2) = \sum_{j=0}^{i} \lambda^j(V_1)\lambda^{i-j}(V_2).$$

This allows one to define a formal power series [9]

$$\lambda_t(V) = \sum \lambda^i(V)t^i$$

with coefficients in $R(G)$ for any $V \in R(G)$. Also

$$\lambda_t(v_1 + v_2) = \lambda_t(v_1)\lambda_t(v_2)$$

for all $v_1, v_2 \in R(G)$.

I learnt the following result from the late Frank Adams.

THEOREM 9.4.

Let G be a finite group and let $V \in R(G)$. The necessary and sufficient condition that V is a representation (rather than a virtual representation) is that the power series $\lambda_t(V)$ is a polynomial.

REMARK 9.5. The remarks of this section constitute a potential alternative method for proving the Shintani correspondence of [117]. To my knowledge no one has attempted this method of proof, although the entry of $\lambda_t(V)$ might appeal to algebraic topologists!

EXAMPLE 9.6. *Galois descent for $GL_2\mathbb{F}_4$ revisited*

For $GL_2\mathbb{F}_4$ the Galois fixed subgroup is $GL_2\mathbb{F}_2 \cong D_6$, the dihedral group of order six. Its subgroups are given by the following table

H	$\hat{H}$	formulae
$D_6 = \langle A, C \rangle$	$1, \tau$	$\tau(A) = -1, \tau(C) = 1$
$C_3 = \langle C \rangle$	$1, \phi \sim \phi^2$	$\phi(C) = \xi_3$
$C_2 = \langle A \rangle$	$1, \tau$	$\tau(A) = -1$
$\{1\}$	1	$-$

The "abelian parts" of $\nu = \operatorname{Ind}_{C_3}^{D_6}(\phi)$ are given by

H	ν
D_6	0
C_3	$\phi + \phi^2$
C_2	$1 + \tau$
$\{1\}$	$2 \cdot 1$

This yields the formula (see Appendix I §5.3)

$$a_{D_6}(\nu) = (C_3, \phi)^{D_6} + (C_2, 1)^{D_6} + (C_2, \tau)^{D_6} - (\{1\}, 1)^{D_6} \in R_+(D_6).$$

We shall now show how the characterisation of the Shintani correspondence (see §4) determines the $\dim_k(\operatorname{Sh}(\nu_4)^{(H,\phi)})$'s and $\dim_k(\operatorname{Sh}(\nu_4)^{(H,\phi)})$'s.

Since, in the bar-monomial resolution for D_6 we have

$$S_{D_6} = 1 \oplus \tau \oplus \underline{\operatorname{Ind}}_{C_3}^{D_6}(1) \oplus \underline{\operatorname{Ind}}_{C_3}^{D_6}(\phi) \oplus \underline{\operatorname{Ind}}_{C_2}^{D_6}(1) \oplus \underline{\operatorname{Ind}}_{C_2}^{D_6}(\tau) \oplus \underline{\operatorname{Ind}}_{\{1\}}^{D_6}(1)$$

we are interested in the following seven dimensions for $i = 4, 5$:

$$\dim_k(\operatorname{Sh}(\nu_i)^{(D_6,1)}), \dim_k(\operatorname{Sh}(\nu_i)^{(D_6,\tau)}), \dim_k(\operatorname{Sh}(\nu_i)^{(C_3,1)}),$$

$$\dim_k(\operatorname{Sh}(\nu_i)^{(C_3,\phi)}), \dim_k(\operatorname{Sh}(\nu_i)^{(C_2,1)}),$$

$$\dim_k(\operatorname{Sh}(\nu_i)^{(C_2,\tau)}), \dim_k(\operatorname{Sh}(\nu_i)^{(\{1\},1)}).$$

We have a formula:

$$\pm\dim_k(\mathrm{Sh}(\nu_4)^{(H,\lambda)})$$

$$= \tfrac{1}{|H|} \sum_{h\in H} \lambda(h)^{-1}\mathrm{Trace}_{\mathrm{Sh}(\nu_4)}(h)$$

$$= \tfrac{1}{|H|} \sum_{h\in H, h=X\sigma(X)} \lambda(h)^{-1}\mathrm{Trace}_{\tilde{\nu}_4}(\sigma, X).$$

When $H = D_6$ and $\lambda = 1$ we have $1 = 1.\sigma(1)$, $A = B\sigma(B)$ and $C = C^2\sigma(C^2)$ we obtain

$$\pm\dim_k(\mathrm{Sh}(\nu_4)^{(D_6,1)})$$

$$= \tfrac{1}{6}(\mathrm{Trace}_{\tilde{\nu}_4}(\sigma,1) + 3\mathrm{Trace}_{\tilde{\nu}_4}(\sigma,B) + 2\mathrm{Trace}_{\tilde{\nu}_4}(\sigma,C^2))$$

$$= 6^{-1}((1-3) + 3 \times (1 + i + i^2 + i^3)$$

$$+ 2 \times (1 - 1 - \xi_3 - \xi_3^2))$$

$$= 6^{-1}(-2 + 2)$$

$$= 0.$$

When $H = D_6$ and $\lambda = \tau$ we obtain

$$\pm\dim_k(\mathrm{Sh}(\nu_4)^{(D_6,\tau)})$$

$$= \tfrac{1}{6}(\mathrm{Trace}_{\tilde{\nu}_4}(\sigma,1) + 3(-1)\mathrm{Trace}_{\tilde{\nu}_4}(\sigma,B) + 2\mathrm{Trace}_{\tilde{\nu}_4}(\sigma,C^2))$$

$$= 6^{-1}((1-3) - 3 \times (1 + i + i^2 + i^3)$$

$$+ 2 \times (1 - 1 - \xi_3 - \xi_3^2))$$

$$= 0.$$

When $H = C_3$ and $\lambda = 1$ we obtain

$$\pm\dim_k(\mathrm{Sh}(\nu_4)^{(C_3,1)})$$

$$= \tfrac{1}{3}(\mathrm{Trace}_{\tilde{\nu}_4}(\sigma,1) + 2\mathrm{Trace}_{\tilde{\nu}_4}(\sigma,C^2))$$

$$= 3^{-1}((1-3) + 2 \times (1 - 1 - \xi_3 - \xi_3^2))$$

$$= 0.$$

When $H = C_3$ and $\lambda = \phi$ we obtain

$$\pm\dim_k(\mathrm{Sh}(\nu_4)^{(C_3,\phi)})$$

$$= \tfrac{1}{3}(\mathrm{Trace}_{\tilde{\nu}_4}(\sigma,1) + (\xi_3 + \xi_3^2)\mathrm{Trace}_{\tilde{\nu}_4}(\sigma,C^2))$$

$$= 3^{-1}((1-3) + \xi_3(1 - 1 - \xi_3 - \xi_3^2))$$

$$+\xi_3^2(1 - 1 - \xi_3 - \xi_3^2))$$

$$= -1.$$

When $H = C_2$ and $\lambda = 1$ we obtain

$$\pm\dim_k(\mathrm{Sh}(\nu_4)^{(C_2,1)})$$

$$= \tfrac{1}{2}(\mathrm{Trace}_{\tilde{\nu}_4}(\sigma,1) + \mathrm{Trace}_{\tilde{\nu}_4}(\sigma,B))$$

$$= -1.$$

When $H = C_2$ and $\lambda = \tau$ we obtain

$$\pm\dim_k(\mathrm{Sh}(\nu_4)^{(C_2,\tau)})$$

$$= \tfrac{1}{2}(\mathrm{Trace}_{\tilde{\nu}_4}(\sigma,1) - \mathrm{Trace}_{\tilde{\nu}_4}(\sigma,B))$$

$$= -1.$$

When $H = \{1\}$ and $\lambda = 1$ we obtain

$$\pm\dim_k(\mathrm{Sh}(\nu_4)^{(\{1\},1)})$$

$$= (\mathrm{Trace}_{\tilde{\nu}_4}(\sigma,1)$$

$$= -2.$$

Next I shall do the same for ν_5.

$$\pm\dim_k(\mathrm{Sh}(\nu_5)^{(H,\lambda)})$$

$$= \tfrac{1}{|H|} \sum_{h\in H} \lambda(h)^{-1}\mathrm{Trace}_{\mathrm{Sh}(\nu_5)}(h)$$

$$= \tfrac{1}{|H|} \sum_{h\in H, h=X\sigma(X)} \lambda(h)^{-1}\mathrm{Trace}_{\tilde{\nu}_5}(\sigma,X).$$

When $H = D_6$ and $\lambda = 1$ we have $1 = 1.\sigma(1)$, $A = B\sigma(B)$ and $C = C^2\sigma(C^2)$ we obtain

$$\pm\dim_k(\mathrm{Sh}(\nu_5)^{(D_6,1)})$$

$$= \tfrac{1}{6}(\mathrm{Trace}_{\tilde{\nu}_5}(\sigma,1) + 3\mathrm{Trace}_{\tilde{\nu}_5}(\sigma,B) + 2\mathrm{Trace}_{\tilde{\nu}_5}(\sigma,C^2))$$

$$= 6^{-1}((3-2) + 3 \times (1 - 2 + i + i^3)$$

$$+ 2(1 + \xi_3 - \xi_3 + \xi_3^2 - \xi_3^2))$$

$$= 6^{-1}(1 - 3 + 2)$$

$$= 0.$$

When $H = D_6$ and $\lambda = \tau$ we obtain

$$\pm\dim_k(\mathrm{Sh}(\nu_5)^{(D_6,\tau)})$$

$$= \tfrac{1}{6}(\mathrm{Trace}_{\tilde{\nu}_5}(\sigma,1) - 3\mathrm{Trace}_{\tilde{\nu}_5}(\sigma,B) + 2\mathrm{Trace}_{\tilde{\nu}_5}(\sigma,C^2))$$

$$= 6^{-1}((3-2) - 3 \times (1 - 2 + i + i^3)$$

$$+ 2(1 + \xi_3 - \xi_3 + \xi_3^2 - \xi_3^2))$$

$$= 6^{-1}(1 + 3 + 2)$$

$$= 1.$$

When $H = C_3$ and $\lambda = 1$ we obtain

$$\pm\dim_k(\mathrm{Sh}(\nu_5)^{(C_3,1)})$$

$$= \tfrac{1}{3}(\mathrm{Trace}_{\tilde{\nu}_5}(\sigma,1) + 2\mathrm{Trace}_{\tilde{\nu}_5}(\sigma,C^2))$$

$$= 3^{-1}((3-2) + 2 \times (1 + \xi_3 - \xi_3 + \xi_3^2 - \xi_3^2))$$

$$= 1.$$

When $H = C_3$ and $\lambda = \phi$ we obtain

$$\pm\dim_k(\mathrm{Sh}(\nu_5)^{(C_3,\phi)})$$

$$= \tfrac{1}{3}(\mathrm{Trace}_{\bar{\nu}_5}(\sigma, 1) + (\xi_3 + \xi_3^2)\mathrm{Trace}_{\bar{\nu}_5}(\sigma, C^2))$$

$$= 3^{-1}((3-2) + \xi_3(1 + \xi_3 - \xi_3 + \xi_3^2 - \xi_3^2)$$

$$+ \xi_3^2(1 + \xi_3 - \xi_3 + \xi_3^2 - \xi_3^2))$$

$$= 3^{-1}(1-1)$$

$$= 0.$$

When $H = C_2$ and $\lambda = 1$ we obtain

$$\pm\dim_k(\mathrm{Sh}(\nu_5)^{(C_2,1)})$$

$$= \tfrac{1}{2}(\mathrm{Trace}_{\bar{\nu}_5}(\sigma, 1) + \mathrm{Trace}_{\bar{\nu}_5}(\sigma, B))$$

$$= 2^{-1}((3-2) - 1)$$

$$= 0.$$

When $H = C_2$ and $\lambda = \tau$ we obtain

$$\pm\dim_k(\mathrm{Sh}(\nu_5)^{(C_2,\tau)})$$

$$= \tfrac{1}{2}(\mathrm{Trace}_{\bar{\nu}_5}(\sigma, 1) - \mathrm{Trace}_{\bar{\nu}_5}(\sigma, B))$$

$$= 2^{-1}((3-2) + 1)$$

$$= 1.$$

When $H = \{1\}$ and $\lambda = 1$ we obtain

$$\pm\dim_k(\mathrm{Sh}(\nu_5)^{(\{1\},1)})$$

$$= \mathrm{Trace}_{\bar{\nu}_5}(\sigma, 1)$$

$$= 1.$$

10. The observation of Digne-Michel [53]

10.1. Let Σ be the Frobenius automorphism topologically generating the absolute Galois group of $\mathbf{F}_q$, $\mathrm{Gal}(\overline{\mathbf{F}}_q/\mathbf{F}_q)$, so that Σ^n generates

$\mathrm{Gal}(\overline{\mathbb{F}}_q/\mathbb{F}_{q^n})$. By Lang's Theorem for any $U \in GL_s\mathbb{F}_{q^n}$ there exists a positive integer t and a matrix $X \in GL_s\mathbb{F}_{q^{nt}}$ for some t such that $U = X^{-1}\Sigma(X)$.

Observe that

$$X^{-1}\Sigma(X) \in GL_s\mathbb{F}_{q^n}$$

$$\Longleftrightarrow \Sigma^n(X^{-1})\Sigma^{n+1}(X) = X^{-1}\Sigma(X)$$

$$\Longleftrightarrow X\Sigma^n(X^{-1}) = \Sigma(X\Sigma^n(X^{-1}))$$

$$\Longleftrightarrow X\Sigma^n(X^{-1}) \in GL_s\mathbb{F}_q.$$

Let $Y \in GL_s\mathbb{F}_{q^{nt}}$ be another matrix and suppose that

$$(\Sigma, X^{-1}\Sigma(X)), \ (\Sigma, Y^{-1}\Sigma(Y)) \in G(\mathbb{F}_{q^n}/\mathbb{F}_q) \propto GL_s\mathbb{F}_{q^n}$$

are conjugate by $(1, V) \in GL_s\mathbb{F}_{q^n}$. Hence, say,

$$(1, V)(\Sigma, X^{-1}\Sigma(X))(1, V^{-1})$$

$$= (\Sigma, VX^{-1}\Sigma(X))(1, V^{-1})$$

$$= (\Sigma, VX^{-1}\Sigma(XV^{-1}))$$

$$= (\Sigma, Y^{-1}\Sigma(Y)).$$

Therefore $YVX^{-1} = \Sigma(YVX^{-1}) \in GL_s\mathbb{F}_q$ and

$$(1, YVX^{-1})(\Sigma^n, X\Sigma^n(X^{-1}))(1, XV^{-1}Y^{-1})$$

$$= (\Sigma^n, YVX^{-1}X\Sigma^n(X^{-1}))(1, XV^{-1}Y^{-1})$$

$$= (\Sigma^n, YVX^{-1}X\Sigma^n(X^{-1}XV^{-1}Y^{-1}))$$

$$= (\Sigma^n, YV\Sigma^n(V^{-1}Y^{-1}))$$

$$= (\Sigma^n, YVV^{-1}\Sigma^n(Y^{-1}))$$

$$= (\Sigma^n, Y\Sigma^n(Y^{-1})).$$

Therefore

$$(\Sigma^n, X\Sigma^n(X^{-1})), \ (\Sigma^n, Y\Sigma^n(Y^{-1})) \in G(\mathbb{F}_{q^n}/\mathbb{F}_q) \times GL_s\mathbb{F}_q$$

are conjugate by an element of $GL_s\mathbb{F}_q$ or, equivalently,

$$X\Sigma^n(X^{-1}), \ Y\Sigma^n(Y^{-1}) \in GL_s\mathbb{F}_q$$

are conjugate in $GL_s\mathbb{F}_q$.

Conversely, if there exists $W \in GL_s\mathbb{F}_q$ such that

$$X\Sigma^n(X^{-1}) = WY\Sigma^n(Y^{-1})W^{-1} \in GL_s\mathbb{F}_q$$

then

$$Y^{-1}W^{-1}X = \Sigma^n(Y^{-1}W^{-1}X) \in GL_s\mathbb{F}_{q^n}.$$

Also

$$(1, Y^{-1}W^{-1}X)(\Sigma, X^{-1}\Sigma(X))(1, X^{-1}WY)$$

$$= (\Sigma, Y^{-1}W^{-1}XX^{-1}\Sigma(X))(1, X^{-1}WY)$$

$$= (\Sigma, Y^{-1}W^{-1}XX^{-1}\Sigma(XX^{-1}WY))$$

$$= (\Sigma, Y^{-1}W^{-1}\Sigma(WY))$$

$$= (\Sigma, Y^{-1}W^{-1}W\Sigma(Y))$$

$$= (\Sigma, Y^{-1}\Sigma(Y))$$

so that

$$(\Sigma, X^{-1}\Sigma(X)), \ (\Sigma, Y^{-1}\Sigma(Y)) \in G(\mathbb{F}_{q^n}/\mathbb{F}_q) \propto GL_s\mathbb{F}_{q^n}$$

are conjugate by an element of $GL_s\mathbb{F}_{q^n}$.

Therefore we have proved the following result.

THEOREM 10.2.

There is a one-one correspondence of the form

$$\left\{ \begin{array}{c} \text{conjugacy} \\[4pt] \text{classes in} \\[4pt] GL_s\mathbb{F}_q \end{array} \right\} \leftrightarrow \left\{ \begin{array}{c} GL_s\mathbb{F}_{q^n} - \text{conjugacy} \\[4pt] \text{classes of} \\[4pt] \text{elements } (\Sigma, A) \text{ in} \\[4pt] G(\mathbb{F}_{q^n}/\mathbb{F}_q) \propto GL_s\mathbb{F}_{q^n} \end{array} \right\}$$

given by

$$X\Sigma^n(X^{-1}) \leftrightarrow (\Sigma, X^{-1}\Sigma(X))$$

for $X \in GL_s\mathbb{F}_{q^{nt}}$ for some t.

PROPOSITION 10.3.

In the situation of §10.1 and Theorem 10.2

$$|GL_s\mathbb{F}_{q^n}| \cdot |GL_s\mathbb{F}_q - \text{conjugacy class of } X\Sigma^n(X^{-1})|$$

$$= |GL_s\mathbb{F}_q|$$

$$\times |GL_s\mathbb{F}_{q^n} - \text{conjugacy class of } (\Sigma, X^{-1}\Sigma(X))|.$$

Proof

The size of the conjugacy class

$$|GL_s\mathbb{F}_{q^n} - \text{conjugacy class of } (\Sigma, X^{-1}\Sigma(X))|$$

is equal to $|GL_s\mathbb{F}_{q^n}|$ divided by the number of $V \in GL_s\mathbb{F}_{q^n}$ such that $(1, V)$ commutes with $(\Sigma, X^{-1}\Sigma(X))$. By the calculation of §10.2 this happens if and only if $VX^{-1}\Sigma(XV^{-1}) = X^{-1}\Sigma(X)$ or equivalently $XVX^{-1} \in GL_s\mathbb{F}_q$. Also in this case $(1, XVX^{-1})$ commutes with $(\Sigma^n, X\Sigma^n(X^{-1}))$.

On the other hand the size of the conjugacy class

$$|GL_s\mathbb{F}_q - \text{conjugacy class of } X\Sigma^n(X^{-1})|$$

is equal to $|GL_s\mathbb{F}_q|$ divided by the number of $W \in GL_s\mathbb{F}_q$ which commute with $X\Sigma^n(X^{-1})$. This happens in and only if $V = X^{-1}WX \in GL_s\mathbb{F}_{q^n}$ and $(1, V)$ commutes with $(\Sigma, X^{-1}\Sigma(X))$.

Fixing X this discussion gives a bijection between the V's and the W's, which immediately yields the result. $\square$

REMARK 10.4. Also note that

$$(\Sigma, X^{-1}\Sigma(X))^n$$

$$= (\Sigma^2, X^{-1}\Sigma(X)\Sigma(X^{-1}\Sigma^2 X)(\Sigma, X^{-1}\Sigma(X))^n$$

$$= (\Sigma^2, X^{-1}\Sigma^2 X)(\Sigma, X^{-1}\Sigma(X))^{n-2}$$

$$= (\Sigma^3, X^{-1}\Sigma^2 X\Sigma^2(X^{-1})\Sigma^3 X))\Sigma(X))^{n-3}$$

$$= (\Sigma^3, X^{-1}\Sigma^3 X))\Sigma(X))^{n-3}$$

$$\vdots \qquad \vdots \qquad \vdots \qquad \vdots$$

$$= (\Sigma^n, X^{-1}\Sigma^n X)).$$

REMARK 10.5. Another way in which to derive the results of this section is to notice that the arguments yield a bijection between double cosets and special pairs (h', h'') of the form

$$GL_s\mathbb{F}_q \backslash GL_s\mathbb{F}_{q^{nt}} / GL_s\mathbb{F}_{q^n} \Longleftrightarrow \{\text{special pairs } (h', h'')\}$$

where a special pair in $(h', h'') \in GL_s\mathbb{F}_q \times GL_s\mathbb{F}_{q^n}$ is a pair of elements constructed from the same double coset in the manner described above but where h' is only defined up to $GL_s\mathbb{F}_q$-conjugacy and (Σ, h'') is only defined up to $GL_s\mathbb{F}_{q^n}$-conjugacy in the semi-direct product.

In the case of the Glauberman correspondence [5] the analogous double coset space has only the identity double coset as G-fixed points. However generally, in the Shintani case,

$$(GL_s\mathbb{F}_q \backslash GL_s\mathbb{F}_{q^{nt}} / GL_s\mathbb{F}_{q^n})^{\mathrm{Gal}(\mathbb{F}_{q^{nt}}/\mathbb{F}_q)} \neq \{*\}$$

since the $GL_s\mathbb{F}_{q^n}$-conjugacy class of (Σ, h'') may contain an element whose second coordinate lies in $GL_s\mathbb{F}_q$.

10.6. *Relation between Shintani descent and Theorem 10.2*

The Shintani correspondence [117] is a bijection of the form

$$\mathrm{Sh} : \mathrm{Irr}(GL_n\mathbb{F}_{q^d})^C \xrightarrow{\cong} \mathrm{Irr}(GL_n\mathbb{F}_q).$$

As explained in §4 of this Appendix, this correspondence is characterised in terms of character functions in the following manner. Let $\chi_{\mathrm{Sh}(\rho)}$ denote the trace function of the irreducible representation $\mathrm{Sh}(\rho)$. Then there exists an irreducible linear representation $\tilde{\rho}$ of the semi-direct product $C \propto GL_n\mathbb{F}_{q^d}$ which is a lift of the projective homomorphism of Proposition 3.2. Let $\chi_{\tilde{\rho}}$ denote the trace function of $\tilde{\rho}$. This linear lift may be chosen in such a way that, for all $g \in GL_n\mathbb{F}_{q^d}$,

$$\chi_{\mathrm{Sh}(\rho)}((\Sigma, g)^d) = \chi_{\tilde{\rho}}(\Sigma, g).$$

The left side of this relation is interpreted in the following manner. By Theorem 10.2, the $C \propto GL_n\mathbb{F}_{q^d}$-conjugacy class of the element

$$(\Sigma, g)^d = (1, g\Sigma(g) \ldots \Sigma^{d-1}(g))$$

intersects $GL_n\mathbb{F}_q$ in a unique $GL_n\mathbb{F}_q$-conjugacy class. The left side of the characterising relation denotes $\chi_{\mathrm{Sh}(\rho)}$ applied to any element of this $GL_n\mathbb{F}_q$-conjugacy class.

Theorem 10.2 guarantees that the character function relation does indeed define a unique virtual representation $\mathrm{Sh}(\rho)$ in $R(GL_n\mathbb{F}_q) \otimes \mathbb{C}$. More careful analysis, carried out in [117] shows that $\mathrm{Sh}(\rho)$ is in fact an irreducible representation.

The remainder of this section contains a recapitulation of some of the main ingredients of Shintani's paper [117].

10.7. *([117] Lemma 2.2)*

Let G be a linear algebraic group defined over $\mathbb{F}_q$ and assume that the centraliser of each $g \in G$ is connected. Then $g, g' \in G(\mathbb{F}_q)$ are conjugate in $G(\mathbb{F}_q)$ if and only if they are conjugate in $G(\mathbb{F}_{q^n})$.

This result applies to the parabolic groups $P_{\mu_1,\mu_2,\ldots}$ of which one is $GL_s\mathbb{F}_{q^n}$.

Therefore for each class function χ on $GL_s\mathbb{F}_q$ we can extend to a class function $\overline{\chi}$ on $GL_s\mathbb{F}_{q^n}$ by the formula

$$\overline{\chi}(g) = \begin{cases} \chi(g') & \text{if } g' \in GL_s\mathbb{F}_q \text{ is } GL_s\mathbb{F}_{q^n} - \text{conjugate to } g \\ \\ 0 & \text{otherwise.} \end{cases}$$

LEMMA 10.8. *([117] Lemma 1.1)*
Denote the cyclic group of order m by C_m acting on G. Let $\tilde{\rho}$ be an irreducible representation of $C_m \propto G$. If $\operatorname{Res}_G^{C_m \propto G}(\tilde{\rho})$ is reducible then $\chi_{\tilde{\rho}}(\Sigma, g) = 0$. If $\operatorname{Res}_G^{C_m \propto G}(\tilde{\rho})$ is irreducible then

$$|G|^{-1} \sum_{l=0}^{m-1} \xi_m^{kl} \sum_{g \in G} |\chi_{\tilde{\rho}}(\Sigma^l, g)|^2 = m \text{ or } 0$$

when $k \equiv 0$ (modulo m) or not, respectively.

LEMMA 10.9. *([117] Lemma 1.2)*
Let $\tilde{\rho}_1$ and $\tilde{\rho}_2$ be two irreducible representations of $C_m \propto G$ whose restrictions to G are irreducible and inequivalent. Then, for $l = 0, 1, \ldots, m - 1$,

$$\sum_{g \in G} \chi_{\tilde{\rho}_1}(\Sigma^l, g) \overline{\chi_{\tilde{\rho}_2}(\Sigma^l, g)} = 0.$$

10.10. *([117] Lemma 1.4)*
Let ρ be a representation of G on a vector space V. Let Σ act on the m-fold tensor product of V with itself by

$$\Sigma(v_1 \otimes v_2 \otimes \ldots \otimes v_m) = v_m \otimes v_1 \otimes v_2 \otimes \ldots \otimes v_{m-1}.$$

Let g act by a formula of the type

$$g(v_1 \otimes v_2 \otimes \ldots \otimes v_m) = \Sigma^{a_1}(g)(v_1) \otimes \Sigma^{a_2}(g)(v_2) \otimes \ldots \otimes \Sigma^{a_m}(g)(v_m).$$

Therefore

$$(\Sigma \cdot g \cdot \Sigma^{-1})(v_1 \otimes v_2 \otimes \ldots \otimes v_m)$$

$$= (\Sigma \cdot g)(v_2 \otimes v_3 \otimes \ldots \otimes v_m \otimes v_1)$$

$$= \Sigma(\Sigma^{a_1}(g)(v_2) \otimes \Sigma^{a_2}(g)(v_3) \otimes \ldots \otimes \Sigma^{a_m}(g)(v_1))$$

$$= \Sigma^{a_m}(g)(v_1) \otimes \Sigma^{a_1}(g)(v_2) \otimes \Sigma^{a_2}(g)(v_3) \otimes \ldots \otimes \Sigma^{a_{m-1}}(g)(v_m))$$

and in order for this to be $\Sigma(g)(v_1 \otimes v_2 \otimes \ldots \otimes v_m)$ we need that

$$a_m = a_1 + 1, a_1 = a_2 + 1, \ldots, a_{m-1} = a_m + 1 \text{ (modulo } m).$$

This works if

$$a_1 = m - 1, a_m = 0, a_{m-1} = 1, a_{m-2} = 2, \ldots, a_2 = m - 2.$$

Therefore, if $\{e_\alpha\}$ is a basis of V, then

$$(\Sigma, g)(e_{\alpha_1} \otimes e_{\alpha_2} \otimes \ldots \otimes e_{\alpha_m})$$

$$= (1, g)(\Sigma, 1)(e_{\alpha_1} \otimes e_{\alpha_2} \otimes \ldots \otimes e_{\alpha_m})$$

$$= \sum_{i_1, i_2, \ldots} \Sigma^{m-1}(g)_{i_1, \alpha_m} e_{i_1} \otimes \Sigma^{m-1}(g)_{i_2, \alpha_1} e_{i_2} \otimes \ldots \otimes g_{i_m, \alpha_{m-1}} e_{i_m}$$

whose trace is given by

$$\sum_{\alpha_1,\alpha_2,\ldots} \Sigma^{m-1}(g)_{\alpha_1,\alpha_m} \Sigma^{m-1}(g)_{\alpha_2,\alpha_1} \cdots g_{\alpha_m,\alpha_{m-1}}$$

$$= \mathrm{Trace}(g \cdot \Sigma(g) \cdot \ldots \cdot \Sigma^{m-2}(g) \cdot \Sigma^{m-1}(g)).$$

The observation of Digne-Michel applies to any parabolic subgroup of GL_s ([117] Lemma 2.6).

10.11. ([117] *Lemma 2.8*)

Let $\tilde{f}$ be a class function on the parabolic group $\tilde{P}_\mu(\mathbb{F}_{q^n})$ and let f be a class function on $P_\mu(\mathbb{F}_q)$ such that $\tilde{f}(\Sigma, g'') = f(g')$ in the Digne-Michel (g', g'') notation used earlier. Then for all $h'' \in GL_s\mathbb{F}_{q^n}$ and $h' \in GL_s\mathbb{F}_q$ we have

$$\mathrm{Ind}_{\tilde{P}_\mu(\mathbb{F}_{q^n})}^{GL_s\mathbb{F}_{q^n}}(\Sigma, h'') = \mathrm{Ind}_{\tilde{P}_\mu(\mathbb{F}_q)}^{GL_s\mathbb{F}_q}(h').$$

Shintani uses J.A. Green's classification of irreducibles of $GL_s\mathbb{F}_q$ and $GL_s\mathbb{F}_{q^n}$ in terms of induction from parabolic groups to make an induction on s.

To handle the representations which are not parabolically induced Shintani uses the elements of $R(GL_s\mathbb{F}_{q^n})$ given by the conjugacy class functions Σ_r^l of [69]. The proofs that these are in $R(GL_s\mathbb{F}_{q^n})$ rather than $R(GL_s\mathbb{F}_{q^n}) \otimes \mathbb{Q}$ uses a criterion for integrality due to Brauer [32].

I shall recall all this below.

DEFINITION 10.12. Choose an injective homomorphism $\theta : \overline{\mathbb{F}}_q^* \longrightarrow \mathbb{C}^*$. Set Σ_r^l equal to the following conjugacy class function on $GL_s\overline{\mathbb{F}}_q$. For $X \in GL_s\overline{\mathbb{F}}_q$ let $\{\lambda_1, \ldots, \lambda_s\}$ denote the set of eigenvalues of X and set

$$\Sigma_r^l(X) = \sum_{1 \leq i_1 < \ldots < i_r \leq s} \theta^l(\lambda_{i_1}, \ldots, \lambda_{i_r}) \in \mathbb{C}.$$

These functions may be collected into polynomials if we set

$$\hat{X} = \begin{pmatrix} \theta(\lambda_1) & 0 & \cdots & \cdots & 0 \\ 0 & \theta(\lambda_2) & 0 & \cdots & 0 \\ \vdots & \vdots & \vdots & \vdots & \vdots \\ \vdots & \vdots & \vdots & \vdots & \vdots \\ 0 & 0 & \cdots & 0 & \theta(\lambda_s) \end{pmatrix}$$

so that the characteristic polynomial of $\hat{X}$ satisfies

$$\det(\hat{X} - tI_s) = t^s - \Sigma_1^1(X)t^{s-1} + \Sigma_2^1((X)t^{s-2} + \ldots$$

$$+ (-1)^i \Sigma_i^1((X)t^{s-i} + \ldots + (-1)^s \Sigma_s^1((X)$$

and, more generally,

$$\det(\hat{X}^l - tI_s) = t^s - \Sigma_1^l(X)t^{s-1} + \Sigma_2^l((X)t^{s-2} + \dots$$

$$+ (-1)^i \Sigma_i^l((X)t^{s-i} + \dots + (-1)^s \Sigma_s^l((X).$$

THEOREM 10.13. *([69] Theorem 1)* Identifying $R(GL_s\mathbb{F}_{q^n}) \otimes \mathbb{C}$, via sending a representation to its trace function, $X \mapsto \det(\hat{X}^l - tI_s)$ lies in $R(GL_s\mathbb{F}_{q^n})[t]$ for all $s, l \geq 1, n \geq 0$.

REMARK 10.14. The Adams operations $\psi^l : R(G) \longrightarrow R(G)$ is a ring homomorphism whose trace function is satisfies $\text{trace}(\psi^l(\rho))(X) = \text{trace}(\rho)(X^l)$ so that Theorem 10.13 is implied by the case in which $l = 1$.

10.15. *Proof of Theorem 10.13 (after [69])*

By Brauer's induction theorem every virtual representation $\lambda \in R(GL_s\mathbb{F}_{q^n})$ may be written as a $\mathbb{Z}$-linear combination of the form

$$\lambda = \sum_{i=1}^{t} \alpha_i \text{Ind}_{J_i}^{GL_s\mathbb{F}_{q^n}}(\phi_i)$$

where $\phi : J_i \longrightarrow \mathbb{C}^*$ is a character and $J_i \cong H_i \times C_i$ with H_i being a p-group for some prime p and C_i is a cyclic group of order prime to p. In particular take $\lambda = 1$ and multiply the relation by a conjugacy class function f to obtain, via Frobenius reciprocity, a relation between class functions

$$f = \sum_{i=1}^{t} \alpha_i f \cdot \text{Ind}_{J_i}^{GL_s\mathbb{F}_{q^n}}(\phi_i) = \alpha_i \text{Ind}_{J_i}^{GL_s\mathbb{F}_{q^n}}(\text{Res}_{J_i}^{GL_s\mathbb{F}_{q^n}}(f) \cdot \phi_i)$$

which shows that if each $\text{Res}_{J_i}^{GL_s\mathbb{F}_{q^n}}(f) \in R(J_i) \subset R(J_i) \otimes \mathbb{C}$ then $f \in R(GL_s\mathbb{F}_{q^n}) \subset R(GL_s\mathbb{F}_{q^n}) \otimes \mathbb{C}$.

Therefore it suffices to prove Theorem 10.13 with $GL_s\mathbb{F}_{q^n}$ replaced by a subgroup of the form $H \times C$ where H is a p-group for some prime p and C is a cyclic group of order prime to p.

Now consider the cyclotomic field $\mathbb{Q}(\xi_{q^m-1})$ where $\xi_a = e^{2\pi\sqrt{-1}/a}$ and m is a multiple of n which is large enough so that $\mathbb{F}_q^m$ contains all the eigenvalues of all the matrices in $H \times C$. Let $\mathcal{P} \lhd \mathbb{Z}[\xi_{q^m-1}]$ be any prime ideal dividing the characteristic of $\mathbb{F}_q$. Then $\mathbb{Z}[\xi_{q^m-1}]/\mathcal{P} \cong \mathbb{F}_q^m$ where the isomorphism sends a $(q^m - 1)$-th root of unity x to its residue class $\bar{x}$.

We may choose θ in §10.12 so that $\theta(\bar{x}) = x$.

Now let l be the characteristic of $\mathbb{F}_q$. If l does not divide the order of $H \times C$ then it is well-known that there is a complex representation of the form $\rho : H \times C \longrightarrow GL_s\mathbb{Z}[\xi_{q^m-1}]$ such that reduction modulo $\mathcal{P}$ gives a representation $\bar{\rho}$ of $H \times C$ over $\mathbb{F}_{q^n}$ which is equivalent to the $\mathbb{F}_{q^n}$-representation

$$H \times C \subseteq GL_s\mathbb{F}_{q^n}.$$

For $g \in H \times C$ the eigenvalues of $\rho(g)$ are precisely the images under θ of the $\overline{\mathbb{F}}_q$ eigenvalues of g. The i-th elementary symmetric function of the θ-values of the $\overline{\mathbb{F}}_q$ eigenvalues of g is equal to the i-elementary symmetric function complex eigenvalues of $\rho(g)$ which is the trace function of the i-th exterior power representation $\lambda^i(\rho)$ of ρ.

Now assume that l does divide the order of $H \times C$. We may write $H \times C \cong H_1 \times H_2$ where the order of H_1 is not divisible by l and H_2 is an l-group. If $g \in H \times C$ corresponds to $(h_1, h_2) \in H_1 \times H_2$ then h_1 and h_2 are two commuting elements of $GL_s\mathbb{F}_{q^n}$ and the $\overline{\mathbb{F}}_q$ eigenvalues of h_2 are all equal to 1. Therefore elementary matrix algebra shows that in $GL_s\mathbb{F}_{q^m}$ we may simultaneously conjugate $(h_1, 1)$ and $(1, h_2)$ to upper triangular matrices of the form

$$
\begin{pmatrix}
\zeta_1 & \cdots & \cdots & \cdots \\
0 & \zeta_2 & \cdots & \cdots \\
\vdots & \vdots & \vdots & \vdots \\
0 & 0 & \cdots & \zeta_s
\end{pmatrix}
\quad \text{and} \quad
\begin{pmatrix}
1 & \cdots & \cdots & \cdots \\
0 & 1 & \cdots & \cdots \\
\vdots & \vdots & \vdots & \vdots \\
0 & 0 & \cdots & 1
\end{pmatrix}
$$

respectively.

Therefore the θ-values of the $\overline{\mathbb{F}}_q$ eigenvalues of $g = (h_1, h_2)$ are exactly those of $(h_1, 1)$. This reduces us, by projection onto H_1 from $H \times C$ to the case in which l does not divide the order of the subgroup, which completes the proof. $\square$

REMARK 10.16. In ([126] Proposition 2.1.17) it is shown by algebraic topological methods that the relation

$$
1 = \sum_{i=1}^{t} \alpha_i \mathrm{Ind}_{J_i}^{GL_s\mathbb{F}_{q^n}}(\phi_i)
$$

which occurs at the beginning of the proof of Theorem 10.13 is also true for a family of J_i's which are M-groups. M-groups are a special type of solvable group and a proof of Theorem 10.13 ought to be possible based on reduction to the case of an M-group.

11. Tables of $(-)^{((H,\lambda))}$ data

11.1. The subgroups of $C_2 \times D_6$ up to conjugation are

H	generators	$\hat{H}$
$C_2 \times D_6$	$(\sigma,1), A, C$	$1, \tau, \phi, \tau\phi$
D_6	A, C	$1, \phi$
$C_2 \times C_3$	$(\sigma,1), C$	$1\ \tau, \phi \sim \phi^2, \tau\phi \sim \tau\phi^2$
$C_2 \times C_2$	$(\sigma,1), A$	$1, \tau, \phi, \tau\phi$
C_3	C	$1, \phi, \phi^2$
C_2	A	$1, \phi$
C_2'	$(\sigma,1)A$	$1, \phi'$
C_2''	$(\sigma,1)$	$1, \tau$
$\{1\}$	1	1

In the following tables $(H,\lambda)^G$ denotes the Line bundle $\underline{\mathrm{Ind}}_H^{C_2 \propto PGL_2\mathbb{F}_4}(k_\lambda)$ for some $H \subseteq C_2 \times D_6 = C_2 \times GL_2\mathbb{F}_2 \subset C_2 \propto PGL_2\mathbb{F}_4$.

With this notation $((H,\lambda)^G)^{((J))} = \sum_{\lambda \in \hat{J}} \underline{\mathrm{Ind}}_H^{C_2 \propto PGL_2\mathbb{F}_4}(k_\lambda)^{((J))}$.

The following tables calculate these invariants when H, J are subgroups (up to conjugation) of $C_2 \times D_6$ and $k = \mathbb{C}$.

$(H,\lambda)^G$	$((H,\lambda)^G)^{((C_2 \times D_6))}$	$((H,\lambda)^G)^{((D_6))}$	$((H,\lambda)^G)^{((C_2 \times C_3))}$
$(C_2 \times D_6, \tau)^G$	τ	1	τ
$(C_2 \times D_6, \phi)^G$	ϕ	ϕ	1
$(C_2 \times D_6, 1)^G$	1	1	1
$(\langle(\sigma,1), C\rangle, \tau\phi)^G$	0	0	$\tau\phi + \tau\phi^2$
$(\langle(\sigma,1), C\rangle, \phi)^G$	0	0	$\phi + \phi^2$
$(\langle(\sigma,1), C\rangle, \tau)^G$	0	0	2τ
$(\langle(\sigma,1), A\rangle, \tau)^G$	0	0	0
$(\langle(\sigma,1), A\rangle, 1)^G$	0	0	0
$(\langle(\sigma,1), A\rangle, \phi)^G$	0	0	0
$(C_3, \phi)^G$	0	0	0
$(C_2, \phi)^G$	0	0	0
$(C_2'', 1)^G$	0	0	0
$(C_2'', \tau)^G$	0	0	0
$(\{1\}, 1)^G$	0	0	0

$(H,\lambda)^G$	$((H,\lambda)^G)^{((C_2 \times C_2))}$	$((H,\lambda)^G)^{((C_3))}$	$((H,\lambda)^G)^{((C_2))}$
$(C_2 \times D_6, \tau)^G$	2τ	1	2
$(C_2 \times D_6, \phi)^G$	$\tau + \tau\phi$	1	2ϕ
$(C_2 \times D_6, 1)^G$	2	1	2
$(\langle(\sigma,1),C\rangle, \tau\phi)^G$	0	$\phi + \phi^2$	0
$(\langle(\sigma,1),C\rangle, \phi)^G$	0	$\phi + \phi^2$	0
$(\langle(\sigma,1),C\rangle, \tau)^G$	0	2	0
$(\langle(\sigma,1),A\rangle, \tau)^G$	2τ	0	2
$(\langle(\sigma,1),A\rangle, 1)^G$	2	0	2
$(\langle(\sigma,1),A\rangle, \phi)^G$	$\phi + \tau\phi$	0	2ϕ
$(C_3, \phi)^G$	0	$2\phi + 2\phi^2$	0
$(C_2, \phi)^G$	0	0	4ϕ
$(C_2'', 1)^G$	0	0	0
$(C_2'', \tau)^G$	0	0	0
$(\{1\}, 1)^G$	0	0	0

$(H,\lambda)^G$	$((H,\lambda)^G)^{((C_2'))}$	$((H,\lambda)^G)^{((C_2''))}$	$((H,\lambda)^G)^{((\{1\}))}$
$(C_2 \times D_6, \tau)^G$	ϕ'	τ	10
$(C_2 \times D_6, \phi)^G$	1	1	10
$(C_2 \times D_6, 1)^G$	1	1	10
$(\langle(\sigma,1),C\rangle, \tau\phi)^G$	$2\phi'$	2τ	20
$(\langle(\sigma,1),C\rangle, \phi)^G$	2	2	20
$(\langle(\sigma,1),C\rangle, \tau)^G$	$2\phi'$	2τ	20
$(\langle(\sigma,1),A\rangle, \tau)^G$	$6\phi'$	6τ	30
$(\langle(\sigma,1),A\rangle, 1)^G$	6	6	30
$(\langle(\sigma,1),A\rangle, \phi)^G$	$3 + 3\phi'$	$3 + 3\tau$	30
$(C_3, \phi)^G$	0	0	40
$(C_2, \phi)^G$	0	0	60
$(C_2'', 1)^G$	6	6	60
$(C_2'', \tau)^G$	$6\phi'$	6τ	60
$(\{1\}, 1)^G$	0	0	120

11.2. Tedious calculations

The following subsection contains the calculations of the entries in the table of §11.1. They are accomplished using the following observation and are included only for completeness.

The Lines of $\underline{\mathrm{Ind}}_H^{C_2 \propto PGL_2 \mathbb{F}_4}(k_\lambda)$ which are stabilised by (J, μ) for some μ must be $g \otimes_H k_\lambda$ where $jg \otimes_H v = g \otimes_H w$ so that $g^{-1} J g \subseteq H$.

Case: $H = C_2 \times D_6 = J$.

Since $C_2 \times D_6$ is its own normaliser in $C_2 \propto GL_2\mathbb{F}_4$ the only possible stabilised Line is $\langle 1 \otimes_{C_2 \times D_6} 1 \rangle$ which J acts on via λ.

Case: $H \neq C_2 \times D_6 = J$.

In this case no conjugate of J lies inside H so there are no Lines stabilised.

Case: $H = C_2 \times D_6, J = D_6$.

D_6 is its own normaliser in $PGL_2\mathbb{F}_4$ so any conjugate of D_6 by $g \notin C_2 \times D_6$ does not lie in $C_2 \times D_6$. Hence the only stabilised Line is $1 \otimes_{C_2 \times D_6} 1 = (\sigma, 1) \otimes_{C_2 \times D_6} 1$ which is acted upon by the restriction of λ to D_6.

Case: $H \neq C_2 \times D_6, D_6, J = D_6$.

No conjugate of D_6 can be a subgroup of these H's.

Case: $H = C_2 \times D_6, J = C_2 \times C_3$.

Any conjugate of $C_2 \times C_3$ contained in $C_2 \times D_6$ has to be the unique cyclic subgroup of order six. Hence g lies in the normaliser of $C_2 \times C_3$ in $C_2 \propto PGL_2\mathbb{F}_4$ which is $C_2 \times D_6$. So the only stabilised Line is $1 \otimes_{C_2 \times D_6} 1$ acted upon by the restriction of λ.

Case: $H = J = C_2 \times C_3$.

The stabiliser of $C_2 \times C_3$ in $C_2 \propto PGL_2\mathbb{F}_4$ is $C_2 \times D_6$ so the stabilised Lines are $1 \otimes_{C_2 \times C_3} 1$ and $A \otimes_{C_2 \times C_3} 1$. The first of these Lines is acted upon by the restriction of λ and the second by

$$(\sigma, 1)A \otimes_{C_2 \times C_3} 1 = A(\sigma, 1) \otimes_{C_2 \times C_3} 1 = A \otimes_{C_2 \times C_3} \lambda(\sigma, 1)$$

and

$$CA \otimes_{C_2 \times C_3} 1 = AC^2 \otimes_{C_2 \times C_3} 1 = A \otimes_{C_2 \times C_3} \lambda(C^2).$$

Case: $H \neq C_2 \times D_6, D_6, C_2 \times C_3$ and $J = C_2 \times C_3$.

No conjugate of $C_2 \times C_3$ can be a subgroup of H so there are no Lines stabilised.

Case: $H = C_2 \times D_6$ and $J = C_2 \times C_2$.

The copies of $C_2 \times C_2$ in $C_2 \times D_6$ are $\langle (\sigma, 1), A \rangle$, $\langle (\sigma, 1), AC \rangle$ and $\langle (\sigma, 1), AC^2 \rangle$ which are all conjugate by powers of $C \in H$. Hence if $g^{-1}C_2 \times C_2 g \subset H$ then g, gC or gC^2 normalises $C_2 \times C_2$. The normaliser of $C_2 \times C_2$ in $C_2 \propto PGL_2\mathbb{F}_4$ is $\langle (\sigma, 1), A, B \rangle$. Hence the stabilised lines are $1 \otimes_{C_2 \times D_6} 1$ and $B \otimes_{C_2 \times D_6} 1$. Now $B(\sigma, 1)B = A(\sigma, 1)$, $BA = AB$ so the first Line is acted upon by the restriction of λ and the second by

$$(\sigma, 1)B \otimes_{C_2 \times D_6} 1 = B \otimes_{C_2 \times D_6} \lambda(A(\sigma, 1))$$

and

$$AB \otimes_{C_2 \times D_6} 1 = B \otimes_{C_2 \times D_6} \lambda(A).$$

Case: $H = C_2 \times C_3$ and $J = C_2 \times C_2$.
No conjugate of J is a subgroup of H.

Case: $H = J = C_2 \times C_2$.
Then g lies in the normaliser of $C_2 \times C_2$ which is $\langle(\sigma,1), A, B\rangle$. So the Lines which are stabilised are $1 \otimes_{C_2 \times C_2} 1$ and $B \otimes_{C_2 \times C_2} 1$. The first is acted on by λ and the second by

$$AB \otimes_{C_2 \times C_2} 1 = B \otimes_{C_2 \times C_2} \lambda(A)$$

and

$$(\sigma,1)B \otimes_{C_2 \times C_2} 1 = B \otimes_{C_2 \times C_2} \lambda(A(\sigma,1)).$$

Case: $H = C_3, C_2, C_2', C_2'', \{1\}$ and $J = C_2 \times C_2$.
No conjugate of J is a subgroup of H.

Case: $H = C_2 \times D_6$ and $J = C_3$.
Any conjugate of C_3 lying in H must be C_3 so g belongs to the normaliser of C_3 which is H. Hence there is only one stabilised line acted upon by the restriction of λ.

Case: $H = C_2 \times C_3$ and $J = C_3$.
Any conjugate of C_3 in H is J so $g \in C_2 \times D_6$. Therefore the stabilised Lines are $1 \otimes_{C_2 \times C_3} 1$ and $A \otimes_{C_2 \times C_3} 1$ which are acted upon via the restriction of λ and λ^2 respectively.

Case: $H = C_2 \times C_2, C_2, C_2', C_2'', \{1\}$ and $J = C_3$.
No conjugate of J is a subgroup of H.

Case: $H = J = C_3$.
The normaliser of C_3 is $C_2 \times D_6$ so there are four stabilised Lines

$$1 \otimes_{C_3} 1, (\sigma,1) \otimes_{C_3} 1, A \otimes_{C_3} 1, (\sigma,1)A \otimes_{C_3} 1.$$

The first two are acted upon via λ and the other two by λ^2.

Case: $H = C_2 \times D_6$ and $J = C_2$.
Any conjugate of A lying in H must be A, AC, AC^2 which are conjugate to each other by powers of C. The centraliser of A in $C_2 \propto PGL_2\mathbb{F}_4$ is $\langle(\sigma,1), A, B\rangle$ so there are two lines stabilised $1 \otimes_{C_2 \times D_6} 1$ and $B \times_{C_2 \times D_6} 1$ which are both acted upon via λ.

Case: $H = C_2 \times C_3$ and $J = C_2$.
Any conjugate of A in H must be $(\sigma,1)$ but these are not conjugate in $C_2 \propto PGL_2\mathbb{F}_4$ so there are no stabilised Lines.

Case: $H = C_2 \times C_2$ and $J = C_2$.

Any conjugate of A lying in H cannot be $(\sigma, 1)$ and $B(\sigma, 1)B = A(\sigma, 1)$, both of which are not conjugate to A. So g commutes with A and therefore lies in $\langle (\sigma, 1), A, B \rangle$. Therefore there are two stabilised Lines $1 \otimes_{C_2 \times C_2} 1$ and $B \times_{C_2 \times C_2} 1$ both acted upon via λ.

Case: $H = C_3, C_2'', \{1\}$ and $J = C_2$.

No conjugate of A lies in H so no Lines are stabilised.

Case: $H = J = C_2$.

The centraliser of A is $\langle (\sigma, 1), A, B \rangle$. Therefore there are two stabilised Lines $1 \otimes_{C_2 \times C_2} 1$ and $B \times_{C_2 \times C_2} 1$ so there are four stabilised Lines

$$1 \otimes_{C_2} 1, B \otimes_{C_2} 1, (\sigma, 1) \otimes_{C_2} 1, B(\sigma, 1) \otimes_{C_2} 1$$

and A acts on each via λ.

Case: $H = C_2 \times D_6$ and $J = C_2'$.

If $A(\sigma, 1)$ is conjugate to an element of H then it is $(\sigma, 1)$ since $B(\sigma, 1)B = A(\sigma, 1)$. Therefore the only Line stablised is $B \otimes_{C_2 \times D_6} 1$ on which $A(\sigma, 1)$ acts like multiplication by $\lambda(\sigma, 1)$.

Case: $H = C_2 \times D_6$ and $J = C_2''$.

There are seven elements of order two in H but $(\sigma, 1)$ is only conjugate to itself. Its centraliser is H so there is only one stabilised Line, acted upon via λ.

Case: $H = C_2 \times C_3$ and $J = C_2''$. $BC_2''B = C_2'$.

If a conjugate of $(\sigma, 1)$ lies in H it must be $(\sigma, 1)$. The centraliser of $(\sigma, 1)$ is $C_2 \times D_6$ so there are two stabilised Lines $1 \otimes_{C_2 \times C_3} 1$ and $A \otimes_{C_2 \times C_3} 1$ both of which are acted upon via λ.

Case: $H = C_2 \times C_2$ and $J = C_2''$. $BC_2''B = C_2'$.

If a conjugate of $(\sigma, 1)$ lies in H it must be $(\sigma, 1)$ or $A(\sigma, 1) = B(\sigma, 1)B$. The centraliser of $(\sigma, 1)$ is $C_2 \times D_6$ so there are six stabilised Lines

$$1 \otimes_{C_2 \times C_2} 1, C \otimes_{C_2 \times C_2} 1, C^2 \otimes_{C_2 \times C_2} 1, B \otimes_{C_2 \times C_2} 1, CB \otimes_{C_2 \times C_2} 1, C^2 B \otimes_{C_2 \times C_2} 1.$$

The first three are acted upon via λ and the last three by multiplication by $\lambda(A(\sigma, 1))$.

Case: $H = C_3, C_2$ and $J = C_2''$. $BC_2''B = C_2'$.

No conjugate of J lies in H.

Case: $H = J = C_2''$. $BC_2''B = C_2'$.

The centraliser of $(\sigma, 1)$ is $C_2 \times D_6$ so there are six stabilised lines corresponding to $g = 1, A, C, AC, C^2, AC^2$ on each of which $(\sigma, 1)$ acts via λ.

Remarks on a paper of Guy Henniart

This appendix contains an account of a calculation, by Deligne and Henniart, of wildly ramified local roots numbers modulo roots of unity. Since this result is relevant to epsilon factors derived from monomial resolutions of GL_n of a local field I have included the account which has been gathering dust on my computer since 2010 or earlier and on my homepage since 2012. This is the homepage version reproduced "as is" — here we go!

Originally, in other lectures in the series which begat this Appendix, one encounters local L-functions, functional equations and local epsilon factors of admissible representations. The p-adic Galois epsilon factors are numbers lying on the unit circle and they are fundamental in the local Langlands correspondence which was proved by Mike Harris and Richard Taylor. Later part of the proof was simplified by Henniart using his "uniqueness theorem", which is characterised in terms of p-adic epsilon factors.

This Appendix, which was formerly a lecture in the above mentioned series, is mainly expository. In it I shall outline the calculation by Deligne and Henniart of the p-adic epsilon factors of wild, homogeneous Galois representations modulo p-primary roots of unity. This formula is an important ingredient in the proofs of the uniqueness theorem. The only novel ingredients in my exposition will be the use of monomial resolutions to reduce to the one-dimensional case and an explicit formulae for the Deligne-Henniart "Gauss sum" (which seems in my opinion to contradict, in the tamely ramified case, one of the lemmas — claimed in general but used by Henniart only in the wild case — at the crux of the proof).

1. The basic ingredients

1.1. These notes are an exposition of the papers [49] and [73] which culminate in the derivation of a formula for Galois local constants (otherwise known as Galois epsilon factors) modulo p-power roots of unity for wildly ramified, homogeneous representations on the Weil group of a p-adic local field.

My account will differ from [49] in §3.3, which I shall derive using monomial resolutions. Furthermore, since it is well-known how to pass to and fro between Galois group and Weil group representations, I shall restrict the discussion to the Galois case.

I posted this on my webpage in this unpolished form, rather than posting it on the Arxiv, because I have been allowing it to languish completely unnoticed for nearly two years. I am very grateful to Paul Buckingham and Guy Henniart for their expert assistance.

1.2. Ramification groups and functions

Let us recall from [114] the properties of the ramification groups $\mathrm{Gal}(K/E)_i$, $\mathrm{Gal}(K/E)^\alpha$ and functions $\phi_{K/E}, \psi_{K/E}$ associated to a Galois extension K/E of local fields with residue characteristic p.

Let v_K denote the valuation on K, $\mathcal{O}_K$ the valuation ring of K and write $G = \mathrm{Gal}(K/E)$. The ramification groups form a finite chain of normal subgroups ([114] p.62 Proposition 1)

$$\{1\} = G_r \subseteq \ldots \subseteq G_{i+1} \subseteq G_i \subseteq \ldots \subseteq G_1 \subseteq G_0 \subseteq G_{-1} = G$$

defined by

$$G_i = \{g \in G \mid v_L(g(x) - x) \geq i + 1 \text{ for all } x \in \mathcal{O}_K\}.$$

The inertia group is G_0 and G_{-1}/G_0 is isomorphic to the Galois group of the residue field extension. If $H = \mathrm{Gal}(K/M) \subseteq G$ then $H_i = G_i \bigcap H$. The quotient G_0/G_1 is cyclic of order prime to p while G_1 is a p-group and each G_i/G_{i+1} with $i \geq 1$ is an elementary abelian p-group.

The function $\phi_{K/E} : [-1, \infty) \longrightarrow [-1, \infty)$ is a piecewise-linear homeomorphism given by

$$\phi_{K/E}(u) = \begin{cases} u & \text{if } -1 \leq u \leq 0, \\[2mm] \dfrac{u|G_1|}{|G_0|} & \text{if } 0 \leq u \leq 1, \\[2mm] \dfrac{|G_1| + \ldots + |G_m| + (u-m)|G_{m+1}|}{|G_0|} & \text{if } m \leq u \leq m+1, \\ & 1 \leq m \text{ an integer.} \end{cases}$$

At a positive integer $i \geq 1$ the slope of $\phi_{K/E}$ just to the left of i equals $\frac{|G_i|}{|G_0|}$ and just to the right it is $\frac{|G_{i+1}|}{|G_0|}$. Therefore the condition that $G_i = G_{i+1}$ is equivalent to $\phi_{K/E}$ being linear at i. If $G_i \neq G_{i+1}$ then $\phi_{K/E}$ is concave downward at i and i is called a "jump" value for the lower filtration. If $G_0 = \ldots = G_r \neq G_{r+1}$ then $\phi_{K/E}(x) = x$ if $-1 \leq x \leq r$ and $\phi_{K/E}(x) < x$ if $r < x$.

If $g \in G_0$ then $g \in G_i$ if and only if $g(\pi_K)/\pi_K \equiv 1$ (modulo $\mathcal{P}_K^i$).

The function $\psi_{K/E} : [-1, \infty) \longrightarrow [-1, \infty)$ is the piecewise-linear homeomorphism given by the inverse of $\phi_{K/E}$. Hence $\psi_{K/E}(i) = \alpha$ is not necessarily an integer. To accommodate this we extend the definition of the

G_i's to G_u for any real number $u \geq -1$ by setting $G_u = G_j$ where j is the smallest integer satisfying $u \leq j$.

Given a chain of fields $E \subseteq M \subseteq K$ there are chain rules

$$\phi_{M/E}(\phi_{K/M}(x)) = \phi_{K/E}(x), \quad \psi_{K/M}(\psi_{M/E}(y)) = \psi_{K/E}(y).$$

The upper numbering of the ramification groups is defined by the relations

$$G^v = G_{\psi_{K/E}(v)} \text{ and } G^{\phi_{K/E}(u)} = G_u.$$

If $H = \mathrm{Gal}(K/M) \lhd G$ is a normal subgroup then $(G/H)^v = G^v H/H$.

In §1.5 we shall utilise the extension of the upper numbering filtration to the case of infinite Galois extensions such as $\overline{F}/E$. Following ([114] Remark 1, p. 75) for K/E an infinite Galois extension we set

$$\mathrm{Gal}(K/E)^v = \varprojlim \mathrm{Gal}(K'/E)^v$$

where K' runs through the set of finite Galois extensions of E contained in K. The filtration $\mathrm{Gal}(K/E)^v$ is left continuous in the sense that

$$\mathrm{Gal}(K/E)^v = \bigcap_{w<v} \mathrm{Gal}(K/E)^w.$$

As we shall see in Proposition 1.3, the upper filtration $\mathrm{Gal}(K/E)^v$ is *not* right continuous. One says that v is a "jump" for the upper numbering filtration if $\mathrm{Gal}(K/E)^v \neq \mathrm{Gal}(K/E)^{v+\epsilon}$ for all $\epsilon > 0$. Even for finite Galois extensions an upper numbering jump need not be an integer ([114] Exercise 2, p. 77).

PROPOSITION 1.3.

(i) Let K/E be a, not necessarily finite, Galois extension of p-adic local fields[1]. Then for any $\alpha \in \mathbb{R}$ there exists $\gamma < \alpha$ such that $\mathrm{Gal}(K/E)^\gamma = \mathrm{Gal}(K/E)^\alpha$.

(ii) Let K/E be an infinite Galois extension of p-adic local fields. Then the filtration $\mathrm{Gal}(K/E)^v$ is not right continuous in the sense that, if v is a jump for the upper numbering filtration,

$$\bigcup_{v<w} \mathrm{Gal}(K/E)^w \overset{\subsetneq}{\neq} \mathrm{Gal}(K/E)^v.$$

Proof

In (i) the set of jumps in the upper ramification filtration is discrete. Suppose that $\{\beta_n\}$ is an increasing sequence of real numbers such that $\beta_n < \alpha$ tending to α from below. Therefore the sequence G^{β_n} will eventually

---

[1]I am very grateful to Paul Buckingham and Guy Henniart for explaining the proof of this proposition to me.

stabilise (i.e. becoming equal for large enough $n < \alpha$). Therefore there is one of these β_n's, say γ, such that

$$\mathrm{Gal}(K/E)^\alpha = \bigcap_{w<\alpha} \mathrm{Gal}(K/E)^w = \bigcap_{w<\gamma} \mathrm{Gal}(K/E)^w = \mathrm{Gal}(K/E)^\gamma,$$

as required.

In (ii) we consider the infimum

$$\underline{\alpha} = \inf\{\alpha \in \mathbb{R} \mid \mathrm{Gal}(K/E)^\alpha \subseteq \mathrm{Gal}(K/E)^v\}.$$

By part (i) there exists $\gamma < \underline{\alpha}$ such that $\mathrm{Gal}(K/E)^\gamma = \mathrm{Gal}(K/E)^{\underline{\alpha}}$. Suppose that $\mathrm{Gal}(K/E)^\gamma \subseteq \mathrm{Gal}(K/E)^v$ then, by definition, $\underline{\alpha} \le \gamma$ which is a contradiction. Therefore

$$\mathrm{Gal}(K/E)^{\underline{\alpha}} = \mathrm{Gal}(K/E)^\gamma \nsubseteq \mathrm{Gal}(K/E)^v.$$

However, if $v < w$ then $\mathrm{Gal}(K/E)^w \subseteq \mathrm{Gal}(K/E)^v$ so that $\underline{\alpha} \le w$ and therefore $\mathrm{Gal}(K/E)^w \subseteq \mathrm{Gal}(K/E)^{\underline{\alpha}}$ which implies that

$$\bigcup_{v<w} \mathrm{Gal}(K/E)^w \subseteq \mathrm{Gal}(K/E)^{\underline{\alpha}} \overset{\subset}{\ne} \mathrm{Gal}(K/E)^v,$$

as required. $\square$

The proof of part (ii) of Proposition 1.3 establishes the following result.

COROLLARY 1.4.

Let K/E be an infinite Galois extension of p-adic local fields. Let v be a jump for the upper numbering filtration and define

$$\underline{\alpha} = \inf\{\alpha \in \mathbb{R} \mid \mathrm{Gal}(K/E)^\alpha \subseteq \mathrm{Gal}(K/E)^v\}.$$

Then $\underline{\alpha}$ is strictly smaller than v.

1.5. *Wild, homogeneous local Galois representations*

All fields are non-Archimedean local containing $F/\mathbb{Q}_p$. Let σ be a nontrivial, continuous, finite-dimensional complex representation of $\mathrm{Gal}(\overline{F}/E)$. The level $\alpha(\sigma)$ is the least α such that σ restricted to $\mathrm{Gal}(\overline{F}/E)^\alpha$ is nontrivial but σ restricted to $\mathrm{Gal}(\overline{F}/E)^{\alpha+\epsilon}$ is trivial for all $\epsilon > 0$. There exists an upper numbering ramification group with this property by left continuity of the $\mathrm{Gal}(\overline{F}/E)^\alpha$'s because there are certainly ramification groups $\mathrm{Gal}(\overline{F}/E)^w$ on which σ is non-trivial. Define $\alpha(\sigma)$ to be the supremum of the set of real numbers v such that σ is non-trivial on $\mathrm{Gal}(\overline{F}/E)^v$. Therefore

$$\mathrm{Gal}(\overline{F}/E)^{\alpha(\sigma)} = \bigcap_{v<\alpha(\sigma)} \mathrm{Gal}(\overline{F}/E)^v.$$

Then σ is wild if $\alpha(\sigma) > 0$. If σ is wild and irreducible then ([**72**] §3)

$$a(\sigma) = \dim(\sigma)(1 + \alpha(\sigma))$$

where $f(\sigma) = \mathcal{P}_E^{a(\sigma)}$ is the Artin conductor of σ (see §1.10 for the definition of $a(\sigma)$).

There exists a finite Galois extension K/E such that σ (faithfully) factors through the finite Galois group $\mathrm{Gal}(K/E)$. Since $\mathrm{Gal}(\overline{F}/E)^{\alpha(\sigma)}$ is the last ramification group (in the upper numbering) on which σ is non-trivial its image in $G = \mathrm{Gal}(K/E)$ is abelian and normal. The representation σ is homogeneous if

$$\mathrm{Res}_{\mathrm{Gal}(\overline{F}/E)^{\alpha(\sigma)}}^{\mathrm{Gal}(\overline{F}/E)}(\sigma) = n\chi_\sigma$$

where $n = \dim(\sigma)$ and $\chi_\sigma : \mathrm{Gal}(\overline{F}/E)^{\alpha(\sigma)} \longrightarrow \mathbb{C}^*$ is a character of finite order.

Suppose that the image of $\mathrm{Gal}(\overline{F}/E)^{\alpha(\sigma)}$ in G is

$$A = \mathrm{Gal}(K/M) = \mathrm{Gal}(K/E)^{\alpha(\sigma)} \lhd G.$$

Suppose that σ is irreducible and wild then it will not necessarily also be homogeneous but let us suppose that it is.

We remark that, if σ is not homogeneous then Clifford theory, which deals with restriction to normal abelian subgroups, implies that

$$\mathrm{Res}_A^G(\sigma) = m(\chi_1 \oplus \chi_2 \oplus \ldots \oplus \chi_t)$$

where the conjugacy G-orbit of χ_1 is $\{\chi_1, \ldots, \chi_t\}$. This means that each of the χ_i's is non-trivial on A.

Furthermore, if σ is irreducible, wild and homogeneous, then χ_σ will be fixed under the conjugation action by G. This is because for $a \in A, g \in G, v \in V$

$$\sigma(a)v = \chi_\sigma(a) \cdot v \text{ and } \sigma(gag^{-1})(v) = \sigma(g)(\chi_\sigma(a) \cdot (\sigma(g)^{-1}(v))) = \sigma(a)v$$

since $\sigma(a)$ is multiplication by a scalar.

Therefore χ_σ corresponds via class field theory to a character $\chi : M^* \longrightarrow \mathbb{C}^*$ which is invariant under the Galois action of G on M. On $1 + \mathcal{P}_M^{a(\chi)-1}$ χ has the form $\chi(1 + x) = \psi_M(gx)$ for $g \in M^*$ such that $g \in M^*/1 + \mathcal{P}_M$ is well-defined. Here ψ_M is a choice of additive character, which depends on the choice of ψ_F and is then defined as $\psi_M = \psi_F \cdot \mathrm{Trace}_{M/F}$. Therefore, defining $C_E = ((E^*/1 + \mathcal{P}_E) \otimes \mathbb{Z}[1/p])$, we have a well-defined element

$$g = g_\sigma \in ((M^*/1 + \mathcal{P}_M) \otimes \mathbb{Z}[1/p])^G \cong ((E^*/1 + \mathcal{P}_E) \otimes \mathbb{Z}[1/p]) = C_E.$$

From the short exact sequence, when p is odd,

$$0 \longrightarrow (\mathcal{O}_E/\mathcal{P}_E)^* \otimes \mathbb{Z}[1/p] \longrightarrow C_E \longrightarrow \mathbb{Z}[1/p] \longrightarrow 0$$

we see that $C_E \otimes \mathbb{Z}/2$ has four elements.

Note that the element g_σ can be defined for any wild, homogeneous representation, irreducible or not.

1.6. Varying M in §1.5

One can vary the choice of M in the above construction. Suppose we take another subfield M' fixed by $A = \mathrm{Gal}(K/M) = \mathrm{Gal}(K/E)^{\alpha(\sigma)}$. Therefore we must have $M' \subseteq M$ and we may as well assume that $M' \neq M$. Set $A' = \mathrm{Gal}(K/M')$ so that $A \subset A'$ is a proper subgroup. Note that A' is not necessarily abelian (it would be if σ were faithful) but A is, in fact it is cyclic because χ is faithful on A.

We shall also assume that σ restricted to A' is equal to $n\chi'$ for a character χ' which must restrict to χ on A. In terms of local class field theory we have

$$\chi = \chi' \cdot N : M^* \xrightarrow{N} (M')^* \xrightarrow{\chi'} \mathbb{C}^*.$$

Consider $\underline{A} = \mathrm{Gal}(\overline{F}/M) = \mathrm{Gal}(\overline{F}/E)^{\alpha(\sigma)}$ and $\underline{A}' = \mathrm{Gal}(\overline{F}/M')$. Let $\underline{\alpha}(M/M')$ denote the real number

$$\underline{\alpha}(M/M') = \inf\{\alpha \in \mathbb{R} \mid (\underline{A}')^{\alpha} \subseteq \underline{A}\}.$$

THEOREM 1.7.
In the notation of §1.6

$$\alpha(M/M') < \alpha(\chi') = a(\chi') - 1.$$

We begin with an intermediate result.

PROPOSITION 1.8.
In the notation of §1.6, $\underline{\alpha}(M/M') \leq \alpha(\chi')$.

Proof

Since the upper ramification index is preserved under passage to quotient Galois groups it will suffice to prove this by studying the finite extension K/E as in §1.6, where we continue to assume that σ is faithful although K/E is not necessarily abelian.

We can show that $\underline{\alpha}(M/M') \leq \alpha(\chi')$ by showing that

$$(A')^{\alpha(\chi')} \subseteq A.$$

By definition ([114] p. 71, Remark 1) $\alpha(\sigma)$ is a jump because $\sigma :$ $\mathrm{Gal}(K/E) \longrightarrow GL_n\mathbb{C}$ is one-one and σ is non-trivial on $A = \mathrm{Gal}(K/M) = \mathrm{Gal}(K/E)^{\alpha(\sigma)}$ but is trivial on $\mathrm{Gal}(K/E)^{\alpha(\sigma)+\epsilon}$ for all $\epsilon > 0$ so

$$\{1\} = \mathrm{Gal}(K/E)^{\alpha(\sigma)+\epsilon} \neq \mathrm{Gal}(K/E)^{\alpha(\sigma)}$$

for all $\epsilon > 0$.

By the theory of the Herbrand functions ϕ and ψ (§1.2; see also [114] Chapter IV §3) there exists a real number $\gamma = \psi_{K/E}(\alpha(\sigma))$ such that

$$\mathrm{Gal}(K/E)^{\alpha(\sigma)} = \mathrm{Gal}(K/E)_{\gamma}.$$

This means that $\text{Gal}(K/E)_\gamma = \text{Gal}(K/E)_i$ where i is the smallest integer satisfying $\psi_{K/E}(\alpha(\sigma)) = \gamma \leq i$. If $\gamma < i$ then

$$\alpha(\sigma) = \phi_{K/E}(\gamma) < \phi_{K/E}(i) = \delta$$

and $\text{Gal}(K/E)^{\alpha(\sigma)} = \text{Gal}(K/E)^\delta$, which is a contradiction. Therefore $\psi_{K/E}(\alpha(\sigma)) = i$, an integer. Furthermore i must be a jump for otherwise $G_i = G_{i+1}$ and $\text{Gal}(K/E)^{\alpha(\sigma)} = \text{Gal}(K/E)^{\phi_{K/E}(i+1)}$. Therefore we have

$$\{1\} = \text{Gal}(K/E)_{i+1} \subset \text{Gal}(K/E)_i = \text{Gal}(K/E)^{\alpha(\sigma)}.$$

Now consider $A' \bigcap A$ which equals, by ([114] p. 62 Proposition 2) since i is an integer,

$$A' \bigcap A = A' \bigcap \text{Gal}(K/E)^{\alpha(\sigma)} = A' \bigcap \text{Gal}(K/E)_i = (A')_i = (A')^\beta$$

for $\beta = \phi_{K/M'}(i)$. Since χ' restricted to A is equal to χ, which is one-one, χ' restricted to $(A')^\beta$ is non-trivial.

Now let j be the largest integer for which the restriction of χ' to $(A')_j$ is non-trivial. Therefore $j \geq i$ and also, by [114] p. 102 Proposition 5), $\phi_{K/M'}(j) = \alpha(\chi')$. Hence ([114] p. 73 Proposition 12)

$$\beta = \phi_{K/M'}(i) \leq \phi_{K/M'}(j) = \alpha(\chi').$$

By definition of $\underline{\alpha}(M/M')$ we have

$$\underline{\alpha}(M/M') \leq \beta \leq \alpha(\chi').$$

$\square$

1.9. *Proof of Theorem 1.7*

Suppose that $\underline{\alpha}(M/M') = \alpha(\chi')$ then, in the notation of the proof of Proposition 1.8, $\underline{\alpha}(M/M') = \beta = \alpha(\chi')$ and so

$$\underline{A}' \bigcap \underline{A} = (\underline{A}')^\beta = (\underline{A}')^{\alpha(\chi')}.$$

Therefore

$$\underline{\alpha}(M/M') = \inf\{\alpha \in \mathbb{R} \mid (\underline{A}')^\alpha \subseteq (\underline{A}')^{\alpha(\chi')}\}.$$

By the proof of Proposition 1.3(ii)

$$(\underline{A}')^{\underline{\alpha}(M/M')} \overset{\subset}{\neq} (\underline{A}')^{\alpha(\chi')},$$

which contradicts the assumption that $(\underline{A}')^{\underline{\alpha}(M/M')} = (\underline{A}')^{\alpha(\chi')}$. $\square$

1.10. *Recap of abelian local root numbers*

Let us recall from ([94] p. 29) the formula for the abelian local roots numbers. Let $\chi : E^* \longrightarrow \mathbb{C}^*$ be a character (i.e. with open kernel). Let $a(\chi) = 0$ if χ is trivial on $\mathcal{O}_E^*$ and otherwise let $a(\chi)$ be the least integer $n \geq 1$ such that χ is trivial on $1 + \mathcal{P}_E^n$. The Artin conductor is given by the ideal $f(\chi) = \mathcal{P}_E^{a(\chi)}$. For example, if the residue field satisfies $\mathcal{O}_E/\mathcal{P}_E \cong \mathbb{F}_{p^d}$ and χ restricted to $\mathcal{O}_E^*$ has the form

$$\mathcal{O}_E^* \longrightarrow (\mathcal{O}_E/\mathcal{P}_E)^* \overset{\text{Norm}_{\mathbb{F}_{p^d}/\mathbb{F}_p}}{\longrightarrow} \mathbb{F}_p^* \subset \mathbb{C}^*$$

then $a(\chi) = 1$.

In each of these cases the local root number is given by the formula ([**94**] p. 29)

$$W_E(\chi) = \frac{1}{\sqrt{N\pi_E}} \sum_{w \in (\mathcal{O}_E/\mathcal{P}_E)^*} \chi(w)\chi(c)^{-1}\psi_E(w/c)$$

where c is a generator of $f(\chi)\mathcal{D}_E$.

1.11. The Gauss sum of [**73**]

Let $\mathcal{P}_E = \pi_E\mathcal{O}_E$ and let ν_E be the E-adic order of ψ_E on so that the inverse different satisfies $\mathcal{D}_E^{-1} = \mathcal{P}_E^{-\nu_E}$.

Suppose that $p \neq 2$ and that $x \in E^*$ satisfies $\nu_E(x) + \nu_E$ is odd. Therefore we have an integer b such that

$$0 = \nu_E(x) + \nu_E + 2b + 1.$$

Hence

$$x\pi_E^{2b}\mathcal{P}_E \subseteq \mathcal{P}_E^{2b-\nu_E-2b-1+1} = \mathcal{D}_K^{-1}$$

so that $\psi_E(x\pi_E^{2b}\xi) = 1$ for all $\xi \in \mathcal{P}_E$.

Consider the Gauss sum

$$\phi(x) = \sum_{\xi \in (\mathcal{O}_E/\mathcal{P}_E)^*} \psi_E(x\pi_E^{2b}\xi^2/2) \in \mathbb{C}^*.$$

Note that there is a misprint[2] in the definition of ϕ in ([**73**] §2).

If we replace ξ by $\xi + \pi_E u$ with $u \in \mathcal{O}_E$ we have

$$\psi_E(x\pi^{2b}(\xi + \pi_E u)^2/2) = \psi_E(x\pi^{2b}\xi^2/2) \cdot \psi_E(x\pi^{2b}(\xi\pi_E u + \pi_E^2 u^2/2)$$

$$= \psi_E(x\pi^{2b}\xi^2/2)$$

so that $\phi(x)$ is well-defined.

If $v \in \mathcal{O}_E^*$ then

$$\phi(xv^2) = \sum_{\xi \in (\mathcal{O}_E/\mathcal{P}_E)^*} \psi_E(x\pi_E^{2b}(v\xi)^2/2) = \phi(x)$$

and for any a

$$\phi(x\pi_E^{2a}) = \sum_{\xi \in (\mathcal{O}_E/\mathcal{P}_E)^*} \psi_E(x\pi_E^{2a}\pi_E^{2b-2a}\xi^2/2) = \phi(x)$$

---

[2]In ([**73**] §2) the sum is taken over *all* the elements of the residue field. The error can be seen by taking $E = \mathbb{Q}_p$ (see §1.12 below).

1. THE BASIC INGREDIENTS

so that we have a function, which is not a homomorphism (see $E = \mathbb{Q}_p$ in §1.12 below),

$$\phi : (E^*/(1 + \mathcal{P}_E)) \otimes \mathbb{Z}/2 \longrightarrow \mathbb{C}^*$$

defined by setting $\phi(x) = 1$ if $\nu_E(x) + \nu_E$ is even. By the usual argument, if q_E is the order of the residue field $\mathcal{O}_E/\mathcal{P}_E$ then $\phi(x)^2 = (-1)^{(q_E-1)/2} q_E$ if $\nu_E(x) + \nu_E$ is odd. Define a map

$$G_E : E^*/(1 + \mathcal{P}_E) \longrightarrow \mu_4$$

by the formula

$$G_E(x) = \begin{cases} \dfrac{\phi(x)}{+\sqrt{q_E}} & \text{if } \nu_E(x) + \nu_E \text{ is odd} \\[2mm] 1 & \text{if } \nu_E(x) + \nu_E \text{ is even.} \end{cases}$$

Since $\phi(x) = \phi(x^p)$ because p is odd we may extend G_E to a non-homomorphic function

$$G_E : C_E = ((E^*/1 + \mathcal{P}_E) \otimes \mathbb{Z}[1/p]) \longrightarrow \mu_4.$$

When $p = 2$ set $G_E(x) = 1$ for all x.

1.12. *The case $E = \mathbb{Q}_p$ and the misprint of ([73] §2)*
When $E = \mathbb{Q}_p$ with $p \neq 2$ we have

$$G_{\mathbb{Q}_p} : \mathbb{Q}_p^*/(\mathbb{Q}_p^{2*}) = \mathbb{Q}_p^* \otimes \mathbb{Z}/2 \longrightarrow \mu_4.$$

Now $\mathbb{Q}_p^*/(\mathbb{Q}_p^{2*}) = \{1, u, p, up\}$ where $u \in \mathbb{Z}_p^*$ and the mod p Legendre symbol satisfies ([125] p. 267)

$$\left(\frac{u}{p}\right) = -1$$

We have $\nu_{\mathbb{Q}_p} = 0$ and $\nu_{\mathbb{Q}_p}(x) + \nu_{\mathbb{Q}_p}$ is even for $x = 1, u$ and $\nu_{\mathbb{Q}_p}(x) + \nu_{\mathbb{Q}_p} + 2(-1) + 1 = 0$ when $x = p, up$. Therefore

$$G_{\mathbb{Q}_p}(1) = 1 = G_{\mathbb{Q}_p}(u)$$

and, if $\xi_p = e^{2\pi\sqrt{-1}/p}$,

$$
\begin{aligned}
G_{\mathbb{Q}_p}(p) &= \tfrac{1}{\sqrt{p}} \sum_{z \in (\mathbb{Z}/p)^*} e^{2\pi\sqrt{-1}pp^{-2}z^2/2} \\[2ex]
&= \tfrac{1}{\sqrt{p}} \sum_{z \in (\mathbb{Z}/p)^*} \xi_p^{z^2/2} \\[2ex]
&= \tfrac{1}{\sqrt{p}} \sum_{z \in (\mathbb{Z}/p)^*} \xi_p^{z/2} + \tfrac{1}{\sqrt{p}} \sum_{w \in (\mathbb{Z}/p)^*} \left(\tfrac{w}{p}\right) \xi_p^{w/2} \\[2ex]
&= \tfrac{1}{\sqrt{p}} \sum_{w \in (\mathbb{Z}/p)^*} \left(\tfrac{w}{p}\right) \xi_p^{w/2} \\[2ex]
&= \left(\tfrac{2}{p}\right) \tfrac{1}{\sqrt{p}} \sum_{w \in (\mathbb{Z}/p)^*} \left(\tfrac{w/2}{p}\right) \xi_p^{w/2} \\[2ex]
&= \left(\tfrac{2}{p}\right) W_{\mathbb{Q}_p}(l(p)) \text{ in the notation of } ([\mathbf{125}] \text{ p. } 267) \\[2ex]
&= \begin{cases} -\left(\tfrac{2}{p}\right)\sqrt{-1} & \text{if } p \equiv 3 \ (\mathrm{mod}\ 4) \\[2ex] \left(\tfrac{2}{p}\right) & \text{if } p \equiv 1 \ (\mathrm{mod}\ 4). \end{cases}
\end{aligned}
$$

Hence $G_{\mathbb{Q}_p}(p)^2 = \left(\tfrac{-1}{p}\right) = (-1)^{(p-1)/2}$.[3]

---

[3]The sum over all the residue field, as in ([**73**] §2), would add $\tfrac{1}{\sqrt{p}}$ to $G_{\mathbb{Q}_p}(p)$ and then its square would not be equal to $(-1)^{(p-1)/2}$ as claimed in ([**73**] §2).

$$G_{\mathbb{Q}_p}(up) = \frac{1}{\sqrt{p}} \sum_{z \in (\mathbb{Z}/p)^*} e^{2\pi\sqrt{-1}upp^{-2}z^2/2}$$

$$= \frac{1}{\sqrt{p}} \sum_{z \in (\mathbb{Z}/p)^*} \xi_p^{uz^2/2}$$

$$= \frac{1}{\sqrt{p}} \sum_{z \in (\mathbb{Z}/p)^*} \xi_p^{z/2} - \frac{1}{\sqrt{p}} \sum_{w \in (\mathbb{Z}/p)^*} \left(\frac{w}{p}\right) \xi_p^{w/2}$$

$$= -\frac{1}{\sqrt{p}} \sum_{w \in (\mathbb{Z}/p)^*} \left(\frac{w}{p}\right) \xi_p^{w/2}$$

$$= -\left(\frac{2}{p}\right) \frac{1}{\sqrt{p}} \sum_{w \in (\mathbb{Z}/p)^*} \left(\frac{w/2}{p}\right) \xi_p^{w/2}$$

$$= -\left(\frac{2}{p}\right) W_{\mathbb{Q}_p}(l(p)) \text{ in the notation of } ([\textbf{125}] \text{ p. 267})$$

$$= \begin{cases} \left(\frac{2}{p}\right)\sqrt{-1} & \text{if } p \equiv 3 \pmod 4 \\[2mm] -\left(\frac{2}{p}\right) & \text{if } p \equiv 1 \pmod 4. \end{cases}$$

Therefore $G_{\mathbb{Q}_p}(up) \neq G_{\mathbb{Q}_p}(u)G_{\mathbb{Q}_p}(p)$ which confirms that $G_{\mathbb{Q}_p}$ is not a homomorphism.

In general, therefore, the formula for $G_{\mathbb{Q}_p}$ is given by ([\textbf{125}] p. 267)

$$G_{\mathbb{Q}_p}(x) = \left(\frac{2^{\nu_{\mathbb{Q}_p}(x)}}{p}\right) W_{\mathbb{Q}_p}(l(x)).$$

1.13. *The formula for G_E in general when $p \neq 2$*

Let $N : \mathbb{F}_{q^d}^* \longrightarrow \mathbb{F}_q^*$ denote the norm. It is a surjective homomorphism, by Hilbert's Theorem 90 and element counting, so that we have a surjection

$$N : \mathbb{F}_{q^d}^*/\mathbb{F}_{q^d}^{2*} = \mathbb{F}_{q^d}^* \otimes \mathbb{Z}/2 \longrightarrow \mathbb{F}_q^*/\mathbb{F}_q^{2*} = \mathbb{F}_q^* \otimes \mathbb{Z}/2$$

which is therefore an isomorphism since both groups have only two elements.

The exact sequence

$$0 \longrightarrow \mathcal{O}_E^* \longrightarrow E^* \longrightarrow \mathbb{Z} \longrightarrow 0$$

yields a short exact sequence

$$0 \longrightarrow \mathcal{O}_E^* \otimes \mathbb{Z}/2 \longrightarrow E^*/E^{2*} \longrightarrow \mathbb{Z}/2 \longrightarrow 0$$

and in the short exact sequence

$$0 \longrightarrow 1 + \mathcal{P}_E \longrightarrow \mathcal{O}_E^* \longrightarrow \mathbb{F}_{q^d}^* \longrightarrow 0$$

the group $1 + \mathcal{P}_E$ is 2-divisible so that we have an isomorphism

$$\mathcal{O}_E^* \otimes \mathbb{Z}/2 \xrightarrow{\cong} \mathbb{F}_q^* \otimes \mathbb{Z}/2.$$

Therefore E^*/E^{2*} has four elements which are $\{1, u, \pi_E, u\pi_E\}$ where $u \in \mathcal{O}_E^*$ maps to a non-square in $\mathbb{F}_{q^d}^*$. If q is a power of p then the condition on u is equivalent to

$$\left(\frac{N_{\mathbb{F}_{q^d}/\mathbb{F}_p}(u)}{p} \right) = -1.$$

Recall that $\mathcal{D}_E^{-1} = (\pi_E)^{-\nu_E}$.
Suppose that $\nu_E + 2b + 1 = 0$ so that

$$\nu_E(1) + \nu_E + 2b + 1 = 0 = \nu_E(u) + \nu_E + 2b + 1.$$

Therefore for $x = 1, u$ we have

$$
\begin{aligned}
G_E(x) &= \frac{1}{\sqrt{N\pi_E}} \sum_{z \in (\mathcal{O}_E/\mathcal{P}_E)^*} \psi_E(x\pi_E^{2b} z^2/2) \\
&= \frac{1}{\sqrt{N\pi_E}} \sum_{z \in (\mathcal{O}_E/\mathcal{P}_E))^*} \psi_E(x\pi_E^{2b} z/2) \\
&\quad + \frac{1}{\sqrt{N\pi_E}} \sum_{w \in (\mathcal{O}_E/\mathcal{P}_E))^*} \left(\frac{N_{\mathbb{F}_{q^d}/\mathbb{F}_p}(w)}{p} \right) \psi_E(x\pi_E^{2b} w/2) \\
&= \left(\frac{N_{\mathbb{F}_{q^d}/\mathbb{F}_p}(2x)}{p} \right) \frac{1}{\sqrt{N\pi_E}} \sum_{w \in (\mathcal{O}_E/\mathcal{P}_E))^*} \left(\frac{N_{\mathbb{F}_{q^d}/\mathbb{F}_p}(w)}{p} \right) \psi_E(\pi_E^{2b} w).
\end{aligned}
$$

It would be nice to be able to apply the Davenport-Hasse theorem ([**87**] p. 20) to this Gauss sum but this is only immediate in the case of $E/\mathbb{Q}_p$ being unramified because the additive character ψ_E involves the trace for $E/\mathbb{Q}_p$ rather than the trace for their residue fields. When ν_E is odd then $G_E(u\pi_E) = 1 = G_E(\pi_E)$.
    Suppose that $\nu_E + 2b + 2 = 0$ so that

$$\nu_E(\pi_E) + \nu_E + 2b + 1 = 0 = \nu_E(u\pi_E) + \nu_E + 2b + 1.$$

Therefore for $x = \pi_E, u\pi_E$ we have

$G_E(x)$

$$= \frac{1}{\sqrt{N\pi_E}} \sum_{z \in (\mathcal{O}_E/\mathcal{P}_E)^*} \psi_E((x/\pi_E)\pi_E^{2b+1}z^2/2)$$

$$= \frac{1}{\sqrt{N\pi_E}} \sum_{z \in (\mathcal{O}_E/\mathcal{P}_E))^*} \psi_E((x/\pi_E)\pi_E^{2b+1}z/2)$$

$$+ \frac{1}{\sqrt{N\pi_E}} \sum_{w \in (\mathcal{O}_E/\mathcal{P}_E))^*} \left(\frac{N_{\mathbb{F}_{q^d}/\mathbb{F}_p}(w)}{p}\right) \psi_E((x/\pi_E)\pi_E^{2b+1}w/2)$$

$$= \left(\frac{N_{\mathbb{F}_{q^d}/\mathbb{F}_p}(2(x/\pi_E))}{p}\right) \frac{1}{\sqrt{N\pi_E}} \sum_{w \in (\mathcal{O}_E/\mathcal{P}_E))^*} \left(\frac{N_{\mathbb{F}_{q^d}/\mathbb{F}_p}(w)}{p}\right) \psi_E(\pi_E^{2b+1}w).$$

1.14. The case when $a(\chi) = 1$

Now suppose that c is a generator of $f(\chi)\mathcal{D}_E$. In the notation of §1.13 the inverse different is given by $\mathcal{D}_E^{-1} = (\pi_E)^{-\nu_E}$. Therefore if $a(\chi) = 1$ and $\nu_E + 2b + 1 = 0$ then $f(\chi)\mathcal{D}_E = \mathcal{P}_E^{1+\nu_E} = \mathcal{P}_E^{-2b}$ and so $c^{-1} = \pi_E^{2b}$. Similarly $a(\chi) = 1$ and $\nu_E + 2b + 2 = 0$ then $f(\chi)\mathcal{D}_E = \mathcal{P}_E^{1+\nu_E} = \mathcal{P}_E^{-2b-1}$ and so $c^{-1} = \pi_E^{2b+1}$.

Therefore, if $\chi_E : E^* \longrightarrow \mathbb{C}^*$ satisfies $\chi_E(\pi_E) = 1$ and

$$\chi_E(z) = \left(\frac{\text{Norm}_{\mathbb{F}_{p^d}/\mathbb{F}_p}(z + \mathcal{P}_E)}{p}\right) \in \{\pm 1\}$$

for $z \in \mathcal{O}_E^*$ and $E^*/E^{2*} = \{1, u, \pi_E, u\pi_E\}$. the formulae of §1.13 become

$$G_E(x) = \begin{cases} \left(\dfrac{N_{\mathbb{F}_{q^d}/\mathbb{F}_p}(2x)}{p}\right) W_E(\chi_E) & x = 1, u \text{ and } \nu_E \text{ odd,} \\[2ex] 1 & x = \pi_E, u\pi_E \text{ and } \nu_E \text{ odd,} \\[2ex] \left(\dfrac{N_{\mathbb{F}_{q^d}/\mathbb{F}_p}(2(x/\pi_E))}{p}\right) W_E(\chi_E) & x = \pi_E, u\pi_E \text{ and } \nu_E \text{ even,} \\[2ex] 1 & x = 1, u \text{ and } \nu_E \text{ even.} \end{cases}$$

2. The formula of ([72] p. 123 (5)) for the biquadratic extension

2.1. Consider the formula of ([73] p. 123 (5)) for an extension N/E

$$W_E(\text{Ind}_{N/E}(1)) = \delta_{N/E}(g)^{-1} G_N(g)^{-1} G_E(g)^{[N:E]} \in \mu_4$$

which holds for all $g \in E^*$ but only depends upon

$$g \in E^*/E^{2*} = \langle 1, u, \pi_E, u\pi_E \rangle.$$

In this subsection we shall examine the formula in the all important case when $N = E(\sqrt{u}, \sqrt{\pi_E})$. Bear in mind that p is odd so that $E(\sqrt{u})/E$ is unramified and $E(\sqrt{\pi_E})/E$ is totally ramified. The extension N/E is the unique biquadratic extension of E.

Recall that (see §3.4)

$$W_E(\mathrm{Ind}_{N/E}(1)) \quad = SW_2(\mathrm{Ind}_{N/E}(1)) \cdot W_E(\mathrm{Det}(\mathrm{Ind}_{N/E}(1)))$$

$$= SW_2(\mathrm{Ind}_{N/E}(1)) \cdot W_E(\delta_{N/E}).$$

The trace form of N/E is represented (in the sense of Galois descent theory ([125] p. 102 Example (2.31)) by the regular representation of $\mathrm{Gal}(N/E) \cong \mathbb{Z}/2 \times \mathbb{Z}/2$

$$\langle N/E \rangle = 1 + l(u) + l(\pi_E) + l(u\pi_E)$$

in the notation of [125]. Therefore the second Hasse-Witt invariant is given in $H^2(E; \mathbb{Z}/2) \cong \{\pm 1\}$ by

$$HW_2(\langle N/E \rangle) \quad = l(u)l(\pi_E) + l(u)l(u\pi_E) + l(\pi_E)l(u\pi_E)$$

$$= l(u)l(\pi_E) + l(u)(l(u) + l(\pi_E)) + l(\pi_E)(l(u) + l(\pi_E))$$

$$= l(u)l(\pi_E) + l(-1)l(u) + l(-1)l(\pi_E).$$

By a formula of Serre ([116]; [120]; [125] p. 95 Corollary 2.8)

$$SW_2(\mathrm{Ind}_{N/E}(1)) = HW_2(\langle N/E \rangle) + l(2)\delta_{N/E} = HW_2(\langle N/E \rangle)$$

since, in cohomological notation

$$\delta_{N/E} = l(u) + l(\pi_E) + l(u\pi_E) \in H^1(E; \mathbb{Z}/2) \cong E^*/E^2$$

which is trivial because $l(u) + l(\pi_E) = l(u\pi_E)$ (cf. [125] p. 102 Example (2.31)).

Therefore the formula under discussion simplifies to the form

$$(l(u)l(\pi_E)) \cdot (l(-1)l(u)) \cdot (l(-1)l(\pi_E)) = W_E(\mathrm{Ind}_{N/E}(1)) = G_N(g)^{-1} \in \{\pm 1\}$$

where the products on the left-hand side are cup-products in mod 2 Galois cohomology.

Next we need to verify that ν_N is even, which will follow from the transitivity formula for discriminants ([114] III §4). Let us recall how that goes.

We have the trace $\mathrm{Tr}_{L/K} : L \longrightarrow K$ for an extension of local fields L/K. The trace form $\langle L/K \rangle : (x, y) \mapsto \mathrm{Tr}_{L/K}(xy)$ is symmetric and non-singular on L. Let $\{e_i\}$ be a choice of $\mathcal{O}_K$-basis for the free module $\mathcal{O}_L$ then the discriminant of L/K is the ideal of $\mathcal{O}_L$ generated by the element $\det(\mathrm{Tr}_{L/K}(e_i e_j)) = (\det(\sigma(e_i))^2$ where σ runs through the set of K-monomorphisms of L into an algebraic closure of K. Set

$$\mathcal{D}_{L/K}^{-1} = \{y \in L \mid \mathrm{Tr}_{L/K}(xy) \in \mathcal{O}_K \text{ for all } x \in \mathcal{O}_L\},$$

which is the inverse different of L/K (or codifferent) and it is the largest $\mathcal{O}_L$-submodule of L whose image under $\mathrm{Tr}_{L/K}$ lies in $\mathcal{O}_K$. The inverse of the codifferent is the different $\mathcal{D}_{L/K}$ which is a non-zero ideal of $\mathcal{O}_L$. The absolute different is the case when $K = \mathbb{Q}_p$, the prime field. Transitivity for the chain of fields $\mathbb{Q}_p \subseteq E \subseteq N$ takes the form

$$\mathcal{D}_{N/\mathbb{Q}_p} = \mathcal{D}_{N/E} \cdot \mathcal{D}_{E/\mathbb{Q}_p}.$$

Also $\mathcal{D}_{E/\mathbb{Q}_p} = \mathcal{D}_E = (\pi_E)^{\nu_E}$ and the ramification index of N/E is 2 so that the order ν_N of the discriminant of N is even unless $E/\mathbb{Q}_p$ is unramified.

Suppose that $E/\mathbb{Q}_p$ is ramified so that ν_N is even, $\pi_N = \sqrt{\pi_E}$ and

$$G_N(x) = \begin{cases} \left(\dfrac{N_{\mathbb{F}_{q^d}/\mathbb{F}_p}(2(x/\pi_N))}{p} \right) W_N(\chi_N) & x = \pi_N, u_N \pi_N \text{ and } \nu_N \text{ even,} \\[4mm] 1 & x = 1, u_N \text{ and } \nu_N \text{ even.} \end{cases}$$

Therefore $G_N(g) = 1$ for $g \in E^*$.

This means that the unique biquadratic of E extends to a Q_8-extension, which is correct because each p-adic local field has a Q_8-extension and each such extension has a unique biquadratic subfield.

It remains to consider the case where $E/\mathbb{Q}_p$ is unramified and so we may assume $\pi_E = p$. In this case, by the ramification criterion of ([**114**] III §5 Theorem 1) $\mathcal{D}_{E/\mathbb{Q}_p} = \mathcal{O}_E$ and $\mathcal{D}_{N/E} = \mathcal{O}_N$ if and only if N/E is unramified. However $E(\sqrt{u})/E$ is unramified but for $N/E(\sqrt{u})$ in fact the inverse different is the fractional ideal generated by $\pi_N^{-1} = \sqrt{\pi_E}^{-1}$ so that $\nu_N = 1$ and

$$G_N(x) = \begin{cases} \left(\dfrac{N_{\mathbb{F}_{q^d}/\mathbb{F}_p}(2x)}{p} \right) W_N(\chi_N) & x = 1, u_N \text{ and } \nu_N \text{ odd,} \\[4mm] 1 & x = \pi_N, u_N \pi_N \text{ and } \nu_N \text{ odd.} \end{cases}$$

Once again, because of the existence of Q_8 extensions of E we must have, for $g \in E^*$,

$$1 = G_N(g) = \left(\frac{N_{\mathbb{F}_{q^d}/\mathbb{F}_p}(2)}{p} \right) W_N(\chi_N).$$

Now

$$\left(\frac{N_{\mathbb{F}_{q^d}/\mathbb{F}_p}(2)}{p} \right) = \left(\frac{2}{p} \right)^{\dim_{\mathbb{F}_p}(\mathcal{O}_N/\mathcal{P}_N)} = (-1)^{\frac{(p^2-1)\dim_{\mathbb{F}_p}(\mathcal{O}_N/\mathcal{P}_N)}{8}}$$

by the second subsidiary law of quadratic reciprocity.

Next we use the Davenport-Hasse theorem to compute the local root number

$$W_N(\chi_N) = \frac{1}{\sqrt{N\pi_N}} \sum_{w \in (\mathcal{O}_N/\mathcal{P}_N)^*} \chi_N(w)\psi_N(w/c),$$

where we have used the fact that $\chi_N(c) = 1$. Also c generates $f(\chi_N)\mathcal{D}_N = \mathcal{P}_N^{1+1} = (\sqrt{p}^2) = (p)$ so that

$$W_N(\chi_N) = \frac{1}{\sqrt{N\pi_N}} \sum_{w \in (\mathcal{O}_N/\mathcal{P}_N)^*} \chi_N(w)\psi_N(w/p).$$

Now let $L = \mathbb{Q}_p(\sqrt{p})$ then

$$W_L(\chi_L) = \frac{1}{\sqrt{p}} \sum_{x \in \mathbb{F}_p^*} \left(\frac{x}{p}\right)\psi_L(x/p)$$

$$= \left(\frac{2}{p}\right)\frac{1}{\sqrt{p}} \sum_{x \in \mathbb{F}_p^*} \left(\frac{2x}{p}\right)\psi_{\mathbb{Q}_p}(2x/p)$$

$$= \left(\frac{2}{p}\right)W_{\mathbb{Q}_p}(l(p))$$

$$= \begin{cases} \left(\frac{2}{p}\right)(-i) & \text{if } p \equiv 3 \ (\text{modulo } 4) \\ \left(\frac{2}{p}\right) & \text{if } p \equiv 1 \ (\text{modulo } 4), \end{cases}$$

by ([125] p. 266).

Since N/L is unramified the Davenport-Hasse theorem ([87] p. 20) implies that

$$W_N(\chi_N) = -(-W_L(\chi_L))^{\dim_{\mathbb{F}_p}(\mathcal{O}_N/\mathcal{P}_N)}.$$

The residue degree is even so set $\dim_{\mathbb{F}_p}(\mathcal{O}_N/\mathcal{P}_N) = 2d$. Then we have[4]

$$G_N(g) = -\left(\frac{2}{p}\right)^{2d}\begin{cases} \left(\frac{2}{p}\right)^{2d}(-1)^d & \text{if } p \equiv 3 \ (\text{modulo } 4) \\ \left(\frac{2}{p}\right)^{2d} & \text{if } p \equiv 1 \ (\text{modulo } 4) \end{cases}$$

$$= \begin{cases} (-1)^{d+1} & \text{if } p \equiv 3 \ (\text{modulo } 4) \\ -1 & \text{if } p \equiv 1 \ (\text{modulo } 4). \end{cases}$$

2.2. *The Davenport-Hasse theorem when $E/\mathbb{Q}_p$ is unramified*

Suppose that $E/\mathbb{Q}_p$ is unramified of degree d and that $p \neq 2$. The restriction of the trace

$$\text{Trace}_{E/\mathbb{Q}_p} : \mathcal{O}_E \longrightarrow \mathbb{Z}_p$$

---

[4]This seems to contradict the formula of ([73] p. 123 (5)). However this is only a tamely ramified extension.

is surjective so that $\nu_E = 0$ and we may choose $\pi_E = p$. Then $G_E(1) = 1 = G_E(u)$ and for $x = p, up$

$$G_E(x)$$

$$= \left(\frac{N_{\mathbb{F}_{p^d}/\mathbb{F}_p}(2(x/p))}{p}\right) \frac{1}{\sqrt{p^d}} \sum_{w \in (\mathcal{O}_E/\mathcal{P}_E))^*} \left(\frac{N_{\mathbb{F}_{p^d}/\mathbb{F}_p}(w)}{p}\right) \psi_E(p^{-1}w)$$

$$= -\left(\frac{N_{\mathbb{F}_{p^d}/\mathbb{F}_p}(2(x/p))}{p}\right) \frac{1}{\sqrt{p^d}}(-1) \sum_{w \in (\mathcal{O}_E/\mathcal{P}_E))^*} \left(\frac{N_{\mathbb{F}_{p^d}/\mathbb{F}_p}(w)}{p}\right) \psi_E(p^{-1}w)$$

$$= -\left(\frac{N_{\mathbb{F}_{p^d}/\mathbb{F}_p}(2(x/p))}{p}\right) \frac{1}{\sqrt{p^d}}\left(-\sum_{v \in \mathbb{F}_p^*}\left(\frac{v}{p}\right)\xi_p^v\right)^d$$

$$= (-1)^{d+1}\left(\frac{N_{\mathbb{F}_{p^d}/\mathbb{F}_p}(2(x/p))}{p}\right)W_{\mathbb{Q}_p}(l(p))^d,$$

by the Davenport-Hasse theorem ([**87**] p. 20).

QUESTION 2.3. Perhaps the proof of the Davenport-Hasse theorem given in ([**87**] p. 20) would evaluate the Gauss sums of §1.13 in general?

3. p-adic epsilon factors modulo p-primary roots of unity

In this section I shall give the proof of the main result of [**73**].

THEOREM 3.1.
Let σ be a wild, homogeneous representation of $\mathrm{Gal}(\overline{F}/E)$ then

$$W_E(\sigma) \equiv \mathrm{Det}(\sigma)(g_\sigma)^{-1}G_E(g_\sigma)^{\deg(\sigma)} \ (\text{modulo } \mu_{p^\infty}).$$

3.2. *How to use the strict inequality of (*[**73**]* p. 121)*
We shall use the strict inequality of §1.6 and Theorem 1.7

$$\alpha(M/M') < \alpha(\chi'),$$

assuming which I can run the following argument from ([**73**] p. 21). In this case, providing that $\alpha(M/M') < n$, we have a commutative diagram ([**49**] Corollary 3.12(i))

$$\mathcal{P}_M^m/\mathcal{P}_M^{m+1} \xrightarrow{\quad x \mapsto 1+x \quad} (1+\mathcal{P}_M^m)/(1+\mathcal{P}_M^{m+1})$$

$$\Big\downarrow \text{Trace}_{M/M'} \qquad\qquad \Big\downarrow \text{Norm}_{M/M'}$$

$$\mathcal{P}_{M'}^n/\mathcal{P}_{M'}^{n+1} \xrightarrow{\quad y \mapsto 1+y \quad} (1+\mathcal{P}_{M'}^n)/(1+\mathcal{P}_{M'}^{n+1})$$

in which the vertical maps are surjective and where $m = \psi_{M/M'}(n)$.

If we have the strict inequality $\alpha(M/M') < \alpha(\chi')$ we may take $n = \alpha(\chi') = a(\chi') - 1$ and $m = \psi_{M/M'}(\alpha(\chi'))$. By [**49**]

$$\alpha(\chi) = \alpha(\chi' \cdot \text{Norm}_{M/M'}) \leq \psi_{M/M'}(\alpha(\chi')) = m$$

and since $\text{Norm}_{M/M'}$ is surjective the composition $\chi = \chi' \cdot \text{Norm}_{M/M'}$ is non-trivial on $(1+\mathcal{P}_M^m)/(1+\mathcal{P}_M^{m+1})$ so that $m \leq \alpha(\chi)$. Therefore

$$m = \alpha(\chi) \text{ and } a(\chi) = m+1.$$

Now suppose that we have $g' \in (M')^*/(1+\mathcal{P}_{M'})$ such that for all $y \in \mathcal{P}_{M'}^{\alpha(\chi')}/\mathcal{P}_{M'}^{\alpha(\chi')+1}$

$$\chi'(1+y) = \psi_{M'}(g'y)$$

and that we have $g \in M^*/(1+\mathcal{P}_M)$ such that for all $x \in \mathcal{P}_M^{\alpha(\chi)}/\mathcal{P}_M^{\alpha(\chi)+1}$

$$\chi(1+x) = \psi_M(gx).$$

Taking $y = \mathrm{Trace}_{M/M'}(x)$ we have

$$\psi_M(gx) \quad = \chi(1+x)$$

$$= \chi'(\mathrm{Norm}_{M/M'}(1+x))$$

$$= \chi(1+y)$$

$$= \psi_{M'}(g' \cdot y)$$

$$= \psi_{M'}(g' \cdot \mathrm{Trace}_{M/M'}(x))$$

$$= \psi_{M'}(\mathrm{Trace}_{M/M'}(g'x))$$

$$= \psi_M(g'x)$$

which shows that choosing χ' instead of χ leads to the construction of the same element

$$g_\sigma \in ((E^*/1 + \mathcal{P}_E) \otimes \mathbb{Z}[1/p]) = C(E).$$

We have

$$C(E) = (E^*/1 + \mathcal{P}_E) \otimes \mathbb{Z}[1/p] \xrightarrow{\cong} (((M')^*/1 + \mathcal{P}_{(M')}) \otimes \mathbb{Z}[1/p])^G$$

$$\xrightarrow{\cong} ((M^*/1 + \mathcal{P}_M) \otimes \mathbb{Z}[1/p])^G$$

and we have just shown that $g_\sigma = g_{\sigma,\chi} = g_{\sigma,\chi'}$ when considered as elements of $C(E)$. This means that, if the elements $\hat{g}_\sigma \in E^*$ and $\hat{g}_{\sigma,\chi'} \in (M')^*$ both represent $g_\sigma \in C(E)$ then, for some positive integer r,

$$(\hat{g}_\sigma/\hat{g}_{\sigma,\chi'})^{p^r} \in 1 + \mathcal{P}_{M'}.$$

Therefore, if $\rho : (M')^* \longrightarrow \mathbb{C}^*$ is a character of finite order then

$$\rho(\hat{g}_\sigma)/\rho(\hat{g}_{\sigma,\chi'}) \in \mu_{p^\infty}.$$

In addition

$$G_{M'}(\hat{g}_\sigma) = G_{M'}(\rho(\hat{g}_{\sigma,\chi'})).$$

These facts are used below in §3.4.

3.3. The improved induction theorem

Here I shall assume a familiarity with monomial resolutions for finite groups.

If σ is a wild and homogeneous representation of $G = \mathrm{Gal}(K/E)$ as in §1.5 then $V^{(A,\chi)} = V$ and the (A,χ)-part of the monomial resolution for V gives another monomial resolution in which every stabilising pair $(\mathrm{Gal}(K/M_j), \chi_j)$ is larger than or equal to (A,χ). Furthermore the Euler characteristic in $R_+(G)$ is well-defined because the Euler characteristic of the whole monomial resolution is well-defined. Therefore we have

$$\sigma = \sum_i n_i \mathrm{Ind}^G_{\mathrm{Gal}(K/M_i)}(\chi_i) \in R(G)$$

where $A \subseteq \mathrm{Gal}(K/M_i)$ and $\chi_i|A = \chi$ for all i. Therefore each $\mathrm{Ind}^G_{\mathrm{Gal}(K/M_i)}(\chi_i)$ restricts to $[K : M_i]\chi$ on A. This means that

$$g_\sigma = g_{\mathrm{Ind}^G_{\mathrm{Gal}(K/M_i)}(\chi_i)}$$

for each i.

Note: There is a subtlety here to be careful of.

The entire representation $\mathrm{Ind}^G_{\mathrm{Gal}(K/M_i)}(\chi_i)$ is wild and homogeneous (with the same associated character as σ) but this does not mean that χ_i is wild. We know that A is a ramification group for $\mathrm{Gal}(K/M_i)$ with the same lower numbering as A has for G but the definition of g_{χ_i} depends on the upper numbering which does not intersect well with subgroups!

3.4. The proof of Theorem 3.1

The result is already known when $p = 2$ [138] so henceforth p is an odd prime.

Write $\lambda_{N/E} = W_E(\mathrm{Ind}_{N/E}(1)) \in \mu_4$. The $\lambda_{N/E}$'s are a very subtle and important family of numbers, to the construction of which approximately 200 pages of the 400 pages essay [89] are devoted. In [48] it is shown that, for any orthogonal Galois representation,

$$W_E(\sigma) = SW_2(\sigma)W_E(\mathrm{Det}(\sigma)).$$

In ([125] p. 274; see also [122]) is given a very quick construction of $W_E(\sigma)$ when σ is orthogonal, which immediately gives the existence and Deligne's formula. The formula is used in §2.1.

Suppose the we are given a wild, homogeneous representation σ as in §1.5. Define

$$\zeta(\sigma) = \mathrm{Det}(\sigma)(g_\sigma)^{-1}G_E(g_\sigma)^{\deg(\sigma)} \in \mathbb{C}^*/\mu_{p^\infty}$$

so that our objective is to show that

$$W_E(\sigma) = \zeta(\sigma) \in \mathbb{C}^*/\mu_{p^\infty}.$$

Notice that if σ and τ are two wild, homogeneous representations such that $g_\sigma = g_\tau$ then $\zeta(\sigma \oplus \tau) = \zeta(\sigma)\zeta(\tau)$. Furthermore when η is one-dimensional we have $W_E(\eta) = \zeta(\eta) \in \mathbb{C}^*/\mu_{p^\infty}$ by ([64] p. 4). Using the inductivity properties of $W_E(-)$ and the induction theorem of §3.3 we shall derive Theorem 3.1 from the one-dimensional case.

Consider the equation of §1.5

$$\sigma = \sum_i n_i \mathrm{Ind}^{\mathrm{Gal}(K/E)}_{\mathrm{Gal}(K/M_i)}(\chi_i) \in R(\mathrm{Gal}(K/E)).$$

By inductivity (in relative dimension zero) of the local constants we have

$$W_E(\sigma) = \prod_i W_E(\mathrm{Ind}_{\mathrm{Gal}(K/M_i)}^{\mathrm{Gal}(K/E)}(\chi_i))^{n_i} = \prod_i \lambda_{M_i/E}^{n_i} W_{M_i}(\chi_i)^{n_i}.$$

Since, by §3.3, for each i

$$g_\sigma = g_{\mathrm{Ind}_{\mathrm{Gal}(K/M_i)}^G(\chi_i)}$$

we find that

$$\zeta(\sigma) = \prod_i \zeta(\mathrm{Ind}_{\mathrm{Gal}(K/M_i)}^{\mathrm{Gal}(K/E)}(\chi_i))^{n_i} \in \mathbb{C}^*/\mu_{p^\infty}.$$

Therefore, if we can show that for each i

$$\frac{W_E(\mathrm{Ind}_{\mathrm{Gal}(K/M_i)}^{\mathrm{Gal}(K/E)}(\chi_i))}{W_{M_i}(\chi_i)} = \frac{\zeta(\mathrm{Ind}_{\mathrm{Gal}(K/M_i)}^{\mathrm{Gal}(K/E)}(\chi_i))}{\zeta(\chi_i)} \in \mathbb{C}^*/\mu_{p^\infty}$$

then the result follows because $W_{M_i}(\chi_i) = \zeta(\chi_i) \in \mathbb{C}^*/\mu_{p^\infty}$.

In the situation of §1.5 we have $\mathrm{Res}_A^{\mathrm{Gal}(K/E)}(\sigma) = n\chi_\sigma$ and on $1 + \mathcal{P}_M^{a(\chi_\sigma)-1}$ we have $\chi_\sigma(1+x) = \psi_M(g_\sigma x)$ and $g_\sigma \in C(E)[1/p]$.
Now suppose (N, χ) is one of the (M_i, χ_i)'s so that

$$(A, \chi_\sigma) \le (H = \mathrm{Gal}(K/N), \chi) \text{ and } N \le M.$$

Therefore, if $\delta_{N/E} = \mathrm{Det}(\mathrm{Ind}_{N/E}(1))$, in $\mathbb{C}^*/\mu_{p^\infty}$ we have

$$W_E(\mathrm{Ind}_{N/E}(\chi)) = \lambda_{N/E} W_N(\chi)$$

$$= \delta_{N/E}(g)^{-1} G_N(g)^{-1} G_E(g)^{[N:E]} \chi(\hat{g})^{-1} G_N(\hat{g}),$$

by Proposition 3.5, for any $g \in E^*$ and where on $1 + \mathcal{P}_N^{a(\chi)-1}$ we have $\chi(1+y) = \psi_N(\hat{g}y)$. Then $\hat{g} = g_\sigma \in C(M)[1/p]$ so that $G_N(g_\sigma) = G_N(\hat{g})$.
Therefore

$$W_E(\mathrm{Ind}_{N/E}(\chi)) = \delta_{N/E}(g_\sigma)^{-1} G_E(g_\sigma)^{[N:E]} \chi(\hat{g})^{-1} \in \mathbb{C}^*/\mu_{p^\infty}$$

but $g_\sigma/\hat{g}$ lies in a pro-p group so $\chi(\hat{g})/\chi(g_\sigma) \in \mu_{p^\infty}$. Hence, by the formula

$$\mathrm{Det}(\mathrm{Ind}_{N/E}(\chi)) = \mathrm{Res}_{E^*}^{N^*}(\chi)\delta_{N/E}$$

of ([47] Proposition 1.2)

$$W_E(\mathrm{Ind}_{N/E}(\chi)) = \mathrm{Det}(\mathrm{Ind}_{N/E}(\chi))^{-1}(g_\sigma) G_E(g_\sigma)^{[N:E]} \in \mathbb{C}^*/\mu_{p^\infty}$$

which implies the general formula for $W_E(\sigma)$ (modulo μ_{p^∞}). $\square$

PROPOSITION 3.5. ([73] Proposition 1 p. 123)
Let K be a finite extension of E in $\overline{F}$ and let $g \in C_E \otimes \mathbb{Z}[1/p]$. Then

$$\frac{W_E(\mathrm{Ind}_{K/E}(\chi))}{W_K(\chi)} = \delta_{K/E}(g)^{-1} G_K(g)^{[K:E]} \in \mathbb{C}^*/\mu_{p^\infty}$$

for every character χ of K^*.

APPENDIX III

Finite general linear and symmetric groups

This Appendix recalls in §1 the characterisation of irreducible complex representations of the symmetric groups and in §2 those for finite general linear groups, together with the construction of their zeta functions. §3 describes how to generalise the Kondo-Gauss sums [**85**] by means of a formula in terms of character values and then derives the functorial properties of the Kondo-style Gauss sums.

1. Symmetric groups

1.1. Following ([**93**] Chapter One, §7) we recall the classification of irreducible representations of the symmetric group Σ_n over an algebraically closed field of characteristic zero. Without loss of generality we shall stick to complex representations in this appendix. Denote by Λ_n the algebra of symmetric polynomials $\mathbb{Z}[x_1, \ldots, x_n]^{\Sigma_n}$ whose homogeneous of degree k part will be written Λ_n^k ([**93**] p. 10) and set $\Lambda = \oplus_{n \geq 0} \Lambda_n$. Hence Λ is a graded algebra.

Define $p_r = \sum x_i^r \in \Lambda$ and if $\underline{\lambda} = (\lambda_1, \lambda_2, \ldots)$ is an integral partition we write ([**93**] p. 15)

$$p_{\underline{\lambda}} = p_{\lambda_1} p_{\lambda_2} \cdots p_{\lambda_i} \cdots .$$

Each permutation $w \in \Sigma_n$ factorises uniquely as a product of disjoint cycles. If the orders of the cycles are $\rho_1, \rho_2, \ldots$ with $\rho_1 \geq \rho_2 \geq \ldots$ we set $\underline{\rho}(w) = (\rho_1, \rho_2, \ldots)$ which is a partition of n called the cycle type of w. It determines the conjugacy class of w in Σ_n giving a bijection between conjugacy classes and partitions.

Define a mapping ([**93**] p. 60)

$$\psi : \Sigma_n \longrightarrow \Lambda_n$$

by $\psi(w) = p_{\underline{\rho}(w)}$ which satisfies the multiplicative relation $\psi(v \times w) = \psi(v)\psi(w)$ where $v \times w$ is the disjoint union of the permutations v and w.

Next consider $R^n = R(\Sigma_n)$, the (complex) representation ring of the symmetric group and $R = \oplus_{n \geq 0} R^n$. This is a graded algebra (in fact, as explained by Zelevinsky [**146**], it is a PSH Hopf algebra). The product is

303

given on $x \in R^n$ and $y \in R^m$ by

$$x \cdot y = \operatorname{Ind}_{\Sigma_n \times \Sigma_m}^{\Sigma_{n+m}} (x \otimes y).$$

Moreover R carries the Schur inner product

$$\left\langle \sum f_n, \sum g_n \right\rangle = \oplus_n \langle f_n, g_n \rangle_{\Sigma_n}.$$

There is also a scalar product on Λ ([93] p. 34). For partitions $\underline{\lambda}, \underline{\mu}$ we define symmetric functions ([93] pp. 11–15)

$$m_{\underline{\lambda}} = \sum x_1^{\alpha_1} x_2^{\alpha_2} \ldots = \sum x^{\underline{\alpha}}$$

where $\underline{\alpha}$ runs through all the permutations of $(\lambda_1, \ldots, \lambda_n)$, n is the length of $\underline{\lambda}$ and for each positive integer r

$$h_r = \sum_{|\underline{\lambda}|=r} m_{\underline{\lambda}} \text{ and } h_{\underline{\lambda}} = h_{\lambda_1} h_{\lambda_2} \ldots .$$

The inner product is characterised by

$$\langle h_{\underline{\lambda}}, m_{\underline{\mu}} \rangle = \begin{cases} 1 & \text{if } \underline{\lambda} = \underline{\mu}, \\ 0 & \text{otherwise.} \end{cases}$$

The characteristic homomorphism is a $\mathbb{Z}$-linear mapping

$$\operatorname{ch} : R \longrightarrow \Lambda \otimes_{\mathbb{Z}} \mathbb{Q}$$

which sends $f \in R^n$ to

$$\operatorname{ch}(f) = \frac{1}{n!} \sum_{w \in \Sigma_n} f(w) \psi(w).$$

If $\underline{\lambda}$ is a partition write $m_i(\underline{\lambda})$ for the number of integers in $\underline{\lambda}$ which are equal to i and define ([93] p. 17)

$$z_{\underline{\lambda}} = \prod_{i \geq 1} i^{m_i} \cdot m_i!.$$

The integer $n!/z_{\underline{\lambda}}$ is equal to the number of elements of Σ_n which have cycle type $\underline{\lambda}$ when $|\underline{\lambda}| = n$. The combinatorial formula for the characteristic homomorphism is

$$\operatorname{ch}(f) = \sum_{|\underline{\rho}|=n} z_{\underline{\rho}}^{-1} f_{\underline{\rho}} p_{\underline{\rho}}$$

where $f_{\underline{\rho}}$ is the value of f at elements of cycle type $\underline{\rho}$.

The following result is important in the classification of irreducible representations of Σ_n.

PROPOSITION 1.2. *([93] p. 61)*
The characteristic homomorphism is an isometric isomorphism of R onto Λ.

Proof

The characteristic map is a ring homomorphism, by Frobenius reciprocity, if $f \in R^m, g \in R^n$ then

$$\mathrm{ch}(f \cdot g) = \langle \mathrm{Ind}_{\Sigma_m \times \Sigma_n}^{\Sigma_{n+m}} (f \otimes g), \psi \rangle_{\Sigma_{m+n}}$$

$$= \langle f \otimes g, \mathrm{Res}_{\Sigma_m \times \Sigma_n}^{\Sigma_{n+m}} (\psi) \rangle_{\Sigma_m \times \Sigma_n}$$

$$= \langle f, \psi \rangle_{\Sigma_m} \langle g, \psi \rangle_{\Sigma_n}$$

$$= \mathrm{ch}(f) \mathrm{ch}(g).$$

Now let $\mathbf{1}_n$ denote the one-dimensional trivial representation of Σ_n. If $\underline{\lambda} = (\lambda_1, \lambda_2, \dots)$ is a partition of n let $\mathbf{1}_{\underline{\lambda}} = \mathbf{1}_{\lambda_1} \mathbf{1}_{\lambda_2} \dots$. Since

$$\mathrm{ch}(\mathbf{1}_n) = \sum_{|\underline{\rho}| = n} z_{\underline{\rho}}^{-1} p_{\underline{\rho}} = h_n$$

we see that $\mathrm{ch}(\mathbf{1}_{\underline{\lambda}}) = h_{\underline{\lambda}}$.

Now, for each partition $\underline{\lambda}$ of n define

$$\chi^{\underline{\lambda}} = \det(\mathbf{1}_{\lambda_i - i + j}) \in R^n$$

where the suffices i, j in the matrix each run through $1, 2, \dots, n$. Therefore $\chi^{\underline{\lambda}}$ is a virtual representation of Σ_n which is characterised by $\mathrm{ch}(\chi^{\underline{\lambda}}) = s_{\underline{\lambda}}$. Since ch is an isometry we obtain

$$\langle \chi^{\underline{\lambda}}, \chi^{\underline{\mu}} \rangle_{\Sigma_n} = \begin{cases} 1 & \text{if } \underline{\lambda} = \underline{\mu}, \\ \\ 0 & \text{otherwise.} \end{cases}$$

A counting argument completes the proof. $\square$

1.3. *The $s_{\underline{\lambda}}$'s ([93] pp. 23-24)*

Writing $x^{\underline{\alpha}} = x_1^{\alpha_1} x_2^{\alpha_2} \dots$ for a strictly non-negative multi-index $\underline{\alpha} = (\alpha_1, \alpha_2, \dots)$, define

$$a_{\underline{\alpha}} = \sum_{w \in \Sigma_n} \mathrm{sign}(w) w(x^{\underline{\alpha}}),$$

the antisymmetrisation of $x^{\underline{\alpha}}$. Since $a_{\underline{\alpha}}$ is skew symmetric it vanishes unless the non-negative integers $\alpha_1, \alpha_2, \dots, \alpha_n$ are all distinct. Hence we may assume $\alpha_1 > \alpha_2 > \dots > \alpha_n \geq 0$. Therefore we may write $\underline{\alpha}$ as the sum of two partitions

$$\underline{\alpha} = \underline{\lambda} + \underline{\delta} = \underline{\lambda} + (n-1, n-2 \dots, 1, 0)$$

where $\alpha_i = \lambda_i + n - i$. In $\mathbb{Z}[x_1, \dots, x_n]$ the polynomial $a_{\underline{\lambda} + \underline{\delta}}$ is divisible by $a_{\underline{\delta}}$ and we define $s_{\underline{\lambda}} = a_{\underline{\lambda} + \underline{\delta}} / a_{\underline{\delta}}$, which is a symmetric polynomial. Note that $a_{\underline{\alpha}}$ is equal to a Vandermond-type determinant whose (i, j)-th entry is $x_i^{\alpha_j}$.

PROPOSITION 1.4. *([93] p. 62)*
The irreducible representations of Σ_n are $\{\chi^{\underline{\lambda}}; |\underline{\lambda}| = n\}$.

Proof

By Proposition 1.2 (proof) we know that each $\chi^{\underline{\lambda}}$ is either an irreducible representation or minus an irreducible representation in R^n. To show that the latter is not the case it suffices to show that $\dim(\chi^{\underline{\lambda}}) = \chi^{\underline{\lambda}}(1) > 0$.
We have $([93]$ p. 61 (7.2))

$$s_{\underline{\lambda}} = \mathrm{ch}(\chi^{\underline{\lambda}}) = \sum_{|\underline{\rho}|=n} z_{\underline{\rho}}^{-1} \chi^{\underline{\lambda}}_{\underline{\rho}} p_{\underline{\rho}}$$

so that $\chi^{\underline{\lambda}}_{\underline{\rho}} = \langle s_{\underline{\lambda}}, p_{\underline{\rho}} \rangle$ and in particular

$$\chi^{\underline{\lambda}}(1) = \chi^{\underline{\lambda}}_{(1,1,\dots,1)} = \langle s_{\underline{\lambda}}, p_1^n \rangle$$

so that

$$h_1^n = p_1^n = \sum_{|\underline{\lambda}|=n} \chi^{\underline{\lambda}}(1) s_{\underline{\lambda}}.$$

Therefore, in the notation for transition matrices of $([93]$ p. 56),

$$\chi^{\underline{\lambda}}(1) = M(h, s)_{(1,1,\dots,1),\underline{\lambda}}$$

and so $\chi^{\underline{\lambda}}(1)$ equals the number of standard tableaux of shape $\underline{\lambda}$, which is a positive integer. $\square$

REMARK 1.5. In the course of the proof of Proposition 1.4 one sees that the transition matrix $M(p, s)$ gives the character table of Σ_n. That is,

$$p_{\underline{\rho}} = \sum_{\lambda} \chi^{\underline{\lambda}}_{\underline{\rho}} s_{\underline{\lambda}}.$$

1.6. *Maximal terms in* $a_{\Sigma_n}(\chi^{\underline{\lambda}})$
In the notation of Appendix I, Section Five the element $a_G(\rho) \in R_+(G)$ is an explicit formula for Brauer's Induction Theorem when G is a finite group and ρ is a representation of G over an algebraically closed field of characteristic zero.
I would like a formula for the coefficient in $a_{\Sigma_n}(\chi^{\underline{\lambda}})$ of $(C_{\underline{\mu}}, \phi_{\underline{\mu}})^{\Sigma_n}$ where $(C_{\underline{\mu}}, \phi_{\underline{\mu}})$ is maximal in $\mathcal{M}_{\Sigma_n}$ of the form

$$C_{\underline{\mu}} = C_{\mu_1} \times C_{\mu_2} \times \dots \times C_{\mu_r} \subseteq \Sigma_{\mu_1} \times \Sigma_{\mu_2} \times \dots \times \Sigma_{\mu_r}$$

and $\phi_{\underline{\mu}} = \phi_{\mu_1} \otimes \phi_{\mu_2} \otimes \dots \otimes \phi_{\mu_r}$ with C_j a cyclic group of order j.
The formula for this is

$$|N_{\Sigma_n}(C_{\underline{\mu}}, \phi_{\underline{\mu}})|^{-1} \sum_{g \in C_{\underline{\mu}}} \chi^{\underline{\lambda}}(g) \overline{\phi}_{\underline{\mu}}(g).$$

In order to benefit from the combinatorial formulae of [**93**] this should be written in terms of the $\chi_{\underline{\rho}}^{\underline{\lambda}}$'s.

However, we shall take an easier route.

Now I want to evaluate the terms involving some maximal pairs $(H,\phi)^{\Sigma_n}$ in $a_{\Sigma_n}(\chi^{\underline{\lambda}})$. To do this is will suffice to replace $\chi^{\underline{\lambda}}$ by $\mathrm{Ind}_{\Sigma_{\underline{\lambda}}}^{\Sigma_n}(1)$ where $\Sigma_{\underline{\lambda}} = S_{\lambda_1} \times \ldots \times \Sigma_{\lambda_d}$. This representation is denoted by $\eta_{\underline{\lambda}}$ in ([**93**] p. 61). As explained in ([**9**] §1) both the $\mathrm{Ind}_{\Sigma_{\underline{\lambda}}}^{\Sigma_n}(1)$'s and the $\chi^{\underline{\lambda}}$'s form a free basis for the abelian group $R(\Sigma_n)$. Since $\mathrm{ch}(\chi^{\underline{\lambda}}) = s_{\underline{\lambda}}$ and $\mathrm{ch}(\eta^{\underline{\lambda}}) = h_{\underline{\lambda}}$ there is an invertible square matrix with integer entries, in the notation of ([**93**] p. 55) it is the transition matrix $M(s,h)$ such that

$$s_{\underline{\lambda}} = \sum_{\underline{\mu}} M(s,h)_{\underline{\lambda},\underline{\mu}}\, \eta_{\underline{\mu}} \text{ and } \chi^{\underline{\lambda}} = \sum_{\underline{\mu}} M(s,h)_{\underline{\lambda},\underline{\mu}}\, \mathrm{Ind}_{\Sigma_{\underline{\mu}}}^{\Sigma_n}(1).$$

Now suppose that $(C_{\underline{\mu}}, \phi_{\underline{\mu}})$ is a maximal pair as in §1.6 and that $\phi_{\underline{\mu}} = \phi_{\mu_1} \otimes \ldots \phi_{\mu_r}$ where χ_{μ_j} is a faithful character on C_{μ_j}. This means that $\phi_{\underline{\mu}}$ is non-trivial on any subgroup of $C_{\underline{\mu}}$. Then

coefficient of $(C_{\underline{\mu}}, \phi_{\underline{\mu}})^{\Sigma_n}$ in $a_{\Sigma_n}(\mathrm{Ind}_{\Sigma_{\underline{\lambda}}}^{\Sigma_n}(1))$

$$= \frac{|C_{\underline{\mu}}|}{|\Sigma_n|} \frac{|\Sigma_n|}{|N_{\Sigma_n}(C_{\underline{\mu}},\phi_{\underline{\mu}})|} \langle \phi_{\underline{\mu}}, \mathrm{Res}_{C_{\underline{\mu}}}^{\Sigma_n} \mathrm{Ind}_{\Sigma_{\underline{\lambda}}}^{\Sigma_n}(1) \rangle_{C_{\underline{\mu}}}$$

$$= \frac{|C_{\underline{\mu}}|}{|N_{\Sigma_n}(C_{\underline{\mu}},\phi_{\underline{\mu}})|} \sum_{z \in C_{\underline{\mu}}\backslash \Sigma_n/\Sigma_{\underline{\lambda}}} \langle \phi_{\underline{\mu}}, \mathrm{Ind}_{C_{\underline{\mu}}\cap z\Sigma_{\underline{\lambda}}z^{-1}}^{C_{\underline{\mu}}}(1) \rangle_{C_{\underline{\mu}}}$$

$$= \frac{|C_{\underline{\mu}}|}{|N_{\Sigma_n}(C_{\underline{\mu}},\phi_{\underline{\mu}})|} |\{z \in C_{\underline{\mu}}\backslash \Sigma_n/\Sigma_{\underline{\lambda}} \mid C_{\underline{\mu}} \cap z\Sigma_{\underline{\lambda}}z^{-1} = \{1\}\}|.$$

1.7. An interesting combinatorial polynomial identity

This section describes a result due to Francesco Mezzadri, my son-in-law. The following sketch of the proof is my responsibility, both for the method and any errors therein!

Let $\underline{\lambda}$ be a partition of n. Francesco was interested in factorising the polynomial

$$f_{\underline{\lambda}}(x) = \frac{1}{\chi^{\underline{\lambda}}(1)} \sum_{g \in \Sigma_n} \chi^{\underline{\lambda}}(g) x^{\mathrm{length}(\underline{\rho}(g))}$$

where $\mathrm{length}(\underline{\rho}(g))$ is the length of the partition $\underline{\rho}(g)$ corresponding to the cycle type of g (see [**83**]). Recall that the length of a permutation whose expression in terms of disjoint cycles has r_i i-cycles is $\sum_i r_i$. Since $n!/z_{\underline{\rho}(g)}$ equals the number of elements in the conjugacy class of g we have

$$f_{\underline{\lambda}}(x) = \frac{n!}{\chi^{\underline{\lambda}}(1)} \sum_{|\underline{\rho}|=n} \frac{\chi_{\underline{\rho}}^{\underline{\lambda}}}{z_{\underline{\rho}}} x^{\mathrm{length}(\underline{\rho})}.$$

On the other hand the characteristic homomorphism satisfies ([**93**] p. 63)

$$s_{\underline{\lambda}} = \mathrm{ch}(\chi^{\underline{\lambda}}) = \sum_{|\underline{\rho}|=n} \frac{\chi^{\underline{\lambda}}_{\underline{\rho}}}{z_{\underline{\rho}}} p_{\underline{\rho}}$$

in the ring of symmetric polynomials in $x_1,\ldots,x_N$ with $N \geq n$. The symmetric polynomials $s_{\underline{\lambda}}$ are, I believe, called the Schur functions.

If we set $1 = x_1 = \ldots = x_N$ we obtain $p_{\underline{\rho}}|_{\{x_i=1,\text{ all }i\}} = N^{\mathrm{length}(\underline{\rho})}$. Therefore

$$f_{\underline{\lambda}}(N) = \frac{n!}{\chi^{\underline{\lambda}}(1)} s_{\underline{\lambda}}|_{\{x_i=1,\text{ all }i\}}.$$

As is well-known, first observed by Weyl I believe, there is an isomorphism of $\Sigma_n \times GL_N\mathbb{C}$-representations of the form ([**9**] §1)

$$(\mathbb{C}^N)^{\otimes n} \cong \sum_{|\underline{\lambda}|=n} \chi^{\underline{\lambda}} \otimes \mathrm{Hom}_{\Sigma_n}(\chi^{\underline{\lambda}}, (\mathbb{C}^N)^{\otimes n}).$$

We write $W_{\underline{\lambda}} = \mathrm{Hom}_{\Sigma_n}(\chi^{\underline{\lambda}}, (\mathbb{C}^N)^{\otimes n})$. Hence

$$N^n = \sum_{|\underline{\lambda}|=n} \chi^{\underline{\lambda}}(1)W_{\underline{\lambda}}(1).$$

The characteristic homomorphism followed by evaluating at $1 = x_1 = \ldots = x_N$ gives a homomorphism depending on N from R to the integers. The theory of polynomial functors, I believe, shows that this homomorphism sends $\chi^{\underline{\lambda}}$ to $W_{\underline{\lambda}}(1)$ so that

$$f_{\underline{\lambda}}(N) = \frac{n!W_{\underline{\lambda}}(1)}{\chi^{\underline{\lambda}}(1)}.$$

The dimensional equation for $(\mathbb{C}^N)^{\otimes n}$ leads to the result, if $\mathrm{length}(\underline{\lambda}) \leq N$, that

$$W_{\underline{\lambda}}(1)\chi^{\underline{\lambda}}(1) = \prod_{1 \leq j \leq k \leq N} \frac{\lambda_j - \lambda_k + k - j}{k - j}$$

which yields

$$f_{\underline{\lambda}}(N) = \prod_{i=1}^{\mathrm{length}(\underline{\lambda})} (N-i+1)(N-i+2)\ldots(N+\lambda_i-i).$$

Varying N yields the following result:

THEOREM 1.8. *(F. Mezzadri 2010; appearing in* [**83**]*)*

$$f_{\underline{\lambda}}(x) = \prod_{i=1}^{\mathrm{length}(\underline{\lambda})} (x-i+1)(x-i+2)\ldots(x+\lambda_i-i).$$

2. Irreducibles for $GL_n\mathbb{F}_q$ and their zeta functions

2.1. *Irreducible representations of* $GL_n\mathbb{F}_q$ *([93] §1)*

Let $\hat{\mathbb{F}}^*_{q^n}$ denote the group of characters of the multiplicative group of the field of q^n elements. When m divides n the (surjective) norm homomorphism induces an injective map

$$N^*_{n,m} : \hat{\mathbb{F}}^*_{q^m} \longrightarrow \hat{\mathbb{F}}^*_{q^n}$$

and we set

$$\Gamma = \lim_{\substack{n \\ \rightarrow}} \hat{\mathbb{F}}^*_{q^n},$$

which is a discrete torsion group non-canonically isomorphic to the multiplicative group of $\overline{\mathbb{F}}_q$, the algebraic closure of $\mathbb{F}_q$. The Frobenius automorphism Fr acts on Γ as the q-th power map. Write Γ_n for the fixed points of Fr^n. We have a pairing

$$\langle -, - \rangle : \hat{\mathbb{F}}^*_{q^n} \times \mathbb{F}^*_{q^n} \longrightarrow \mathbb{C}^*$$

given by $\langle \gamma, x \rangle = \gamma(x)$.

If f is a Frobenius orbit in Γ denote by $d(f)$ the number of elements in f, called the degree of f. The irreducible representations of $GL_n\mathbb{F}_q$ are described in the following manner, which originally appeared in [69]. To each Frobenius orbit $f \in \Gamma$ with $d(f) = n$ there corresponds a unique irreducible cuspidal[1] representation of $GL_n\mathbb{F}_q$ and the remaining irreducible representations of $GL_n\mathbb{F}_q$ are constructed from these representations of $GL_m\mathbb{F}_q$ for $m < n$.

Let $n_1, \dots, n_r$ be positive integers whose sum is equal to n. Let $P_{n_1,\dots,n_r} \subset GL_n\mathbb{F}_q$ be the standard parabolic subgroup consisting of matrices of blocks of the form

$$\begin{pmatrix} Y_{1,1} & Y_{1,2} & Y_{1,3} & \cdots & \cdots & \cdots \\ 0 & Y_{2,2} & Y_{2,3} & \cdots & \cdots & \cdots \\ 0 & 0 & Y_{3,3} & \cdots & \cdots & \cdots \\ 0 & \vdots & \vdots & \vdots & \vdots & \vdots \\ 0 & \vdots & \vdots & \vdots & \vdots & \vdots \\ 0 & 0 & 0 & \cdots & \cdots & Y_{r,r} \end{pmatrix}$$

where $Y_{i,i} \in GL_{n_i}\mathbb{F}_q$ and all blocks below the diagonal are zero. The parabolic subgroup is a semi-direct product $P_{n_1,\dots,n_r} = U_{n_1,\dots,n_r} D_{n_1,\dots,n_r}$

---

[1]Cuspidal representations are also sometimes called discrete series representations.

where $U_{n_1,\dots,n_r}$ is the subgroup of $P_{n_1,\dots,n_r}$ in which the diagonal blocks are identity matrices and $D_{n_1,\dots,n_r}$ is the subgroup in which all off-diagonal blocks are zero. Hence

$$D_{n_1,\dots,n_r} \cong GL_{n_1}\mathbb{F}_q \times GL_{n_2}\mathbb{F}_q \times \dots \times GL_{n_r}\mathbb{F}_q.$$

Suppose we are given irreducible representations π_i of $GL_{n_i}\mathbb{F}_q$ for $1 \leq i \leq r$ then the iterated tensor product $\pi_1 \otimes \pi_2 \otimes \dots \otimes \pi_r$ is an irreducible representation of $D_{n_1,\dots,n_r}$. The "induction product" of $\pi_1, \pi_2, \dots, \pi_r$ is denoted by $\pi_1 \circ \pi_2 \circ \dots \circ \pi_r$ and defined by

$$\pi_1 \circ \pi_2 \circ \dots \circ \pi_r = \operatorname{Ind}_{P_{n_1,\dots,n_r}}^{GL_n\mathbb{F}_q} \operatorname{Inf}_{D_{n_1,\dots,n_r}}^{P_{n_1,\dots,n_r}} (\pi_1 \otimes \pi_2 \otimes \dots \otimes \pi_r).$$

The induction product is part of the PSH algebra structure used in the approach of [146] to the results of [69].

Suppose for the moment that $\pi_1 = \pi_2 = \dots = \pi_r = (f)$, which is a cuspidal representation of $GL_d\mathbb{F}_q$ with $d = d(f)$ and $n = rd$. The commuting algebra of $(f) \circ (f) \circ \dots \circ (f) = (f)^{\circ r}$ is the group-ring of the symmetric group Σ_r. Therefore the irreducible $GL_n\mathbb{F}_q$-components of $(f)^{\circ r}$ are given by $(f^\lambda) = \operatorname{Hom}_{\Sigma_r}(\lambda, (f)^{\circ r})$ as λ runs through the irreducible representations of Σ_r. From Proposition 1.4 the irreducible representations λ are indexed by partitions of r.

More generally suppose that λ_i is a partition of $|\lambda_i|$ for $1 \leq i \leq m$ and that $f_1, \dots, f_m$ are distinct Frobenius orbits in Γ with $n = \sum_{i=1}^m d(f_i)|\lambda_i|$ then $(f_1^{\lambda_1}) \circ (f_2^{\lambda_2}) \circ \dots \circ (f_m^{\lambda_m})$ is an irreducible representation of $GL_n\mathbb{F}_q$ and all irreducible representations are obtained in this way, without repetition.

If we think of the above data as a function λ from Frobenius orbits in Γ to partitions given by $f_i \mapsto \lambda_i$ for $1 \leq i \leq m$ and sending every other Frobenius orbit to zero then we shall denote the resulting irreducible representation by π_λ. In other words $\lambda \mapsto \pi_\lambda$ gives a canonical bijection between partition-valued functions on Frobenius orbits of Γ and irreducible representations of $GL_n\mathbb{F}_q$ as n varies.

2.2. Zeta functions for $GL_n\mathbb{F}_q$

Let $\Psi : M_n\mathbb{F}_q \longrightarrow \mathbb{C}^*$ be the usual additive character, as used in the section on Kondo-Gauss sums (Definition 3.1). Let $C(M_n\mathbb{F}_q)$ denote the space of complex-valued functions on $M_n\mathbb{F}_q$, the set of $n \times n$ matrices with entries in $\mathbb{F}_q$.

As we shall soon see as we recapitulate the process of ([93] §2), the finite field case is precisely taking the top level in the Godement-Jacquet approach to zeta functions for GL_n of a local field (see [40] pp. 147–155).

For $\Phi \in C(M_n\mathbb{F}_q)$ the Fourier transform of Φ is $\hat{\Phi} \in C(M_n\mathbb{F}_q)$ given by

$$\hat{\Phi}(X) = q^{-n^2/2} \sum_{Y \in M_n\mathbb{F}_q} \Phi(Y)\Psi(XY).$$

Hence

$$\hat{\hat{\Phi}}(X)$$

$$= q^{-n^2/2} \sum_{Y \in M_n\mathbb{F}_q} \hat{\Phi}(Y)\Psi(XY)$$

$$= q^{-n^2/2} \sum_{Y \in M_n\mathbb{F}_q} q^{-n^2/2} \sum_{Z \in M_n\mathbb{F}_q} \Phi(Z)\Psi(YZ)\Psi(XY)$$

$$= q^{-n^2} \sum_{Y \in M_n\mathbb{F}_q} \sum_{Z \in M_n\mathbb{F}_q} \Phi(Z)\Psi(Y(Z+X)).$$

Fixing Z and hence fixing $Z + X$ we see that $Y \mapsto \Psi(Y(Z + X))$ is a non-trivial character unless $X + Z = 0$ and so this subsum over Y vanishes. Therefore

$$\hat{\hat{\Phi}}(X) = q^{-n^2} \sum_{Y \in M_n\mathbb{F}_q} \Phi(-X) = \Phi(-X),$$

which is the usual relation satisfied by the Fourier transfer.

Let π be a finite-dimensional complex representation of $GL_n\mathbb{F}_q$. Then the zeta function associated to $\Phi \in C(M_n\mathbb{F}_q)$ and π is defined by

$$\zeta(\Phi, \pi) = \sum_{X \in GL_n\mathbb{F}_q} \Phi(X)\pi(X) \in M_{\dim_{\mathbb{C}}(\pi)}(\mathbb{C}).$$

Define the normalised trace of the zeta function to be

$$\zeta_{\mathrm{tr}}(\Phi, \pi) = \frac{1}{\dim_{\mathbb{C}}(\pi)} \sum_{X \in GL_n\mathbb{F}_q} \Phi(X)\chi_\pi(X) \in \mathbb{C}.$$

Hence

$$\mathrm{Trace}(\zeta(\Phi, \pi)) = \dim_{\mathbb{C}}(\pi) \cdot \zeta_{\mathrm{tr}}(\Phi, \pi).$$

Define

$$W(\pi, \Psi, X) = q^{-n^2/2} \sum_{Y \in GL_n\mathbb{F}_q} \pi(Y)\Psi(YX)$$

and

$$W_{\mathrm{tr}}(\pi, \Psi, X) = \frac{q^{-n^2/2}}{\dim_{\mathbb{C}}(\pi)} \sum_{Y \in GL_n\mathbb{F}_q} \chi_\pi(Y)\Psi(YX).$$

For each $Z \in GL_n\mathbb{F}_q$ we have

$$W(\pi, \Psi, XZ) = q^{-n^2/2} \sum_{Y \in GL_n\mathbb{F}_q} \pi(Y)\Psi(YXZ)$$

$$= q^{-n^2/2} \sum_{Y \in GL_n\mathbb{F}_q} \pi(Y)\Psi(ZYX)$$

$$= q^{-n^2/2} \sum_{Y \in GL_n\mathbb{F}_q} \pi(Z)^{-1}\pi(ZY)\Psi(ZYX)$$

$$= \pi(Z)^{-1}W(\pi, \Psi, X)$$

and

$$W(\pi, \Psi, ZX) = W(\pi, \Psi, X)\pi(Z)^{-1}.$$

PROPOSITION 2.3.

Let π be an irreducible representation of $GL_n\mathbb{F}_q$ such that $\langle \pi, 1\rangle_{GL_n\mathbb{F}_q} = 0$. Then $W(\pi, \Psi, X) = 0$ for all singular matrices X.

Proof

Let $H(X) = \{Y \in GL_n\mathbb{F}_q \mid YX = X\}$. For $Y \in H(X)$ we have

$$W(\pi, \Psi, X) = W(\pi, \Psi, YX) = W(\pi, \Psi, X)\pi(Y)^{-1}$$

so that

$$W(\pi, \Psi, X) = W(\pi, \Psi, X)\frac{1}{|H(X)|}\sum_{U\in H(X)} \pi(U).$$

Therefore it suffices to show that $\sum_{U\in H(X)} \pi(U)$ is the zero matrix. If the rank of X is r with $0 \le r \le n-1$ then for suitable $A, B \in GL_n\mathbb{F}_q$ we have $X = Ae_rB$ where $\begin{pmatrix} 1_r & 0 \\ 0 & 0 \end{pmatrix}$ we have $H(X) = AH(e_r)A^{-1}$ so it is sufficient to assume $X = e_r$ in which case

$$H(e_r) = \{Y = \begin{pmatrix} 1_r & * \\ 0 & * \end{pmatrix}\}.$$

If the matrix $\sum_{U\in H(e_r)} \pi(U)$ is non-zero then the map it induces on the representation space for π maps it non-trivially into the subspace fixed by the action of $H(e_r)$. Therefore it suffices to show that there is no non-zero vector in the representation space which is fixed by $H(e_r)$, which contains the subgroup $H(e_{n-1})$, which is a normal subgroup of the parabolic subgroup $P_{n-1,1}$.

Since

$$\langle \operatorname{Ind}_{H(e_{n-1})}^{GL_n\mathbb{F}_q}(1), \pi\rangle_{GL_n\mathbb{F}_q} = \langle 1, \operatorname{Res}_{H(e_{n-1})}^{GL_n\mathbb{F}_q}(\pi)\rangle_{H(e_{n-1})} \ne 0$$

so that the irreducible π is a component of

$$\operatorname{Ind}_{H(e_{n-1})}^{GL_n\mathbb{F}_q}(1) = \operatorname{Ind}_{P_{n-1,1}^{GL_n\mathbb{F}_q}}(\operatorname{Ind}_{H(e_{n-1})}^{P_{n-1,1}}(1)).$$

But $P_{n-1,1}/H(e_{n-1}) \cong GL_{n-1}\mathbb{F}_q$ so that $\operatorname{Ind}_{H(e_{n-1})}^{P_{n-1,1}}(1)$ is obtained by inflating $\operatorname{Ind}_{\{1\}}^{GL_{n-1}\mathbb{F}_q}(1)\otimes 1$ from $GL_{n-1}\mathbb{F}_q \times GL_1\mathbb{F}_q$ so that, by the classification of irreducibles of $GL_n\mathbb{F}_q$, π is an irreducible component of $\pi_1 \circ (1)$ for some irreducible representation π_1 of $GL_{n-1}\mathbb{F}_q$. This implies that $\langle \pi, 1\rangle_{GL_n\mathbb{F}_q} \ne 0$, which is a contradiction. $\square$

2.4. Next we observe that

$$\sum_{X \in M_n\mathbb{F}_q} \hat{\Phi}(-X)W(\pi, \Psi, X)$$

$$= \sum_{X \in M_n\mathbb{F}_q} W(\pi, \Psi, X)\, q^{-n^2/2} \sum_{Y \in M_n\mathbb{F}_q} \Phi(Y)\Psi(-XY)$$

$$= \sum_{X \in M_n\mathbb{F}_q} q^{-n^2/2} \sum_{Z \in GL_n\mathbb{F}_q} \pi(Z)\Psi(ZX)\, q^{-n^2/2}$$

$$\times \sum_{Y \in M_n\mathbb{F}_q} \Phi(Y)\Psi(-YX)$$

$$= \sum_{Z \in GL_n\mathbb{F}_q} \pi(Z)\Phi(Z)$$

$$= \zeta(\Phi, \pi).$$

Taking traces

$$\mathrm{Trace}(\zeta(\Phi, \pi))$$

$$= \dim_{\mathbb{C}}(\pi) \cdot \zeta_{\mathrm{tr}}(\Phi, \pi)$$

$$= \dim_{\mathbb{C}}(\pi) \sum_{X \in M_n\mathbb{F}_q} \hat{\Phi}(-X)W_{\mathrm{tr}}(\pi, \Psi, X)$$

or

$$\zeta_{\mathrm{tr}}(\Phi, \pi) = \sum_{X \in M_n\mathbb{F}_q} \hat{\Phi}(-X)W_{\mathrm{tr}}(\pi, \Psi, X).$$

THEOREM 2.5.
Let π be an irreducible representation of $GL_n\mathbb{F}_q$ such that $\langle \pi, 1\rangle_{GL_n\mathbb{F}_q} = 0$. Then

$$\zeta(\hat{\Phi}, \pi^\vee)^{\mathrm{Transpose}} = W(\pi^\vee, \Psi, 1)\zeta(\Phi, \pi).$$

Proof
By Proposition 2.3 and §2.4

$$\zeta(\hat{\Phi}, \pi^\vee)^{\mathrm{Transpose}}$$

$$= \sum_{X \in M_n\mathbb{F}_q} \hat{\hat{\Phi}}(-X)W(\pi^\vee, \Psi, X)^{\mathrm{Transpose}}$$

$$= \sum_{X \in GL_n\mathbb{F}_q} \Phi(X)W(\pi^\vee, \Psi, X)^{\mathrm{Transpose}}$$

$$= W(\pi^\vee, \Psi, 1)^{\mathrm{Transpose}} \sum_{X \in GL_n\mathbb{F}_q} \Phi(X)\pi(X)$$

$$= W(\pi^\vee, \Psi, 1) \sum_{X \in GL_n\mathbb{F}_q} \Phi(X)\pi(X)$$

$$= W(\pi^\vee, \Psi, 1)\zeta(\Phi, \pi)$$

because $W(\pi^\vee, \Psi, 1)$ is a scalar matrix. $\square$

COROLLARY 2.6.

In the situation of Theorem 2.5

$$\zeta_{\mathrm{tr}}(\hat{\Phi}, \pi^\vee) = q^{-n^2/2} W_{GL_n\mathbb{F}_q}(\pi^\vee) \zeta_{\mathrm{tr}}(\Phi, \pi)$$

where $W_{GL_n\mathbb{F}_q}(\pi^\vee)$ is as in §2.2.

2.7. $W(\pi, \Psi, X)$ when π is reducible

In §2.2 we defined

$$W(\pi, \Psi, X) = q^{-n^2/2} \sum_{Y \in GL_n\mathbb{F}_q} \pi(Y)\Psi(YX)$$

which is an endomorphism of the representation space of π. Suppose that

$$\pi = \pi_1 \oplus \pi_2 \oplus \ldots \oplus \pi_r$$

where the π_i's are irreducible representations (possibly with repetitions). Suppose that π_i is represented by a homomorphism

$$\rho_i : GL_n\mathbb{F}_q \longrightarrow GL_{d_i}\mathbb{C}$$

for $1 \leq i \leq r$. Therefore π is represented by the matrix homomorphism into $GL_{d_1+\ldots+d_r}\mathbb{C}$

$$X \mapsto \begin{pmatrix} \rho_1(X) & 0 & 0 & \cdots & \cdots & 0 \\ 0 & \rho_2(X) & 0 & \cdots & \cdots & 0 \\ \vdots & \vdots & \vdots & \vdots & \vdots & \vdots \\ \vdots & \vdots & \vdots & \vdots & \vdots & \vdots \\ 0 & 0 & 0 & \cdots & 0 & \rho_r(X) \end{pmatrix}.$$

Therefore the formula for $W(\pi, \Psi, X)$ in terms of a matrix is given by

$$\begin{pmatrix} W(\rho_1, \Psi, X) & 0 & 0 & \cdots & \cdots & 0 \\ 0 & W(\rho_2, \Psi, X) & 0 & \cdots & \cdots & 0 \\ \vdots & \vdots & \vdots & \vdots & \vdots & \vdots \\ \vdots & \vdots & \vdots & \vdots & \vdots & \vdots \\ 0 & 0 & 0 & \cdots & 0 & W(\rho_r, \Psi, X) \end{pmatrix}.$$

By Proposition 2.3 if X is singular and $\langle \pi_i, 1 \rangle_{GL_n\mathbb{F}_q} = 0$ then $W(\pi_i, \Psi, X) = 0$. If

$$0 = \langle \pi, 1 \rangle_{GL_n\mathbb{F}_q} = \sum_{i=1}^{r} \langle \pi_i, 1 \rangle_{GL_n\mathbb{F}_q}$$

then each positive integer $\langle \pi_i, 1 \rangle_{GL_n\mathbb{F}_q}$ is zero and so $W(\pi, \Psi, X)$ is the zero matrix. In other words, we may remove the adjective "irreducible" from the statement of Proposition 2.3.

PROPOSITION 2.8.

Let π be a finite-dimensional complex representation of $GL_n\mathbb{F}_q$ such that $\langle \pi, 1 \rangle_{GL_n\mathbb{F}_q} = 0$. Then $W(\pi, \Psi, X) = 0$ for all singular matrices X.

2.9. *Theorem 2.5 when π is reducible*

The formula of §2.4 does not require that π be irreducible so that Proposition 2.8, §2.4 and the proof of Theorem 2.5 establish the following result.

THEOREM 2.10.

Let π be a finite-dimensional complex representation of $GL_n\mathbb{F}_q$ such that $\langle \pi, 1 \rangle_{GL_n\mathbb{F}_q} = 0$. Then

$$\zeta(\hat{\Phi}, \pi^\vee)^{\text{Transpose}} = W(\pi^\vee, \Psi, 1)\zeta(\Phi, \pi).$$

COROLLARY 2.11.

In the situation of Theorem 2.10

$$\zeta_{\text{tr}}(\hat{\Phi}, \pi^\vee) = q^{-n^2/2} W_{GL_n\mathbb{F}_q}(\pi^\vee)\zeta_{\text{tr}}(\Phi, \pi)$$

where $W_{GL_n\mathbb{F}_q}(\pi^\vee)$ is as in §2.2. In particular, if $\pi = \text{Ind}_H^{GL_n\mathbb{F}_q}(\rho)$ with $\langle \rho, 1 \rangle_H = 0$ then

$$\zeta(\hat{\Phi}, \text{Ind}_H^{GL_n\mathbb{F}_q}(\rho^\vee))^{\text{Transpose}} = q^{-n^2/2} W_H(\rho^\vee)\zeta(\Phi, \text{Ind}_H^{GL_n\mathbb{F}_q}(\rho)).$$

3. Kondo-Gauss sums for $GL_n\mathbb{F}_q$

DEFINITION 3.1. Let $\rho : H \longrightarrow GL_n\mathbb{C}$ denote a representation of a subgroup H of $GL_n\mathbb{F}_q$. If q is a power of the prime p we have the (additive) trace map

$$\text{Tr}_{\mathbb{F}_q/\mathbb{F}_p} : \mathbb{F}_q \longrightarrow \mathbb{F}_p.$$

In addition we have the matrix trace map

$$\text{Trace} : GL_n\mathbb{F}_q \longrightarrow \mathbb{F}_q.$$

Define a measure map Ψ on matrices $X \in GL_n\mathbb{F}_q$ by

$$\Psi(X) = e^{\frac{2\pi\sqrt{-1}\text{Tr}_{\mathbb{F}_q/\mathbb{F}_p}(\text{Trace}(X))}{p}}$$

which is denoted by $e_1[X]$ in [85]. Let χ_ρ denote the character function of ρ which assigns to X the trace of the complex matrix $\rho(X)$.

Define a complex number $W_H(\rho)$ by the formula

$$W_H(\rho) = \frac{1}{\dim_\mathbb{C}(\rho)} \sum_{X \in H} \chi_\rho(X)\Psi(X).$$

When $H = GL_n\mathbb{F}_q$ and ρ is irreducible $W_{GL_n\mathbb{F}_q}(\rho) = w(\rho)$, the Kondo-Gauss sum which is introduced and computed in [85].

THEOREM 3.2.

Let σ be a finite-dimensional representation of $H \subseteq GL_n\mathbb{F}_q$. Then for any subgroup J such that $H \subseteq J \subseteq GL_n\mathbb{F}_q$

$$W_H(\sigma) = W_J(\text{Ind}_H^J(\sigma)).$$

Proof

Set $\rho = \text{Ind}_H^J(\sigma)$. By definition

$$W_J(\rho) = \frac{|H|}{|J| \cdot \dim_\mathbb{C}(\sigma)} \sum_{X \in J} \chi_\rho(X)\Psi(X)$$

$$= \frac{1}{|J| \cdot \dim_\mathbb{C}(\sigma)} \sum_{X \in J} \sum_{Y \in J,\ YXY^{-1} \in H} \chi_\sigma(YXY^{-1})\Psi(X)$$

by the character formula for an induced representation ([**126**] Theorem 1.2.43). Consider the free action of J on $J \times J$ given by $(X,Y)Z = (Z^{-1}XZ, YZ)$ for $XY, Z \in J$. The map from $J \times J$ to J sending (X,Y) to YXY^{-1} is constant on each J-orbit. Therefore

$$W_J(\rho) = \frac{1}{|J| \cdot \dim_\mathbb{C}(\sigma)} \sum_{X \in J} \sum_{Y \in J,\ YXY^{-1} \in H} \chi_\sigma(YXY^{-1})\Psi(YXY^{-1})$$

$$= \frac{1}{|J| \cdot \dim_\mathbb{C}(\sigma)} |J| \sum_{U \in H} \chi_\sigma(U)\Psi(U)$$

$$= W_H(\sigma).$$

LEMMA 3.3.

$(\dim_\mathbb{C}(\sigma_1) + \dim_\mathbb{C}(\sigma_2))W_H(\sigma_1 \oplus \sigma_2) = \dim_\mathbb{C}(\sigma_1)W_H(\sigma_1) + \dim_\mathbb{C}(\sigma_2)W_H(\sigma_2)$.

EXAMPLE 3.4. *The Weil representation $r(\Theta)$ of $GL_2\mathbb{F}_q$*

The Weil representation is a very ingenious construction of a $(q-1)$-dimensional irreducible complex representation of $GL_2\mathbb{F}_q$. It is constructed from scratch in ([**126**] Chapter Three). However there is a very simple description of $r(\Theta)$ in terms of induced representation, which may be verified (for example) using the character formulae of ([**126**] Chapter Three).

There is a copy of $\mathbb{F}_{q^2}^*$, unique up to conjugation, embedded in $GL_2\mathbb{F}_q$. For example, if σ is a non-square in $\mathbb{F}_q^*$ then sending $a + b\sqrt{\sigma}$ to

$$\begin{pmatrix} a & b\sigma \\ b & a \end{pmatrix}$$

gives such an embedding. Let F denote the generator of $\text{Gal}(\mathbb{F}_{q^2}/\mathbb{F}_q)$ and suppose that $\Theta : \mathbb{F}_{q^2}^* \longrightarrow \mathbb{C}^*$ is a character such that $F(\Theta) \neq \Theta$.

Let H denote the "top line" subgroup consisting of matrices of the form

$$\begin{pmatrix} a & b \\ 0 & 1 \end{pmatrix}$$

so that $H \cong \mathbb{F}_q^* \times \mathbb{F}_q$ by sending the above matrix to $(a, b/a)$. Therefore we have a character on H given by

$$(\Theta \otimes \Psi) \begin{pmatrix} a & b \\ 0 & 1 \end{pmatrix} = \Theta(a)\Psi(b/a).$$

Here Ψ is the additive measure defined in Definition 3.1 on $n \times n$ matrices in the case $n = 1$.

There is a (split) short exact sequence of complex $GL_2\mathbb{F}_q$-representations

$$0 \longrightarrow \mathrm{Ind}_{\mathbb{F}_{q^2}^*}^{GL_2\mathbb{F}_q}(\Theta) \longrightarrow \mathrm{Ind}_H^{GL_2\mathbb{F}_q}(\Theta \otimes \Psi) \longrightarrow r(\Theta) \longrightarrow 0.$$

To establish this result I first used the fact that ([**126**] Chapter Three) provides an easy description of the right-hand map together with a complicated argument to show that the left-hand representation was inside the kernel. Then, smugly pleased with the discovery, I check it using the character values of ([**126**] Chapter Three) only to find the same result appears in ([**40**] p. 47)!

By Theorem 3.2 and Lemma 3.3 we have

$$W_{GL_2\mathbb{F}_q}(r(\Theta)) + W_{\mathbb{F}_{q^2}^*}(\Theta) = W_H(\Theta \otimes \Psi).$$

However

$$W_H(\Theta \otimes \Psi) = \sum_{(a,b) \in H} \Theta(a)\Psi(b/a)\Psi(a+1) = 0$$

since the sum of the values of a non-trivial character over a finite abelian group ($\mathbb{F}_q$ in this case) is zero. Therefore

$$W_{GL_2\mathbb{F}_q}(r(\Theta)) = -W_{\mathbb{F}_{q^2}^*}(\Theta)$$

where the right side in the familiar Gauss sum over a finite field.

PROPOSITION 3.5.

For $i = 1, 2$ let $\sigma_i : H_i \longrightarrow GL_{n_i}\mathbb{C}$ be a representation of $H_i \subseteq GL_{s_i}\mathbb{F}_q$. Then we have a representation of $H_1 \times H_2$ (embedded into $GL_{s_1+s_2}$ by direct sum of matrices) given by the tensor product $\sigma_1 \otimes \sigma_2$ and

$$W_{H_1 \times H_2}(\sigma_1 \otimes \sigma_2) = W_{H_1}(\sigma_1)W_{H_2}(\sigma_2).$$

Proof

We have

$$W_{H_1 \times H_2}(\sigma_1 \otimes \sigma_2)$$

$$= \tfrac{1}{s_1 \cdot s_2} \sum_{X_1 \oplus X_2 \in H_1 \times H_2} \chi_{\sigma_1 \otimes \sigma_2}(X_1 \oplus X_2)\Psi(X_1 \oplus X_2)$$

$$= \tfrac{1}{s_1 \cdot s_2} \sum_{X_1 \oplus X_2 \in H_1 \times H_2} \chi_{\sigma_1}(X_1)\chi_{\sigma_2}(X_2)\Psi(X_1)\Psi(X_2)$$

$$= W_{H_1}(\sigma_1)W_{H_2}(\sigma_2).$$

$\square$

REMARK 3.6. In [85] Kondo gives formulae for his Gauss sums on the irreducible complex representations of $GL_n\mathbb{F}_q$. The calculations of [85] do not use the function $W_H(\rho)$ but stick to the case of an irreducible ρ and $H = GL_n\mathbb{F}_q$. The greater freedom and generality of $W_H(\rho)$ should make the calculations much simpler.

To obtain Kondo's formulae and (elsewhere in this monograph) to generalise to the case in which $\mathbb{F}_q$ is replaced by a local field we need to understand the behaviour of Kondo-Gauss sums under taking spaces of homomorphisms of irreducibles of Σ_n into ρ's.

3.7. Towards Kondo's formulae

The semi-direct product $\Sigma_n \int H_1$ consists of elements $(\tau, (X_1, X_2, \ldots, X_n)) \in \Sigma_n \times H_1^n$ with multiplication defined by

$$(\tau, (X_1, X_2, \ldots, H_n)) \cdot (\tau', (X_1', X_2', \ldots, X_n'))$$
$$= (\tau\tau', (X_1 X_{\tau(1)}', X_2 X_{\tau(2)}', \ldots, X_n X_{\tau(n)}')).$$

If $H_1 \subseteq GL_r\mathbb{F}_q$ then $\Sigma_n \int H_1 \subseteq GL_{nr}\mathbb{F}_q$. If σ is a representation of H_1 then $\Sigma_n \int H_1$ acts on $\sigma \otimes \ldots \otimes \sigma$ (the n-fold tensor product) by the formula

$$(\tau, (X_1, X_2, \ldots, H_n))(v_1 \otimes \ldots \otimes v_n) = \sigma(X_1)(v_{\tau(1)}) \otimes \ldots \otimes \sigma(X_n)(v_{\tau(n)}).$$

This defines a left action because

$$(\tau, (X_1, X_2, \ldots, H_n)) \cdot (\tau', (X_1', X_2', \ldots, X_n'))(v_1 \otimes \ldots \otimes v_n)$$

$$= (\tau, (X_1, X_2, \ldots, H_n))(\sigma(X_1')(v_{\tau'(1)}) \otimes \ldots \otimes \sigma(X_n')(v_{\tau'(n)}))$$

$$= \sigma(X_1)(\sigma(X_{\tau(1)}')(v_{\tau\tau'(1)})) \otimes \ldots \otimes \sigma(X_n)(\sigma(X_{\tau(n)}')(v_{\tau\tau'(n)}))$$

$$= \sigma(X_1 X_{\tau(1)}')(v_{\tau\tau'(1)})) \otimes \ldots \otimes \sigma(X_n X_{\tau(n)}')(v_{\tau\tau'(n)}))$$

$$= (\tau\tau', (X_1 X_{\tau(1)}', X_2 X_{\tau(2)}', \ldots, X_n X_{\tau(n)}'))(v_1 \otimes \ldots \otimes v_n)$$

as required. These formulae define the exterior n-fold tensor representation $\sigma^{\bigcirc^n}$ of $\Sigma_n \int H_1$.

In order to calculate the Kondo Gauss sum for the representation given by

$$\text{Hom}_{\Sigma_d}(\text{Ind}_{\Sigma_{a_1} \times \cdots \times \Sigma_{a_r}}^{\Sigma_d}(1), \pi^{\otimes d}) = \otimes_{i=1}^{r} \text{Hom}_{\Sigma_{a_i}}(1, \pi^{\bigcirc^{a_i}})$$

we can get the Kondo-Gauss sum of $\text{Hom}_{\Sigma_{a_i}}(1, \pi^{\bigcirc^{a_i}})$ by Möbius inversion from the Kondo Gauss sums of $\pi^{\bigcirc^{a_i}}$.

Given a representation λ of Σ_n we may inflate it to give a representation $\text{Inf}(\lambda)$ of $\Sigma_n \int H_1$. Set $H_n = \Sigma_n \int H_1$. We shall now examine how to calculate $W_H(\text{Inf}(\lambda)^\vee \otimes \sigma^{\bigcirc^n})$ where $\text{Inf}(\lambda)^\vee$ is the contragredient (equals dual in this case) of $\text{Inf}(\lambda)$. It will suffice to calculate this in case of the free generators $\lambda = \text{Ind}_{\Sigma_{u_1} \times \Sigma_{u_2} \times \cdots \times \Sigma_{u_m}}^{\Sigma_n}(1)$ where $\underline{u} = (u_1, \ldots, u_m)$ is a partition of n by positive integers (i.e. $u_i > 0$ and $\sum_j u_j = n$). Since this representation is self-dual we shall consider

$$W_{H_n}(\text{Inf} \cdot \text{Ind}_{\Sigma_{u_1} \times \Sigma_{u_2} \times \cdots \times \Sigma_{u_m}}^{\Sigma_n}(1) \cdot \sigma^{\bigcirc^n})$$

$$= W_{H_n}(\text{Ind}_{H_{u_1} \times H_{u_2} \times \cdots \times H_{u_m}}^{H_n}(\sigma^{\bigcirc^n}))$$

$$= W_{H_{u_1} \times \cdots \times H_{u_m}}(\sigma^{\bigcirc^n})$$

$$= \prod_{j=1}^{m} W_{H_{u_j}}(\sigma^{\bigcirc^{u_j}}).$$

Therefore it suffices to calculate $W_{H_n}(\sigma^{\bigcirc^n})$.

3.8. $W_{H_n}(\sigma^{\bigcirc^n})$

Let $v_1, v_2, \ldots, v_d$ be a basis for the vector space underlying σ. Suppose that $X_i \in H$ acts via σ according to the matrix formula

$$\sigma(X_i)(v_s) = \sum_{t=1}^{d} X_{i;t,s} v_t.$$

If $\tau \in \Sigma_n$ then

$$(\tau, (X_1, X_2, \ldots, X_n))(v_{s_1} \otimes \ldots \otimes v_{s_n})$$

$$= \sigma(X_1)(v_{s_{\tau(1)}}) \otimes \ldots \otimes \sigma(X_n)(v_{s_{\tau(n)}})$$

$$= \sum_{t_1=1}^{d} \sum_{t_2=1}^{d} \cdots \sum_{t_n=1}^{d} X_{1;t_1,s_{\tau_1}} X_{1;t_2,s_{\tau_2}} \cdots X_{1;t_n,s_{\tau_n}} v_{t_1} \otimes v_{t_2} \otimes \ldots \otimes v_{t_n}.$$

Therefore the trace of $(\tau, (X_1, X_2, \ldots, X_n)) \in H_n$ acting on $\sigma^{\bigcirc^n}$ is equal to

$$\sum_{1 \leq s_1, s_2, \ldots, s_n \leq d} X_{1;s_1,s_{\tau_1}} X_{2;s_2,s_{\tau_2}} \cdots X_{n;s_n,s_{\tau_n}}.$$

The formula for $W_{H_n}(\sigma^{\bigcirc^n})$ is equal to

$$W_{H_n}(\sigma^{\bigcirc^n})$$

$$= \tfrac{1}{\dim(\sigma)^n} \sum_{(\tau,(X_1,\dots))\in H_n} \chi_{\sigma^{\bigcirc^n}}((\tau,(X_1,\dots)))\Psi((\tau,(X_1,\dots,X_n))).$$

Now consider the expression for $\Psi((\tau,(X_1,\dots,X_n)))$. When τ is the transposition $(1,2)$, in terms of matrices in H_n the element $((1,2),(X_1,\dots,X_n))$ corresponds to the product of matrices of $d \times d$ blocks

$$\begin{pmatrix} X_1 & 0 & 0 & 0 & \cdots & \cdots \\ 0 & X_2 & 0 & 0 & \cdots & \cdots \\ 0 & 0 & X_3 & 0 & \cdots & \cdots \\ \vdots & \vdots & \vdots & \vdots & \vdots & \vdots \end{pmatrix} \begin{pmatrix} 0 & 1 & 0 & 0 & \cdots & \cdots \\ 1 & 0 & 0 & 0 & \cdots & \cdots \\ 0 & 0 & 1 & 0 & \cdots & \cdots \\ 0 & 0 & 0 & 1 & \cdots & \cdots \\ \vdots & \vdots & \vdots & \vdots & \vdots & \vdots \end{pmatrix}$$

$$= \begin{pmatrix} 0 & X_1 & 0 & 0 & \cdots & \cdots \\ X_2 & 0 & 0 & 0 & \cdots & \cdots \\ 0 & 0 & X_3 & 0 & \cdots & \cdots \\ 0 & 0 & 0 & X_4 & \cdots & \cdots \\ \vdots & \vdots & \vdots & \vdots & \vdots & \vdots \end{pmatrix}.$$

The behaviour for general τ is clear and we obtain the formula

$$\Psi((\tau,(X_1,\dots,X_n)))$$

$$= e^{\frac{2\pi\sqrt{-1}\mathrm{Tr}_{\mathbb{F}_q/\mathbb{F}_p}(\mathrm{Trace}((\tau,(X_1,\dots,X_n))))}{p}}$$

$$= \prod_{\tau(i)=i} \Psi(X_i).$$

Therefore we have the formula

$$W_{H_n}(\sigma^{\bigcirc^n})$$

$$= \tfrac{1}{\dim(\sigma)^n} \sum_{(\tau,(X_1,\dots))\in H_n}$$

$$\times \sum_{1\le s_1,s_2,\dots,s_n\le d} X_{1;s_1,s_{\tau_1}} \cdots X_{n;s_n,s_{\tau_n}} \prod_{\tau(i)=i} \Psi(X_i).$$

Now let us evaluate some small examples of the above formula.

3.9. *Small examples of* $W_{H_n}(\sigma^{\bigcirc^n})$ *when* $\langle\sigma,1\rangle_H = 0$

In this subsection $\langle\sigma,1\rangle_H$ denotes the Schur inner product, which is zero when the trivial representation does not occur in σ.

When $n = 2$ we have

$$W_{H_2}(\sigma^{\bigcirc^2})$$

$$= \tfrac{1}{\dim(\sigma)^2} \sum_{(1,(X_1,X_2))} \sum_{1 \le s_1, s_2 \le d} X_{1;s_1,s_1} X_{2;s_2,s_2} \Psi(X_1)\Psi(X_2)$$

$$+ \tfrac{1}{\dim(\sigma)^2} \sum_{((1,2),(X_1,X_2))} \sum_{1 \le s_1, s_2 \le d} X_{1;s_1,s_2} X_{2;s_2,s_1}$$

$$= W_H(\sigma)^2 + \tfrac{1}{\dim(\sigma)^2} \sum_{(X_1,X_2) \in H^2} \mathrm{Trace}(X_1 X_2).$$

However, as (X_1, X_2) runs through H^2 the element $X_1 X_2$ runs through H precisely $|H|$ times. Therefore we have

$$W_{H_2}(\sigma^{\bigcirc^2}) = W_H(\sigma)^2 + \tfrac{1}{\dim(\sigma)^2} \langle \sigma, 1 \rangle_H = W_H(\sigma)^2.$$

When $n = 3$ consider the contribution to the formula of §3.8 from a 2-cycle, which may as well be $\tau = (1,2)$, and a 3-cycle, which may as well be $\tau = (1,2,3)$. We have

$$\tfrac{1}{\dim(\sigma)^3} \sum_{((1,2),(X_1,X_2,X_3))} \sum_{1 \le s_1, s_2, s_3 \le d} X_{1;s_1,s_2} X_{2;s_2,s_1} X_{3;s_3,s_3} \Psi(X_3)$$

$$= \tfrac{1}{\dim(\sigma)^2} \sum_{(X_1,X_2) \in H^2} \mathrm{Trace}(X_1 X_2) W_H(\sigma)$$

$$= 0$$

and

$$\tfrac{1}{\dim(\sigma)^3} \sum_{((1,2,3),(X_1,X_2,X_3))} \sum_{1 \le s_1, s_2, s_3 \le d} X_{1;s_1,s_2} X_{2;s_2,s_3} X_{3;s_3,s_1}$$

$$= \tfrac{|H|^2}{\dim(\sigma)^3} \sum_{X \in H} \mathrm{Trace}(X)$$

$$= 0.$$

Therefore we obtain the formula

$$W_{H_3}(\sigma^{\bigcirc^3}) = W_H(\sigma)^3.$$

In general we have the following result.

THEOREM 3.10.
If the Schur inner product $\langle \sigma, 1 \rangle_H$ vanishes then

$$W_{H_n}(\sigma^{\bigcirc^n}) = W_H(\sigma)^n.$$

PROPOSITION 3.11.
In the notation of §2.1

$$W_{P_{n_1,\dots,n_r}}\left(\mathrm{Inf}_{D_{n_1,\dots,n_r}}^{P_{n_1,\dots,n_r}}(\rho)\right) = |U_{n_1,\dots,n_r}| W_{D_{n_1,\dots,n_r}}(\rho).$$

Also

$$|U_{n_1,\dots,n_r}| = q^{n^2 - n_1^2 - n_2^2 - \dots - n_r^2}.$$

Proof

We have

$$W_{P_{n_1,\dots,n_r}}\left(\mathrm{Inf}_{D_{n_1,\dots,n_r}}^{P_{n_1,\dots,n_r}}(\rho)\right)$$

$$= \frac{1}{\dim(\rho)} \sum_{A\in U_{n_1,\dots,n_r},\, B\in D_{n_1,\dots,n_r}} \chi_{\mathrm{Inf}_{D_{n_1,\dots,n_r}}^{P_{n_1,\dots,n_r}}(\rho)}(AB)\Psi(AB)$$

$$= \frac{1}{\dim(\rho)} \sum_{A\in U_{n_1,\dots,n_r},\, B\in D_{n_1,\dots,n_r}} \chi_{\mathrm{Inf}_{D_{n_1,\dots,n_r}}^{P_{n_1,\dots,n_r}}(\rho)}(B)\Psi(B)$$

$$= |U_{n_1,\dots,n_r}|\, W_{D_{n_1,\dots,n_r}}(\rho).$$

The order of $|U_{n_1,\dots,n_r}|$ follows from the fact that there are $(n - n_i) \times n_i$ arbitrary entries from $\mathbb{F}_q$ to the right of the diagonal block given by the $n_i \times n_i$ identity matrix for $1 \le i \le r$. $\square$

3.12. *Computing* $W_{GL_n\mathbb{F}_q}((f_1^{\lambda_1}) \circ (f_2^{\lambda_2}) \circ \dots \circ (f_m^{\lambda_m}))$

Let $f_1,\dots,f_m$ be distinct Frobenius orbits in Γ in the notation of §2.1 with degrees $d(f_i) > 1$. Let $\lambda_1,\dots,\lambda_m$ be partitions such that $n = \sum_{t=1}^m d(f_t)|\lambda_t|$. We shall use the results of this section to calculate

$$W_{GL_n\mathbb{F}_q}((f_1^{\lambda_1}) \circ (f_2^{\lambda_2}) \circ \dots \circ (f_m^{\lambda_m})).$$

Since

$$(f_1^{\lambda_1}) \circ (f_2^{\lambda_2}) \circ \dots \circ (f_m^{\lambda_m})$$

$$= \mathrm{Ind}_{P_{d(f_1)|\lambda_1|,\dots,d(f_m)|\lambda_m|}}^{GL_n\mathbb{F}_q}\left(\mathrm{Inf}_{D_{d(f_1)|\lambda_1|,\dots,d(f_m)|\lambda_m|}}^{P_{d(f_1)|\lambda_1|,\dots,d(f_m)|\lambda_m|}}((f_1^{\lambda_1}) \otimes (f_2^{\lambda_2}) \otimes\right.$$

$$\left.\dots \otimes (f_m^{\lambda_m}))\right)$$

we have

$$W_{GL_n\mathbb{F}_q}((f_1^{\lambda_1}) \circ (f_2^{\lambda_2}) \circ \dots \circ (f_m^{\lambda_m}))$$

$$= W_{P_{d(f_1)|\lambda_1|,\dots,d(f_m)|\lambda_m|}}\left(\mathrm{Inf}_{D_{d(f_1)|\lambda_1|,\dots,d(f_m)|\lambda_m|}}^{P_{d(f_1)|\lambda_1|,\dots,d(f_m)|\lambda_m|}}((f_1^{\lambda_1}) \otimes (f_2^{\lambda_2}) \otimes\right.$$

$$\left.\dots \otimes (f_m^{\lambda_m}))\right)$$

$$= q^{n^2 - d(f_1)^2|\lambda_1|^2 - \dots - d(f_m)^2|\lambda_m|^2} W_{D_{d(f_1)|\lambda_1|,\dots,d(f_m)|\lambda_m|}}((f_1^{\lambda_1}) \otimes (f_2^{\lambda_2}) \otimes$$

$$\dots \otimes (f_m^{\lambda_m}))$$

$$= q^{n^2 - d(f_1)^2|\lambda_1|^2 - \dots - d(f_m)^2|\lambda_m|^2} \prod_{t=1}^m W_{GL_{d(f_t)|\lambda_t|}}((f_t^{\lambda_t})).$$

Now let χ_{λ_t} denote the irreducible representation of $\Sigma_{|\lambda_t|}$ of Proposition 1.4. There exist unique integers a_{μ_t} such that

$$\chi_{\lambda_t} = \sum_{\mu_t \text{ a partition of } |\lambda_t|} a_{\mu_t} \operatorname{Ind}_{\Sigma_{\mu_t}}^{\Sigma_{|\lambda_t|}}(1)$$

where $\Sigma_{(\mu_{t,1},\mu_{t,2},\ldots,\mu_{t,u})} = \Sigma_{\mu_{t,1}} \times \ldots \times \Sigma_{\mu_{t,u}}$.
Therefore

$$W_{GL_{d(f_t)|\lambda_t|}}((f_t^{\lambda_t}))$$

$$= \prod_{\substack{\mu_t \text{ a partition of } |\lambda_t| \\ \mu_t = \mu_{t,1},\ldots,\mu_{t,u}}} \prod_{v=1}^{u} W_{d(f_t)}((f_t))^{\mu_{t,v} a_{\mu_t}}$$

$$= \prod_{\substack{\mu_t \text{ a partition of } |\lambda_t| \\ \mu_t = \mu_{t,1},\ldots,\mu_{t,u}}} W_{d(f_t)}((f_t))^{|\lambda_t| a_{\mu_t}}$$

and from [93]

$$W_{d(f_t)}((f_t)) = (-1)^{d(f_t)} q^{-d(f_t)/2} \tau(f_t, \Psi_{d(f_t)}).$$

Due to laziness I have not attempted to compare these formulae with those of [85]!

4. The symmetric group's PSH algebra and Theorem 1.8

4.1. In §3 of [83] a Schur inner-product of some character functions on the symmetric group are computed. I mentioned (and reproved) the result in 1.8. To recapitulate the result we need to recall that the length of a permutation in $\sigma \in \Sigma_n$, the symmetric group on n objects, is $\sum_i r_i$ where $l(\sigma) = r_i$ is the number of i-cycles in the expression for σ are the product of disjoint cycles.

Let N be a positive integer. The function $l_n : \sigma \mapsto l(\sigma)$ is a function on the conjugacy classes of permutations in Σ_n and so also is $\rho_{N,n} : \sigma \mapsto N^{l(\sigma)}$. There exist irreducible representations V_j of Σ_n and complex numbers α_j such that the character function of $\sum_j \alpha_j V_j$ is $\rho_{N,n}$ or, equivalently,

$$\sum_j \alpha_j \chi_{V_j}(\sigma) = \rho_{N,n}(\sigma)$$

for all $\sigma \in \Sigma_n$. Here χ_V is the trace function (i.e. the character) of a representation V.

Recall (see also Chapter Nine) the PSH-algebra for the symmetric groups $R = \oplus_{n \geq 0} R(\Sigma_n)$. This Hopf algebra is the unique PSH-algebra over the integers with a unique primitive irreducible. The coproduct

$$R(\Sigma_n) \longrightarrow \oplus_k R(\Sigma_k) \otimes R(\Sigma_{n-k})$$

has $(k, n-k)$-component given by the restriction to $\Sigma_k \times \Sigma_{n-k}$. That is, a representation maps via

$$V \mapsto \mathrm{Res}^{\Sigma_n}_{\Sigma_k \times \Sigma_{n-k}}(V) \in R(\Sigma_k \times \Sigma_{n-k}) \cong R(\Sigma_k) \otimes R(\Sigma_{n-k}).$$

The dual Hopf algebra may be identified with complex class functions.

The product in the PSH algebra is given by mapping

$$V_1 \otimes V_2 \in R(\Sigma_k) \otimes R(\Sigma_{n-k})$$

to $\mathrm{Ind}^{\Sigma_n}_{\Sigma_k \times \Sigma_{n-k}}(V_1 \otimes V_2) = m(V_1 \otimes V_2)$. Let us calculate $\chi_{m(V_1 \otimes V_2)}$. The trace formula for this character is

$$\chi_{m(V_1 \otimes V_2)}(\hat{\sigma}) = \frac{1}{k!(n-k)!} \sum_{\tau \in \Sigma_n, \tau\hat{\sigma}\tau^{-1} \in \Sigma_k \times \Sigma_{n-k}} \chi_{V_1 \otimes V_2}(\tau\hat{\sigma}\tau^{-1})$$

which is zero unless $\hat{\sigma}$ is conjugate in Σ_n to an element $(\sigma, \sigma') \in \Sigma_k \times \Sigma_{n-k}$.

Two elements of $\Sigma_k \times \Sigma_{n-k}$ are conjugate in Σ_n if and only if they are conjugate in $\Sigma_k \times \Sigma_{n-k}$, because cycle shape determines conjugacy.

Suppose that $\hat{\sigma}$ is conjugate in Σ_n to an element $(\sigma, \sigma') \in \Sigma_k \times \Sigma_{n-k}$. Let r_i, r_i' be the numbers of i-cycles in the cycle decomposition for σ, σ' respectively. The number of distinct elements in the Σ_n-conjugacy class of (σ, σ') is

$$\frac{n!}{1^{r_1+r_1'}(r_1+r_1')! 2^{r_2+r_2'}(r_2+r_2')! \dots}$$

while the number in the Σ_k-conjugacy class of σ is

$$\frac{k!}{1^{r_1}(r_1)! 2^{r_2}(r_2)! \dots}$$

and the number in the Σ_{n-k}-conjugacy class of σ' is

$$\frac{(n-k)!}{1^{r_1'}(r_1')! 2^{r_2'}(r_2')! \dots}.$$

Therefore the denominators are the orders of the relevant centralisers.

Next we want to simplify the formula for $\chi_{m(V_1 \otimes V_2)}(\hat{\sigma})$. If $\tau\hat{\sigma}\tau^{-1} \in \Sigma_k \times \Sigma_{n-k}$ there exists $\mu \in \Sigma_k \times \Sigma_{n-k}$ such that $\mu\tau\hat{\sigma}\tau^{-1}\mu^{-1} = \hat{\sigma}$. Setting $\tau_1 = \mu\tau$ we see that $\tau_1 \in Z_{\Sigma_n}(\hat{\sigma})$, the centraliser of $\hat{\sigma}$ in Σ_n. Therefore $\tau = \mu^{-1}\tau_1$ and we have a surjective map

$$(\Sigma_k \times \Sigma_{n-k}) \times Z_{\Sigma_n}(\hat{\sigma}) \longrightarrow \{\tau \in \Sigma_n, \tau\hat{\sigma}\tau^{-1} \in \Sigma_k \times \Sigma_{n-k}\}$$

sending (ν, τ_1) to $\nu\tau_1$. Also $\nu\tau_1 = \nu'\tau_1'$ if and only if

$$(\nu')^{-1}\nu = \tau_1'\tau_1^{-1} \in Z_{\Sigma_k \times \Sigma_{n-k}}(\sigma, \sigma').$$

Therefore if $\lambda = (\nu')^{-1}\nu$ then $(\nu, \tau_1) = (\nu'\lambda, \lambda^{-1}\tau')$. Therefore there is a bijection

$$(\Sigma_k \times \Sigma_{n-k}) \times_{Z_{\Sigma_k \times \Sigma_{n-k}}(\sigma,\sigma')} Z_{\Sigma_n}(\hat{\sigma}) \leftrightarrow \{\tau \in \Sigma_n, \tau\hat{\sigma}\tau^{-1} \in \Sigma_k \times \Sigma_{n-k}\}.$$

Therefore the number of τ's in the formula for $\chi_{m(V_1 \otimes V_2)}(\hat{\sigma})$ is

$$\frac{k!(n-k)!1^{r_1+r_1'}(r_1+r_1')!2^{r_2+r_2'}(r_2+r_2')!\ldots}{1^{r_1}(r_1)!2^{r_2}(r_2)!\ldots 1^{r_1'}(r_1')!2^{r_2'}(r_2')!\ldots}$$

and each $\tau\hat{\sigma}\tau^{-1}$ gives an element in the $\Sigma_k \times \Sigma_{n-k}$-conjugacy class of $\hat{\sigma} = (\sigma, \sigma')$ we find that

$$\chi_{m(V_1 \otimes V_2)}(\hat{\sigma}) = \frac{1^{r_1+r_1'}(r_1+r_1')!2^{r_2+r_2'}(r_2+r_2')!\ldots}{1^{r_1}(r_1)!2^{r_2}(r_2)!\ldots 1^{r_1'}(r_1')!2^{r_2'}(r_2')!\ldots}\chi_{V_1}(\sigma)\chi_{V_2}(\sigma')$$

if $\hat{\sigma}$ is Σ_n-conjugate to $(\sigma, \sigma') \in \Sigma_k \times \Sigma_{n-k}$ and zero otherwise.

DEFINITION 4.2. Following ([83] Corollary 3.1), if V is a representation (not necessarily irreducible) of the symmetric group Σ_n and N is a strictly positive integers define

$$f_{V,N} = \frac{1}{\dim_{\mathbb{C}}(V)} \sum_{\hat{\sigma} \in \Sigma_n} \chi_V(\hat{\sigma})N^{\text{length}(\hat{\sigma})}.$$

Up to a scalar factor this is the Schur inner product of V with the virtual representation (with complex coefficients) whose character function takes the form $\hat{\sigma} \mapsto N^{\text{length}(\hat{\sigma})}$. This function is the analogue for symmetric groups of the additive character Ψ for finite general linear groups and the function $f_{V,N}$ is the anaolgue of the Kondo-Gauss sum. The Schur inner product is the non-generate bilinear form of the PSH-algebra structure of $R = \oplus_{t \geq 0} R(\Sigma_t)$ [146]. This suggests examining the properties of $f_{V,N}$ with respect to the Hopf algebra structure of R. The following result gives the behaviour under the product in R.

PROPOSITION 4.3.
If V_1 and V_2 are representations of Σ_k and Σ_{n-k} respectively and $m(V_1 \otimes V_2)$ is the product in R then

$$f_{m(V_1 \otimes V_2),N} = f_{V_1,N}f_{V_2,N}.$$

Proof
From the preceding discussion about the character values of

$$\text{Ind}_{\Sigma_k \times \Sigma_{n-k}}^{\Sigma_n}(V_1 \otimes V_2) = m(V_1 \otimes V_2)$$

we see that

$f_{m(V_1 \otimes V_2),N}$

$$= \frac{k!(n-k)!}{n!\dim_{\mathbb{C}}(V_1)\dim_{\mathbb{C}}(V_2)} \sum_{\hat{\sigma} \in \Sigma_n} \chi_{m(V_1 \otimes V_2)}(\hat{\sigma})N^{\text{length}(\hat{\sigma})}$$

$$= \frac{k!(n-k)!}{n!\dim_{\mathbb{C}}(V_1)\dim_{\mathbb{C}}(V_2)} \sum_{\hat{\sigma} \in \Sigma_n} \frac{1^{r_1+r_1'}(r_1+r_1')!2^{r_2+r_2'}(r_2+r_2')!\ldots}{1^{r_1}(r_1)!2^{r_2}(r_2)!\ldots 1^{r_1'}(r_1')!2^{r_2'}(r_2')!\ldots}\chi_{V_1}(\sigma)$$

$$\times \chi_{V_2}(\sigma')N^{\text{length}(\hat{\sigma})}$$

where the sum is taken over those $\hat{\sigma}$ in Σ_n which are conjugate to some (σ, σ') in $\Sigma_k \times \Sigma_{n-k}$. Now if $\hat{\sigma} = (\sigma, \sigma')$ with legnths as in the preceding discussion then the ratio of conjugacy class sizes satisfies

$$\frac{|\{\Sigma_n-\text{conjugates of } (\sigma,\sigma')\}|}{|\{\Sigma_k \times \Sigma_{n-k}-\text{conjugates of } (\sigma,\sigma')\}|}$$

$$= \frac{n!}{1^{r_1+r_1'}(r_1+r_1')!2^{r_2+r_2'}(r_2+r_2')!...} \frac{1^{r_1}(r_1)!2^{r_2}(r_2)!...}{k!} \frac{1^{r_1'}(r_1')!2^{r_2'}(r_2')!...}{(n-k)!}.$$

Therefore we may re-write the function as a sum over $\Sigma_k \times \Sigma_{n-k}$ in the form

$$f_{m(V_1 \otimes V_2),N}$$

$$= \frac{1}{\dim_{\mathbb{C}}(V_1)\dim_{\mathbb{C}}(V_2)} \sum_{(\sigma,\sigma')\in\Sigma_k\times\Sigma_{n-k}} \chi_{V_1}(\sigma)\chi_{V_2}(\sigma')N^{\text{length}(\hat{\sigma})}$$

$$= f_{V_1,N}f_{V_2,N},$$

as required. $\square$

EXAMPLE 4.4. Let $m, k_1, k_2, \ldots, k_t \geq 1$ be positive integers such that

$$m = \sum_{i=1}^{t} k_i.$$

Let $1_{k_i} \in R(\Sigma_{k_i})$ denote the class of the one-dimensional trivial representation. Then the iterated PSH-algebra product

$$m(1_{k_1} \otimes 1_{k_2} \otimes \ldots \otimes 1_{k_t}) = \text{Ind}_{\Sigma_{k_i} \times \ldots \times \Sigma_{k_t}}^{\Sigma_m}(1) \in R(\Sigma_m).$$

Therefore

$$f_{\text{Ind}_{\Sigma_{k_i} \times \ldots \times \Sigma_{k_t}}^{\Sigma_m}(1),N} = \prod_{i=1}^{t} f_{1_{k_i},N}$$

and

$$f_{1_{k_i},N} = \sum_{\sigma\in\Sigma_{k_i}} N^{l(\sigma)} = N^{k_i} + a_1 N^{k_i-1} + \ldots + N \in \mathbb{Z}[N].$$

If $\underline{k} = (k_1, \ldots, k_t)$ is a partition of m let us denote $\text{Ind}_{\Sigma_{k_i} \times \ldots \times \Sigma_{k_t}}^{\Sigma_m}(1)$ by $m_{\underline{k}}$. According to [9] as $\underline{k}$ runs through the partitions of m the $m_{\underline{k}}$'s run through a $\mathbb{Z}$-basis for the free abelian group $R(\Sigma_m)$. On the other hand, as explained in (Appendix III, Proposition 1.4), the irreducible representations of Σ_m are given by $\chi^{\underline{\lambda}}$ as $\underline{\lambda}$ runs through the partitions of m. Hence there are integers $t_{\underline{k},\underline{\lambda}}$, indexed by ordered pairs of partitions of m such that

$$\chi^{\underline{\lambda}} = \sum_{\underline{k}} t_{\underline{k},\underline{\lambda}} \cdot m_{\underline{k}} \in R(\Sigma_m).$$

Consequently we obtain, for all values of N,

$$f_{\chi^{\underline{\lambda}},N} = \sum_{\underline{k}} t_{\underline{k},\underline{\lambda}} \cdot f_{m_{\underline{k}},N}.$$

Locally p-adic Lie groups

This Appendix assures the reader, without going into a single detail, that replacing GL_nK and its Bruhat-Tits building by any locally p-adic Lie group and its Baum-Connes space $\underline{E}(G,\mathcal{C})$, where $\mathcal{C}$ is the family of compact open modulo the centre subgroups $H \subseteq G$, results in a construction of functorial monomial resolutions for any admissible representation V of G with a fixed central character $\underline{\phi}$. The construction is accomplished by a direct imitation of that of Chapter Four.

1. Monomial resolutions for arbitrary locally p-adic Lie groups

1.1. Let G be a locally p-adic Lie group and let V be a smooth representation defined on a k-vector space with central character $\underline{\phi}$. A functorial monomial resolution may be constructed for V by the same method as that used in Chapter Four for GL_nK.

In order to accomplish this construction one requires a canonical simplicial complex $\underline{E}(G,\mathcal{C})$ on which G acts simplicially in such a way that, for every compact open modulo the centre subgroup $H \subseteq G$, the H-fixed subcomplex $\underline{E}(G,\mathcal{C})^H$ is non-empty and contractible. One then replaces the Bruhat-Tits building by $\underline{E}(G,\mathcal{C})$ in the construction of Chapter Four.

The construction of $\underline{E}(G,\mathcal{C})$ is due to Tammo tom Dieck ([135], [136]) and is discussed in ([100] pp. 6–7). Here is a sketch of the construction.

Let M be a zero-dimensional simplicial complex with a G-action. For example, if $\mathcal{C}$ is a family of subgroups of G which is closed under conjugation and passage to subgroups then we may take M equal to the disjoint union of all the cosets G/H as H varies throughout $\mathcal{C}$. Form the iterated n-fold join

$$M(n) = M * M * \ldots * M \qquad (n \text{ copies of } M).$$

Set

$$\underline{E}(G,\mathcal{C}) = \bigcup_{1 \leq n} M(n)$$

with the weak topology ($A \subseteq \underline{E}(G,\mathcal{C})$ is closed if and only if $A \bigcap M(n)$ is closed in $M(n)$ for all n).

If H is a subgroup of G then $\underline{E}(G, \mathcal{C})^H$ is empty unless $H \in \mathcal{C}$ in which case it is non-empty and contractible because $M(n)$ is an $(n-2)$-connected space.

This G-space is very large but it is canonical and would suffice to give functorial monomial resolutions for admissible G-representations, as in the case of $GL_n K$.

Presumably for many families of classical p-adic Lie groups Bruhat-Tits buildings provide canonical, finite dimensional complexes which are G-homotopy equivalent to $\underline{E}(G, \mathcal{C})$. This is what happens in the case of $GL_n K$.

Bibliography

[1] *Automorphic forms, Representations and L-functions*; (eds. A. Borel and W. Casselman) A.M. Soc. Proc. Symp. Pure Math. XXXIII (Parts 1 and 2) (1979).

[2] *An Introduction to the Langlands Program*; (eds. J. Bernstein and S. Gelbart) Birkhäuser (2004).

[3] Modular Forms and Fermat's Last Theorem (eds. G. Cornell, J.H. Silverman and G. Stevens) Springer Verlag (1997).

[4] P. Abaramenko and K.S. Brown: *Buildings: Theory and Applications*; Springer Grad. Texts in Math. #248 (2008).

[5] A. Case and V. Snaith: Explicit Brauer Induction and the Glauberman Correspondence; J. Alg. and Appl. (2) 1 (2003) 105-118.

[6] J.F. Adams: *Stable Homotopy Theory*; Lecture Notes in Math. #3 (1966) Springer Verlag.

[7] J. Arthur and L. Clozel: Simple Algebras, Base Change and the Advanced Theory of the Trace Formula; Annals of Math. Study #120. Princeton Univ. Press (1989).

[8] M.F. Atiyah: Characters and cohomology of finite groups; Pub. Mth. IHES (Paris) 1960.

[9] M.F. Atiyah: Power operations in K-theory; Quart. J. Math. Oxford (2) 17 (1966) 165-193.

[10] M.F. Atiyah: Bott periodicity and the index of elliptic operators; Quart. J. Math. Oxford (2) 19 (1968) 113-140.

[11] A-M. Aubert, S. Hasan and R. Plymen: Cycles in the chamber homology of $GL(3)$; http://arxiv.org/abs/math.KT/0410147 (22 April 2006).

[12] H. Bass: *Algebraic K-theory*; Benjamin (1968).

[13] P. Baum, A. Connes and N. Higson: Classifying spaces for proper actions and K-theory of group C^*-algebras; Contemp. Math. 167 (1994) 241-291.

[14] J-N. Bernstein, P. Deligne, D. Kazhdan and M-F. Vigneras: *Représentations des groupes rëductifs sur un corps local*; Hermann Trav. en Cours (1984).

[15] I. Bernstein and A. Zelevinsky: Representations of the groups $GL_n(F)$ where F is a nonarchimedean local field; Russian Math. Surveys 31 (1976), 1-68.

[16] I. Bernstein and A. Zelevinsky: Induced representations of the group $GL(n)$ over a p-adic field; Functional analysis and its applications 10 (1976), no. 3, 74-75.

[17] R. Boltje: A canonical Brauer induction formula; Astérisque 181-182 (1990) 31-59.

[18] R. Boltje: A general theory of canonical induction formulae; J. Alg. 206 (1998) 293-343.

[19] R. Boltje: Monomial resolutions; J. Alg. 246 (2001) 811-848.

[20] R. Boltje, V. Snaith and P. Symonds: Algebraicisation of Explicit Brauer Induction; J. Alg. (2) 148 (1992) 504-527.

[21] A. Borel: *Linear algebraic groups*; Benjamin (1969).

[22] A. Borel and J. Tits: Groupes Rëductifs; Publ. Math. I. H. E. S. 27 (1965) 55-150.

[23] A. Borel and J. Tits: Complements á l'article: Groupes Réductifs; Publ. Math. I. H. E. S. 41 (1972) 253-276.

[24] A. Borel and J. Tits: Homomorphismes abstraits de groupes algébriques simples, Ann. Math. 97 (1973), 499-571.

[25] R. Bott: The stable homotopy of the classical groups; Annals of Math. (2) 70 (1959) 313-337.

[26] N. Bourbaki: Algèbre; Hermann (1958).

[27] N. Bourbaki: Mésures de Haar; Hermann (1963).

[28] N. Bourbaki: Variétés différentielles et analytiques; Hermann (1967).

[29] N. Bourbaki: Groupes et algèbres de Lie; Hermann (1968).

[30] N. Bourbaki: Groupes er algèbres de Lie; Chapters IV-VI (1968) Hermann Paris.

[31] R. Brauer: Bezeihungen zwischen Klassenzahlen von Teilkörpern eines Galoischen Körpers; Math. Nachr. 4 (1951) 158-174.

[32] R. Brauer: A characterization of the characters of groups of finite order; Annals of Math. (2) 57 (1953) 357-377 (also R. Brauer: Collected Papers (vol I), MIT Press (1980) (eds. Paul Fong and Warren J. Wong) 588-608.

[33] R. Brauer and J. Tate: On the characters of finite groups; Ann. Math. 62 (1955) 1-7.

[34] G.E. Bredon: Equivariant cohomology; Lecture Notes in Math. #34, Springer Verlag (1967).

[35] K.S. Brown: Buildings; Springer Verlag (1989).

[36] F. Bruhat: Distributions sur un groupe localement compact et applications á l' étude des représentations des groupes p-adiques; Bull. Soc. Math. France 89 (1961) 43-75.

[37] F. Bruhat and J. Tits: Groupes algébriques simples sur un corps local; Proceedings of a Conference on Local Fields (Driebergen, 1966) Springer Verlag (1967) 23-36.

[38] F. Bruhat and J. Tits: Groupes réductifs sur un corps local I: Données radicielles valuées; Publ. Math. I. H. E. S. 41 (1972) 5-251.

[39] D. Bump: Automorphic Forms and Representations; Cambridge Uiversity Press (1997).

[40] C.J. Bushnell and G. Henniart: The Local Langlands Conjecture for GL(2); Grund. Math. Wiss. #335; Springer Verlag (2006).

[41] R.W. Carter: Simple Groups of Lie Type; Wiley (1989).

[42] W. Casselman: The Steinberg character as a true character; Harmonic analysis on homogeneous spaces (Calvin C. Moore, ed.), Proceedings of Symposia in Pure Mathematics, vol. 26, American Mathematical Society (1973) 413-417.

[43] W. Casselman: An assortment of results on representations of $GL_2(k)$; Modular functions of one variable II (P. Deligne and W. Kuijk, eds.), Lecture Notes in Mathematics, vol. 349, Springer (1973) 1-54.

[44] C. Chevalley: The Theory of Lie Groups; Princeton Univ. Press (1962).

[45] J.H. Conway, R.T. Curtis, S.P. Norton and R.A. Wilson: Atlas of Finite Groups; Clarendon Press, Oxford (1985).

[46] P. Deligne: Formes modulaires et représentations de $GL(2)$; Modular functions of one variable II, Lecture Notes in Mathematics, vol. 349, Springer Verlag (1973).

[47] P. Deligne: Le support du caractère d'une représentation supercuspidale; C. R. Acad. Sci. Paris 283 (1976) 155-157.

[48] P. Deligne: Les constantes locales de l' équation fonctionelle de la fonction L d'Artin d'une représentation orthogonale; Inventiones Math. 35 (1976) 299-316.

[49] P. Deligne and G. Henniart: Sur la variation, par torsion, des constantes locales d'équations fonctionelles de fonctions L; Inventiones Math. 64 (1981) 89-118.

[50] M. Demazure and A. Grothendieck: *Schémas en groupes*; Lecture Notes in Mathematics vol. 151-153 Springer Verlag (1970).

[51] F. Diamond and J. Im: Modular forms and modular curves; Can. Math. Soc. Conference Proceedings vol. 17 (1995).

[52] F. Diamond and J. Shurman: *A First Course in Modular Forms*; Springer Grad. Texts in Math. #228 (2005).

[53] F. Digne and J. Michel: Fonctions-L des variétés de Deligne-Lusztiget descente de Shintani; Mem. Math. Soc. France #20 (1985).

[54] F. Digne and J. Michel: *Representations of Finite Groups of Lie Type*; London Math. Soc. Students Texts #2, Cambridge Univ. Press (1991).

[55] Doi and Naganuma: On the algebraic curve uniformized by automorphic functions: Ann. Math. 86 (1967) 449-460.

[56] Doi and Naganuma: On the functional equation of certain Dirichlet series; Invent. Math. 9 (1969) 1-14.

[57] B. Dwork: On the Artin root number; Amer. J. Math. 78 (1956) 444-472.

[58] G. van Dijk: Computation of certain induced characters of p-adic groups; Math. Ann. 199 (1972) 229-240.

[59] F.G. Friedlander: *Introduction to the Theory of Distributions*; Cambridge Univ. Press (1982).

[60] P. Garrett: Buildings and classical groups; Chapman and Hall (1997).

[61] P. Garrett: Buildings, BN-pairs, Hecke algebras, classical groups; Building notes (http://www.math.umn.edu/ garrett).

[62] Stephen S. Gelbart: *Automorphic Forms on Adéle Groups*; Ann. Math. Study #83, Princeton University Press (1975).

[63] P. Gérardin and J-P. Labesse: The solution of a basechange problem for $GL(2)$ (following Langlands, Saito, Shintani); Amer. Math. Soc. Proc. Symp. Pure Math. XXXIII, *Automorphic Forms,Representations and L-Functions* (1979) 115–134.

[64] P. Gérardin and P. Kutzko: Facteurs locaux pour $GL(2)$; Ann. Scient. E.N.S. t. **13** (1980) 349-384.

[65] G. Glauberman: Correspondences of characters for relatively prime operator groups; Can. J. Math. 20 (1968) 1465-1488.

[66] Roger Godement and Hervé Jacquet: Zeta functions of simple algebras; Lecture Notes in Math. #260, Springer-Verlag (1981).

[67] Dorian Goldfeld and Joseph Hundley: *Automorphic Representations and L-functions for the General Linear Group* vols. I and II; Cambridge studies in advanced mathematics #129 and #130, Cambridge University Press (2011).

[68] D. Goss: *Basic Structures in Function Field Arithmetic*; Ergebnisse der Math. vol. 35 (1996) Springer-Verlag (ISBN 3-540-61087).

[69] J.A. Green: The characters of the finite general linear groups; Trans. Amer. Math. Soc. 80 (1955) 402-447.

[70] P. Griffiths and J. Harris: *Principles of Algebraic Geometry*; Wiley (1978).

[71] Harish-Chandra: Harmonic analysis on reductive p-adic groups: Harmonic analysis on homogeneous spaces (Calvin C. Moore, ed.), Proceedings of Symposia in Pure Mathematics, vol. 26, American Mathematical Society (1973) 167-192.

[72] G. Henniart: Représentations du groupe de Weil d'un corps local; L'Enseign. Math. t.26 (1980) 155-172.

[73] G. Henniart: Galois ε-factors modulo roots of unity; Inventiones Math. 78 (1984) 117-126.

[74] F. Hirzebruch: *Topological Methods in Algebraic Geometry* (3rd edition); Grund. Math. Wiss. #131 Springer Verlag (1966).

[75] R. Howe: Dual pairs in physics: harmonic oscillators, photons, electrons and singletons; Applications of group theory in physics and mathematical physics (Chicago 1982) 179-207, Lectures in Appl. Math. 21 A.M. Soc. (1985).

[76] D. Husemoller: *Fibre Bundles*; McGraw-Hill (1966).

[77] I. M. Isaacs: *Character Theory of Finite Groups* Academic Press (1976).

[78] N. Iwahori and H. Matsumoto: On some Bruhat decomposition and the structure of the Hecke rings of p-adic Chevalley groups; Publ. Math. I. H. E. S. 25 (1965) 5-48.

[79] H. Jacquet: Représentations des groupes linéaires p-adiques; Theory of Group Representations and Fourier Analysis (Proceedings of a conference at Montecatini, 1970), Edizioni Cremonese, 1971.

[80] H. Jacquet and R. P. Langlands: *Automorphic forms on GL(2)*; Lecture Notes in Mathematics, # 114, Springer Verlag (1970).

[81] H. Jacquet: *Automorphic Forms on GL_2 II*; Springer Verlag Lecture Notes in Math. #278 (1972).

[82] J. Kaminker and C. Schochet: K-theory and Steenrod homology; Trans. A.M.Soc. 227 (1977) 63-107.

[83] J. P. Keating, F. Mezzadri and B. Singphu: Rate of convergence of linear functions on the unitary group; J. Phys. A. Math. Theor. 44 (2011) o35204.

[84] B. Kendirli: On representations of $SL(2, F)$, $GL(2, F)$, $PGL(2, F)$; Ph.D. thesis, Yale University (1976).

[85] T. Kondo: On Gaussian sums attached to the general linear groups over finite fields; J. Math. Soc. Japan vol. 15 #3 (1963) 244-255.

[86] S.S. Kudla: Tate's thesis; *An Introduction to the Langlands Program* Birkhäuser (2004) 109-132.

[87] S. Lang: *Cyclotomic Fields*; Grad. Texts in Math. #59 Springer-Verlag (1978).

[88] S. Lang: *Algebra*; (2nd edition) Addison-Wesley (1984).

[89] R. P. Langlands: On the functional equation of the Artin L-functions; Yale University preprint (unpublished).

[90] R. P. Langlands: *On the functional equations satisfied by Eisenstein series*; Lecture Notes in Mathematics, vol. 544, Springer Verlag (1976).

[91] R. P. Langlands: *Base Change for GL(2)*; Annals of Math. Study #96. Princeton Univ. Press (1980).

[92] E. Lluis-Puebla and V. Snaith: On the homology of the symmetric group; Bol. Soc. Mat. Mex. (2) 27 (1982) 51-55.

[93] I.G. Macdonald: Zeta functions attached to finite general linear groups; Math. Annalen 249 (1980) 1-15.

[94] J. Martinet: Character theory and Artin L-functions; London Math. Soc. 1975 Durham Symposium, *Algebraic Number Fields* Academic Press (1977) 1-87.

[95] H. Matsumoto: Analyse harmonique dans certains systèmes de Coxeter et de Tits; Analyse harmonique sur les groupes de Lie, Lecture Notes in Mathematics, vol. 497, Springer-Verlag (1975) 257-276.

[96] H. Matsumoto: Analyse harmonique dans les systèmes de Tits bornologiques de type affine; Lecture Notes in Mathematics, vol. 590, Springer-Verlag (1977).

[97] J.P. May, V. Snaith and P. Zelewski: A further generalisation of the Segal conjecture; Quart. J. Math. Oxford (2) 40 (1989) 457-473.

[98] J. W. Milnor: On the Steenrod homology theory; *Novikov conjectures, index theorems and rigidity* vol. 1 (Oberwolfach 1993), London Math. Soc. Lecture Notes Ser. #226, Cambridge Univ. Press (1995) 79-96.

[99] J.W. Milnor and J.C. Moore: On the structure of Hopf algebras; Annals of Math. (2) 81 (1965) 211-264.

[100] Guido Mislin and Alain Valette: *Proper group actions and the Baum-Connes conjecture*; Advanced Courses in Math., CRM Barcelona (2003).

[101] M. Nakaoka: Homology of the infinite symmetric group; Annals of Math. (2) 73 (1961) 229-257.

[102] O. Neisse and V.P. Snaith: Explicit Brauer Induction for symplectic and orthogonal representations; Homology, Homotopy Theory and Applications 7 (3) 2005, Conference edition "Applications of K-theory and Cohomology"(ed. J.F. Jardine).

[103] G. Olshanskii: Intertwining operators and complementary series in the class of representations induced from parabolic subgroups of the general linear group over a locally compact division algebra; Math. USSR Sbornik 22 (1974) no. 2, 217-255.

[104] D.G. Quillen: Higher algebraic K-theory I; Lecture Notes in Math. #341 (1973) Springer-Verlag.

[105] D.G. Quillen: On the cohomology and K-theory of the general linear groups over a finite field; Ann. Math. 96 (1972) 552-586.

[106] A. Robert: Modular representations of the group $GL(2)$ over a p-adic field; Journal of Algebra 22 (1972) 386-405.

[107] H. Saito: *Automorphic Forms and Algebraic Extensions of Number Fields*; Lecture in Math. Kyoto Univ. (1975).

[108] P. Schneider and U. Stuhler: Resolutions for smooth representations of the general linear groups over a local field; J. Reine Angew Math. 436 (1993) 19-32.

[109] P. Schneider and U. Stuhler: Representation theory and sheaves on the Bruhat-Tits building; IHES Publ. Math. #85 (1997) 97-191.

[110] P. Schneider and J. Teitelbaum: $\mathcal{U}(g)$-finite locally analytic representations; Represent. Theory 5 (2001) 111-128 (electronic).

[111] P. Schneider and J. Teitelbaum: Duality for admissible locally analytic representations; Represent. Theory 5 (2005) 297-326 (electronic).

[112] J-P. Serre: *Cohomologie Galoisienne*; Lecture Notes in Mathematics #5 (1973) Springer-Verlag.

[113] J-P. Serre: *Linear Representations of Finite Groups*; Grad. Texts in Math. # 42 (1977) Springer-Verlag.

[114] J-P. Serre: *Local Fields*; Grad. Texts in Math. # 67 (1979) Springer-Verlag.

[115] J-P. Serre: *Trees*; Springer-Verlag (1980).

[116] J-P. Serre: L'invariant de Witt de la forme $\text{Tr}(x^2)$; Comm. Math. Helv. 59 (1984) 651-676.

[117] T. Shintani: Two remarks on irreducible characters of finite general linear groups; J. Math. Soc. Japan 28 (1976) 396-414.

[118] T. Shintani: On liftings of holomorphic cusp forms; A.M.Soc. Proc. Symp. Pure Math. 33 Part 2 (1979) 79-110.

[119] A. Silberger: On work of Macdonald and $L_2(G/B)$ for a p-adic group; Harmonic analysis on homogeneous spaces (Calvin C. Moore, ed.) Proceedings of Symposia in Pure Mathematics, vol. 26, American Mathematical Society (1973) 387-393.

[120] V.P. Snaith: Stiefel-Whitney classes of symmetric bilinear forms - a formula of Serre; Can. Bull. Math. (2) 28 (1985) 218-222.

[121] V.P. Snaith: Applications of Explicit Brauer Induction; A.M.Soc. Proc. Symp. Pure Math. 47 Part 2 (1987) 495-531.

[122] V.P. Snaith: Explicit Brauer induction; Inventiones Math. 94 (1988) 455-478.

[123] V.P. Snaith: A construction of the Deligne-Langlands local root numbers of orthogonal Galois representations; Topology 27 (2) 119-127 (1988).

[124] V.P. Snaith: Invariants of representations; NATO ASI series C #279 Proc. Lake Louise Algebraic K-theory Conf. Kluwer (1989) 445-508.

[125] V.P. Snaith: *Topological Methods in Galois Representation Theory*, C.M.Soc Monographs, Wiley (1989) (republished by Dover in 2013).

[126] V.P. Snaith: *Explicit Brauer Induction (with applications to algebra and number theory)*; Cambridge studies in advanced mathematics #40,Cambridge University Press (1994).

[127] V.P. Snaith: Base Change Yoga; McMaster University Notes (c. 1994).

[128] V.P. Snaith: Monomial resolutions for admissible representations of GL_2 of a local field; preprint 2011 (revised September 2013).

[129] V.P. Snaith: Remarks on a paper by Guy Henniart (an essay concerning Galois epsilon factors modulo roots of unity), posted on my Sheffield webpage April 2012.

[130] V.P. Snaith: Galois descent of representations, posted on my Sheffield webpage January 2013.

[131] V. Snaith and P. Zelewski: Stable maps into classifying spaces of compact Lie groups; U.W.O. Preprint (1986).

[132] E.H. Spanier: *Algebraic Topology*; McGraw-Hill (1966).

[133] T.A. Springer: Characters of special groups; Lecture Notes in Math. #131, Springer-Verlag (1970).

[134] P. Symonds: A splitting principle for group representations; Comm. Math. Helv. 66 (1991) 169-184.

[135] T. tom Dieck: Orbitten und äquivariante Homologie I; Arch. Math. 23 (1972) 307-317.

[136] T. tom Dieck: Transformation Groups; Studies in Math. # 8, de Gruyter (1987).

[137] J.T. Tate: Duality theorems in Galois cohomology over number fields; Proc. Int. Cong. Math. Stockholm (1962).

[138] J.T. Tate: Local Constants; London Math. Soc. 1975 Durham Symposium, *Algebraic Number Fields* (ed. A. Fröhlich) 89-131 (1977) Academic Press.

[139] J. Tits: Buildings and their applications; Proc. I.C.M. Vancouver (1974) Part 1 209-220.

[140] J. Tits: Reductive groups over local fields; Proc. Symp. Pure Math. A.M. Soc. #XXXIII Part I (1977) (eds. A. Borel and W. Casselman) 29-69.

[141] C.A. Weibel: The K-book: An Introduction to Algebraic K-theory; A.M. Soc. Grad. Studies in Math. # 145 (2013).

[142] A. Weil: Fonction zeta et distributions; Séminaire Bourbaki III (1966) in Collected Papers vol. III; Springer-Verlag New York (1979) 158-163.

[143] A. Weil: *Basic Number Theory*; Springer-Verlag New York (1967).

[144] G. D. Williams: The principal series of a p-adic group; Ph.D. thesis, Oxford University, (1974).

[145] N. Winarsky: Reducibility of principal series representations of p-adic groups; Ph.D. thesis, University of Chicago (1974).

[146] A.V. Zelevinsky: *Representations of finite classical groups - a Hopf algebra approach*; Lecture Notes in math. #869, Springer-Verlag (1981).

Index

Printed in the United States
By Bookmasters